Enhancing Data-Driven Electronics Through IoT

Bhagwan Das
School of Information Technology and Engineering, Melbourne Institute of Technology, Australia

Muhammad Zakir Shaikh
Mehran University of Engineering, Pakistan

Samreen Hussain
Dawood University of Engineering and Technology, Pakistan

Enrique Nava Baro
Universidad de Málaga, Spain

IGI Global
Scientific Publishing
Publishing Tomorrow's Research Today

Vice President of Editorial — Melissa Wagner
Managing Editor of Acquisitions — Mikaela Felty
Managing Editor of Book Development — Jocelynn Hessler
Production Manager — Mike Brehm
Cover Design — Phillip Shickler

Published in the United States of America by
IGI Global Scientific Publishing
701 East Chocolate Avenue
Hershey, PA, 17033, USA
Tel: 717-533-8845
Fax: 717-533-7115
Website: https://www.igi-global.com E-mail: cust@igi-global.com

Library of Congress Cataloging-in-Publication Data

LCCN: 2024049273 (CIP Data Pending)
ISBN: 979-8-3693-5448-3
EISBN: 979-8-3693-5450-6
British Cataloguing in Publication Data
A Cataloguing in Publication record for this book is available from the British Library.

This book is dedicated to all the visionaries, researchers, and innovators who strive to bridge the gap between data-driven electronics and the limitless possibilities of the Internet of Things.

To our family, whose unwavering support and encouragement have been the cornerstone of our journey.

To our mentors and colleagues, whose guidance and wisdom have shaped our path in research and academia.

To our students, who constantly inspire us with their curiosity and passion for learning.

To our friends and peers, whose thought-provoking discussions and collaborations have enriched this journey.

To those who dare to push the boundaries of knowledge, may this work serve as a stepping stone toward a smarter, more connected world.

And to the future generations of technologists, may this work inspire new ideas and ground-breaking innovations.

With gratitude and hope,
Bhagwan Das
Muhammad Zakir Shaikh
Samreen Hussain
Enrique Nava Baro

Table of Contents

Detailed Table of Contents

Internet of Medical Things is application of Internet of Things to medical and health care fields. Internet of Medical Things has created a revolution in the health care industry by integrating technology in the medical devices, enabling remote monitoring, data collection and real time data analytics. IOMT make use of variety of sensors for collection of patient data along with different communication devices. Upcoming technologies such as artificial intelligence and edge computing are being exploited to enhance IOMT capabilities. The research work focusses on sensors, protocols, and communication technologies in IOMT. Real-time case studies highlighting the practical applications of IoMT and showcasing its ability to monitor vital signs, manage chronic conditions, and improve patient outcomes are also explored. While IoMT has an enormous potential to transform healthcare, addressing integration challenges is crucial for understanding its benefits. The work discusses the challenges related to cybersecurity and interoperability, to ensure integration of IOMT into clinical practice.

Internet of Things (IoT) technology has revolutionized the way devices interact and communicate, enabling seamless connectivity and automation across various domains including computer systems, networking and embedded systems. Communication is one of the most important fundamental aspects of any digital system. In IoT system, apart from modern paradigms, many classical computer systems communication technologies are used to achieve the main purpose of information acquisition and processing. The actual proposed chapter aims to provide and in-depth and comprehensive exploration of communication technologies and protocols used in IoT environments according to different context and use cases. This exploration will also take into account the data acquisition, storage, processing and transmission technics, internal and external electronic architecture and communication technics.

The integration of Internet of Things (IoT) into the agricultural sector exhibits the potential for modernising conventional farming and tackling various obstacles faced by the community. This chapter focuses on the transformative capacity of the IoT in the agricultural sector, stressing significant insights and discoveries obtained from several case studies. The IoT enables farmers to empower for efficient farming operations and risk reduction by monitoring and analysing data in real-time. Real world applications show IoT's precise agriculture solutions, which contribute to sustainable farm mechanisation, enhanced livestock management, food safety, blockchain technology, and supply chain visibility. Collaboration among stakeholders, giving priorities for research and innovation, and encouraging technology adoption are essential for progress. By adopting the IoT and incorporating research inputs, the agricultural industry can initiate an approach towards efficient and sustainable automation, thereby guaranteeing a prosperous future for farmers and communities on a global scale.

Understanding our world's condition and human activity's effects on ecosystems depends on environmental monitoring. This chapter examines how sensor network integration with Deep Belief Networks (DBNs) might improve environmental monitoring. The intricacy and noise of ecological data provide challenges for traditional approaches, but real-time information from sensor networks is invaluable. DBNs are perfect for deciphering sensor data because they extract intricate patterns from high-dimensional data. Environmental monitoring is now more precise and flexible thanks to this connection. Case studies in climate change forecasting and air and water quality monitoring demonstrate the useful advantages. The chapter highlights the promise of merging deep learning and environmental monitoring for sustainable management while addressing integration issues and outlining future developments.

Apart from powering robots with innovative energy harvesting, the domain of robotics relies on such integration to bring about autonomy and efficiency into systems. This chapter describes state-of-the-art energy harvesting techniques designed for use in robotics, focusing particularly on the aspects of sustainability and lifetime of operation. Some of the techniques considered in this work are kinetic energy recovery, thermoelectric generators, and solar cells, which are analyzed with regard to their potential applications for autonomous power supply. Such synergy between energy harvesting and the Internet of Things is also discussed, showing how IoT connectivity can make its contribution to the optimality of energy management and distribution within robotic networks. The chapter covers technical and practical aspects of how advancements such as this can drive the future of robotics toward more self-sufficient and intelligent systems.

This chapter examines the collaboration between machine learning and remote sensing technology to improve air quality and environmental monitoring. It applies advanced techniques and extensive datasets to examine, forecast, and manage environmental elements in real time. The chapter explores the significance of monitoring air quality and the environment for public health and sustainable development. It explains the fundamentals of remote sensing and machine learning and how they can join forces. This research delves into how satellite imagery, sensor networks, and data fusion methods can provide a comprehensive view of the environment. It investigates their successful applications in various regions with diverse climates. The chapter highlights the challenges in this field such as obtaining quality data, addressing intensive computational requirements, and fostering interdisciplinary collaboration. It also outlines future directions and emerging opportunities, with a focus on promising technologies like deep learning and cloud computing.

Preface

In a world that is increasingly driven by data and connectivity, the role of the Internet of Things (IoT) has become pivotal in shaping the future of electronic devices and systems. The convergence of IoT with electronics engineering is revolutionizing how we design, interact with, and optimize technology across every sector of society—from smart homes and autonomous vehicles to healthcare and agriculture. However, this rapid evolution presents a pressing challenge: how do we seamlessly integrate IoT technologies into electronic devices to enhance their functionality and performance? This is the key question that Enhancing Data-Driven Electronics Through IoT aims to address.

This book is a comprehensive exploration of how IoT can be harnessed to make electronic devices smarter, more efficient, and more connected. It bridges the gap between theory and practice, offering readers both a solid foundation in IoT concepts and practical insights into real-world applications. In doing so, it provides engineers, researchers, and students with the tools they need to innovate and stay ahead in this fast-paced field. Throughout the chapters, we delve into a wide range of critical topics, from the underlying communication technologies and protocols that make IoT possible to the ethical considerations that must be addressed in its deployment. Emerging technologies such as 5G connectivity, machine learning, and quantum computing are also discussed, providing a forward-looking perspective on how these advancements will shape the future of IoT-enabled electronics. We examine how IoT can be applied in specific domains, such as healthcare, agriculture, and smart cities, showcasing its transformative potential across industries. Moreover, the book addresses the critical challenges of IoT integration, including security, privacy, and scalability. With the rise of connected devices, ensuring that they remain secure and reliable is of paramount importance. Each chapter offers valuable insights into these issues, presenting practical solutions and case studies that illustrate how they can be tackled.

The goal of Enhancing Data-Driven Electronics Through IoT is not only to provide a roadmap for understanding the technology but also to inspire innovation. By highlighting the latest trends and offering guidance on the practical implementation of IoT in electronics, this book aims to equip readers with the knowledge and confidence to push the boundaries of what is possible in their respective fields. As the demand for smarter, more interconnected devices continues to grow, it is my hope that this book will serve as a valuable resource for anyone involved in the world of electronics engineering and IoT. Whether you are a seasoned professional or a student eager to learn, this book will guide you through the intricacies of IoT integration and help you unlock new possibilities for the future.

Welcome to the exciting world of data-driven electronics, where the possibilities are as limitless as the technology itself.

In today's rapidly evolving technological landscape, the integration of the Internet of Things (IoT) with electronics has become a key driver of innovation. We live in a world where everything—from our homes and cars to our healthcare systems and cities—is becoming smarter and more connected. At the heart of this transformation is IoT, a technology that enables devices to collect, share, and act on data in real time, creating an ecosystem of interconnected systems that communicate seamlessly.

The topic of IoT-enabled electronics fits squarely at the intersection of modern convenience and cutting-edge engineering. As consumer expectations for smarter, more efficient devices grow, industries are racing to develop solutions that improve performance, reduce energy consumption, and enhance user experience. From smart home systems that adjust to our preferences, to healthcare devices that monitor patient vitals in real time, IoT is reshaping the way we interact with the world around us. However, beyond its convenience, IoT is also paving the way for significant advancements in fields like agriculture, where data-driven systems can optimize water usage and crop management, and in urban infrastructure, where smart cities are using IoT to enhance traffic flow, reduce energy consumption, and improve quality of life.

In this context, the challenge for professionals and researchers today is not just to understand IoT as a concept, but to master its integration into electronic devices in a way that drives meaningful change. As the world becomes more interconnected, the role of IoT in electronics will only continue to expand, making it a crucial area of focus for engineers, technologists, and innovators aiming to shape the future.

The target audience for Enhancing Data-Driven Electronics Through IoT includes a diverse group of professionals, scholars, and students who are passionate about the future of electronics and connectivity. This book is designed for engineers and technologists who are working at the forefront of electronics design, looking to integrate IoT technology into their devices and systems. It offers practical guidance for those seeking to enhance performance, streamline operations, and create smarter, more efficient devices. Academics and researchers, particularly those focused on IoT, machine learning, data analytics, and electronics engineering, will find this book a valuable resource. It bridges the gap between theory and real-world application, offering insights that are both academically rigorous and practically relevant. Whether you are conducting research in emerging technologies or seeking inspiration for innovative solutions, this book provides the tools and knowledge needed to stay ahead in the field. Students studying electronics, computer science, or IoT-related fields will also benefit from the comprehensive overview and deep dives into specific technologies. The content is accessible enough to provide a clear understanding of IoT's impact on electronics, while offering the depth necessary for more advanced learners who want to explore cutting-edge applications and future trends.

Overall, this book is for anyone interested in how IoT is reshaping electronics today and how they can be a part of this exciting journey toward a more connected and intelligent world.

The selected book chapter's summary is explained to give an overview and explain the state of the art in IoT-driven electronics, providing readers with insights into the latest developments, challenges, and innovations in the field. Each chapter serves as a focused exploration of a specific area, offering both foundational knowledge and advanced concepts, ensuring that readers are equipped to understand the current trends and future directions of IoT integration in electronics engineering. Through this, the book enables a deeper understanding of how IoT is shaping the future of technology and its diverse applications.

Chapter 1- Unveiling The Role of Diversity Measure in Concept Drift Detection - A Comparative Analysis and Future Prospects

This chapter dives into the critical challenge of concept drift in machine learning, particularly in large-scale data processing. By comparing various diversity measures for detecting concept drift, the authors present a thorough analysis of techniques that can help IoT systems handle dynamic environments and changing data patterns. The insights gained here are crucial for developing adaptive IoT applications that remain effective over time.

Chapter 2- RPL and IoT Security: A Holistic Examination of Vulnerabilities

Security in IoT networks is a significant concern, and this chapter explores the vulnerabilities in Routing Protocol for Low-power and Lossy Networks (RPL)-based systems. It delves into weaknesses at both device and network levels, providing solutions and mitigation techniques to fortify IoT networks. Understanding these security challenges is essential for anyone designing IoT systems in critical sectors such as healthcare or industrial automation.

Chapter 3- IoT Security and Privacy

As IoT systems become more widespread, privacy concerns grow. This chapter outlines the core architecture of IoT systems, highlighting potential security and privacy vulnerabilities. It also discusses ethically sound deployment strategies, offering a balanced perspective on securing human-to-machine and machine-to-machine communications. This is an invaluable read for professionals focused on maintaining trust and security in IoT implementations.

Chapter 4- IoT Protocols for Healthcare: Enhancing Connectivity and Efficiency in Medical Systems

This chapter provides an in-depth look at how IoT protocols are transforming healthcare by improving connectivity, real-time data management, and patient monitoring. It analyzes key protocols such as MQTT, CoAP, and HTTP, offering a clear understanding of their role in advancing medical IoT applications. This chapter is essential for anyone interested in IoT's growing role in improving healthcare efficiency and patient outcomes.

Chapter 5- AI and IoT in Edge Computing for Climate Analysis and Sustainability

The integration of AI, IoT, and edge computing offers a powerful tool for addressing climate challenges. This chapter explains how real-time data collection and local processing can enhance sustainability and environmental monitoring. It highlights the potential of IoT-driven solutions for climate analysis, making it particularly relevant for environmental engineers and sustainability experts.

Chapter 6-5G and Cloud Computing Synergy in Smart City Networks: A Review

With the rise of smart cities, the synergy between 5G and cloud computing is pivotal for managing the massive amounts of data generated by IoT devices. This chapter reviews the fundamental characteristics of 5G technology and its integration with cloud systems, providing insights into how these technologies are transforming urban environments. City planners and IoT developers will find this chapter crucial for understanding the future of smart infrastructure.

Chapter 7-A Hybrid Approach for Smart Home Activity Recognition using Sensor Data and Deep Learning Techniques

This chapter addresses the challenge of recognizing human activities in smart homes by introducing a hybrid approach combining deep learning with sensor data. It showcases how natural language processing and time series classification can improve smart home automation, providing a pathway for developers to create more intuitive and adaptive living environments.

Chapter 8- Achieving Smart Connectivity and Automation through Data-Driven Applications of IoT in Electronics

In this chapter, the authors discuss the IoT's role in revolutionizing electronics by offering seamless connectivity and automation. The chapter explores areas like predictive maintenance, energy management, and cost reduction, showing how IoT is reshaping the way electronics interact with their surroundings. It is an essential read for anyone involved in designing data-driven electronic devices.

Chapter 9-Advancements in Internet of Medical Things: Technology, Integration, and Challenges

This chapter focuses on the Internet of Medical Things (IoMT), a growing field that integrates IoT technologies with healthcare systems. It explores the use of sensors and AI in remote patient monitoring and healthcare data management, providing insights into the challenges of integrating these technologies into clinical practices. Healthcare professionals and medical technologists will benefit from its in-depth analysis.

Chapter 10- Communication Technologies and Protocols in IoT Systems: Principles, Standards, and Security

A deep dive into the communication technologies that power IoT systems, this chapter covers the key principles and standards that enable seamless data transmission. It discusses various protocols in relation to their security implications and practical applications across different sectors. IoT developers and system architects will find this chapter invaluable for building robust and secure IoT networks.

Chapter 11- Field to Cloud: Unveiling IoT Triumphs in Agricultural Evolution

Agriculture is increasingly adopting IoT for smart farming practices, and this chapter explores how IoT can enhance efficiency in farming operations. It provides real-world case studies that demonstrate the power of IoT in improving crop management, livestock tracking, and food safety. This chapter is essential for those in the agricultural technology field seeking sustainable and data-driven farming solutions.

Chapter 12- Hybrid Models for Environmental Monitoring by Integrating Deep Belief Networks with Sensor Networks

This chapter examines how deep learning, particularly Deep Belief Networks (DBNs), can be integrated with IoT sensor networks to enhance environmental monitoring. By addressing the challenges of ecological data processing, the authors present a cutting-edge approach that can improve climate monitoring and environmental protection efforts.

Chapter 13- Innovative Energy Harvesting Techniques for Robotics: Enhancing Autonomous Systems with IoT

Energy efficiency is critical for autonomous systems, and this chapter covers innovative energy harvesting techniques that can sustain IoT-powered robots. The authors discuss applications of technologies such as solar cells and thermoelectric generators, offering insights into how IoT can optimize energy use in robotic networks. This chapter is crucial for developers of autonomous IoT systems.

Chapter 14- Machine Learning Integrated Air Quality and Environmental Monitoring Processes

Focusing on air quality and environmental monitoring, this chapter demonstrates how machine learning and remote sensing technologies can work together to analyze environmental data. It explores how real-time data collection and analysis can enhance public health and sustainability efforts, making it a vital resource for environmental scientists and public health officials.

Chapter 15- Security in Internet of Things Applications using Network Attached Storage for Resource Usage Optimization

IoT applications often face resource limitations that impact security. This chapter introduces a novel approach using Network Attached Storage (NAS) to optimize resource use while enhancing security. It provides a practical solution for securing data in low-capability IoT devices, making it highly relevant for cybersecurity professionals working on IoT systems.

Chapter 16- Sustainable Solutions for Autonomous Vehicles using Deep Learning and IoT

As autonomous vehicles become more prevalent, ensuring they are eco-friendly is a growing concern. This chapter discusses how IoT and deep learning can be integrated to reduce energy consumption and optimize vehicle efficiency. It offers valuable insights for automotive engineers and researchers focused on creating sustainable transportation solutions.

Chapter 17- Wireless Sensor Networks (WSNs) Integrated Machine Learning Algorithms for Water Resource Management

Water management is a critical global issue, and this chapter explores how WSNs combined with machine learning algorithms can optimize water resource management. It highlights the potential of IoT-driven solutions to monitor water quality, predict floods, and improve irrigation systems, providing practical applications for environmental engineers and water management specialists.

CONCLUSION

As we look to the future of electronics engineering, Enhancing Data-Driven Electronics Through IoT stands as a timely and impactful contribution to the field. The book offers a fresh perspective on how IoT is reshaping the way we design and interact with electronic devices, while also underscoring the importance of data-driven decision-making in modern engineering practices. The diverse range of topics covered throughout the chapters reflects the complexity of IoT integration, from the technicalities of communication protocols to the broader ethical and security implications. By addressing the interconnectedness of these subjects, this book provides a holistic understanding that transcends traditional approaches to electronics engineering.

Moreover, this publication fosters a deeper appreciation of the role interdisciplinary collaboration plays in driving innovation. As IoT continues to bridge electronics, machine learning, data analytics, and beyond, the importance of cross-field expertise becomes clearer. This book contributes meaningfully by encouraging a broad, inclusive dialogue around these emerging opportunities. The rapidly evolving landscape of IoT calls for resources that not only document present challenges but also anticipate future ones. In this way, Enhancing Data-Driven Electronics Through IoT doesn't just serve the current needs of its audience but also prepares them for the future, offering thought leadership in a field that is poised for exponential growth.

In essence, this book stands as a testament to the transformative potential of IoT and its far-reaching effects on electronic systems. It serves as a foundational resource for those who seek to be at the forefront of IoT-driven innovation, contributing to both academic excellence and practical breakthroughs in electronics design. With this in mind, the publication holds the promise of being a lasting reference point for those dedicated to advancing the field of electronics through data-driven insights and IoT technologies.

We encourage you to engage with this book as a resource for expanding your knowledge and exploring new dimensions of research. May it serve as a valuable tool in deepening your understanding and fostering innovation in the rapidly evolving field of IoT-driven electronics. We wish you a productive and insightful learning experience as you embark on this academic journey.

Bhagwan Das
Muhammad Zakir Shaikh
Samreen Hussain
Enrique Nava Baro

Introduction

In today's digital age, the fusion of electronics and the Internet of Things (IoT) has transformed how we interact with technology, heralding a new era of intelligent, interconnected devices. For professionals in electronics engineering, this presents an exciting, yet complex challenge—how to effectively incorporate IoT technologies into their designs to create smarter, more efficient systems. With IoT now at the heart of innovations across industries, from healthcare to agriculture and smart cities, the need for practical, well-rounded resources is more critical than ever. *Enhancing Data-Driven Electronics Through IoT* addresses this growing demand. This book is not just a collection of theories and academic discussions; it is a guide that bridges the gap between conceptual knowledge and hands-on implementation. It empowers scholars and engineers alike to fully harness the capabilities of IoT and create designs that meet the increasing demand for data-driven, connected devices. From protocols and communication technologies to data analytics and security measures, the book offers a comprehensive exploration of the IoT landscape.

With each chapter offering a deep dive into a specific aspect of IoT integration—whether it be machine learning, 5G connectivity, or quantum computing—the book becomes an indispensable resource for those seeking to remain at the forefront of technological advancements in electronics. The inclusion of real-world case studies provides a practical context, showcasing the transformative power of IoT in diverse applications such as healthcare, agriculture, and smart homes. In an era where the pace of technological innovation is relentless, the importance of staying informed and equipped with cutting-edge knowledge cannot be overstated. *Enhancing Data-Driven Electronics Through IoT* offers that much-needed edge. Whether you are an academic, a researcher, or a practitioner, this book will guide you through the complexities of IoT, offering insights that are as forward-thinking as they are practical.

As we continue to move toward a future defined by smarter devices and more efficient systems, this book will serve as a crucial reference, helping professionals not only keep pace with change but lead it. The IoT revolution is here, and this book is your key to unlocking its full potential.

With that, we invite you to delve into the pages of this book and discover the opportunities that await in the ever-evolving world of IoT-enabled electronics.

Bhagwan Das

Centre for Artificial Intelligence Research and Optimization (AIRO), Torrens University Australia, Australia

Muhammad Zakir Shaikh

Universidad de Málaga, Spain

Samreen Hussain

Dawood University of Engineering and Technology, Pakistan

Enrique Nava Baro

Universidad de Málaga, Spain

Chapter 1
Unveiling the Role of Diversity Measure in Concept Drift Detection:
A Comparative Analysis and Future Prospects

Osamah A. Mahdi
Melbourne Institute of Technology, Australia

Nawfal Ali
Monash University, Australia

Eric Pardede
La Trobe University, Australia

Bhagwan Das
https://orcid.org/0000-0002-0731-6353
Melbourne Institute of Technology, Australia

ABSTRACT

Concept drift, marked by changes in the statistical properties of a target variable, is a notable challenge in machine learning, data mining and applications involving big data and large-scale data processing. Addressing this, the employment of diversity measure has emerged as an effective strategy. This chapter examine and investigate the role of the diversity measure in detecting concept drift and compare and analysis four different ways of using them: DMDDM for drift` detection in a fully supervised binary classification context, DMDDM-S in a semi-supervised context, DMODD for online drift detection in a fully supervised multi-classification context,

DOI: 10.4018/979-8-3693-5448-3.ch001

and HBBE, a hybrid block-based ensemble designed for addressing different types of concept drifts. The comparative analysis evaluates the efficacy of these methods and the results confirm their effectiveness within their respective settings. Overall, this chapter explores advancements in using diversity measures for concept drift detection in big data, emphasizing their importance and future research in machine learning contexts.

INTRODUCTION

The constant transformation or evolution of data is a crucial concern in dynamic settings and applications, including aviation, autonomous vehicles, nuclear reactors, healthcare, the military, smart cities, and the aerospace sector. Essentially, the critical characteristics of these environments are subject to change, potentially leading to negative consequences, such as endangering human lives, if not adequately addressed (Gama 2010). Consequently, learning methods must employ sophisticated algorithms to monitor these changes and adapt accordingly. Furthermore, the effectiveness of learning algorithms may vary due to the changing nature of incoming data, meaning that an algorithm that is effective today may become outdated following changes in the environment or data.

The literature on learning from data streams identifies the phenomenon of class distribution changes within data streams as concept drift (Mahdi, Ali, Pardede, Alazab, et al. 2024). In the context of machine learning, concept drift describes a situation where the statistical properties of the target variable, which the model is designed to forecast, shift over time (Prasad and Agarwal 2016). This implies that the original input data's relevance to the model has altered significantly, yet the model remains oblivious to these modifications and, as a result, fails to make precise predictions. Thus, it is imperative for learning algorithms to detect concept drift in dynamic data streams and accordingly adjust or renew their prediction models. To address this challenge, adaptive learning models are developed, employing drift detection methods to pinpoint the instances of drift in changing environments (Mahdi, Ali, Pardede, Alazab, et al. 2024).

The rapid alterations in human behavior triggered by the COVID-19 pandemic serve as a stark illustration of concept drift. To explore the impact of pandemic-induced concept drift on predictive models, consider Melbourne City's scenario, where quarantine measures have influenced various aspects of daily life, including shopping habits, electricity consumption, and traffic accidents. For instance, analyzing predictions for future traffic accidents, VicRoads, which documents all incidents involving injuries, fatalities, or property damage exceeding $1,000, noted a marked decrease in annual accident figures during the pandemic due to lockdowns and the shuttering of businesses. Additionally, electricity consumption patterns shifted significantly as routines gradually returned to normal post-pandemic. Predictive models established before the pandemic, predicated on certain input-output relationships, are likely to underperform owing to these shifts in underlying data trends.

Concept drift is a phenomenon that impacts a broad range of applications, recognized and tackled across various domains including medicine, industry, education, and commerce. For instance, in the medical sector, the diminishing efficacy of antibiotics over time due to microbial resistance highlights concept drift; misuse of antibiotics can lead to resistance, compromising their effectiveness when critically needed. This reflects how alterations in medication usage can influence disease progression. In the financial sector, areas like bankruptcy forecasting or credit evaluation, traditionally viewed as stable, may experience concept drift due to underlying shifts in social trends and behaviors. In the realm of industrial monitoring, shifts in production processes, service monitoring, or consumer actions can lead to concept drift. Similarly, in transportation, traffic management systems that utilize data mining to assess traffic conditions, such as vehicle density and accident rates, need to adapt to concept drift caused by seasonal or long-term changes in traffic flows. This phenomenon has been extensively studied across fields like data mining, machine learning, statistics, and information retrieval, with the term "concept drift" being referred to by various names in different research areas, as indicated in Table 1, showcasing the wide-ranging terminologies used to describe concept drift across distinct disciplines (Gama et al. 2014; Mahdi, Ali, Pardede, Alazab, et al. 2024; Tsymbal 2004).

Table 1. Concept Drift Terminology

Field	**Terminology**
Data Mining	Concept Drift
Machine Learning	Concept Drift, Covariate Shift
Intrusion Detection Systems (IDS)	Model Drift
Evolutionary Computation	Changing Environment
AI and Robotics	Dynamic Environment
Statistics, Time Series	Non-Stationary
Databases	Concept Drift, Load Shedding
Information Retrieval	Temporal Evolution

Among the variety of concept drift techniques that have been proposed so far, the diversity measure has been used as a promising method for detecting concept drifts (Mahdi et al. 2020a, 2020b; Mahdi, Pardede, and Ali 2021), however, there is a necessity for a comparative analysis to ascertain their efficiency, identify the most effective approach, and delineate a trajectory for their sustained evolution in machine learning contexts. In this paper we examine the role of the diversity measure in detecting concept drift and compare four different ways of using them: DMDDM for drift detection in a fully supervised binary classification context (Mahdi et al. 2020a), DMDDM-S in a semi-supervised framework (Mahdi et al. 2020b), DMODD for online drift detection in a fully supervised multi classification context(Mahdi, Ali, Pardede, and Al-Quraishi 2024), and HBBE, a hybrid block-based ensemble designed for addressing various forms of concept drifts (Mahdi et al. 2021).

This study aims to address the following research questions:

> **Research Question 1**: How do different methodologies, such as DMDDM in a fully supervised context, DMDDM-S in a semi-supervised framework, DMODD as online drift detection, and HBBE as a hybrid block-based ensemble, utilize diversity measures effectively for concept drift detection?
> **Research Question 2**: In a comparative analysis, which method among DMDDM, DMDDM-S, DMODD, and HBBE performs the best in terms of detecting concept drift and enhancing model performance?
> **Research Question 3**: What are the potential future directions and research opportunities in using diversity measures for concept drift detection?

To answer these research questions, we present three comparative analyses of the four methods, evaluating their effectiveness in detecting concept drift and their ability to improve model performance.

The rest of the report is structured as follows. We present an overview and notation on concept drift in Section. Section describes the literature review that will be used for the analysis. Insights into diversity measure and its role in concept drift detection are presented in Section, followed by the experimental results and analysis of DMDDM, DMDDM-S, DMODD and HBEE in Section. Finally, the conclusion is presented in Section.

OVERVIEW OF CONCEPT DRIFT

This section presents an overview of concept drift, including its definition, origins, varieties, and the methodologies for adapting to concept drift (Mahdi, Ali, Pardede, Alazab, et al. 2024; Tsymbal 2004). The concept of concept drift was initially introduced by (Schlimmer and Granger 1986), primarily highlighting the phenomenon where the distribution of underlying data streams evolves over time.

Concept Drift Phenomenon

When it comes to non-stationary environments, the underlying distribution might change $D_i \neq D_j$, for any two time points i and j. Accordingly, the concept of the two points becomes unstable and the model will not be able to approximate the recent incoming data distribution. Consequently, the essential task of streaming data analytics is finding any significant changes in incoming data. Furthermore, the problem of changing definitions and distributions of a learner of incoming data will influence the classification accuracy of a model that is trained on data used previously. One of the most important properties of data streams is that they can change over time. Therefore, classifiers for data streams need to be capable of predicting, detecting, and adapting to concept changes. In order to do so, the nature of changes needs to be studied, including their rate, cause, predictability and severity.

According to Bayesian decision theory, a classification model can be described by the prior probabilities of classes $p(y)$ and class conditional probabilities $p(y|x)$, for all classes y K_1, ...,Kc, where c is the number of predefined classes. The dynamic nature of data streams is reflected by changes in these probability distributions in an event called concept drift. In practical terms, concept drift means that the concept about which data is being collected may shift from time to time after a minimal stability period. Depending on the research area, concept drift can sometimes be referred to as temporal evolution, population drift, covariate shift, or non-stationarity. Most studies assume that concept drift occurs unexpectedly and is unpredictable, in contrast to seasonal changes. However, concept drift adaptation mechanisms often entail solutions for cases where changes can be anticipated in correlation with environmental events.

The Formal Definition of Concept Drift

Assuming P_{t0} denotes the joint probability distribution of the input variable x and the target variable y at time t_0, and P_{t1} denotes the joint probability distribution of x and y at time t_1, concept drift is said to occur if Equation (1) is satisfied when transitioning from t_0 to t_1.

$$\exists x : P_{t0}(x,y) \neq P_{t1}(x,y) \tag{1}$$

Currently, the distribution of the underlying data has shifted away from concept C_1 to a new concept C_2. This shift is attributed to the dynamics of joint probability $P_t(x, y) = P_t(x)\ P_t(y|x)$, and if Equation (2) is fulfilled as time progresses from t_0 to t_1, a concept drift is observed. Variations in either $P_t(x)$ or $P_t(y|x)$ are capable of inducing concept drift.

$$\exists x : P_{t0}(x)P_{t0}(x|y) \neq P_{t1}(x)P_{t1}(x|y) \quad (2)$$

Therefore, concept drift can be formally defined as follows:

Definition 1. For a given data stream S, concept drift may occur between two points in time, t and $t + \Delta$, *iff* x: $p^t(x, y) \neq p^{t+\Delta}(x, y)$ where *pt* refers to the joint distribution at time t between the set of input attributes and the class label.

By considering this, any changes in incoming data can be characterized by changes in the components of Bayesian decision theory:

- Prior probabilities $p(y)$ are prone to changes.
- Probabilities $p(X|y)$ of class conditional are also prone to changes.
- Consequently, posterior probabilities $p(y|X)$ may/may not change.

Based on the cause and effect of these changes, two types of drift are identified: real drift and virtual drift.

1. ***Real Concept Drift*:** When the probability of y given x undergoes a change, the probability of x remains the same, i.e., $P_{t0}(x|y) \neq P_{t1}(x|y)$ and $P_{t0}(x) = P_{t1}(x)$. This scenario significantly affects the prediction model, representing a genuine concept drift that alters both the feature space and its decision-making boundary.
2. ***Virtual Concept Drift*:** When there is a change in the probability of x, while the probability of y given x remains unchanged, i.e., $P_{t0}(x) \neq P_{t1}(x)$ and $P_{t0}(x|y) \neq P_{t1}(x|y)$. This case belongs to virtual concept drift, which does not affect its decision boundary and only changes the feature space.

Also, in line with Bayesian decision theory(Hart, Stork, and Duda 2000), Equation (3) is derived:

$$P(yx)\frac{P(y)*P(xy)}{P(x)} \quad (3)$$

To make the distinction between real and virtual drift clearer, consider the following example, detailed in **Table 2.** Imagine a company that developed a machine learning model to forecast customer churn, utilizing data on customer demographics and buying patterns. A few months later, the company launches a new range of products, leading to alterations in the buying behaviors of certain customers. Consequently, the model's predictive accuracy begins to diminish. The company is now faced with the task of determining if this decrease in performance is attributed to real or virtual drift.

Table 2. Real Drift VS Virtual Drift

Type of Drift	Example	Description
Real Drift	Changes in purchasing behavior	The introduction of a new product alters customer purchasing behaviors, which changes the previously understood relationship between demographics and churn. This shift in the conditional probability distribution $P(y|x)$ necessitates model retraining to adapt to these new behavioral patterns.
Virtual Drift	Change in feature space	Adding a demographic feature like occupation to the model, which doesn't affect churn, exemplifies virtual drift. Here, the input feature space expands, but since occupation doesn't influence churn, the underlying relationship and $P(y|x)$ remain stable, and model performance is unaffected.

The Types of Concept Drifts

Beyond the origins and effect of concept changes, research identifies several patterns in how these changes occur, as depicted in **Figure 1**. An Abrupt Drift occurs when, at a specific time *t*, the original source distribution S_t is instantly replaced by a new distribution S_{t+1}. Such abrupt changes can significantly impair a classifier's performance since the classifier was trained on a different class distribution. In contrast, Gradual Drifts involve a more moderate pace of change, characterized by a period during which examples from two distinct distributions, S_t and S_{t+1}, are intermingled. Over time, the likelihood of encountering examples from S_t diminishes, while the probability of encountering examples from S_{t+1} rises. Another form of drift is Recurrent Concepts, which refers to concepts that have been active in the past and may re-emerge after a period of absence.

LITERATURE REVIEW

Numerous forecast models have been put forward to tackle the issue of concept drifts. In this section, we examine the renowned drift detectors from past studies:

The FHDDM, or Fast Hoeffding Drift Detection Method (Pesaranghader and Viktor 2016), leverages Hoeffding's inequality within a window of a specified size, n, to identify drifts. If a notable change occurs between the present probabilities and the peak of accurate forecasts, the window indicates a drift.

The DDM, or Drift Detection Method (Gama et al. 2004) stands out as a prominent method in its domain. It hinges on classifier error; a rise in this error alongside an uptick in the data stream's training samples may signal a potential sudden shift or drift. Recognizing this, DDM sets a cautionary threshold. When model error rates hit this mark, incoming samples are set aside in a designated window. If the error rate escalates to a preset drift threshold, the classifier undergoes a reconfiguration process, drawing insights from the samples in this specific window.

The ADWIN, or Adaptive Sliding Window approach (Bifet and Gavalda 2007) evaluates two sub-windows by shifting a window, w, across prediction outcomes. If a significant difference in the averages of these sub-windows arises, ADWIN flags a concept drift. Upon identifying this drift, ADWIN removes elements from the window's end until the substantial variation disappears.

The HDDM_A test and HDDM_W test (Frias-Blanco et al. 2014) both employ Hoeffding's bounds for drift identification. The former method contrasts moving averages, while the latter examines the weight of such averages. The weighting process incorporates the EMWA forgetting scheme (Ross et al. 2012). Nonetheless, the creators observed that HDDM_A test is better suited for immediate shifts, whereas the HDDM_W test excels in detecting slow, gradual drifts.

The PH Test technique, as discussed in(Gama, Sebastiao, and Rodrigues 2013), is commonly applied in signal processing applications for drift detection. In this approach, a cumulative discrepancy is calculated between observed values and their average up to the present moment T is captured by a variable m_T. Additionally, the minimum value of M_T, denoted as M_T, is updated over time. A concept drift is indicated when a significant discrepancy between m_T and M_T is observed.

The SegDrift2 approach, referenced in (Pears, Sakthithasan, and Koh 2014), employs two storage mechanisms for incoming data: the first repository amalgamates new with old data entries, while the second exclusively houses new entries. SegDrift2 evaluates the mean values of the entries in both repositories. A concept drift is identified when the discrepancy between these two mean values exceeds a predefined threshold.

SEED (Huang et al. 2014) employs a window approach, contrasting two sub-windows within this main window. When a noticeable difference in averages is observed, the earlier segment of the window is discarded. Moreover, SEED applies the Hoeffding Inequality combined with the Bonferroni correction to determine the test statistic and to execute block compression, eliminating extraneous cut points.

RDDM (Barros et al. 2017) was introduced to address the reduced sensitivity issue that negatively impacts DDM's performance. RDDM incorporates a process to discard older data points and periodically refreshes the RDDM statistics essential for drift detection. The creators highlighted that, in many instances, RDDM tends to outperform DDM in terms of accuracy and promptness in drift identification, though it may result in more false positives and greater memory usage.

EDDM, developed by the authors in (Baena-Garca et al. 2006), modifies the DDM approach. It employs the same warning alarm system as DDM, but instead of tracking the error rate, EDDM focuses on the distance between consecutive errors. Thus, when the concepts within incoming data remain stable, the gap between sequential errors narrows; if not, warnings and indications of drift are triggered.

The Accuracy Updated Ensemble (AUE2), referenced in (Krawczyk and Woźniak 2015), employs an online classifier that enables direct updates to individual learning models, in contrast to the Adaptive Weighted Ensemble (AWE), which only adjusts weights. When drift is not anticipated, the classifiers enhance their performance as though they were trained on a single, extensive dataset. This means that the block size can be reduced without compromising the ensemble's accuracy.

Accuracy Weighted Ensemble (AWE) (Wang et al. 2003), introduces a method where, upon receiving a new data block, a fresh classifier is trained utilizing a conventional static learning algorithm, like Naive Bayes, C4.5, or RIPPER. Post-training, the ensemble's current classifiers are assessed using the latest data block. Following this evaluation, which employs mean square error, the algorithm identifies and selects the top n classifiers to refresh the ensemble.

The Dynamic Weighted Majority (DWM), introduced in (Kolter and Maloof 2001), operates by weighting a collection of incremental classifiers based on their accuracy after each example. For every error detected by a component classifier, its weight is decreased by a predetermined factor. Moreover, following periodic predictions, the entire ensemble undergoes evaluation, and a new classifier may be added if necessary. However, employing numerous examples for training tends to produce a significant number of components, suggesting that future efforts could explore the possibility of pruning some classifiers.

Learn++.NSE (.NSE), described in (Elwell and Polikar 2011), adopts a block-based ensemble strategy for the incremental learning of concept drift. With each new block of data, the .NSE approach trains a new classifier and integrates these classifiers through a dynamically weighted majority voting system. The innovation in this work lies in the calculation of voting weights, which are determined by the time-adjusted accuracy of each classifier across both current and past environments.

Type-LDA, introduced in (Li et al. 2024), a novel framework for concept drift adaptation that focuses on identifying both the timing and nature of drifts to enhance model re-learning. Unlike traditional methods, Type-LDA does not rely on human-set statistical tests but uses an end-to-end model to automatically detect concept drift. The framework includes a drift type identifier and a drift point locator, which guide the adaptation process by leveraging drift type information. Experiments demonstrate that Type-LDA outperforms existing techniques, and future work aims to incorporate automatic active learning to reduce manual labeling requirements.

CDDA, introduced in (Karimian and Beigy 2023), presents a novel method for handling concept drift in data streams using a domain adaptation approach called CDDA. This method effectively manages concept drift by leveraging multiple sources, reducing both execution time and memory usage. Extensive experiments on synthetic and real-world data streams demonstrate that CDDA outperforms existing algorithms in terms of accuracy, kappa statistics, execution time, and model size. The study highlights the importance of selecting an appropriate window size to balance accuracy and computational efficiency, especially in data streams with fast drifts.

Group Drift Detection Method (GDDM), introduced in (Yu et al. 2023), to address the challenges of detecting concept drift across multiple data streams. GDDM leverages error rates as input variables, allowing it to handle streams with varying numbers and scales of features. It is the first method designed to detect group concept drift, where changes occur collectively across streams. Future work will focus on adapting to these drifts to enhance classifier performance, potentially by integrating dynamic ensemble selection techniques. The proposed method has been validated through experiments with both synthetic and real-world datasets, demonstrating its effectiveness

INSIGHTS INTO DIVERSITY MEASURE AND ITS ROLE IN CONCEPT DRIFT DETECTION

Within non-stationary environments characterized by concept drift, existing methodologies are categorized into *Statistical-based Methods*, *Windows-based Methods*, and *Ensemble-based Methods* (Mahdi, Ali, Pardede, Alazab, et al. 2024). Statistical-based Methods focus on monitoring the progression of the learning process by observing changes in the online error rates of the base learners. A decline in model performance that exceeds a predetermined significance threshold suggests the occurrence of concept drift. Windows-based methods utilize a static reference window to encapsulate historical data, alongside a dynamic sliding window to capture the most recent information. A significant discrepancy between the data distributions of these two windows signals concept drift. Ensemble-based strategies segment data streams into blocks, frequently substituting the least effective ensemble member with a new classifier after routine evaluations. This approach is adept at identifying gradual concept drifts while sustaining high accuracy levels. These evaluation techniques serve as benchmarks for assessing the efficacy of learning algorithms in managing concept drift.

In the realm of ensemble learning within standard, static data environments, diversity among ensemble members is recognized as a critical attribute. The evaluation of diversity can provide insights into the efficiency of methods designed to foster diversity. Consequently, a considerable number of studies have explored the concept of diversity as a criterion for pruning ensemble components (Banfield et al. 2003; Giacinto and Roli 2001; Margineantu and Dietterich 1997). Moreover, there have been limited efforts aimed at enhancing diversity; for instance, research identified in (Minku, White, and Yao 2009) examines the impact of diversity on the performance of online ensemble learning and modifies the Poisson distribution traditionally utilized in online bagging to better respond to concept drift. Nonetheless, this adaptation was applied with the goal of improving accuracy rather than directly assessing diversity. To the best knowledge of the authors, employing a diversity measure to directly evaluate the diversity among component classifiers and utilizing this as a criterion for drift detection represents a novel approach. This contrasts with prior methods that primarily relied on classification accuracy as a means of identifying drifts.

This innovation incorporates a diversity metric known as the *disagreement measure* alongside the *PH test* to create four distinct algorithms, each with its unique objective (Mahdi 2020). The disagreement measure quantifies diversity as the proportion of discordant decisions to the total observations, essentially reflecting the variance in performance between two classifiers on identical training sets. This can also be interpreted as the fraction of instances where one classifier is correct while the other is not, thereby indicating the classifiers' varied responses to changes in data streams. The disagreement measure is considered one of the most straightforward indicators of diversity between two classifiers (Kuncheva 2014). Thus, the four algorithms that integrate diversity measures for adapting to drift in dynamic settings mark a significant departure from existing approaches. They offer rapid response to drift with minimal time and memory requirements compared to other drift detection techniques. The subsequent subsections (A, B, C, and D) will delve into the primary contributions of each method.

A. DMDDM as a Drift Detector in a Fully Supervised Binary Classification Context.

To calculate the diversity between component classifiers on a pairwise basis, consider $X = x_1, \ldots, x_n$ as a labeled dataset and y' v = [y' v ($x1$), . . ., y' v (xn)] as an n-dimensional binary vector representing the outputs of classifier h_v. In this vector, $y'_v(x_j) = 1$ indicates a correct prediction of the class label by h_v, and 0 indicates an incorrect prediction. **Table 3** showcases (referred to as oracle outputs) all possible prediction outcomes for a pair of classifiers h_u and h_v, with the condition that $h_u =$ h_v. Here, N_{ab} denotes the count of instances $x_j \in X$ where $y'_u(x_j) = a$ and $y'_v(x_j) =$ b. Consequently, the probability values for N_{ab} are outlined as follows:

- N_{10} represents the count of instances where classifier C_i forecasts class 1 while classifier C_j forecasts class 0.
- N_{01} represents the count of instances where classifier C_j forecasts class 1 while classifier C_i forecasts class 0.

- N_{11} represents the count of instances where both C_i and C_j forecast class 1.
- N_{00} represents the count of instances where both C_i and C_i forecast class 0.

Table 3. The association between two classifiers (2 × 2).

$h_u = h_v$	h_u **corrects (1)**	h_u **incorrect (0)**
h_v corrects (1)	N^{11}	N^{10}
h_v incorrect (0)	N^{01}	N^{00}

Calculating the diversity between two base classifiers (h_u and h_v) through the disagreement measure is quantified by Equation (4):

$$D_{u:v} = N^{10} + N^{01}$$

As noted previously and referenced in [18], the PH test utilizes a variable m_T to track the cumulative differences in observed values e (error estimates). To calculate these values in a prequential (incremental with forgetting) manner, two primary methods are employed: *sliding windows and fading factors*.

The fading factor method is applied across the four strategies. This method systematically removes outdated information by applying a factor to the previous summary, followed by the addition of a new value derived from recent data. Conversely, alternative strategies employ sliding windows to maintain a collection of the d most recent examples at any given time, thereby constraining the sample size for analysis. Consequently, the *fading sum*$S_{x,\alpha}$ and the fading increment N_α at time t for a sequence of objects x are computed as follows:

$$S_{x:\alpha}(t) = x^t + \alpha + S_{x,\alpha}\left(t - 1\right)$$

$$N_\alpha(t) = 1 + \alpha * N_\alpha\left(t - 1\right)$$

Then, the *fading average* is computed at observation i:

$$M_\alpha(t) = \frac{S_\alpha(t)}{N_\alpha(t)}$$

Per (Gama et al. 2013), the fading factors method is more efficient in terms of time and memory usage than the sliding windows technique. Hence, across the four methodologies, the focus is on diversity measures observed through the fading factor, rather than on error rates. By incorporating the diversity value from Equation (4) into Equation (5), we derive the subsequent equations.

$$S_{u:v,\alpha}(t) = D_{u:v} + \alpha * S_{u:v,\alpha}(t-1)$$

$$M_{\alpha}(t) = \frac{S_{u:v,\alpha}(t)}{N_{\alpha}(t)}$$

Equation (9) is applicable alongside the PH test for tracking the diversity between two classifiers. Through this approach, the PH test triggers a drift alert when the predictions of the components (h_u and h_v) begin to diverge in an unexpected manner, effectively recognizing a notable rise in diversity. Equation (10) is then utilized to compute the cumulative difference m_T, representing the discrepancy between observed values and their average up to the present time t:

$$m_T = \sum_{t=1}^{T} \left(x_t - \bar{x}_T - \delta\right)$$

Where $\bar{x}_T = \frac{1}{T} \sum_{t=1}^{t} x_t$ and δ corresponds to the magnitude of changes that are allowed. Additionally, the minimum value of this variable is calculated as follows:

$$M_T = \min\left(m_T, t = 1 \ldots T\right)$$

In the final step, the test observes the disparity between M_T and m_T: as follow:

$$PH_T = m_T - M_T.$$

A drift is indicated when this discrepancy exceeds a specified threshold (λ).

Algorithm 1 presents the DMDDM method as cited in reference (Mahdi et al. 2020a), beginning with the initial processing of data stream samples to assess the predictions made by pairs of classifiers (lines 1-3). It proceeds to create an oracle output table, as mentioned in Table 3, which aids in identifying instances (N^{10} and N^{01}) where pairs of classifiers produce differing outcomes using the same training data x^t. Upon identifying these discrepancies, the algorithm tallies occurrences where one classifier is correct while the other is incorrect (lines 4-5). The disagreement metric is subsequently derived by aggregating these tallies and normalizing against the total count of classifiers (line 6). Following this, the fading factor technique is applied (lines 7-9), calculating the fading sum and increment (lines 7-8) and leveraging the diversity score from line 6 as a substitute for error predictions initially derived from the PH test. The fading average is calculated in line 9. For the purpose of monitoring diversity among classifier pairs, the PH test, outlined in lines 10-13, employs a variable m_T to evaluate the cumulative deviation between the observed diversity measure and its ongoing average. This test compares the current m_T value against the lowest M_T value recorded, determining if the difference exceeds a pre-determined threshold. Surpassing this threshold signals a drift. Required: *S*: Data Stream of Examples with Fully Labelled Data, Forgetting factor α: $0 < \alpha < -1$, Admissible change: $\delta = 0.1$, Drift threshold: $\lambda = 100$, M_t: 1.0D

Algorithm 1. Pseudocode of Drift Detection Method (DMDDM)

Output: Ensure: Drift {TRUE, FALSE}
For each example $x^t \in S$ DO
C_v_*Prediction* = get prediction using x_t
C_u_*Prediction* = get prediction using x_t
IF C_v_*Prediction* = 0.0 and C_u_*Prediction* = 1.0 THEN b++
IF C_u_*Prediction* = 0.0 and C_v_*Prediction* = 1 THEN c++
The Disagreement, $D_{u,v} = b + c / L$
$S_{u:v,\alpha}(t) = D_{u:v} + \alpha \times S_{u:v,\alpha}(t - 1)$
$N_\alpha(t) = 1 + \alpha \times N_\alpha\ (t - 1)$
$M\alpha(t) = S_{u:v,\alpha(t)} / N\alpha(t)$
SumDiversity = SumDiversity + $M\alpha(t)$
$m_T = (m_T + M_\alpha(t)$ (*SumDiversity* / instancesSeen) - δ)
$M_T = GettheMin\ (M_T, m_T)$
$PH^{test} = m_T - M_T$
IF $PH^{test} > \lambda$ THEN
Return True and incrementally train classifier C_v and C_u with x^t
Else Return False
ENDFOR

B. DMDDM-S as a Drift Detector in a Semi-Supervised Binary Classification Context.

DMDDM-S (Mahdi et al. 2020b) builds upon the concept of DMDDM by addressing situations where class labels for the incoming data stream are unavailable, particularly in the context of concept drift challenges. In a binary classification scenario involving a pair of classifiers, each component can predict an example as either 0 (negative class) or 1 (positive class). By analyzing the predictions from each classifier, we can determine the level of disagreement between their predictions.

Thus, this method focuses on quantifying the discrepancy in predictions between pairs of classifiers without considering the actual class labels.

Initially, the DMDDM strategy is employed for identifying concept drift within a semi-supervised setting (DMDDM-S), where its primary benefit in binary classification is the ability to detect disagreements between component predictions without needing the actual labels. Secondly, this newly devised drift detection mechanism is utilized to pinpoint abrupt drifts in scenarios lacking class labels for incoming data. To our knowledge, this constitutes the inaugural attempt at leveraging such a method for concept drift detection. Thirdly, we integrate k-prototype clustering to assign labels to unlabeled data, subsequently retraining the model with both newly labeled and previously labeled data to align with the prevailing concept. Our findings indicate that the proposed drift detector, even when only 50% of the data is labeled, identifies drifts more promptly and efficiently in terms of runtime and memory compared to traditional methods that rely on fully labeled datasets.

Algorithm 2 introduces the DMDDM-S technique, which commences by processing each data stream example to obtain predictions from a pair of classifiers (line 1). It then identifies examples where these classifiers disagree, constructing outputs reflecting these differences. Observations where one classifier is correct and the other is incorrect are tallied (lines 1-3). This approach utilizes a disagreement measure, combined with the PH test, for detecting concept drift (line 5). The disagreement measure is calculated by summing these observations and dividing by the total number of classifiers. Subsequently, the fading factor method is applied (lines 6-8), where the diversity value from line 5 replaces error estimates used in the original PH test. To track classifier pair diversity (lines 9-13), the modified PH test uses variable m_T, to gauge the accumulated disparity between the observed diversity value and its average. If the difference between m_T and the minimum M_T exceeds a certain threshold, a drift is indicated.

Algorithm 2. Pseudocode of Diversity Measure as a Drift Detection Method in a Semi-Supervised Environment (DMDDM-S)

Required: S: data stream of examples (labeled), α: forgetting factor (0 < α < 1), δ: admissible change (e.g., δ = 0.01), λ: drift threshold, Classifiers (Hoeffding Tree, Perceptron): L = 2, Buffer sizes: $\|W_{ld}\|$ = 100, $\|W_{uld}\|$ = 100, Ensemble components: b,c = 0, M_T: 1.0D, Output: Drift ∈ {TRUE, FALSE}	
1	For each example $x^t \in$ S do
2.	Obtain predictions C_v and C_u using x^t
3.	If predictions differ ($C_v = 0$ *and* $C_u = 1$ or $C_v = 1$ *and* $C_u = 0$):
4.	Increment disagreement counter b_c

continued on following page

Algorithm 2. Continued

5.	Calculate disagreement ratio $D_{u,v}$ = b + c /l
6.	$S_{u:v,\alpha}(t) = D_{u,v} + \alpha + S_{u:v,\alpha}(t\text{-}1)$
7.	$N_{\alpha}(t) = 1 + \alpha + N_{\alpha}(t\text{-}1)$
8.	$M_{\alpha}(t) = S_{u:v,\alpha(t)} / N_{\alpha}(t)$
9.	$Sumdiversity = SumDiversity + M_{\alpha}(t)$
10.	Calculate current $m_T = m_T + M_{\alpha} + -$ (SumDiversity / instancesSeen) - δ)
11.	$M_T = min(M_T, m_T)$
12.	Compute $PH_{test} = m_T - M_T$
13.	if $PH_{test} > \lambda$ (Return TRUE)
14.	Manage buffer W_{ld} by updating with W_{uld}
15.	$N_{ld\,=}$ $K - PrototypeClustering$ (W_{ld})
16.	Incrementally train classifiers C_v and C_u with x^t
17.	If buffer W_{uld} is full, manage by removing oldest instances
18.	End for

In the event of a drift, the algorithm labels current unlabeled data for model retraining, maintaining its accuracy. It merges windows of labeled (W_{ld}) and unlabeled data (W_{uld}) employing K-prototype clustering to label the unlabeled data. Upon drift detection, the drift detector is evaluated (lines 14-15). The model is then incrementally trained with the newly labeled data (N_{ld}) (line 16), adjusting the dynamic window of labeled data (W_{ld}) If this window reaches a predefined size, the oldest data is replaced with new; otherwise, x^t is added to (W_{ld}).

C. DMODD as an Online Drift Detector in a Fully Supervised K-Class Problem Context.

In the initial iterations of this research (DMDDM, DMDDM-S), for binary classification, the disagreement measure $D_{v,u}$ (Eq. 1) was employed. Yet, for multi-label classification challenges, relying solely on the outcomes outlined in **Table 3** to distinguish discrepancies between classifier pairs that misclassify the same instance with different labels proves ineffective. As an advancement from these early efforts, this study introduces a refined approach for meticulously tracking the precise predictions of classifiers beyond the binary correct/incorrect assessment. To accurately capture the variance between classifier pairs, we introduce a contingency table $C_{i,j}$. This table records the count of instances $x \in X$ at the intersection of row i and column j, where classifier $h_v(x) = i$ and classifier $h_u(x) = j$, as demonstrated in **Table 4** for a multi-label issue. In this table, the aligned decisions of classifier pairs are cataloged along the diagonal of matrix $C_{i,j}$. To assess their similarity, Equation 13 calculates the sum of these diagonal values, dividing it by the total instance count n. Moreover, to indicate potential drifts, we apply the PH test, as delineated in Equations 10 and 12 from our preliminary findings. Consequently, a drift is suggested when this disparity surpasses a predetermined threshold (λ).

Table 4. Output of a Pair of Classifiers for the Multi-class Classification Problem

	h_u (x) = 0	**h_v (x) = 0**	**...**	**h_u (x) = (k-1)**
h_u (x) = 0	**C_{00}**	**C_{01}**	**...**	**$C_{0(k-1)}$**
h_u (x) = 1	C_{10}	C_{11}	...	$C_{1(k-1)}$
...	...	...	...	...
h_u (x) = (k-1)	$C_{(k-1)0}$	$C_{(k-1)1}$	...	$C_{(k-1)(k-1)}$

$$\theta = \frac{1}{n}\sum_{i=0}^{K}\left(C_{i,j}\right)$$

The DMODD, referenced in (Mahdi 2020; Mahdi et al. 2021)and detailed in **Algorithm 3**, processes data stream examples and evaluates classifier predictions. It employs a contingency table to tally occurrences where classifier decisions diverge (lines 1-3). This approach quantifies classifier decisions, aggregating these tallies and normalizing them by the total count of instances (lines 4-5). It then calculates the fading sum and increment to determine the fading average (lines 6-8). The method tracks classifier diversity, utilizing the PH test to monitor the cumulative variation in diversity (lines 9-14). A drift is identified and addressed when the disparity between the current and minimum cumulative variations surpasses a predefined threshold. The algorithm concludes with the incremental update of the model (line 15), ensuring it remains responsive to new data and identified drifts.

Algorithm 3. DMODD: Diversity Measure as Online Drift Detector

Required: ***S:** A stream of labeled examples, Forgetting factor* **α: 0 < α < -1,** ***Admissible change*: δ = 0.1,** ***Threshold*: λ = 100, M_T: 1.0D**		
Output: Drift can be either True or False.		
1	**for***each example* x^t in **Sdo**	
2	*Cv* = obtain prediction using x^t;	
3	*Cu* = obtain prediction using x^t;	
4	*P* [*Cv prediction*] [*Cu prediction*] + +	

continued on following page

Algorithm 3. Continued

$Sum+ = P[i][i]$	
$Su{:}v{,}\alpha(t) = Sum + \alpha \times Sum\ (t-1);$	
$N\alpha(t) = 1 + \alpha \times N\alpha\ (t-1);$	
$\mathrm{M}\alpha(\mathrm{t}) = \frac{S_{u:v,\alpha}(t)}{N_\alpha(t)}$	
$SumDiversity = SumDiversity + M_\alpha(t);$	
$m_T = (m_T + M_\alpha(\mathrm{t}) - (SumDiversity/instancesSeen) - \delta);$	
$M_T = Min\ (M_T, m_T);$	
$PHtest = m_T - M_T;$	
if$PHtest > \lambda$**then**	
	Return TRUE
incrementally train Cv and Cu with x_t;	
end	

D. HBBE as a Hybrid Block-Based Ensemble Designed for Addressing Different Types of Concept Drifts.

The examination of block-based ensembles and strategies for detecting drifts enriches the comprehension of adaptive block-based ensembles alongside online drift detection broadly, and specifically, the interplay among their mechanisms for responding to concept drift. Consequently, a Hybrid Block-Based Ensemble (HBBE) framework has been developed and empirically proven (Mahdi et al. 2021). This proposed model outperforms the predictive accuracy of other leading adaptive learning algorithms across various scenarios, including sudden, gradual, recurring, and multi-class challenges. The HBBE strategy merges an online drift director (DMODD), tailored for K-class problems and capable of processing instances in real-time (one instance at a time), with block-based weighting to effectively address diverse forms of drifts. The operational mechanism of the block-based ensemble framework is characterized as follows:

- An Online Drift Detector, tailored for K-class problems, processes data instance by instance to promptly enhance the ensemble's responsiveness to sudden drifts.
- In a Block-Based manner, following every *d* examples, an evaluation is conducted alongside incremental updates to the ensemble's components, plus

the integration of a new component. This process is aimed at bolstering the ensemble's ability to adapt to gradual drifts.

- Upon drift detection by the online detector, a new classifier (nominee) is constructed using the latest instances. This nominee is weighted and incorporated into the ensemble based on a specific criterion $\theta()$. Post-drift, existing components of the ensemble are re-adjusted in terms of their weight.
- During periods of stability, where no drifts are identified, the framework operates akin to a conventional block-based ensemble.
- The decision outputs from both the online drift detector and batch learners are amalgamated through a weighted majority vote, employing a measure of suitability to guide the weighting.

Algorithm 4 integrates a block-based ensemble approach with an online drift detector. It processes data stream examples on an instance-by-instance basis (line 1) and accumulates each example in a buffer with a defined capacity *d* (line 2). The online drift detector, as specified in Algorithm 1, is then applied (line 3). Upon detecting a drift or when the buffer is full, a new classifier is constructed using the most recent examples from the buffer and is assigned a weight (line 4). Subsequently, weights are assigned to all components of the ensemble based on the buffer (line 5). If the number of classifiers in the ensemble is less than *k*, the newly built classifier is incorporated into the ensemble (lines 7-8); if the ensemble already contains *k* classifiers, the least effective classifier is replaced with the new one (lines 9-10). This methodology is designed to swiftly adapt to both sudden and gradual changes in the data stream.

Algorithm 4. Pseudocode of the Block-Based Ensemble and Drift detector (HBBE)

Required: S: data stream of examples (labelled), D: drift detector, k: number of ensemble members, B: example buffer of size d, (Q): classifier quality measure, t: example number M_T: 1.0D	
Output: *E*: ensemble of *k* weighted classifiers and 1 classifier with a drift detector	
1	**for** each example $x_t \in S$**do**
2	*B B* $\{x_t\}$
3	**if***drift detected (Algorithm 1) OR* $\lvert B \rvert = d$**then**
4	build and weight candidate classifier *C′* using B and Q (6);
5	weight all classifiers C_i in ensemble *E* using B and Q (8);
6	**end**
7	**if**$\lvert E \rvert < k$**then**
8	*E E {C′}*

continued on following page

Algorithm 4. Continued

else

if$Q()' > Q()$**then**

replace weakest ensemble member with $C\_$;

end

end

end

DISCUSSION

This section discusses the experimental results and analysis of DMDDM, DMDDM-S, DMODD and HBEE. The performance of each of the four methods is assessed through a range of experiments with concept drift detection methods such as FHDDM, DDM, ADWIN, Wtest, PH Test, SeqDrift, Atest, STEPD, SEED, RDDM, EDDM. In addition, HBBE is evaluated against well-known ensemble approaches such as AUE2, AWE, DWM, and Learn++.NSE. For full details of the literature and experiments settings of each method refer to the original works in (Mahdi 2020; Mahdi et al. 2020a, 2020b, 2021) . The subsequent section elaborates on these experiments, evaluations, and findings. To simplify the process of comparative analysis of each method, we will use the multi-criteria decision-making method, the Weighted Sum Model (WSM) (Alshamsi et al. 2023; Chakraborty et al. 2023; Daghouri, Mansouri, and Qbadou 2018; Ma et al. 2022). With WSM, we'll transform and weight each metric, then sum them up for each method to great a single score. This score will represent the balance between all metrics. We can then visualize this score for each approach using a simple bar chart.

To achieve this:

1. Each metric will be normalized between 0 and 1.
2. Weights will be assigned to each metric based on their importance. For this illustration, we assume equal weights.
3. A single score will be computed for each method.

For a better visualization and to easily identify which method offers the best balance across the metrics. The method with the longest bar is the one that optimally balances all the metrics. The length of each bar reflects the score, where a longer bar indicates a better balance across all metrics.

Table 5. Abbreviations of Measures Names and Definition.

Metric	Abbreviation	Definition
Average Delay Detection	ADD	Average time taken to detect a drift after it occurs
Average True Detection	ATD	Average rate of correctly identifying true drifts.
Average False Alarm	AFA	Average rate of incorrect warnings or alerts
Average False Negative	AFN	Average rate of missed detections or unflagged true drifts.
Average Detection Runtime (MS)	DRMS	Average time taken to complete a detection process, measured in milliseconds
Average Memory Usage (Byte)	MUB	Average amount of memory used during a process, measured in bytes.
Average Accuracy	ACC	Average rate at which a system correctly identifies or predicts both true and false events.

Methods on top have better-balanced scores. In addition, **Table 5** shows the abbreviations of the measures that will be used during the analysis.

Discussion Related to RQ1.

This section presents a comparative analysis of DMDDM, DMDDM-S, DMODD, and HBEE in respect to our research questions. To address RQ1, we must evaluate each method in relation to its counterparts. The next paragraphs show the results of the experiments and the analyses of each drift detector.

In Relation to DMDDM

In the comprehensive comparative visualization across the SEA and Sine1 datasets, detectors showcase varied performance across several metrics, with DMDDM emerging as a notable detector, as shown in **Figure 2**. DMDDM consistently exhibits superior performance in both datasets, particularly shining in the Average True Detection (ATD) and Average False Negative (AFN) metrics, where it consistently scores optimally. It ensures swift event detection, evidenced by its low Average Delay Detection (ADD) values, especially in the SEA dataset. Despite its stellar detection capabilities, DMDDM demonstrates a relatively higher Average False Alarm (AFA) in SEA, warranting attention to false alert minimization. Comparatively, other detectors like FHDDM and ADWIN exhibit trade-offs between true detection and false alarms. For instance, ADWIN, while maintaining a low ADD in Sine1, contends with higher false alarms, impacting its overall accuracy. In essence, while DMDDM stands out in rapid, true event detection, a holistic approach to enhancing detection algorithms may consider amalgamating the strengths across detectors, ensuring timely, accurate event detection with minimized false alerts.

Analyzing the performance across the two datasets, DMDDM consistently emerges as a notable method, achieving the highest Weighted Sum Model (WSM) scores, indicating a balanced performance across all metrics, as shown in **Figure. 4**. Its lower Average Delay Detection (ADD) and minimal false alarms/negatives (AFA and AFN) contribute significantly to this superior score. While DMDDM excels in rapid, accurate change detection, other methods like FHDDM and Atest also exhibit commendable performance, particularly in the Sine1 dataset, signifying their potential efficacy in certain contexts. However, methods like ADWIN and PH Test generally reflect lower WSM scores, indicating possible areas for improvement in detection speed, accuracy, or resource efficiency. The tailored optimization of these models could further enhance their practical applicability across varied data-streaming environments.

In Relation to DMDDM -S

The visualization provides a comprehensive comparative analysis of different drift detectors, including DMDDM-S, across two datasets: Sine1 and Mixed. **Figure. 2** focusing on DMDDM-S, it consistently demonstrates competitive performance across most metrics in both datasets, signifying its robustness in diverse contexts. Particularly, DMDDM-S boasts the lowest delay in detection and maintains minimal memory usage, essential for real-time applications and scenarios with memory constraints. Although SEED and FHDDM also exhibit low delay, DMDDM-S is distinguishable with its balance between low delay and low memory usage, which is pivotal for optimizing computational resources. While SEED outperforms in Mean Accuracy, DMDDM-S shows appreciable accuracy, especially in the Sine1 dataset. This balanced, efficient performance underscores DMDDM-S's viability as a key drift detector in varied applications, providing a blend of timely detection, memory efficiency, and respectable accuracy, crucial for maintaining model performance in dynamic environments.

Further examinations into other specific use-cases will bolster understanding of DMDDM-S's applicability in varied contexts. In the visualized data, DMDDM-S consistently demonstrates superior performance, securing the highest WSM scores across both the “Sine1” and “Mixed” datasets, as shown in **Figure 4**. Its low delay, efficient time usage, and commendable accuracy underscore its robustness and applicability in diverse contexts. Notably, DMDDM-S achieves this with modest memory usage, signifying an optimal balance between performance and computational resource allocation. While other methods like SEED and RDDM exhibit competitive performance, especially in the “Sine1” dataset, they are sometimes encumbered by significantly higher memory usage. Conversely, methods like PHTest and EDDM generally display lower overall performance, hampered by their less impressive accuracy and, in certain instances, suboptimal delay and time metrics. These insights spotlight DMDDM-S as a notably effective and efficient detection method amidst the evaluated options.

In Relation to DMODD

The presented visualization in **Figure. 6** elucidates a comprehensive comparison of several drift detectors, with a focal point on DMODD, across two distinct datasets: Wave and SEA, under four pivotal metrics: ADD, DRMS, MUB, and ACC. DMODD demonstrates exemplary performance, especially in the ADD metric, where it showcases the lowest delay in detection in both datasets. This indicates its proficiency in promptly identifying concept drifts, which is imperative to maintain model reliability in dynamic environments. Although DMODD does not secure the lead in the ACC metric, its balanced performance across all metrics, especially its minimal DRMS and MUB, underlines its effectiveness in resource optimization while maintaining commendable accuracy. While other detectors like MDDM-A also exhibit low ADD in the Wave dataset, DMODD maintains consistent low ADD across both datasets, bolstering its versatility and reliability in varied scenarios. Consequently, DMODD emerges as a robust, resource-efficient drift detector that assures timely and accurate drift detection in diverse contexts.

> **Figure 6** represents the balanced score for each drift detection method based on the four metrics. We can clearly see the differences in performance for each detection method across the two datasets. The length of each bar reflects the score, where a longer bar indicates a better balance across all metrics. Methods on top have better-balanced scores. From this visualization, we can easily identify which method offers the best balance across the four metrics. The method with the longest bar is the one that optimally balances all the metrics. The Balanced Score evaluations for both the SEA and Wave datasets provide a comprehensive perspective on the performance of various drift detection methods across critical metrics. When assessing the SEA Dataset, DMODD emerges prominently, showcasing its adeptness at balancing swift drift detection, efficient runtime, judicious memory utilization, and high predictive accuracy. This method's harmonious performance makes it a prime contender for applications demanding an equilibrium of these attributes.

Contrastingly, in the Wave Dataset, the competitive landscape is more diversified. While DMODD still showcases commendable performance, other methods also manifest their strengths, leading to a more varied distribution of scores. However, DMODD's consistent top-tier performance across both datasets underscores its versatility and adaptability to different data scenarios. In essence, while certain methods may be tailored to specific datasets, DMODD's resilience and balanced performance across both SEA and Wave datasets highlight its potential as a go-to choose for diverse drift detection challenges. Its ability to consistently strike an optimal balance amidst varying data dynamics is a testament to its robust design and efficiency.

In Relation to HBEE

Figure. 8 analyzing the performance of the algorithms AUE2, DWM, HBBE, AWE, NSE, and DWM across various datasets reveals distinct behaviors in terms of accuracy, runtime, and memory usage. Focusing on HBBE, it generally demonstrates competitive accuracy across all datasets, often surpassing other algorithms, particularly noticeable in the 'Wave' and 'Sea' datasets. This implies a reliable prediction capability in various contexts, making it a versatile choice for different data types. However, this does not translate into efficiency in terms of computational resources. HBBE tends to consume more runtime and memory compared to its peers, especially evident in the 'Airline' dataset where it has one of the highest usages among the algorithms. Such trade-offs between predictive accuracy and computational efficiency require careful consideration in practical applications.

In contrast, DWM consistently showcases lower memory usage and, in most cases, a more favorable runtime, albeit occasionally at the cost of accuracy. NSE stands out for its sporadically high runtime and memory usage in certain datasets (e.g., 'RBFGR'), demonstrating that its applicability might be constrained in scenarios with limited computational resources. In summary, while HBBE might be preferred for its generally robust accuracy across varied datasets, its computational demands necessitate a balanced evaluation based on the specific use-case requirements.

Figure 9 visualized Weighted Sum Model (WSM) scores elucidate varying performances of five algorithms across multiple datasets. HBBE, while not the highest-scoring algorithm, demonstrates a compelling equilibrium among accuracy, runtime, and memory usage, yielding a WSM score of 0.550. This reflects a balanced, stable performance across diverse datasets and criteria, showcasing its potential applicability in assorted contexts, especially where balanced performance in all three metrics is requisite. DWM excels with the highest WSM score of 0.930, indicating an optimal balance across all considered metrics. Conversely, NSE, with the lowest WSM score, might warrant optimization or be considered for use cases where its specific strengths align with the task's demands. In scenarios where particular metrics dominate the decision-making process, a deeper dive into individual metric performances would be pivotal.

Findings Related to RQ1

Each methodology displays a nuanced approach in leveraging diversity measures for concept drift detection. DMDDM, in fully supervised contexts, performs exceptionally in detecting drifts, focusing on high precision and operational efficiency. The analysis revealed its robustness and reliability, especially in environments with pronounced drifts, maintaining a well-rounded performance with high WSM scores, illustrating its versatility in binary classification scenarios. DMDDM-S, operating in semi-supervised frameworks, exhibits adaptability and efficacy, especially in mixed datasets. The analysis underlined its balanced performance, managing to reconcile accuracy with resource efficiency in situations with limited labeled data. It's a viable option for environments where acquiring labeled instances is challenging, maintaining competitive WSM scores and demonstrating robustness in less supervised scenarios. Lastly, HBBE, designed for K-class challenges, highlights adaptability to various drift types and exceptional accuracy across different datasets. Its methodology is beneficial in multi-class scenarios where understanding the complex interactions and dynamics between classes is crucial. However, the analysis also pointed out the need for a judicious evaluation of its applicability, particularly in resource-constrained environments, due to its moderate balanced performance in terms of runtime and memory usage. In conclusion, the in-depth analysis of each methodology accentuates their distinctive strengths, adaptabilities, and applicabilities in different contexts, showcasing the versatility and effectiveness of diversity measures in detecting concept drifts.

Discussion Related to RQ2

To address RQ2, the comparative bar chart reflects diverse WSM scores for DMDDM, DMDDM-S, DMODD, and HBBE across distinct datasets and scenarios. DMODD, with a WSM score of 20.72, ostensibly outperforms the others, suggesting a superior balance of accuracy, runtime, and memory usage in its evaluated contexts. However, it's pivotal to note the scores' inherent disparities due to originating from different analyses, potentially involving varied metrics and weights. Both DMDDM-S and HBBE exhibit WSM scores of 0.55, indicating similar balanced performances across their respective evaluated metrics, while DMDDM trails slightly at 0.30.

This underscores the necessity to consider the specific analytical contexts when interpreting these scores and deploying algorithms in practical scenarios, ensuring alignment with the demands and characteristics of the data being processed. **Figure 10** provides a succinct, albeit contextually varied, overview of the algorithms' performances, warranting a nuanced interpretation.

Discussion Related to RQ3.

To address RQ3, this subsection addresses the advantages as well as the assumptions and constraints of the diversity measures used during concept drift handling.

The diversity measure is notably efficient in terms of computational time and memory consumption. This efficiency primarily stems from the disagreement measure's reliance on fewer variables, leading to reduced memory demands and shorter update times for these variables, in contrast to methods that employ sliding windows, which necessitate more memory and time for drift detection. Regarding time complexity, the majority of drift detectors, including the one discussed here, operate with constant time complexity. However, exceptions like ADWIN and SeqDrift exhibit logarithmic complexity.

The diversity measure records the highest rates of false alarms, attributed to the sensitivity of the parameter values utilized. For instance, in the DMDDM approach, it was observed that escalating γ (100-200-300) reduces false alarms but may introduce delays in detecting changes or potentially miss them altogether, presenting a balance between minimizing false alarms and avoiding delayed detection. Despite these adjustments, DMDDM maintains a position of having one of the highest rates of detection delay while still being among the lowest in terms of time and memory requirements. It's important to note that this compromise arises from employing the PH test; one potential strategy to mitigate this balance involves managing the diversity of the disagreement measure with dual fading factors instead of a single one, as implemented in this study. This approach remains a consideration for future research. Another possible solution is to integrate fuzzy decision-making concept drift detection. The idea is to introduce fuzzy rules that guide the decision-making process to determine whether a drift has genuinely occurred, based on the output from the drift detector. Integrating fuzzy decision-making could offer several advantages:

- Instead of relying on a strict threshold, we can establish fuzzy membership functions to translate the drift detector's output into a degree of belief that a drift has occurred.

- False alarms are a significant issue in concept drift detection. Fuzzy decision-making could potentially mitigate this by allowing for more nuanced interpretations of the detector's output.

Conversely, the Hybrid Block-Based Ensemble (HBBE) demonstrates superior accuracy particularly in scenarios involving sudden drift, primarily because of the incorporation of an online drift detector that responds more swiftly to abrupt concept changes than many block-based ensemble methods. Additionally, it was observed that handling the ensemble in an online manner yields the highest classification accuracy, yet this approach compromises performance in terms of runtime. This performance trade-off stems from HBBE's integration of an online drift detector alongside a block-based ensemble framework. Finally, regarding future directions, we identify the following areas for exploration:

- Traditional Intrusion Detection Systems (IDS) rely heavily on fixed rules and patterns, which limits their flexibility and prevents them from recognizing new types of attacks(Mahdi et al. 2023; Wahab 2022). Moreover, as the statistical characteristics of data change over time, IDS may become ineffective or obsolete, a situation often described as concept or model drift. Implementing drift detection algorithms within IDS is crucial to address these issues. These algorithms analyze incoming data for changes in distribution, allowing the IDS to dynamically adjust and refresh its model to detect and classify new anomalies or threats effectively. The inclusion of drift detection algorithms, particularly those based on the diversity measure, can maintain high levels of detection accuracy, reduce false positives and negatives, and improve the overall effectiveness and security of the system.
- Within the realm of Big Data and Large-Scale Data Processing Applications, detecting various types of concept drift in massive data streams is a significant challenge. There is a critical need to enhance accuracy simultaneously decreasing computational time and memory consumption.
- For scalability in large-scale deployments, such as Internet of Things (IoT) systems that generate large quantities of data, drift detection approaches must be scalable. Future studies should concentrate on creating distributed and parallel algorithms capable of processing substantial amounts of IoT data and identifying concept drift in real time, thereby guaranteeing efficient data handling and system reactivity.

CONCLUSION

The diversity measure offers adaptability to different drift types and robustness across varied datasets, reinforcing its universal applicability and reliability in detecting concept drifts. In conclusion, diversity measures are notably effective in detecting concept drift, offering a harmonious amalgamation of accuracy, adaptability, and resource efficiency across different scenarios and classification contexts. Their balanced proficiency and versatility make them a preferable choice compared to other methods, especially in applications necessitating a nuanced understanding of evolving data patterns and concept dynamics. The Weighted Sum Model (WSM) scores illustrate that algorithms leveraging diversity measures can achieve a balanced performance concerning accuracy, runtime, and memory usage. For instance, DMDDM depicted well-rounded performance with the highest WSM scores in specific scenarios, emphasizing its reliability and versatility in binary classification scenarios.

REFERENCES

Alshamsi, A. M., El-Kassabi, H., Serhani, M. A., & Bouhaddioui, C. (2023). A Multi-Criteria Decision-Making (MCDM) Approach for Data-Driven Distance Learning Recommendations. *Education and Information Technologies*, *28*(8), 1–38. DOI: 10.1007/s10639-023-11589-9 PMID: 36718426

Baena-Garcıa, M., del Campo-Ávila, J., Fidalgo, R., Bifet, A., Gavalda, R., & Morales-Bueno, R. (2006, September). Early drift detection method. In *Fourth international workshop on knowledge discovery from data streams* (Vol. 6, pp. 77-86).

Banfield, R. E., Hall, L. O., Bowyer, K. W., & Kegelmeyer, W. P. (2003). A new ensemble diversity measure applied to thinning ensembles. In Multiple Classifier Systems: 4th International Workshop, MCS 2003 Guildford, UK, June 11–13, 2003 Proceedings 4 (pp. 306-316). Springer Berlin Heidelberg.

Barros, R. S. M., Cabral, D. R. L., Gonçalves, P. M.Jr, & Santos, S. G. T. C. (2017). RDDM: Reactive Drift Detection Method. *Expert Systems with Applications*, *90*, 344–355. DOI: 10.1016/j.eswa.2017.08.023

Bifet, A., & Gavalda, R. (2007). "Learning from Time-Changing Data with Adaptive Windowing." Pp. 443–48 in *Proceedings of the 2007 SIAM international conference on data mining*. DOI: 10.1137/1.9781611972771.42

Chakraborty, S., Raut, R. D., Rofin, T. M., & Chakraborty, S. (2023). A Comprehensive and Systematic Review of Multi-Criteria Decision-Making Methods and Applications in Healthcare. *Healthcare Analytics*, *4*, 100232. DOI: 10.1016/j.health.2023.100232

Daghouri, A., Mansouri, K., & Qbadou, M. (2018). "Multi Criteria Decision Making Methods for Information System Selection: A Comparative Study." Pp. 1–5 in *2018 International Conference on Electronics, Control, Optimization and Computer Science (ICECOCS)*. DOI: 10.1109/ICECOCS.2018.8610579

Elwell, R., & Polikar, R. (2011). Incremental Learning of Concept Drift in Nonstationary Environments. *IEEE Transactions on Neural Networks*, *22*(10), 1517–1531. DOI: 10.1109/TNN.2011.2160459 PMID: 21824845

Frias-Blanco, I., del Campo-Ávila, J., Ramos-Jimenez, G., & Morales-Bueno, R. (2014). Online and Non-Parametric Drift Detection Methods Based on Hoeffding's Bounds. *IEEE Transactions on Knowledge and Data Engineering*, *27*(3), 810–823. DOI: 10.1109/TKDE.2014.2345382

Gama, J. (2010). *Knowledge Discovery from Data Streams*. CRC Press. DOI: 10.1201/EBK1439826119

Gama, J., Medas, P., Castillo, G., & Rodrigues, P. (2004). "Learning with Drift Detection." Pp. 286–95 in *Advances in Artificial Intelligence–SBIA 2004: 17th Brazilian Symposium on Artificial Intelligence, Sao Luis, Maranhao, Brazil, September 29-Ocotber 1, 2004.Proceedings 17.*

Gama, J., Sebastiao, R., & Rodrigues, P. P. (2013). On Evaluating Stream Learning Algorithms. *Machine Learning*, *90*(3), 317–346. DOI: 10.1007/s10994-012-5320-9

Gama, J., Žliobaitė, I., Bifet, A., Pechenizkiy, M., & Bouchachia, A. (2014). A Survey on Concept Drift Adaptation. *ACM Computing Surveys*, *46*(4), 1–37. DOI: 10.1145/2523813

Giacinto, G., & Roli, F. (2001). An Approach to the Automatic Design of Multiple Classifier Systems. *Pattern Recognition Letters*, *22*(1), 25–33. DOI: 10.1016/S0167-8655(00)00096-9

Hart, P. E., Stork, D. G., & Duda, R. O. (2000). *Pattern Classification*. Wiley Hoboken.

Huang, D. T. J., Koh, Y. S., Dobbie, G., & Pears, R. (2014). "Detecting Volatility Shift in Data Streams." Pp. 863–68 in *2014 IEEE International Conference on Data Mining*. DOI: 10.1109/ICDM.2014.50

Karimian, M., & Beigy, H. (2023). Concept Drift Handling: A Domain Adaptation Perspective. *Expert Systems with Applications*, *224*, 119946. DOI: 10.1016/j.eswa.2023.119946

Kolter, J., & Maloof, M. (2001). "Dynamic Weighted Majority: A New Ensemble Method for Tracking Concept Drift." in *Int'l Conf. Data Mining (ICDM).*

Krawczyk, B., & Woźniak, M. (2015). "Reacting to Different Types of Concept Drift with Adaptive and Incremental One-Class Classifiers." Pp. 30–35 in *2015 IEEE 2nd International Conference on Cybernetics (CYBCONF)*. DOI: 10.1109/CYBConf.2015.7175902

Kuncheva, L. I. (2014). *Combining Pattern Classifiers: Methods and Algorithms*. John Wiley & Sons. DOI: 10.1002/9781118914564

Li, J., Yu, H., Zhang, Z., Luo, X., & Xie, S. (2024). Concept Drift Adaptation by Exploiting Drift Type. *ACM Transactions on Knowledge Discovery from Data*, *18*(4), 1–22. DOI: 10.1145/3638777

Ma, W., Du, Y., Liu, X., & Shen, Y. (2022). Literature Review: Multi-Criteria Decision-Making Method Application for Sustainable Deep-Sea Mining Transport Plans. *Ecological Indicators*, *140*, 109049. DOI: 10.1016/j.ecolind.2022.109049

Mahdi, O. A. (2020). *Diversity Measures as New Concept Drift Detection Methods in Data Stream Mining*. La Trobe University Melbourne.

Mahdi, O. A., Alazab, A., Bevinakoppa, S., Ali, N., & Khraisat, A. (2023). "Enhancing IoT Intrusion Detection System Performance with the Diversity Measure as a Novel Drift Detection Method." Pp. 50–54 in *2023 9th International Conference on Information Technology Trends (ITT)*. DOI: 10.1109/ITT59889.2023.10184268

Mahdi, O. A., Ali, N., Pardede, E., & Al-Quraishi, T. (2024). Online Concept Drift Detector: Optimally Balancing Delay Detection, Runtime, Memory, and Accuracy. *Procedia Computer Science*, *237*, 559–567. DOI: 10.1016/j.procs.2024.05.140

Mahdi, O. A., Ali, N., Pardede, E., Alazab, A., Al-Quraishi, T., & Das, B. (2024). Roadmap of Concept Drift Adaptation in Data Stream Mining, Years Later. *IEEE Access : Practical Innovations, Open Solutions*, *12*, 21129–21146. DOI: 10.1109/ACCESS.2024.3358817

Mahdi, O. A., Pardede, E., & Ali, N. (2021). A Hybrid Block-Based Ensemble Framework for the Multi-Class Problem to React to Different Types of Drifts. *Cluster Computing*, *24*(3), 2327–2340. DOI: 10.1007/s10586-021-03267-7

Mahdi, O. A., Pardede, E., Ali, N., & Cao, J. (2020a). Diversity Measure as a New Drift Detection Method in Data Streaming. *Knowledge-Based Systems*, *191*, 105227. DOI: 10.1016/j.knosys.2019.105227

Mahdi, O. A., Pardede, E., Ali, N., & Cao, J. (2020b). Fast Reaction to Sudden Concept Drift in the Absence of Class Labels. *Applied Sciences (Basel, Switzerland)*, *10*(2), 606. DOI: 10.3390/app10020606

Margineantu, D. D., & Dietterich, T. G. (1997). "Pruning Adaptive Boosting." Pp. 211–18 in *ICML*. Vol. 97.

Minku, L. L., White, A. P., & Yao, X. (2009). The Impact of Diversity on Online Ensemble Learning in the Presence of Concept Drift. *IEEE Transactions on Knowledge and Data Engineering*, *22*(5), 730–742. DOI: 10.1109/TKDE.2009.156

Pears, R., Sakthithasan, S., & Koh, Y. S. (2014). Detecting Concept Change in Dynamic Data Streams: A Sequential Approach Based on Reservoir Sampling. *Machine Learning*, *97*(3), 259–293. DOI: 10.1007/s10994-013-5433-9

Pesaranghader, A., & Viktor, H. L. (2016). "Fast Hoeffding Drift Detection Method for Evolving Data Streams." Pp. 96–111 in *Machine Learning and Knowledge Discovery in Databases: European Conference, ECML PKDD 2016, Riva del Garda, Italy,September 19-23, 2016,Proceedings, Part II 16*. DOI: 10.1007/978-3-319-46227-1_7

Prasad, B. R., & Agarwal, S. (2016). Stream Data Mining: Platforms, Algorithms, Performance Evaluators and Research Trends. *International Journal of Database Theory and Application*, *9*(9), 201–218. DOI: 10.14257/ijdta.2016.9.9.19

Ross, G. J., Adams, N. M., Tasoulis, D. K., & Hand, D. J. (2012). Exponentially Weighted Moving Average Charts for Detecting Concept Drift. *Pattern Recognition Letters*, *33*(2), 191–198. DOI: 10.1016/j.patrec.2011.08.019

Schlimmer, J. C., & Granger, R. H.Jr. (1986). Incremental Learning from Noisy Data. *Machine Learning*, *1*(3), 317–354. DOI: 10.1007/BF00116895

Tsymbal, A. (2004). The Problem of Concept Drift: Definitions and Related Work. *Computer Science Department, Trinity College Dublin*, *106*(2), 58.

Wahab, O. A. (2022). Intrusion Detection in the Iot under Data and Concept Drifts: Online Deep Learning Approach. *IEEE Internet of Things Journal*, *9*(20), 19706–19716. DOI: 10.1109/JIOT.2022.3167005

Wang, H., Fan, W., Yu, P. S., & Han, J. (2003). "Mining Concept-Drifting Data Streams Using Ensemble Classifiers." Pp. 226–35 in *Proceedings of the ninth ACM SIGKDD international conference on Knowledge discovery and data mining*. DOI: 10.1145/956750.956778

Yu, H., Liu, W., Lu, J., Wen, Y., Luo, X., & Zhang, G. (2023). Detecting Group Concept Drift from Multiple Data Streams. *Pattern Recognition*, *134*, 109113. DOI: 10.1016/j.patcog.2022.109113

Chapter 2
RPL and IoT Security: A Holistic Examination of Vulnerabilities

Mohamed Hechmi Jeridi
National Engineering School of Tunis, Tunisia

Tarek Azzabi
Tunisian Naval Academy, Tunisia

jalel Dziri
National Engineering School of Tunis, Tunisia

Tahar Ezzedine
National Engineering School of Tunis, Tunisia

ABSTRACT

The Internet of Things (IoT) represents a groundbreaking advancement in connectivity across homes, industries, healthcare, and transportation. It has introduced a multitude of possibilities for innovation, yet simultaneously raised critical concerns about the security of this interconnected landscape. This research provides an in-depth exploration of IoT, focusing on its core components and the pivotal protocols that enable seamless communication. Central to our analysis is the paramount concern of IoT security. This work meticulously scrutinizes vulnerabilities stemming from device weaknesses, network-related threats, and communication breaches, with a particular emphasis on fortifying IoT security. Furthermore, this study dissects the security of RPL-based IoT networks, identifying vulnerabilities and suggesting potential solutions. Subsequently, it delves into the pivotal role of Intrusion Detection Systems (IDSs) in bolstering IoT network defenses, offering insights into the deployment of IDSs within RPL-based IoT networks.

DOI: 10.4018/979-8-3693-5448-3.ch002

1. INTRODUCTION

The advent of the Internet of Things (IoT) has triggered a profound revolution in connectivity, facilitating seamless communication among devices in various domains, including homes, industries, healthcare, transportation, and more. This technology, driven by a variety of IoT devices, not only holds the potential for numerous opportunities but also introduces new security challenges. In this research, we initiate an exhaustive study of IoT, examining the essential protocols to address security concerns. Our work examines carefully vulnerabilities originating from device weaknesses, network threats, and communication breaches, with a particular emphasis on enhancing IoT security. Our research initiates at the foundational level, where IoT devices gather data, progressing to the network layer where we meticulously assess threats and vulnerabilities within the Routing Protocol for Low-Power and Lossy Networks (RPL). Our examination extends to the application layer. This work, adopts a dual-pronged approach, first analyzing the security of RPL-based IoT networks to unveil vulnerabilities and explore potential solutions. Subsequently, we delve into the role of Intrusion Detection Systems (IDSs) in fortifying IoT networks, offering insights into deploying IDSs within RPL-based IoT networks.

Figure 1. Research challenges in Internet of Things (Nivedita & Sharnil,2021)

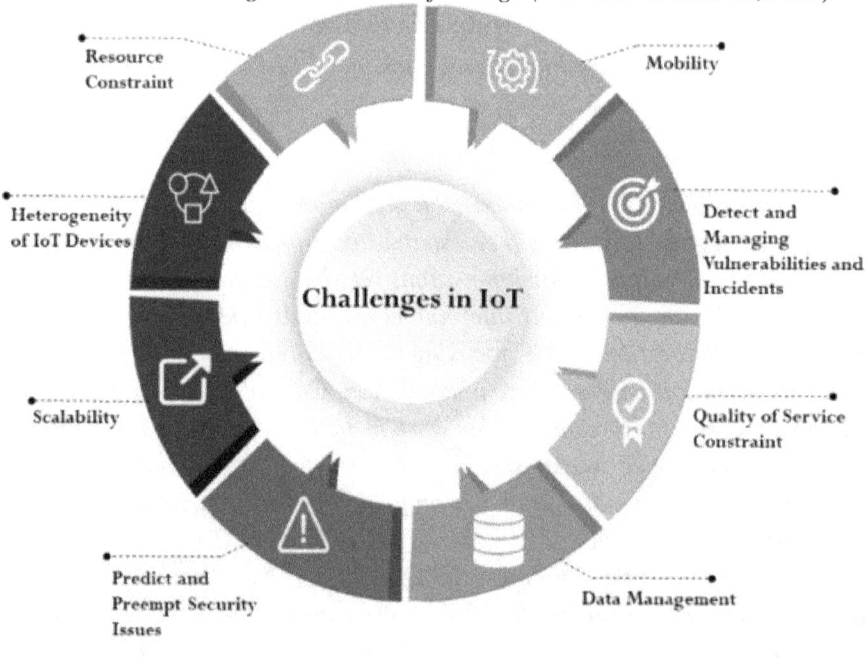

The growing significance of mobility in the IoT realm, we also address security issues related to it. Our objective is to enhance comprehension of IoT security issues and the efforts made to ensure its robust security. To achieve our research objectives, we have organized this article into a structured framework comprising distinct and interconnected sections. In Section II, we establish the fundamental principles of IoT, providing readers with a solid foundation for comprehending IoT security challenges. Transitioning to Section III, we explore the intricate realm of IoT protocols, including 6LoWPAN, IEEE 802.15.4, RPL, and CoAP, demystifying the technical intricacies that facilitate IoT communication. Section IV is dedicated to dissecting the multi-faceted security considerations within the IoT ecosystem, where we examine associated challenges and outline mitigation strategies. Finally, in Section V, we conduct a thorough examination of threats and attacks across various IoT layers, spanning from the perception/sensing layer to the network and application layers, granting readers a comprehensive understanding of the complex IoT security landscape. This structured approach guides readers from fundamental IoT concepts to security intricacies, equipping them with the knowledge required to enhance and protect IoT environments in an evolving digital landscape.

2. IOT FEATURES AND PROPERTIES

In order to efficiently institute an Internet of Things (IoT) network, several important technical considerations demand meticulous attention and resolution. These considerations encompass the management of device and technology heterogeneity, the assurance of scalability to accommodate the prospering count of networked entities, the amelioration of challenges linked to wireless connectivity and ubiquity, the facilitation of self-organization and self-healing capabilities within IoT entities, the optimization of resource allocation in light of the varied energy, storage, and computational capacities of these devices, the establishment of interoperability standards to facilitate communication among diverse vendor-sourced IoT devices, adaptation to the mobility inherent in IoT devices, and the implementation of robust security and privacy protocols to safeguard data and communications, all while concurrently balancing resource consumption through the utilization of lightweight protocols. Each of these considerations is elaborated upon below:

Heterogeneity: IoT devices and technologies exhibit a spectrum of disparate hardware platforms, computational capabilities, and communication protocols. Scalability: The architecture of the IoT system necessitates the capacity for expansion to accommodate an escalating number of networked entities, ensuring that it remains operationally efficient and effective.

Connectivity and Ubiquity: IoT objects predominantly rely on wireless connectivity to access the network and exchange data ubiquitously, necessitating solutions to mitigate associated challenges and complexities.
Self-Organization and Self-Healing: IoT objects are endowed with embedded intelligence, affording them the capability to autonomously react to environmental changes and self-organize into ad hoc networks based on contextual parameters. Resource Limitations: IoT objects are endowed with varying energy, storage, and computational capacities, demanding the development of platforms that judiciously optimize resource utilization to ensure sustainable and efficient operation.
Interoperability: The IoT ecosystem encompasses devices and services sourced from diverse vendors, mandating the establishment of interoperable standards to facilitate seamless communication and collaboration among these disparate entities.
Mobility: A substantial proportion of IoT devices exhibit mobility, thereby subjecting network conditions to dynamic alterations and necessitating adaptive communication strategies to maintain connectivity and data exchange.
Security and Privacy: IoT devices generate copious amounts of data and must be equipped with robust security and privacy measures to safeguard not only hardware integrity but also data, communication channels, and endpoint devices. Achieving this balance requires the deployment of lightweight protocols that minimize resource consumption while fortifying security measures.

3. INTERNET OF THINGS PROTOCOLS

This section explores the IoT paradigm, where diverse entities are interconnected via the Internet to collaboratively achieve shared objectives. This interconnected network creates a varied landscape, encompassing a wide range of devices and technologies. The overarching goals of scalability and universal connectivity in IoT present multifaceted challenges that require meticulous consideration and resolution. Notably, IoT entities possess inherent embedded intelligence, enabling them to autonomously self-organize and self-heal in response to dynamic environmental stimuli. While these autonomous capabilities enhance operational efficiency, they concurrently escalate the security risks associated with IoT deployments. Common Iot—such as energy, storage, and computation—with security preservation stands as a paramount objective for IoT platforms.

Figure 2. Internet of Things Characteristics

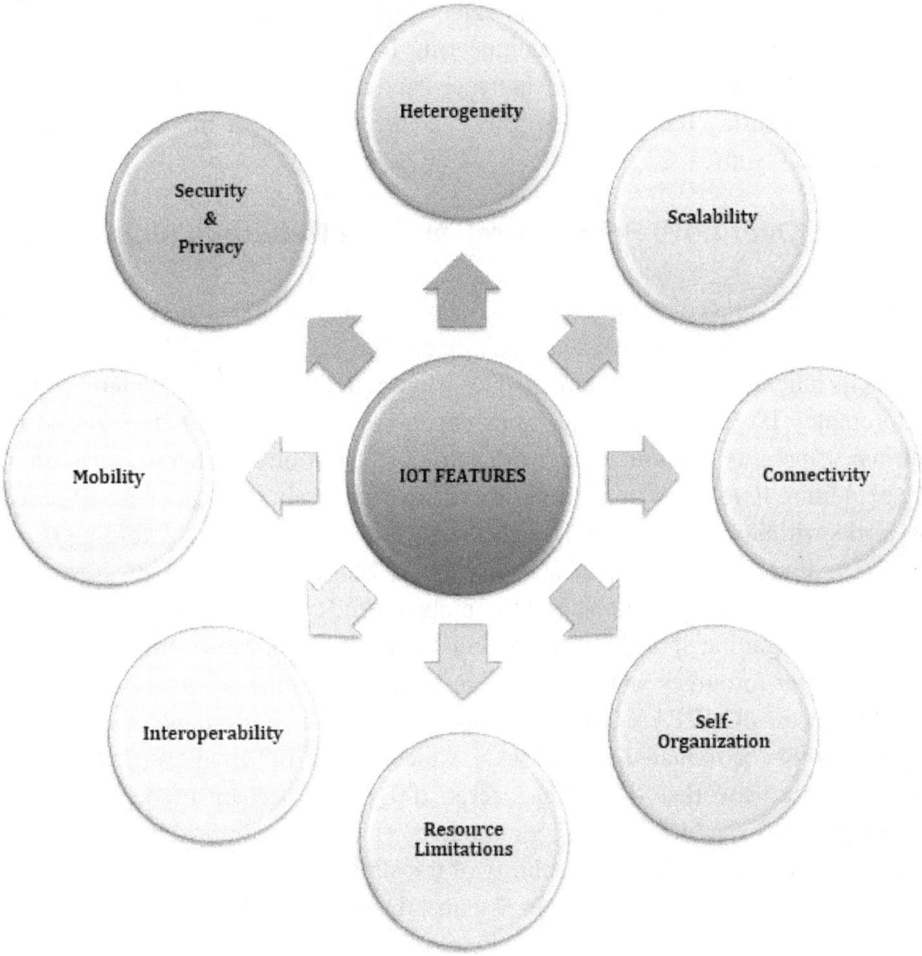

Moreover, achieving seamless interoperability amidst the diversity and heterogeneity of IoT entities assumes vital significance. Achieving these objectives depends on the widespread adoption of standardized protocols and communication frameworks, en- abling harmonious interaction among disparate IoT devices and systems. The dynamic nature of IoT, underscored by the inherent mobility of IoT devices, requires a strategic reassessment of network architecture and communication paradigms to accommodate evolving network conditions and transparent data transfer. Alongside these consider- ations, robust security and privacy measures must be firmly established to safeguard the integrity of IoT ecosystems. These measures encompass comprehensive protocols and policies designed to fortify hardware, data, communication channels, and end- point devices against potential threats. To underpin these pivotal characteristics and requirements, a myriad of protocols and standards are currently under development and standardization by reputable organizations, most notably the Internet Engineering Task Force (IETF). These protocols are tailored to address the unique challenges and demands inherent to the Internet of Things, ensuring the continued growth and maturation of this transformative technological landscape.

3.1. IPv6 Over LoW Power Wireless Area Networks (6LoWPAN)

Given the context of the Internet of Things (IoT) and the focus on security topics, the widespread proliferation of connected objects has made the use of IPv4 unfeasible due to its inherently limited address space. Consequently, a more pragmatic choice is to employ IPv6, distinguished by its expansive 128-bit address format and its intrinsic capability to facilitate network auto-configuration and stateless operation. By exploiting IPv6, each intelligent object can seamlessly integrate with IP-based networks without necessitating intermediary gateways or proxies, affording these objects a remarkable degree of autonomy in defining their own addresses. This autonomy significantly diminishes the complexity and cost associated with configuration. In recognition of the IoT's unique constraints, including restricted packet size, limited power resources, and other inherent limitations, the research community, specifically the 6LoWPAN IETF Working Group, has devised a concise adaptation of IPv6, aptly denominated 6LoW- PAN, which stands for" IPv6 over Low-Power Wireless Area Networks", (Kushalnagar et al.,2007). This mechanism adeptly reduces the length of IPv6 addresses to accommodate resource-constrained devices, while border routers assume the responsibility of translating these compressed addresses into conventional IPv6 addresses. In essence, the 6LoWPAN standard extends the applicability of IPv6 to the realm of wireless embedded environments.

Given the resource-constrained nature of IoT devices or "things," 6LoWPAN net- works leverage a compressed IPv6 protocol for networking, predominantly relying on the IEEE 802.15.4 standard as the underlying data-link and physical layer protocol. The architecture of 6LoWPAN incorporates an adaptation layer positioned between IPv6 and IEEE 802.15.4 (MAC/PHY layer). This adaptation layer efficiently com- presses IPv6 headers, eliminating superfluous overhead, and delimits packet routing within wireless networks employing the IEEE 802.15.4 protocol, architectural components and protocol stack are represented in Figures 3. Additionally, 6LoWPAN defines mechanisms for fragmenting IPv6 datagrams when their size exceeds the Maximum Transmission Unit (MTU) of 127 bytes as stipulated by the IEEE 802.15.4 standard. Transmission of IPv6 Packets over IEEE 802.15.4 Networks defines the original IPv6 over Low-Power Wireless Personal Area Network (6LoWPAN) datagram fragmentation mechanism for LLNs. One critical issue with this original design is that routing an IPv6 packet across a route-over mesh requires the reassembly of the packet at each hop. "An Architecture for IPv6 over the TSCH mode of IEEE 802.15.4", 6TiSCH indicates that this may cause latency along a path and impact critical resources such as memory and battery; to reduce those undesirable effects, it recommends using a 6LoWPAN Fragment Forwarding (6LFF) technique (Thubert P.,2020).

3.2. IEEE 802.15.4

The IEEE 802.15.4 standard delineates the specifications for low-power wireless embedded radio communications operating across three distinct frequency bands: 2.4 GHz, 915 MHz, and 868 MHz. These frequency bands serve as the designated spec- tral regions for facilitating wireless communication among embedded devices with an emphasis on minimizing power consumption, making them particularly suitable for battery-operated and energy-constrained applications within the field of wireless sensor networks and the Internet of Things (IoT). The standard's provisions encompass protocols and modulation schemes optimized for efficient and reliable data exchange in these frequency bands, thus establishing a robust foundation for wireless communication systems operating in diverse industrial, commercial, and residential contexts (Figure4). The IEEE 802.15.4 standard establishes a comprehensive framework comprising a Medium Access Control (MAC) sub-layer and a physical layer (PHY) tailored specifically for low-rate wireless private area networks (LR-WPAN). Its pivotal role in the domain of wireless communication, especially within the context of the Internet of Things (IoT), is underscored by its foundational role in the development of 6LoW- PAN communication protocols.

Figure 3. 6LoWPAN Architecture and protocol stack (PADMA, Bhukya & ERUKALA, Suresh Babu, 2023)

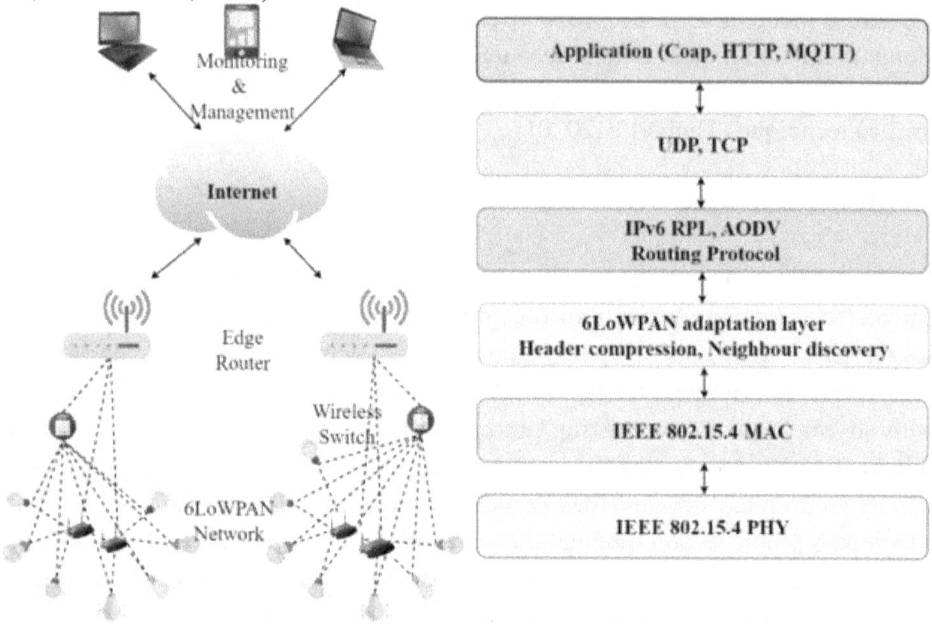

The IEEE 802.15.4 standard has several important features, including low power consumption, modest data rates, cost-effectiveness, and robust message throughput capabilities. As a result, it has found widespread adoption as the core technology underpinning various IoT protocols such as ZigBee, Wireless HART, MiWi, and ISA 100.11a. Over the years, IEEE 802.15.4 has undergone iterative enhancements to meet the evolving demands of embedded networking applications. It was initially released in 2003, followed by revisions in 2006, 2011, and 2015, with the 2.4 GHz frequency band being the predominant choice today due to its favorable attributes, including reasonable data rates, reliable communication, scalability for accommodating numerous nodes, and global usability. IEEE 802.15.4 networks exhibit versatility in forming different topological configurations, including star, cluster-tree, and mesh (peer-to-peer) arrangements, thereby accommodating a spectrum of deployment scenarios. Within these networks, the standard defines two categories of devices: Full Function Devices (FFD) and Reduced Function Devices (RFD). FFDs can as sume the roles of common nodes or serve as coordinators for Personal Area Networks (PANs).

Conversely, RFDs are characterized by their simplicity, possessing limited resources and communication capabilities, rendering them unsuitable for coordinator roles. Notably, the IEEE 802.15.4 standard imposes a maximum transmission unit

(MTU) size of 127 bytes, and frames are equipped with a 16-bit cyclic redundancy check (CRC) for error detection and correction (Figure 5), ensuring data integrity and reliability in wireless communication scenarios (IEEE Std. 802.15.4-2011.2011; Shelby & Bor- mann,Shelby & Bormann,).

3.3. Routing Protocol for Low-Power and Lossy Networks (RPL)

The focal point of this work revolves around the Routing Protocol for Low-Power and Lossy Networks (RPL) and its significant impact on securing the IoT. This protocol serves as a specialized routing protocol designed for 6LoWPAN (IPv6 over Low-Power Wireless Personal Area Networks) networks, as previously explained. These networks pose distinctive challenges, primarily due to their inherent resource constraints, notably limited energy and computational capabilities. RPL crosses these constraints by fashioning a conceptual model of the network's structure in the form of Destination Oriented Directed Acyclic Graphs (DODAGs). These DODAGs underpin the routing infrastructure for conveying data packets within the network. Within this context, a DODAG assumes the role of a hierarchical structure interconnecting nodes, wherein each DODAG designates a 6LoWPAN Border Router (6LBR) as its edge router. These 6LBRs, in turn, serve as vital gateways, bridging the 6LoWPAN network with both the wider Internet and other 6LBRs via a backbone link, as visually depicted in Figure 3. The establishment and ongoing maintenance of these DODAGs hinge on the exchange of control messages, namely DODAG Information Objects (DIO), DODAG Information Solicitations (DIS), and DODAG Destination Advertisement Objects (DAO), coupled with the implementation of a Trickle timer mechanism. The Trickle algorithm assumes a pivotal role in expediting the identification and rectification of routing in- consistencies, thereby bolstering network stability. The nodes within the network can acknowledge DAO messages through the emission of DAO-ACK messages. Further- more, RPL incorporates an Objective Function (OF) in conjunction with optional node and link metrics and constraints, which are transmitted via DIO and DAO messages. This collective framework serves to optimize routing paths, ensuring the selection of the most efficient routes. Within the RPL protocol, each network object is ascribed an IPv6 address, coupled with a Rank value that delineates its positional context relative to its parent node within the DODAG.

The Rank metric adheres to a monotonic principle, ascending from the 6LBR to- wards the network's leaf nodes and descending from the leaf nodes towards the 6LBR. In response to topology modifications and failures, RPL exploits two complementary mechanisms: global repair and local repair. Global repair, instigated by the 6LBR, entails the updating of the DIO DODAG version number field, leading to the comprehensive reconstruction of the entire network topology.

Figure 4. IEEE 802.15.4 characteristics. (Nazabal et al.,2013)

PHY (MHz)	Frequency Band (MHz)	Coverage	Data Rate (Kbps)	Number of Channels	RX Sensitivity	Modulation
868/915	868-868.6 902-928	ISM Europe ISM America	20 40	1 10	- 92 dBm - 92 dBm	BPSK BPSK
868/915 (optional)	868-868.6 902-928	ISM Europe ISM America	250 250	1 10	- 85 dBm - 85 dBm	ASK ASK
868/915 (optional)	868-868.6 902-928	ISM Europe ISM America	100 250	1 10	- 85 dBm - 85 dBm	O-QPSK O-QPSK
2450	2400-2483.5	ISM Worldwide	250	16	- 85 dBm	O-QPSK

Figure 5. IEEE 802.15.4 Frame details. (Mahmud et al.,2018)

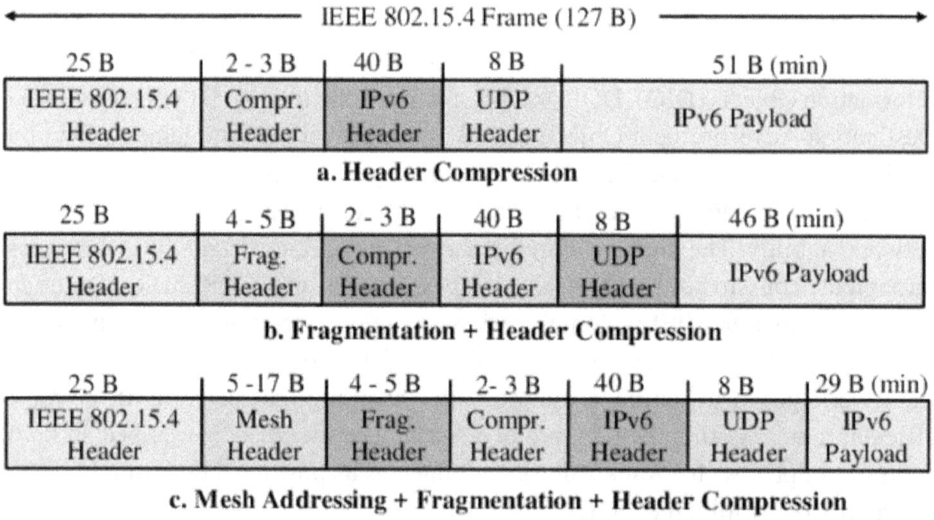

Upon receiving these updated DIO messages, the constituent nodes reset their Trickle timers, revise their parent node listings, and recalibrate their Ranks accordingly. In contrast, local repair mechanisms come into play when specific inconsistencies are detected. These mechanisms afford nodes the capability to reset their Trickle timers independently and to select new preferred parent nodes. This autonomy avoids waiting for global repairs or the arrival of fresh DIO messages. RPL lends support to three distinct traffic topologies: Multipoint-to-Point (MP2P), Point-to-Multipoint (P2MP), and Point-to-Point (P2P). In the MP2P configuration, traffic flows from in-network nodes toward the 6LBR, while in P2MP, traffic is routed from the 6LBR towards in-network nodes, making it apt for applications requiring control of actuating devices. The P2P topology involves the transmission of packets from source nodes to a common ancestor and subsequently cascading downward to the destination node in a storing mode. Conversely, packets are routed from source nodes to the 6LBR and then relayed downward to the destination node in a non-storing mode. Facilitating these diverse traffic patterns necessitates the discovery of both upward and downward routes within the network. (Thubert et al.,2012; Vasseur et al.,2012).

3.4. Constrained Application Protocol (CoAP)

In the context of Internet of Things (IoT) protocols, the prevalent choice for the trans- port layer in IoT networks is the connection-less User Datagram Protocol (UDP). This preference arises from the inherent challenges of maintaining continuous connections between low-powered devices operating over lossy links. In response to these challenges, the Constrained Application Protocol (CoAP) was introduced by the Internet Engineering Task Force (IETF) Constrained RESTful Environments (CoRE) working group.

Figure 6. RPL Functional Model (Badach A.,2018)

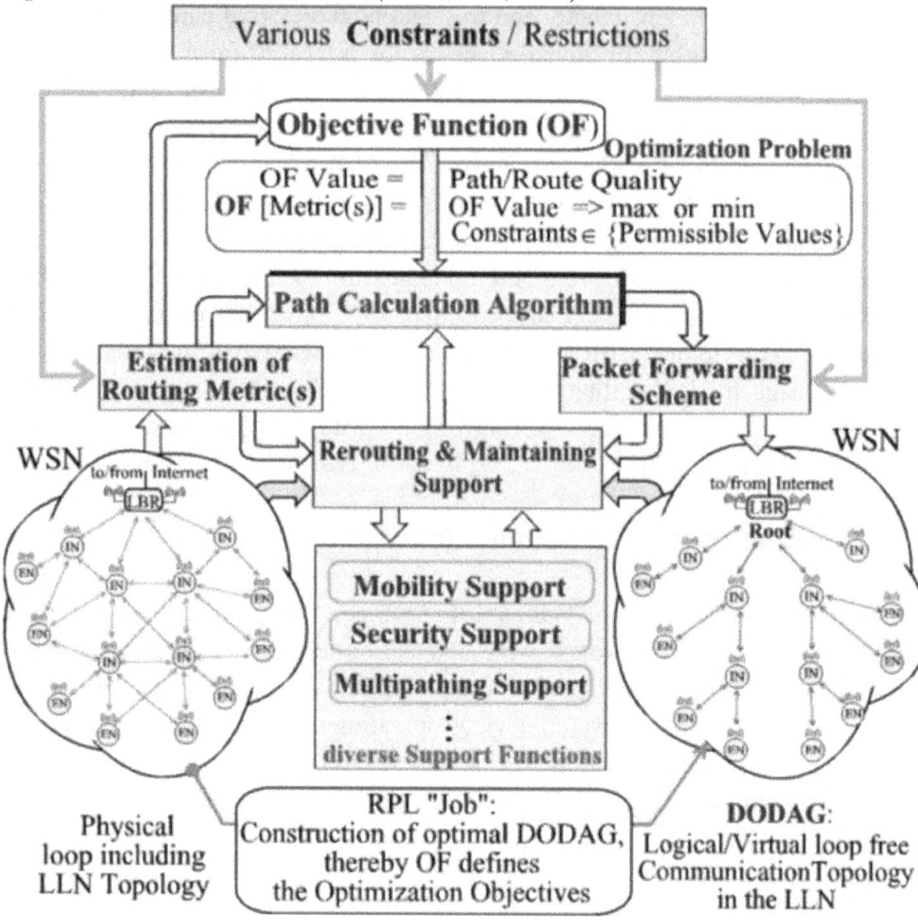

CoAP serves as an application layer protocol tailored for IoT applications, adapting certain HTTP functionalities to align with the power-efficiency and the characteristic challenges of lossy and noisy links in IoT environments. It is a standardized, lightweight, and efficient web transfer protocol explicitly designed for low-power net- works, which often feature high packet error rates and relatively limited throughput, as exemplified by 6LoWPAN networks. CoAP operates on constrained devices atop the unreliable UDP transport layer, offering a well-suited interface for standard Inter- net services. When employed in conjunction with 6LoWPAN, as defined in RFC4944, CoAP messages are crafted to fit within a single IEEE 802.15.4 frame, minimizing the need for fragmentation (Shelby et al.,2014; Al-Fuqaha et al.,2015).

A prominent attribute of the Constrained Application Protocol (CoAP) is its exceptional capacity to harmoniously coexist with prevailing application layer protocols and applications, all while sidestepping the need for extensive and resource-intensive reengineering efforts. This remarkable feature is underpinned by CoAP's adept implementation of mechanisms designed to condense metadata associated with application layer protocols. In doing so, CoAP finds a delicate balance, simultaneously optimizing its own performance and remaining compliant with the principles of Representational State Transfer (REST) architecture, which is foundational to the World Wide Web. By striking this balance, CoAP ensures the seamless interoperability of applications, irrespective of the diverse and sometimes resource-constrained environments within the Internet of Things (IoT). This proficiency is particularly advantageous as it facili tates the swift and efficient integration of CoAP into existing IoT ecosystems and the broader Internet landscape, fostering cohesive communication and interaction between diverse entities, devices, and systems (Ishaq et al.,2013; Hummen et al.,2011). The diagram in Figure 7 illustrates the comprehensive functionality of the CoAP protocol.

Figure 7. Structure of CoAP Protocol (C̦ orak et al.,2018)

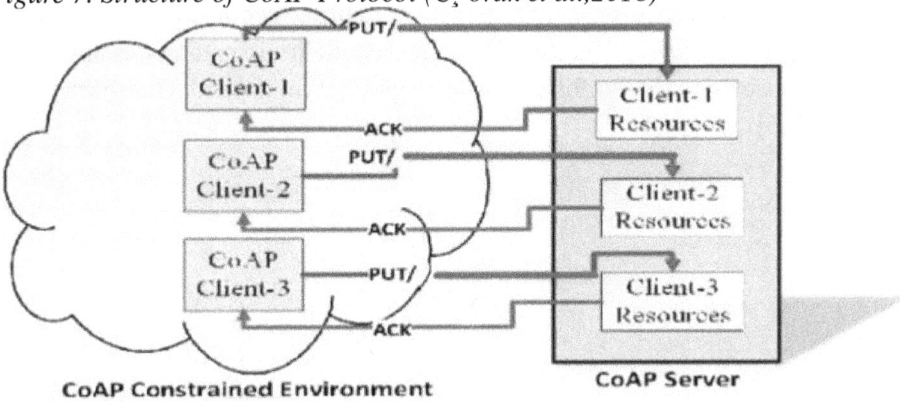

The advent of the Constrained Application Protocol (CoAP) marked the start of significant advancement in the field of networking protocols, particularly for resource- constrained devices and environments. CoAP, as detailed by Ishaq et al.,(2013) in their work, forms a pivotal component of a comprehensive networking stack tailored to meet the unique demands of such constrained devices and environments. Moreover, CoAP serves a crucial role as an application layer protocol within the Internet of Things (IoT) ecosystem. One of its notable features is the facilitation of end-to-end security between two applications through the utilization of Datagram Transport Layer Security (DTLS). DTLS, an adaptation of TLS for datagram-based communication, provides robust security mechanisms for data exchanged between IoT applications.

Within the context of CoAP, the secure iteration known as CoAPs employs a stream- lined version of DTLS. This optimization serves to efficiently safeguard CoAP messages during transmission between two IoT applications, ensuring data confidentiality and integrity, even in resource-constrained environments.

Additionally, CoAP ensures reliability in data exchange through the implementation of acknowledgment messages. As outlined in the works of Hummen et al., (2011), as well as Raza et al., (2012a), this mechanism guarantees that transmitted data is received and processed successfully, minimizing the risk of data loss or corruption within the CoAP protocol.

4. INTERNET OF THINGS SECURITY REQUIREMENTS

Securing Internet of Things (IoT) networks is imperative due to the profound im- pact IoT technologies have the potential to exert across diverse industries. These transformative capabilities, while promising, introduce new and complex security challenges. Essential security requirements are multifaceted and encompass various facets of IoT security (Figure8). Data confidentiality is paramount, involving the protection of sensitive information through encryption and Transport Layer Security (TLS).

Figure 8. Key Security Requirements for IoT

Data integrity ensures that information remains unaltered during transmission, monitored through Message Integrity Codes (MICs). Privacy controls must strike a delicate balance between anonymity and tracking, particularly important in healthcare and location-based services. Identity and endpoint authentication using unique device identification and certificate-based methods are essential to thwart unauthorized ac- cess. Dynamic authorization and access control adapt to the evolving IoT landscape. Availability is guaranteed through the use of Intrusion Detection Systems (IDSs) and firewalls to counteract attacks and ensure uninterrupted services. Trustworthiness, es- pecially in healthcare, demands that devices and data be consistently reliable and accurate. Fault tolerance mechanisms ensure IoT systems can withstand changes and failures, while continuous monitoring proactively identifies security threats. These se- curity requirements collectively guide the design and deployment of IoT systems, safe- guarding sensitive data and applications amidst the transformative potential of IoT technologies across industries. Here is a more in-depth look at each requirement:

Data Confidentiality: Ensuring that only authorized entities can access and mod- ify data is a critical security concern in IoT. This includes both the messages exchanged between source and destination and the data stored in IoT devices. To protect data from unauthorized access, encryption and decryption mechanisms should be used, and Transport Layer Security protocols like TLS or VPN should be implemented.

Data Integrity: It is important to detect any changes to messages or stored data that may have been made by an attacker or intermediary. Message Integrity Codes (MICs) can be used to provide this service. For example, in a healthcare system, MICs can ensure that medical data is not altered during transmission.

Privacy: IoT devices and the wireless medium used for exchanging data raise privacy concerns. To protect user privacy, access to data should be controlled through rules that specify who is allowed to access it. Balancing the need for anonymity with localization and tracking requirements is a challenge in this area. Identity and Endpoint Authentication: Each IoT device must be uniquely iden- tified and mutual authentication between users and devices is required to ensure the security of IoT data. Certificate-based authentication is recommended for end-to-end secure communication between nodes. For example, in a smart home system, a certificate-based authentication mechanism can ensure that only au- thorized users can access and control devices.

Authorization and Access Control: Authorization mechanisms must be in place to manage user access rights and permissions, ensuring that only authorized users can access IoT resources. With the scale and heterogeneity of IoT networks, these mechanisms must be dynamic and scalable. In a smart city system, for example, a context-based access control mechanism can protect the privacy of citizens by only disclosing their data to authorized parties.

Availability: Devices, networks, and services must be always available and work- ing properly, especially given the sensitive nature of many IoT applications. In- trusion Detection Systems (IDSs) and firewalls can be used to detect intrusions and ensure the availability of IoT systems. For example, in a military system, IDSs and firewalls can prevent Distributed Denial of Service (DDoS) attacks and ensure the continued operation of critical systems.

Trust: In many sensitive applications, such as healthcare, the trustworthiness of devices and data must be assessed. In IoT, trust must be consistently safeguarded and maintained, as non-trustworthy behavior can arise from intentional misbehavior or unintentional errors. For example, in a health monitoring sys- tem, a trustworthiness assessment feature can ensure that the data from medical devices is accurate and trustworthy.

Fault Tolerance: IoT systems must be designed to be robust and able to adapt to changing conditions. Intrusion detection and prevention mechanisms can help protect devices from failure, and recovery services should be in place to redirect functionality in case of an attack. In a smart energy system, for example, fault tolerance mechanisms can ensure the continued operation of the system even in the event of a network failure.

The incorporation of these security requirements is essential in the entire lifecycle of Internet of Things (IoT) systems, from their initial design phase to their actual deployment and operation. Their significance lies in the fact that IoT environments are susceptible to a wide array of security threats due to their interconnected and often resource-constrained nature.

In the design phase, engineers and architects need to carefully consider these security measures as integral components of the system's architecture. This involves integrating encryption, access controls, and authentication mechanisms into the very fabric of the IoT solution. By doing so, the system can inherently withstand poten- tial attacks and breaches. During deployment, these security requirements continue to play a pivotal role. They guide the setup of robust access controls, the configuration of intrusion detection systems, and the establishment of secure communication channels. By adhering to these principles, IoT operators can ensure that the deployed systems are resilient to both known and emerging threats. Moreover, these security measures are not static; they require continuous monitoring and adaptation to address evolving security landscapes. Regular audits, updates, and patches are necessary to stay ahead of new vulnerabilities and emerging attack vectors.

Ultimately, the enforcement of these security requirements serves a dual purpose: it not only shields IoT networks and systems from compromise but also upholds the privacy, trust, and integrity of sensitive data and applications. By integrating these security principles into the very core of IoT design and operations, organizations can unlock the full potential of IoT technologies while mitigating the associated security risks.

5. INTERNET OF THINGS THREATS

In this segment of our comprehensive examination, considering the significant evolu- tion of IoT threats in complexity and scale, presenting substantial security challenges, we will delve into the various threats. One of the most notable threats is the ex- ploitation of IoT devices to create extensive botnets, which can execute devastating distributed denial-of-service (DDoS) attacks (Kumari & Kumar,2023). IoT devices are particularly vulnerable due to firmware and software vulnerabilities, often increased by their limited computational capabilities, making them susceptible to attacks like re- mote code execution and privilege escalation (Raizada et al.,2023; Classen et al.,2022). The proliferation of IoT devices also raises grave concerns about data privacy, as these devices collect and transmit vast amounts of sensitive information, opening the door to unauthorized access and potential data breaches. Additionally, inadequate authen- tication methods, default or weak passwords, and suboptimal network security further compound IoT vulnerabilities Jeridi et al.2021. Supply chain attacks, insecure com- munication protocols, and the intricate challenge of managing numerous devices in various settings contribute to the multifaceted IoT threat landscape. Physical attacks, such as tampering and theft, represent tan- gible risks to IoT device integrity. Further- more, as IoT technology continues to advance, the emergence of new and unforeseen threats becomes increasingly likely, potentially exploiting novel attack vectors. Effec- tively addressing these challenges necessitates a comprehensive and proactive security approach, encompassing robust security protocols, regular software updates, secure device management practices, and vigilant monitoring. These measures are essential for protecting IoT ecosystems and safeguarding the integrity and confidentiality of the data they handle, particularly as IoT adoption continues to expand across industries.

5.1. IoT architectures

The exponential growth of Internet of Things (IoT) architectures in scholarly litera- ture is indicative of the rich landscape of approaches and frameworks devised to tackle the complexities of IoT systems. These architectural paradigms span a spec- trum of de- sign philosophies, with notable examples including middleware-based, service-oriented architecture (SOA)-based, and intricate multi-layered models, such as the five-layer architecture depicted in Figure 9 (Ant˜ao et al.,2018).

Figure 9. The five-layer IoT architecture (Khan et al.,2012)

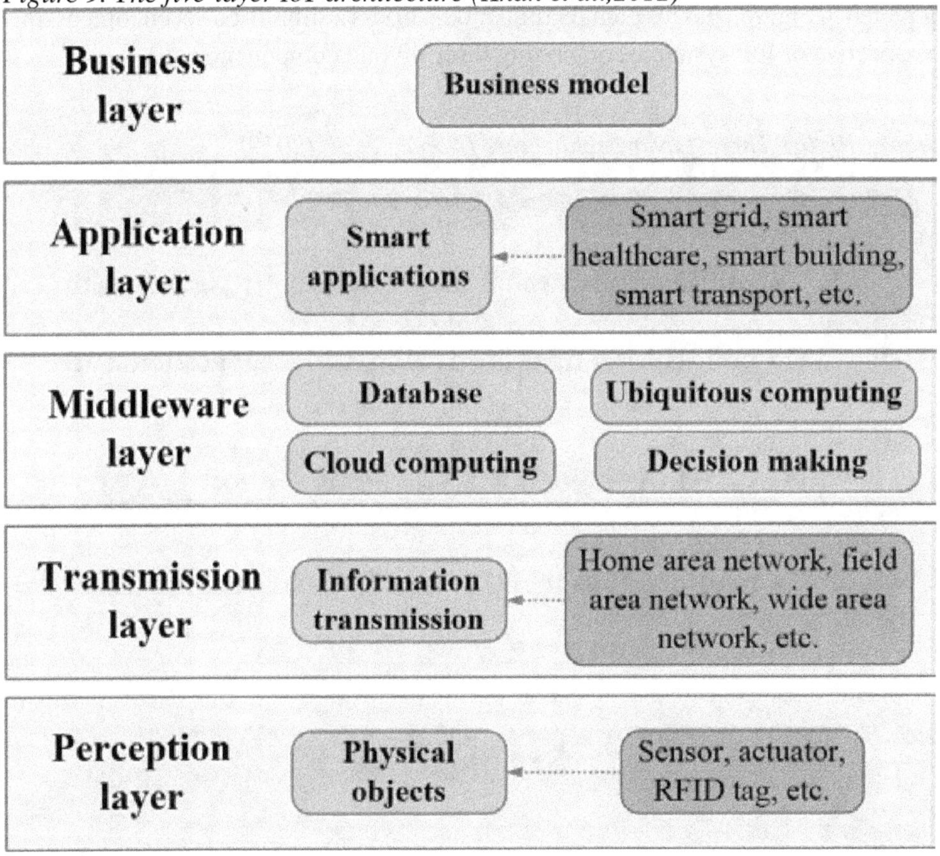

This proliferation of diverse architectural models underscores the dynamism and versatility of IoT ecosystems. Each of these architectural approaches brings its own set of principles and advantages to the table, contributing to the ongoing discourse on the best practices and strategies for structuring and conceptualizing IoT systems, as articulated by Al-Fuqaha et al., in their 2015 work.

These various architectural paradigms not only provide blueprints for the organization and operation of IoT deployments but also serve as a testament to the adaptability of IoT technology in addressing a wide range of application domains and use cases. As IoT continues to evolve, the exploration and refinement of these architectural frame- works play a pivotal role in shaping the future landscape of interconnected devices and systems.

Within the realm of IoT architectural paradigms, each presents a distinctive set of design principles and considerations, attuned to the intricacies and specific require- ments of various IoT applications. Amidst this rich diversity of architectures, one fun- damental and widely recognized IoT model is the three-layer architecture, depicted in Figure 10. This architectural model offers a simplified yet encompassing perspective of IoT systems, comprising three pivotal components:

Figure 10. IoT Three-Layer Architecture (Abbasi et al.,(2019)

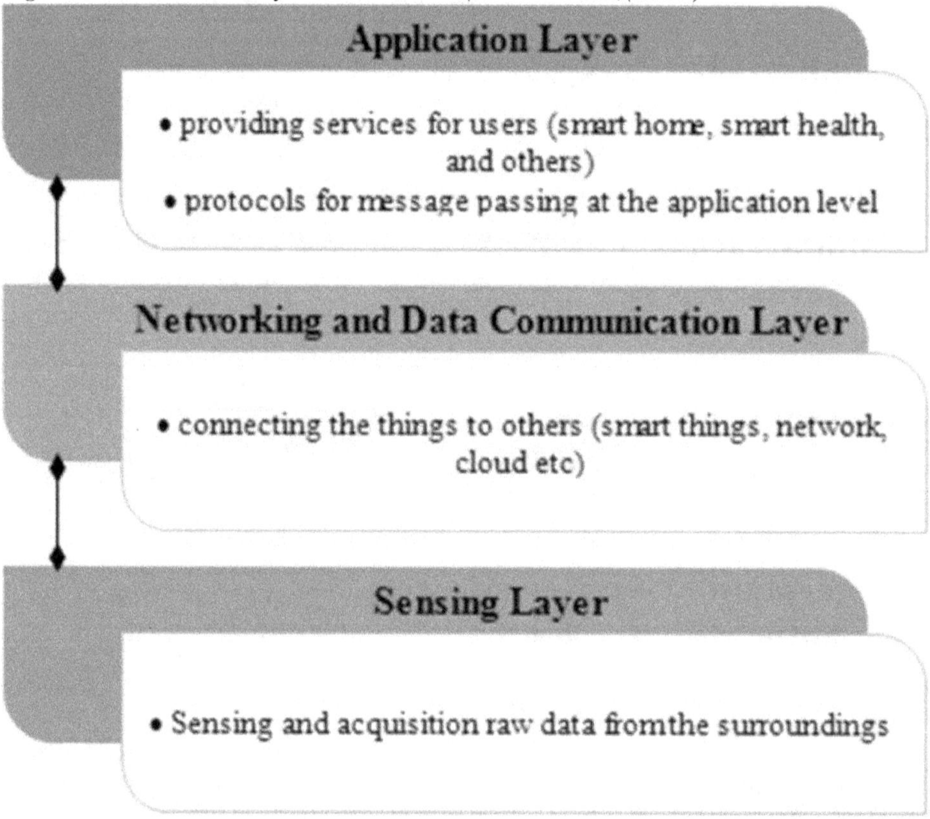

Perception Layer: This foundational layer represents the frontlines of IoT sys- tems, where data is acquired from the physical world through various sensors, actuators, and devices. These devices serve as the sensory input points, collect- ing data on variables like temperature, humidity, motion, or any other relevant envi- ronmental or contextual information. The Perception Layer acts as the in- terface between the physical environment and the digital realm, serving as the initial data source for IoT applications.

Network and Communication Layer: Positioned as an intermediary layer, the Network and Communication Layer plays a critical role in facilitating the seamless transfer of data collected from the Perception Layer to the higher layers of the IoT architecture. This layer is responsible for managing the connectivity, data routing, and communication protocols necessary to transmit data from sensors and devices to the subsequent layers, often leveraging wired or wireless networks, including technologies like Wi-Fi, cellular networks, or even specialized IoT protocols like LoRaWAN.

Application and Service Layer: Situated at the top of the hierarchy, the Application and Service Layer represents the highest level of IoT functionality. It encompasses the software applications, analytics engines, and services that process, interpret, and act upon the data received from the lower layers. This layer is where IoT applications are developed to derive meaningful insights, make informed decisions, and trigger actions based on the data acquired. It often includes cloud-based platforms, edge computing solutions, and specialized IoT applications tailored to specific use cases.

The three-layer IoT architecture model encapsulates the core components required for IoT systems to function cohesively. While it offers a simplified representation, it serves as a foundational framework upon which more intricate and specialized IoT architectures can be built. This model underscores the importance of the Perception Layer's data collection, the crucial role of the Network and Communication Layer in data transport, and the ultimate value extraction achieved by the Application and Service Layer in delivering actionable insights and functionality within IoT ecosystems. Comprehending the intricacies of these architectural layers is imperative, as they constitute the fundamental building blocks upon which the edifice of IoT systems is constructed. Each layer plays a unique and indispensable role, and the security considerations associated with them exhibit nuances that demand careful attention. Vulnerabilities and threats have the potential to manifest at any of these layers, ne- cessitating a global security approach capable of addressing the distinct challenges presented by each.

5.2. Common IoT attack factors

The rapid proliferation of IoT technology has concurrently exposed these systems to an ever-expanding spectrum of security threats. Several contributing factors amplify this vulnerability, including the proliferation of connected devices, the complexities surrounding the accurate identification and securing of IoT nodes, and the intrinsic resource constraints inherent to many IoT devices.

Among the prevalent threats faced by IoT systems, two noteworthy adversaries stand out: unauthorized data access and disruptive denial of service (DoS) attacks. These adversaries have the potential to wreak havoc on IoT services, threaten the integrity of data streams, and, in severe cases, gain unauthorized control of IoT devices from their rightful owners. To effectively mitigate these threats, it becomes imperative to systematically categorize IoT attacks. A well-structured taxonomy of IoT attacks, as exemplified in Figure 11, must take into account a multitude of factors, including:

Device Properties: Attacks may selectively target either low-end or high-end device classifications, each exhibiting its own set of vulnerabilities and capabilities. These distinctions necessitate tailored security measures to safeguard against potential exploits.
Access Level: Adversaries may employ passive or active attack strategies, each entailing varying degrees of interaction with the target IoT system. Understand- ing the nature of these interactions is crucial for devising appropriate counter- measures.

Figure 11. Factors in IoT Attack

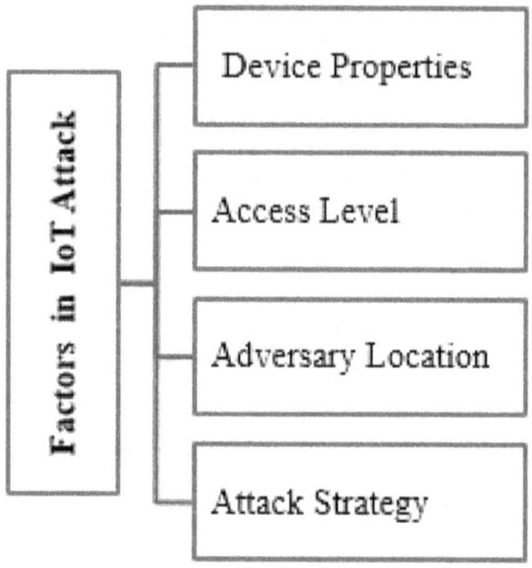

Adversary Location: Threats can originate from within the confines of the IoT network, constituting internal threats, or emanate from external sources, representing external threats. The source of the attack informs the defensive strategies required to thwart it effectively.
Attack Strategy: IoT attacks can assume diverse forms. Some may take on physical manifestations, involving tampering with physical hardware components, while others may exploit logical vulnerabilities inherent in software and communication protocols. Acknowledging these varied attack vectors is instru- mental in crafting a comprehensive defense strategy.

This study places a specific emphasis on attacks targeting the network layer within IoT systems, with a strong focus on vulnerabilities and threats against the Routing Protocol for Low-Power and Lossy Networks (RPL). The rationale for this focus is rooted in the understanding that the choice of the layer to be targeted in an IoT architecture depends on the attacker's objectives and capabilities. By narrowing the scope to network layer security and examining RPL-specific vulnerabilities, this re- search aims to shed light on the critical role of the network layer in IoT security assessments. It recognizes that securing the network layer is paramount for the over- all security and resilience of IoT ecosystems, as it serves as the backbone for data transmission and communication among IoT devices and applications. This research seeks to contribute valuable insights into IoT security, ultimately helping to fortify the integrity, confidentiality, and availability of IoT systems in the face of evolving threats and challenges.

5.3. Perception/Sensing Layer

Regarding the perception layer, and as for the network layer, often referred to as the transmission layer, it serves as the conduit for data transmission between the perception layer and the upper layers of the IoT architecture. This crucial layer is vul- nerable to various types of attacks, including eavesdropping, spoofing, and tampering. These vulnerabilities result from the absence of robust encryption and authentication mechanisms within the network layer, allowing attackers to intercept and manipulate transmitted data. An overview is given in Figure 12.

Figure 12. Overview of the IoT perception layer (Mamdouh et al.,2021)

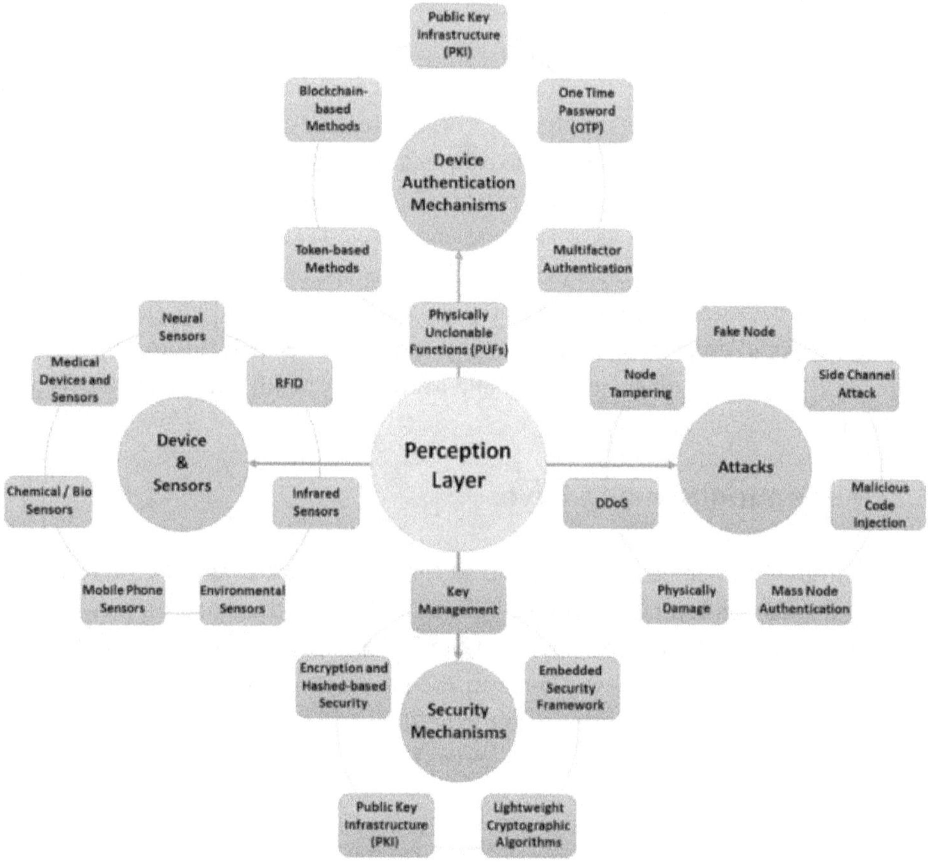

Moreover, if an attacker gains control over the network layer, they have the capabil- ity to launch disruptive denial of service (DoS) attacks, potentially causing the entire IoT system to fail. At the highest tier of the IoT architecture, the application layer assumes the responsibility of providing a wide range of services and applications to end-users. However, this layer is exposed to vulnerabilities such as cross-site scripting (XSS) as documented by Rodr´ıguez et al.,(2020), cross-site request forgery (CSRF) as reported by Barabanov et al.,(2018), and SQL injection Sadiq et al.,(2021). Malicious actors, as discussed by Gowtham M. & Pramod B.(2017), can exploit these weaknesses to gain unauthorized access to sensitive data or manipulate the behavior of the IoT system.

The three-tier IoT architecture remains susceptible to a multitude of attacks, encompassing unauthorized data access, denial of service attacks, and assaults targeting distinct layers of the architecture. These attacks can manifest based on various parameters, including device attributes, access levels, adversary location, strategy, in- formation damage severity, host-based characteristics, protocol-based intricacies, or communication stack protocols. To strengthen security of IoT systems, the implementation of robust security mechanisms at each architectural layer is imperative to prevent and mitigate the impact of these attacks. Additionally, it is worth noting that certain attacks, like replay attacks, can transcend all layers. Depending on the specific communication technology employed, an array of other attacks may also be instigated against these layers. In fact, some attacks can be triggered against all layers such as replay attacks. Besides unauthorized access to data, DoS and DDoS attacks, several other attacks can be triggered against this layer depending on the communication technology, for instance: (Nawir et al.,2016; Mendez et al.,2017; Azzabi et al.,2017).

The Perception/Sensing Layer is susceptible to various forms of attacks, including eavesdropping, spoofing, and tampering due to the absence of robust encryption and authentication mechanisms. Attackers can intercept and manipulate data, potentially leading to disruptive denial of service (DoS) attacks that can cripple the entire IoT system. At the highest application layer, vulnerabilities like cross-site scripting (XSS), cross-site request forgery (CSRF), and SQL injection expose the system to unauthorized access and data manipulation. These attacks can vary based on factors like device attributes, access levels, adversary strategies, and communication proto- cols. To enhance IoT security, robust security measures at each architectural layer are essential, given that certain attacks, including replay attacks, can transcend all lay- ers. Moreover, depending on communication technologies, IoT systems are susceptible to a range of attacks, including side-channel attacks, brute force attacks, man-in-the- middle (MITM) attacks, unfairness attacks, masquerading attacks, and various others, which can disrupt network functionality or compromise sensitive data:

Side-channel attacks: Side-channel attacks focus on exploiting information that is leaked through the physical implementation of a device, such as its power consumption or electromagnetic emanations. These attacks can reveal sensitive information such as encryption keys, and are becoming increasingly relevant in the context of IoT where many devices are physically accessible and have limited computing resources (Gnad et al.,2019). Examples of side-channel attacks include timing attacks, power analysis attacks, electromagnetic analysis attacks, fault induction attacks, optical side-channel attacks, traffic analysis attacks, acoustic attacks, and thermal imaging attacks.

Brute force attack: A brute force attack is a type of cryptanalytic attack that involves trying every possible key or combination of keys until the correct one is found. In the context of IoT, brute force attacks have been demonstrated to be particularly effective against some smart home devices that use weak encryption or no encryption at all (Raikar & Meena.,2021).

Man-in-the-middle (MITM) attacks: MITM attacks occur when an attacker intercepts and modifies the communication between two devices, making the two devices believe that they are communicating directly with each other. MITM at- tacks can be used to steal sensitive information, modify data, or prevent two devices from communicating altogether. In the context of IoT, MITM attacks have been demonstrated to be effective against some IoT hubs and other net- working devices.

Unfairness attacks: Unfairness attacks occur when malicious nodes violate the standard communication protocols of an IoT network, giving them an unfair advantage over other nodes in terms of channel utilization. For example, an attacker might manipulate the back off procedure used in the IoT IEEE 802.15.4 MAC layer to capture the channel more frequently than other nodes. Unfairness attacks can significantly disrupt the normal functioning of an IoT network.

Masquerading attacks: Masquerading attacks occur when an attacker steals the identity of an authorized node in the network, allowing them to impersonate that node and potentially access sensitive information or perform unauthorized actions. These attacks can be triggered at the network layer, such as through Sybil or ClonID attacks.

Other attacks: There are many other types of attacks that can be carried out against IoT networks, including node tampering and physical attacks, collision attacks, jamming attacks, battery exhaustion attacks, replay attacks, traffic sniffing/eavesdropping attacks, data-corruption/message-alteration attacks, key sniffing attacks, proof-of-concept attacks, tag modification and tag cloning at- tacks, RFID authentication attacks, and others. These attacks can have a wide range of effects, from disrupting the normal functioning of the network to com- promising sensitive information.

5.4. Network Layer

In this context of data transmission within the IoT architecture, the network layer serves as an intermediate stratum responsible for consolidating and relaying sensory data from the perception layer to the application layer. This process involves the uti- lization of a diverse range of communication networks, encompassing both wired and wireless mediums. These include Local Area Networks (LAN), for instance, WiFi; Per- sonal Area Networks (PAN), like ZigBee; and expansive Wide Area Networks (WAN) such as LTE and GPRS. According to Kumar et al.,(2014), it is the "Central Nervous System" of the IoT. WSNs are the most popular networks for IoT regarding their ability to cover large areas of things and to retain adequate consumption of energy. It is important to pinpoint that the network layer should support the communication requirement for latency, bandwidth and security. Nevertheless, the characteristics of IoT environments cause several security and privacy concerns, especially associated to the network layer. It should be pointed out that in 6LoWPAN networks, this layer is composed of two sub-layers: the 6LoWPAN adaptation layer and the RPL routing protocol. Because RPL support MP2P and P2MP traffic patterns makes 6LoWPAN networks more vulnerable to routing attacks as summarized in Figure 13. A variety of attacks targeting the two sub-layers of the network layer have been identified. In the following, authors present attacks related to each sub-layer.

5.4.1. Adaptation Layer Attacks

The adaptation layer is implemented at the border router for translating packets be- tween the 6LoWPAN network and Internet. The adaptation layer is mandatory as the size of IEEE 802.15.4 frames (limited to 127 bytes) do not permit to use conventional IPv6 packets (1280 bytes). In this condition, header compression, fragmentation and reassembling are handled by the adaptation layer. Verma & Ranga,(2020) presented a survey on RPL based 6LoWPAN network and related attacks.

5.4.1.1. Fragmentation attacks:.

Figure 13. Example Network Layer Attacks (Gupta, N. & Garg, U.,2022)

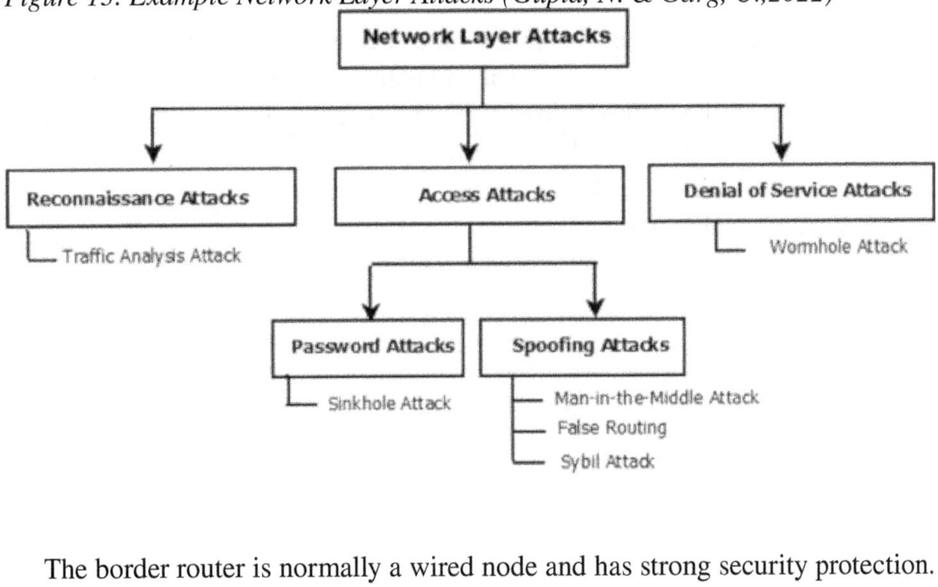

The border router is normally a wired node and has strong security protection. However, the packet fragmentation and reassembly progress still have some vulnerability. In these attacks the attacker can modify or reconstruct the packet fragmentation fields like datagram size, datagram tag or datagram offset. These attacks can cause critical damage to a node, for instance, reassemble buffer overflow because of packet re-sequence, exhausting resource because of processing unnecessary fragmentation, or shutting down and rebooting. Also, as there is no authentication mechanism at receiver side for checking that received fragment is not a spoofed or duplicated one, the attacker may put his own fragments in frag- mentation chain. The fact that integrity checksums and signatures are calculated over whole packets instead of over intermediate fragments, the validity of fragmented packets cannot be verified before packet reassembly. For this reason, an attacker can fill up the limited buffer space of IoT devices with invalid fragments by flooding the resource constrained objects with few large packets. In Figure 14 an overview of fragmentation attacks and countermeasures is given (Alyami et al.,2022).

Authentication attacks: Unfortunately, there are no mechanisms for 6LoWPAN nodes to authenticate before joining the network. So, it is obvious that malicious nodes can easily join the network and trigger other internal attacks, which is very harmful for IoT applications. Many authentication protocols have been proposed in the literature. Tanveer et al.,(2020) proposed a lightweight schema for the Authentication and Key the Exchange (AKE) based on an encryption algorithm and hash mechanism. Authors in Alrababah et al.,(2017) surveyed authentication protocols for IoT according to their mechanisms. The proposed authentication protocols for 6LoWPAN networks are used to check the identity of each device in the network and authenticate it. Providing strong authentication mechanisms can help to mitigate several attacks such as Sybil and CloneID attacks . The Figure 15 provides an overview of the technical methods of the authentication mechanism in IoT.

Confidentiality attacks: Only legal and authorized nodes can access, watch and control data in the network. Providing confidentiality in 6LoWPAN can help to miti- gate various attacks such as eavesdropping, MITM, spoofing attacks, and so on. As for authentication, identity management represents a key factor to assure confidentiality Mendez et al.,(2017).

Figure 14. Taxonomy of fragmentation attacks and countermeasures. (Alyami et al.,2022)

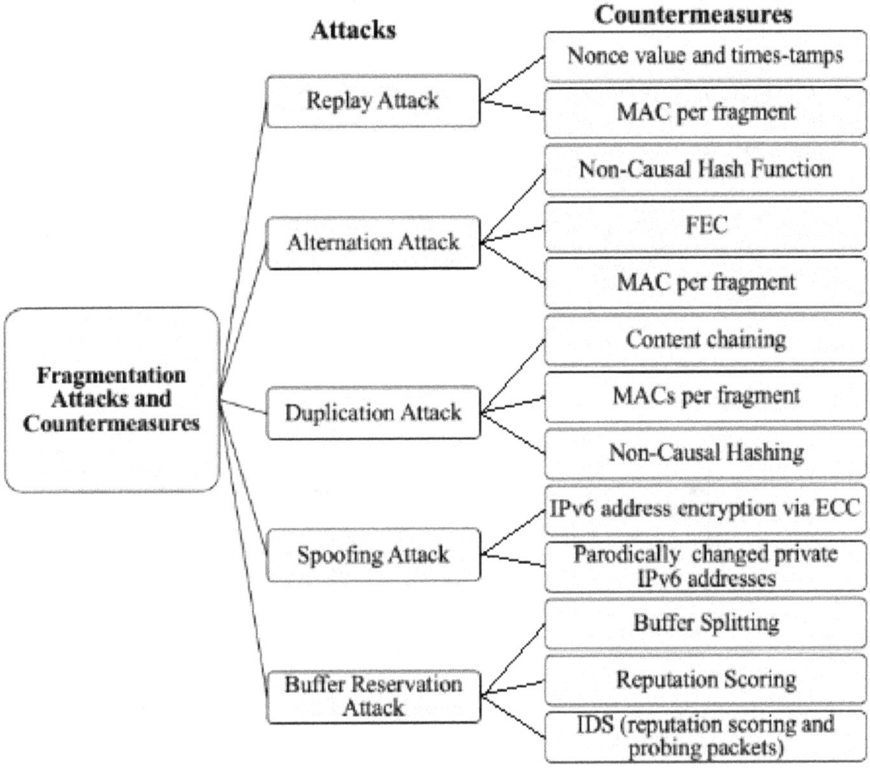

Besides, cryptography is considered the first line for solving the confidentiality and authentication issues. In fact, Internet Protocol Security (IPSec) provides end-to-end network layer security by enabling the authentication and encryp- tion of exchanged IP packets, using Authentication Header (AH) to provide integrity and authentication, and Encapsulating Security Payload (ESP) headers to provide integrity and confidentiality. Actually, since IPSec is very greedy and supplies energy and space, some research works proposed compressed version of IPsec headers to secure the 6LoWPAN adaptation layer (Raza et al.,2011; Varadarajan & Crosby,2014). Valea et al., (2019) propose SCSU, a plug-and-play SoC module for secure context saving, ensuring confidentiality with encryption and integrity with a MAC.

Internet attacks: 6LoWPAN IoT devices and Internet hosts differ strongly regarding their available resources. Indeed, objects in 6LoWPAN networks have limited mem- ory, computational power and very limited security provision. Whereas Internet hosts are equipped with CPUs in the GHz range and huge memory. These capacities dif- ferences, in addition to the openness of IoT make 6LoWPAN networks vulnerable to several attacks from Internet (Figure 16. For avoiding such attacks a firewall could be installed on the 6LBR to control the malicious packets from Internet.

Figure 15. Technical methods of the authentication mechanism in IoT. (Nandy et al.,2019)

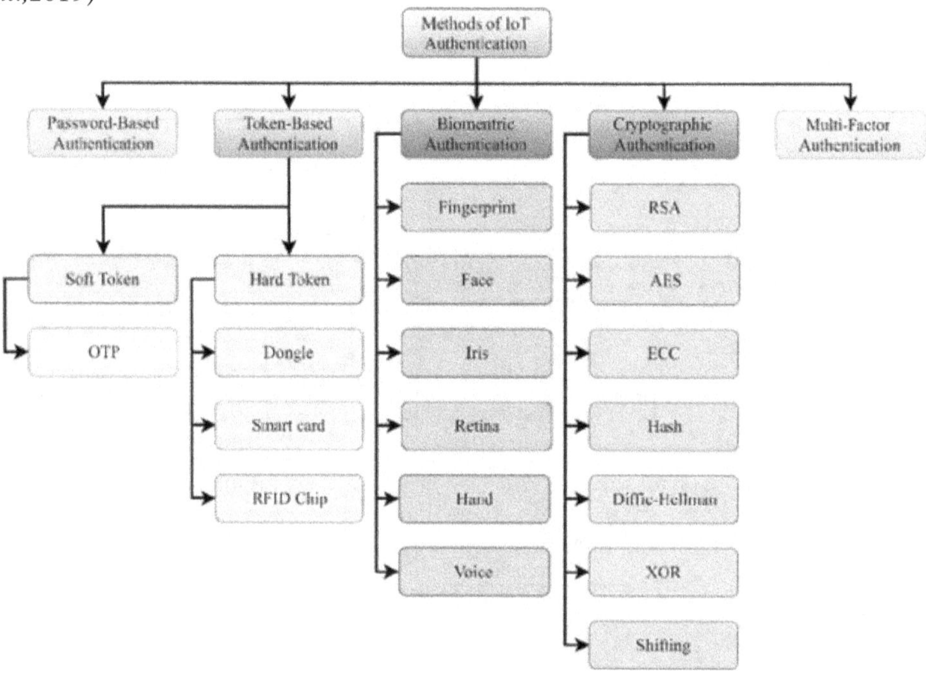

Hummen et al.,(2011) proposed an adaptation layer-based approach to enabling security boot- strapping between the IoT domain and the Internet with existing IP security protocols such as DTLS. Pongle & Chavan,(2015a) classified authentication, confidentiality, and internet attacks as adaptation layer attacks, however, authors believe that these at- tacks can target any layer of the three-layer architecture. Wei et al.,(2022) proposed a K-nearest neighbor based (KNN) intrusion detection mech- anism and design a secure network framework IPv6-based WSNs.

5.4.2. Routing Attacks - RPL Threats

The RPL specification defines some fault tolerance mechanisms, such as local and global repairs, and loops detection and avoidance. Furthermore, it defines secure ver- sions of the different RPL control messages (i.e. control messages encryption) and some security modes and mechanisms. The defined security mechanisms ensure control messages integrity and confidentiality against outsider attackers (Thubert et al.,2012). However, from one end, the proposed security mechanisms are not clearly defined; for instance, it is not specified how asymmetric cryptography may be employed to support node authentication and key retrieval. From the other end, an insider attacker can bypass these mechanisms by gaining access to shared keys and trigger several attacks against RPL in order to break routing operation rules, and cause network disruption for different IoT applications. In fact, malicious nodes may falsify or lie the advertisement of link and node routing metrics to disturb traffic routing.

Figure 16. Internet attacks in the RPL-connected 6LoWPAN Networks (Shreenivas et al.,2017)

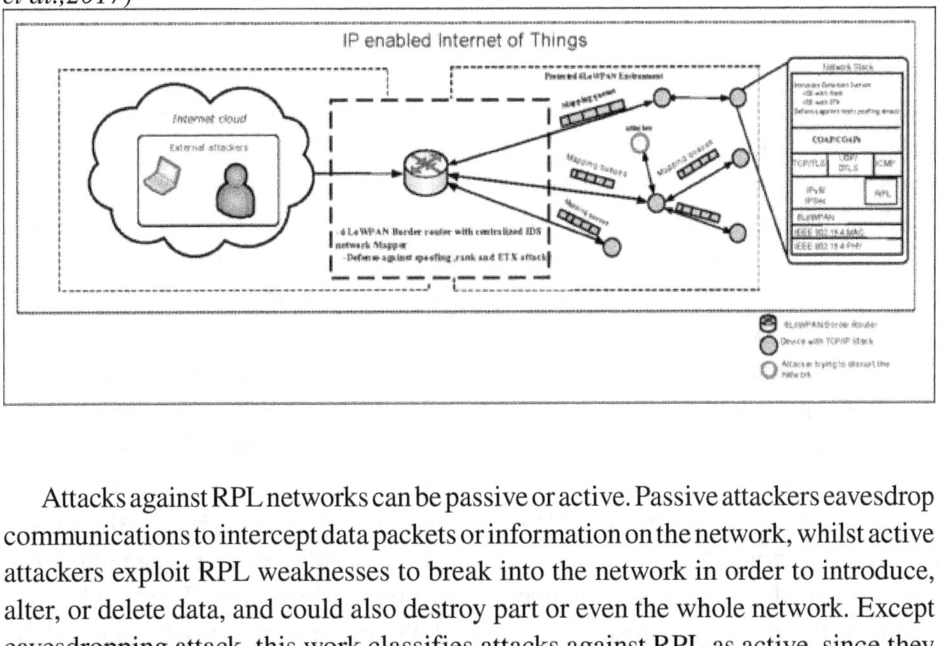

Attacks against RPL networks can be passive or active. Passive attackers eavesdrop communications to intercept data packets or information on the network, whilst active attackers exploit RPL weaknesses to break into the network in order to introduce, alter, or delete data, and could also destroy part or even the whole network. Except eavesdropping attack, this work classifies attacks against RPL as active, since they aim to disrupt and/or destroy the network.

Figure 17. Taxonomy of RPL attacks. (Al-Amied et al.,2022)

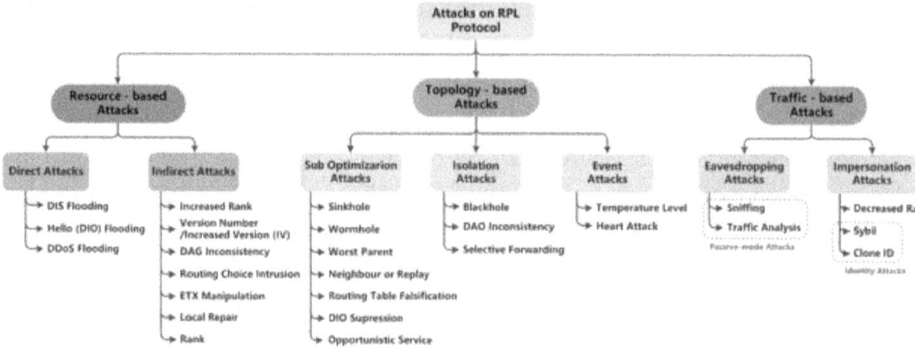

Several classifications for RPL threats exist in the literature. Pongle & Chavan,(2015a) put all attacks in a same category, namely attacks on RPL topology. Mayzaud et al.,(2016) classified RPL attacks on three categories: attacks on resources, attacks on topology and attacks on traffic (Figure 17). Based on the resource constrained nature of IoT nodes, in this work the authors believe that both attacks on topology and traffic impact directly or indirectly the network resources which means that all attacks can be seen as attacks on resources. Tsao et al.,(2015) proposed four categories of RPL attacks under the ISO7498-2 model: attacks due to failures to authenticate, attacks due to failures to keep routing information confidential, attacks on integrity, and attacks on availability. Like in Tsao et al.,(2015), Airehrour et al.,(2016a) presented a summary of attacks against RPL and classified these attacks as con- fidentiality, integrity and/or availability attacks. Verma & Ranga,(2020), categorize defense mechanisms against RPL attacks into two groups: Secure Protocol and Intru- sion Detection System (IDS). Secure Protocol enhances RPL with security features: Cryptography, Trust, and Threshold-based solutions. Cryptography uses traditional methods. Trust assesses node trustworthiness for routing. Threshold-based improves the reset mechanism. Traditional IDS isn't suitable for IoT; lightweight IDS with min- imal resource usage is needed. In RPL, IDS detects anomalies using Signature-based, Anomaly-based, Specification-based, or Hybrid-based methods.

From the perspective presented in this section, vulnerabilities in the RPL (Routing Protocol for Low-Power and Lossy Networks) can be categorized into two primary classes. Firstly, there are Novel RPL Specification-based attacks, which include attacks targeting RPL-specific attributes like rank, neighbor relationships, and version numbers. Secondly, there are Existing routing attacks adapted to exploit RPL's unique characteristics, encompassing spoofing, hello flood, homing, selective forwarding, sybil, wormhole, acknowledgment flooding, eavesdropping, impersonation, relay, and replay attacks. In the subsequent sections, the authors provide a comprehensive examination of the various attacks that have been identified and documented within the context of RPL.

A. RPL Specification-Based Attacks

This class includes new threats exploiting some functioning rules in the RPL specification. This kind of attacks may cause control messages overhead, discard downward routing state, and exhaust nodes resources.

Rank attacks: In the literature there are several variants of Rank attack Nandhini et al.,(2021). These variants are termed Rank attacks, Increased Rank attacks, Decreased Rank Attacks, and Worst Parent Attacks. In the opinion of the au- thors, each of these variants can be simply called Rank attack because they are based on the malicious manipulation and/or exploitation of the Rank field and/or rules. For instance, in (Le, A. et al.,2011; Le, A. et al.,2012) Rank attack, two or more malicious nodes may misbehave and cooperate on skipping the Rank checking function. This leads to break the Rank rule and thus to create unopti- mized routes, or undetectable loops, or even never discovering existing optimized paths (Figure 18). In (Le, A. et al.,2013) Rank attack, the attacker chooses a random parent, which has a Rank higher than the Rank of the preferred parent (worst parent attack). This attack creates unoptimized routes which leads to poor performance. In Dvir et al.(2011) Rank attack, an adversary node illegit- imately advertises a better Rank equal to a lower Rank value (decreased Rank attack).

Rehman et al.,(2016) introduced a new variant of this attack termed the Rank Attack using Objective Function (RAOF attack), where the advertised Rank value is bounded. In this attack attacker announces a Rank value less than the minimum Rank among its neighbors and greater than its preferred parent. In addition, because RPL suggests no measures to monitor the change in routing metric values, attacker announces in the DIO message a drastically lowered value of the routing metric compared to the minimum observed among its neighbors. In both attacks the honest neighbors will select the attacker as their new pre- ferred parent, and thus allow it to manage and manipulate more network traffic. As consequence, they allow malicious node to trigger other attacks (e.g. eaves- dropping, deleting and modifying data). In Alsukayti & Alreshoodi.,(2023), the attacker voluntarily increases its Rank value (i.e. increased Rank attack). This leads to generate loops in the network, to exhaust node resources, and to congest the network.

Figure 18. Rank Attack on RPL-Based 6LoWPAN (Bang et al.,2023)

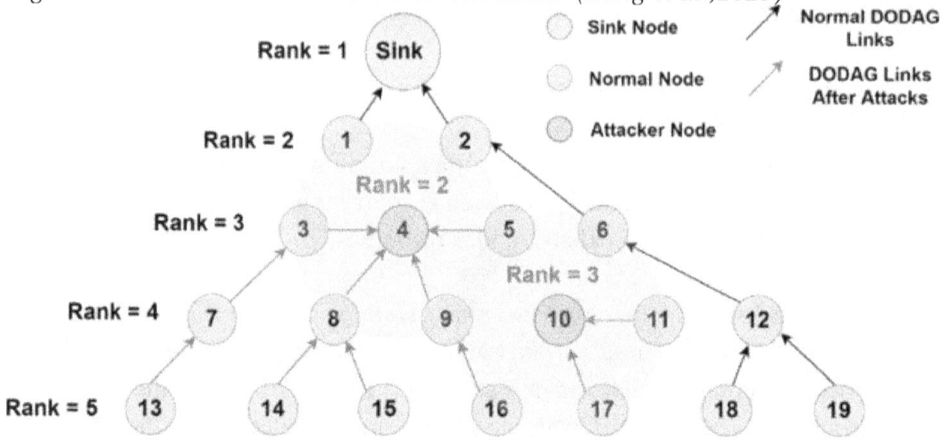

Neighbor attack: In Neighbor attack, misbehaving nodes resend replicated DIO messages (without updates) to honest nodes (Le, A. et al.,2013). Hence, honest nodes can consider the messages owners as neighbors. Consequently, they can update their routes to the out range neighbors. This leads to create false route, to disrupt the network, or to consume more resources. As described, Neighbor attack is more like a Routing information replay attack where the attacker for- wards outdated DIO messages inducing honest nodes to update their routing tables with stale routes Alsukayti & Alreshoodi.,(2023). This attack has also been termed DIO replay attack (Figure 19).

Figure 19. DIO replay attack. (Verma & Ranga,2020)

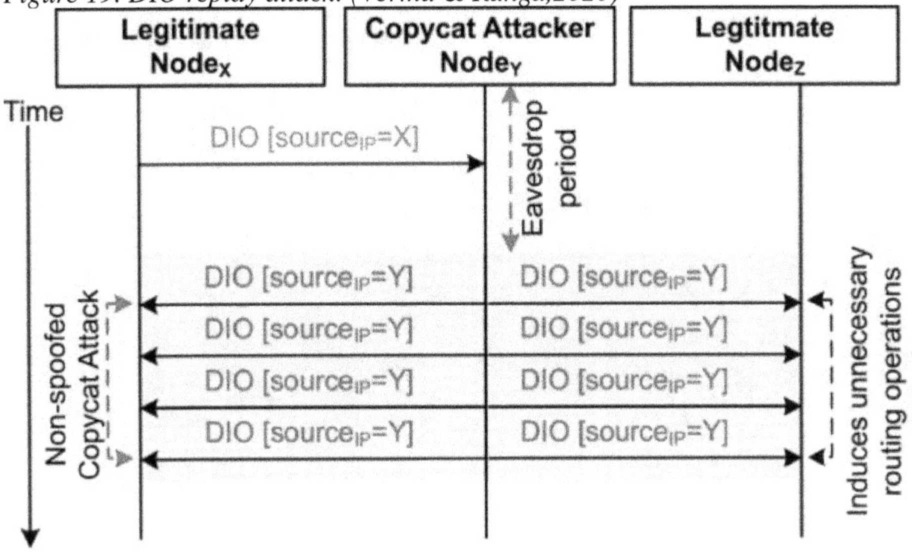

Routing Choice Intrusion attack: Nodes use DIO messages metrics and objective function rules to decide whether to join the DODAG or not. Wallgren et al.,(2013), showcases new IPv6 security features for routing choice intrusion at- tack. Zhang et al.,(2015) defined a new RPL internal attack where the attacker learns RPL's routing rules (choices to route packets), captures control messages and broadcasts fake ones. According to the authors, this attack is hard to detect because the intruder have only to ignore the legitimate RPL internal detection by itself, and work as normal. One or more attackers may participate in the attack. This attack can generate un-optimized routes, create loops, and exhaust resources. In a variant of this attack, attackers tamper the DIO metric value and ignore inside legitimate metric detection by itself. This variant is identical to the rank attack.

Figure 20. DAO attacks (Sheibani et al.,2022)

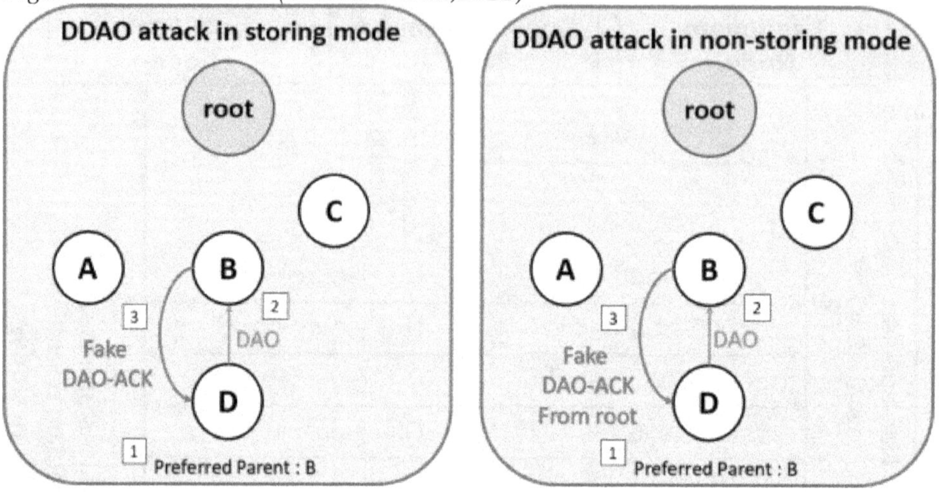

DAO attacks: The DAO message is used to maintain downward routes in RPL storing mode. Thus, after a node joins a DODAG, it advertises a DAO message for its neighbors to update their routing tables. In addition, parents can use DIO messages to request DAO from sub-DODAG. Attackers can exploit RPL storing mode to trigger DAO-related attacks called in the literature: DAO Inconsistency Attack, Routing Table Overload Attack, and Routing Table Falsification Attack (Figure 20. In the first attack, attackers can use the Forwarding-Error F flag to make a node discard available downward routes Hui J.W.,(2012). This attack aims to make the topology of the DODAG suboptimal, and to isolate the sub- DODAG bound the attacker. The second and the third attacks can be grouped. Indeed, in both attacks malicious nodes announce fake routes by modifying or forging DAO messages Mayzaud et al.,(2016). This leads to build false downward routes and overload the targeted nodes' routing tables with these false paths. Consequently, honest nodes will be prevented of building new legitimate routes. Additionally, used paths can be longer resulting delay, packet drops and/or net- work congestion.

Figure 21. DIS flooding attacks. Verma & Ranga,(2020)

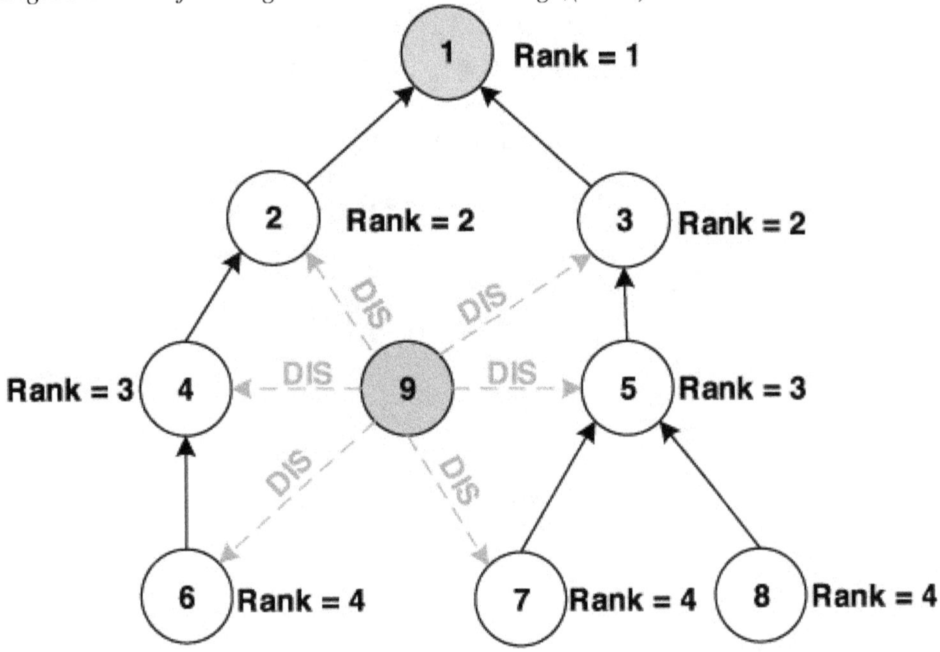

DIS attacks: DIS messages are sent by new nodes wishing to join a DODAG. These messages allow nodes to request DIO messages from the in-network in range nodes (neighbors). There exist two types of DIS attack. In the first one malicious node unicast periodically fake DIS messages to a node which will replay by a DIO message. In the second one, malicious node multicast periodically DIS messages to its neighbors which have to reset their trickle timers, and thus to send more DIO messages. Both attacks lead to generate more control messages overhead, flood the network with fake messages and thus, disrupt the network operation and exhaust resources Le, A. et al., (2013). These attacks were defined in Mayzaud et al.,(2016) as flooding attacks (Figure 21.

Figure 22. Local repair attack (Bang et al.,2023)

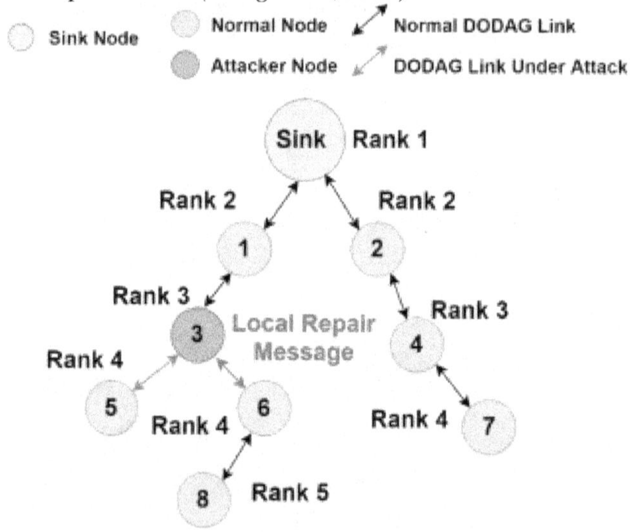

Local repair attack (Figure 22: An attacker can repeatedly trigger local repair by changing the DODAG ID field or broadcasting infinite rank, which leads to update the network topology, and thus to consume more resources Farzaneh et al.,(2020). Furthermore, to trigger a local repair attack, compromised nodes can modify the Down 'O' flag and Sender-Rank field. Indeed, this attack may be triggered just by modifying flags or adding new flags in the header. This last attack is defined in Ghaleb et al.,(2023) as DAG Inconsistency Attack. As defined, the local repair attack is more like a poisoning or detaching attack.

Version number attack (Figure 23): The DIO message version number field is set and updated only by the 6LBR. Each time a rebuilding of the DODAG is necessary, the DODAG version number is incremented (by the 6LBR) and propagated unchanged down the DODAG graph. This process is known as global repair. As there is no mechanism in RPL to protect the version number field from modifications, an adversary can illegitimately increase the version number of the DOGAG which triggers the global repair mechanism and thus the reconstruction of the RPL topology from scratch. This attack leads to increase control messages overhead, generate loops and unoptimized topology, bring inconsistencies, and hence exhaust resources. Also, the version number attack is more effective when the attacker is located far from the 6LBR (Dvir et al.2011; Mayzaud et al.,2014; Aris et al.,2016; Alsukayti & Alreshoodi.,2023).

Figure 23. Version number attack (Mayzaud et al.,2016)

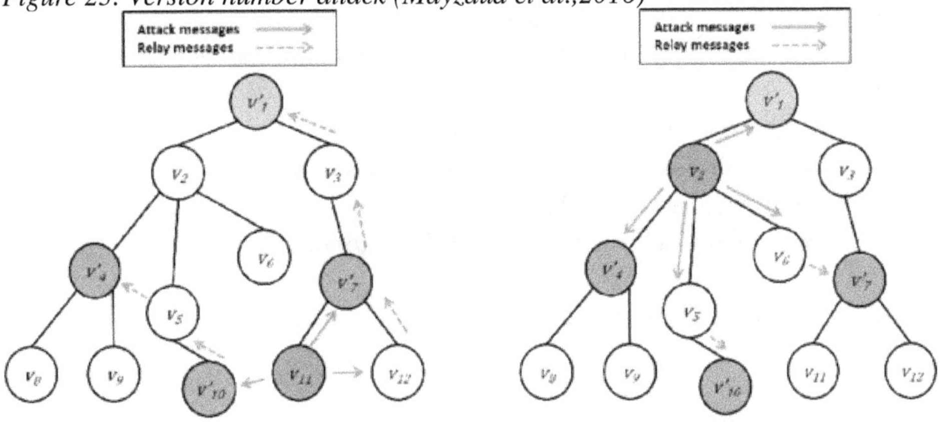

(a) Attacker located at position 11. (b) Attacker located at position 2.

Resource depletion attacks: Cong Pu. (2019) defined Resource depleting attacks as being the one triggered when an adversary initiates greedy activities aiming to exhaust the nodes resources. As already stated, authors in this work believe that all attacks against RPL impact directly or indirectly nodes resources. The fact that attacks create unoptimized routes, generate loops, or congest the network, all these factors lead honest nodes to consume more resources (Figure 24).

Figure 24. An example of energy depletion attack and the performance impact of energy depletion attack. (Cong Pu., 2019)

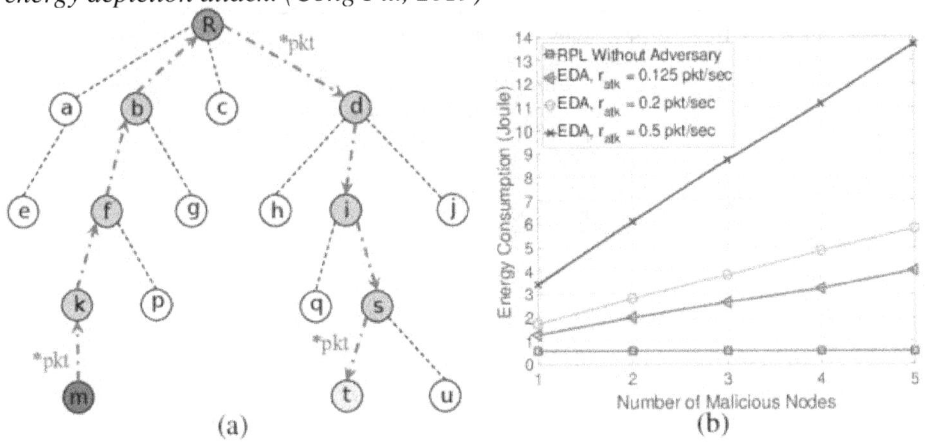

(a) A malicious node “nm” intentionally generates and sends a large number of packets “*pkt” to a destination node “nt”.
(b) The energy consumption against number of malicious nodes and attack rate (ratk).

Green node is the DODAG root
Red dash-dotted lines represent forwarding path
Network area 200×200 (m2),
Normal packet injection rate is 0.1 pkt/sec.

The attacks presented previously are summarized in Table 1.

Table 1. RPL Specification-Based Attacks

Attack Type	Description	Attack Goal
Rank attacks	Malicious manipulation and/or exploitation of the Rank field and/ or rules	Unoptimized routes
Neighbor attack	Malicious node replays a replicated DIO messages to create a false route	Loops in the network exhaust node resources congest the network
Routing Choice Intrusion attack	According to RPL’s routing rules, in order to disrupt optimal route choice, an attacker broadcasts a previously captured counterfeit control message.	Unoptimized routes create loops exhaust resources
DAO attacks	Malicious nodes announce fake routes by modifying or forging DAO messages	Build false routes overload the targeted nodes’ routing tables
DIS attacks	The attacker periodically sends (unicast/multicast) fake DIS messages to reset the timers of the target nodes	Generate control messages overhead Flood the network with fake messages
Version number attack	An adversary can illegitimately increase the version number of the DOGAG (due to the lack of version number field protection), triggering the global repair mechanism and consequently leading to the reconstruction of the RPL topology from scratch	increase control messages overhead generate loops unoptimized topology

continued on following page

Table 1. Continued

Attack Type	Description	Attack Goal
Local repair attack	An attacker initiates the update of the network topology by repeatedly triggering local repairs through changes in the DODAG ID field or by broadcasting an infinite rank	exhaust node resources
Resource depletion attacks	Encompass attacks that result in high energy consumption	lead nodes to consume more resources

B. Attacks Tailored to The Context of RPL

This class includes well known routing attacks which have already been studied by the research communities, and have been adjusted to the context of RPL.

HELLO Flood Attacks (Figure 25): In the context of an RPL network, threat actors can assume the guise of neighboring nodes by disseminating DIO messages under the pretense of HELLO messages, utilizing amplified signal power and favorable routing metrics. Consequently, when legitimate nodes dispatch data packets to these deceptive entities, the messages may experience loss due to the attackers potentially residing outside the communication range Wallgren et al.,(2013). This attack exhibits similarities to the Neighbor attack. The re- semblance arises from the fact that both attacks involve deceptive actions aimed at disrupting the normal operation of network nodes, albeit with distinct strate- gies and objectives. To mitigate this threat, Cakir et al., (2020) introduced a GRU-based deep learning model, which has demonstrated remarkable efficacy in forecasting and thwarting HF attacks on the RPL protocol within IoT networks, surpassing conventional methods while maintaining lower error rates in attack detection.

Figure 25. HELLO Flood Attacks (Bang et al., 2023)

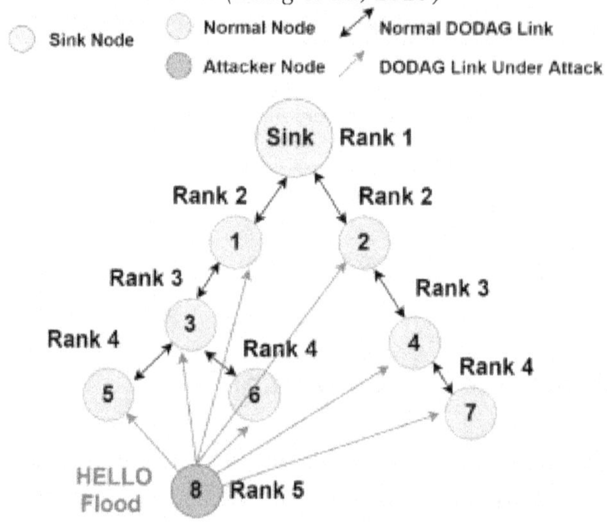

Sinkhole attack: In this attack, the malicious node advertises itself as the best path in order to be chosen as preferred parent by its neighbors, and thus to route traffic through it. In opinion of the authors, this attack is similar to Rank attack, where a malicious node advertises an artificial beneficial Rank to be selected as preferred parent. As it is, this attack does not appear to be harmful (passive at- tack), however it become harmful (active attack) if combined with other attacks (Chugh et al.,2012; Weekly & Pister,2012; Haas et al.2014). Zaminkar & Fotohi, (2020), focuses on the development of a defense mechanism against sinkhole at- tacks in Internet of Things (IoT) devices using node rating and ranking in a real-world enterprise setting, where IoT devices are evaluated for security and integrity in the face of potential attacks, with a particular emphasis on securing end-to-end connections in a heterogeneous IoT environment. Black-hole attacks: An intruder trigger a black-hole attack by dropping all data packets routed through it. This attack can be considered as a DoS Attack. In- deed, the black-hole attack is more dangerous if combined with Rank or Sink- hole attacks since the attacker is in a position where huge traffic is routed through it. This attack increases the number of exchanged DIO messages which leads to instability of the network, data packets delay, and thus resources exhausting (Chugh et al.,2012; Weekly & Pister,2012; Haas et al.2014; Kumar et al.,2016). Rani et al., (2020) presents a defense mechanism against dual-threat attacks (Black Hole Attack and Gray Hole Attack) in Mobile Ad Hoc Networks (MANETs) using swarm-based Artificial Bee Colony (ABC) optimization technique, improving the selection of optimal nodes for data packet transmission. Dhanaraj et al., (2021) presents a hybrid detection framework for countering black hole and sink hole attacks by combining Proportional Coinciding Score (PCS) and the MK-Means algorithm, which reduces feature count, enhances attack detection accuracy, and mitigates computational challenges.

Figure 26. Sinkhole node in RPL network (Alzubaidi et al., 2018)

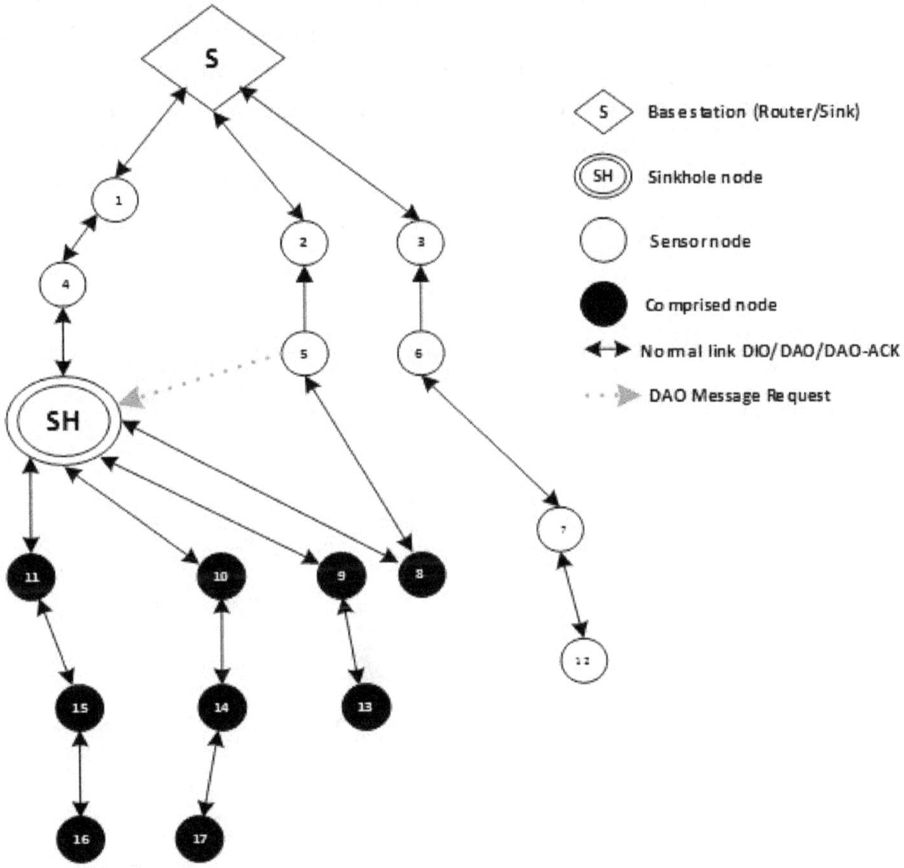

Selective-forwarding attacks: In the Selective-forwarding attacks, a misbehaving node can either aggressively filter RPL control messages or drop data packets and forward only the control messages traffic. The first attack affects negatively the topology construction and the network functions which leads to disrupt rout- ing. While the second attack leads to a DoS attack because no data will be transmitted to destination nodes. These attacks are also known as gray-hole at- tacks which are special case of black-hole attack Chugh et al., (2012). These attacks are more dangerous and cause great harm if they are combined with other attacks such as Sinkhole attack or Rank attack (Dvir et al.2011; Weekly & Pister,2012; Wallgren et al.,2013; Raza et al.,2013). Airehrour et al., (2017) intro- duces a novel trust-based Routing Protocol for Low-Power and Lossy Networks designed to counter selective forwarding attacks in IoT networks by implement- ing feedback-driven trust-aware security mechanisms while minimizing network traffic overhead. Jiang, J., & Liu, Y., (2022) presents an advanced selective for- warding attack model in Routing Protocol for Low Power and Lossy Networks (RPL) that offers attackers the ability to flexibly drop packets, dynamically adjust packet forwarding rates, and engage in se- lective bad-mouthing attacks, thereby enabling stealthy, sophisticated threats, and it proposes a centralized trust-based defense mechanism deployed at the root node to effectively detect and mitigate these selective forwarding attacks.

Figure 27. Selective forwarding attack. (Choudhary S. & Kesswani N., 2019)

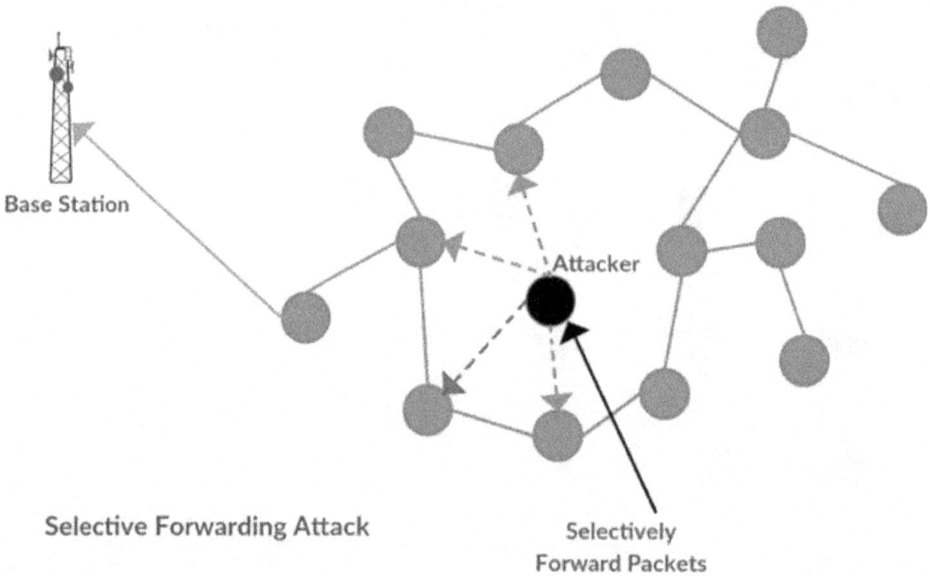

Figure 28. Wormhole attack. (Sharma et al.,, 2023)

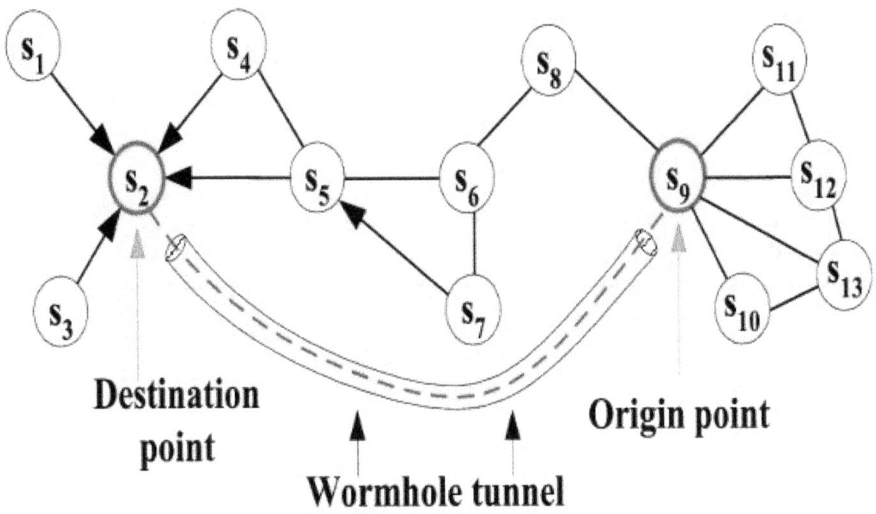

Wormhole attack: To trigger a wormhole attack, two or multiple attackers have to connect via wired or wireless links called tunnels. Wormhole attack permits an attacker to replay the network traffic in the other ends of the tunnels. In the case of RPL, some attackers can be outside the 6LoWPAN, and thus can bypass the 6LBR. In addition, if control messages are replayed to another part of the net- work, nodes which are actually distant see each other's as if they are neighbors which leads to distorts routing paths and create unoptimized routes Mayzaud et al., (2016). The Wormhole attack is highly harmful especially combined with other attacks and its detection is considered as a research challenge (Wallgren et al.,2013; Khan et al.,2013). Lai G.H;, (2016) presents a distributed wormhole detection mechanism for wireless sensor networks, based on the IPv6 Routing Protocol for Low-Power and Lossy Networks (RPL), which utilizes node-defined RPL ranks to measure distance, effectively identifying malicious wormhole nodes by detecting unreasonable rank values. Perazzo et al., (2018) presents a practical implementation of a wormhole attack, leveraging an IEEE 802.15.4-based Wire- less Sensor and Actuator Network, and measures its impact on various parame- ters, while also examining existing countermeasures, ultimately suggesting that the most effective defense against this attack may involve precluding subsequent malevolent activities such as traffic eavesdropping and selective packet dropping. Zahra & al., (2022) presents a lightweight multiclass classification-based model for the detection of specific security attacks, including rank and wormhole at- tacks, aimed at the IoT infrastructure and the RPL protocol. The model utilizes optimized feature selection strategies, and leverages a lightweight gradient boost- ing machine algorithm for multiclass classification-based attack detection. The evaluation encompasses cross-entropy, Cohn's kappa, and Matthews correlation coefficient.

Figure 29. Classification of wormhole attacks based on hidden and participating nodes (Hanif, M. et al., 2022)

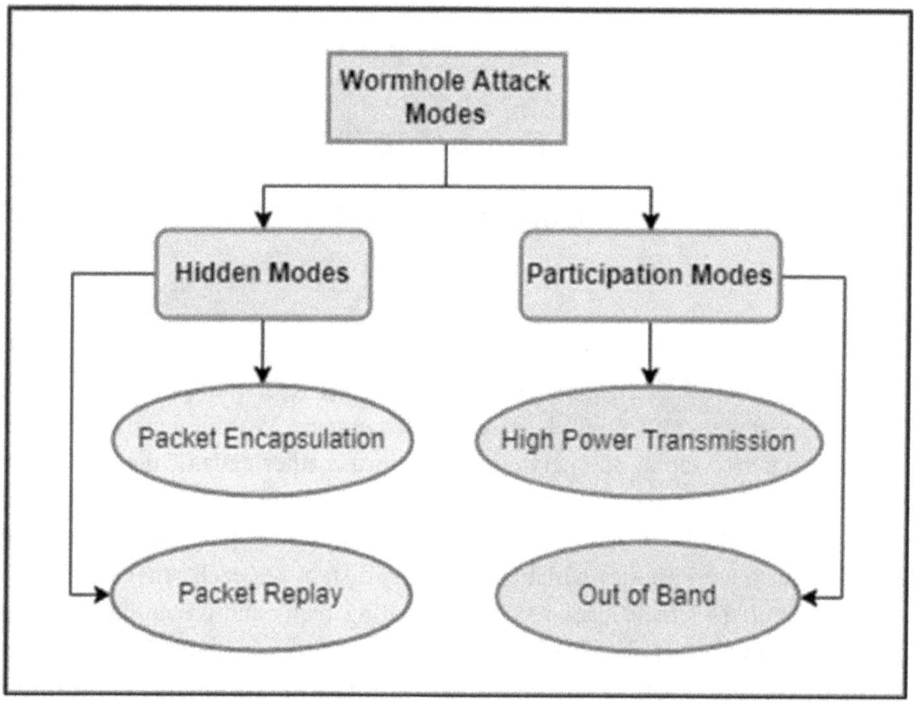

Sybil and CloneID attacks: Sybil and CloneID attacks are similar and known as Identity attacks Tsao et al., (2015). In a CloneID attack (spoofing attack), an attacker copies the same logical identity on several physical nodes. In a Sybil attack (impersonation attack), an attacker copies several logical identities on one physical node. A malicious node can trigger these attacks to access the traffic, take the control of the network, or to overcome a voting scheme (Wallgren et al.,2013; Raza et al.,2012a). Sybil attack can be combined with other attacks to affect harmfully the network operations. For instance, in SybM attack, at- tackers use periodically new fabricated identities (Sybil identities)

and trigger a Sybil-mobile attack in order to overload the network with fake messages and thus exhaust nodes resources (Medjek et al.2015; Medjek et al.,2017b). Tandon & Srivastava, (2019), present a context-aware trust-based approach, to secure RPL routing in IoT networks by simultaneously detecting Rank and Sybil at- tacks, outperforming the SecTrust-RPL protocol in terms of detection accuracy, energy efficiency, and network throughput. The proposed scheme offers several key contributions, including direct and indirect trust calculations for parent and child nodes, leveraging factors relevant to detecting these attacks.

Figure 30. Types of Sybil attack (Arshad et al., 2022)

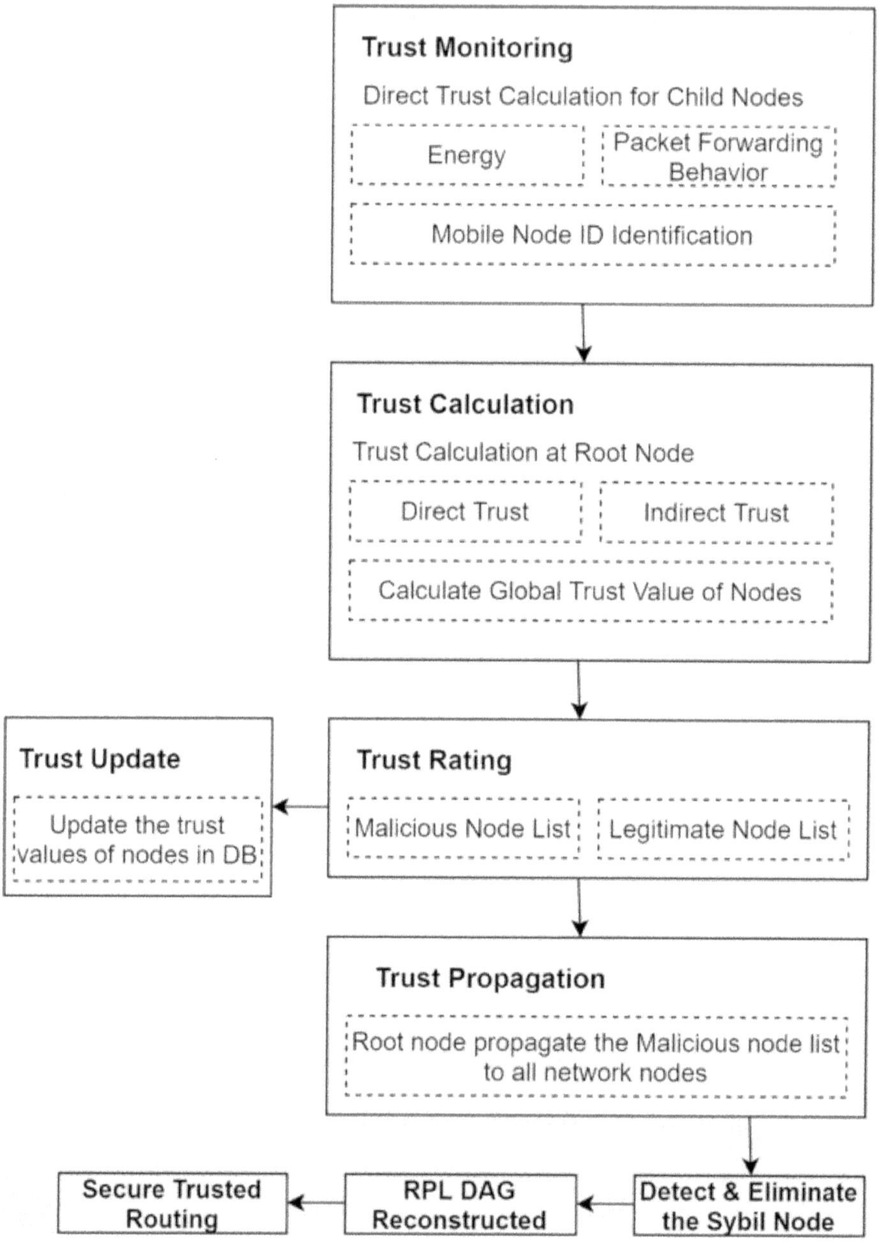

Groves & Pu, (2019), introduce a Gini index-based method to detect and mitigate Sybil attacks in RPL-based LLNs by evaluating RPL vulnerabilities and proposing a counter- measure based on identity dispersity in received DIS messages, demonstrating its effectiveness in improving detection accuracy, energy efficiency, and detection latency. Morales-Molina et al., (2021), address Sybil and Clone ID attacks by proposing pre-training techniques to select key characteristics from RPL net- work samples, followed by training a Dense Neural Network to maximize deep feature engineering, with the aim of improving classification results to protect against malicious identity impersonation attacks in the context of IoT and RPL protocol. Hassan, M. et al. (2023), address Sybil attacks in RPL-enabled Smart Grid Advanced Metering Infrastructures by introducing a novel fog-enabled GINI index-based trust mechanism (GITM) that utilizes the forwarding behavior of legitimate member nodes to detect and isolate Sybil attacks, outperforming ex- isting methods in terms of detection rate, energy consumption reduction, and isolation latency while maintaining a low increase in end-to-end delay and re- ducing the number of control messages.

Figure 31. Clone ID attack simulation. Morales-Molina et al., (2021)

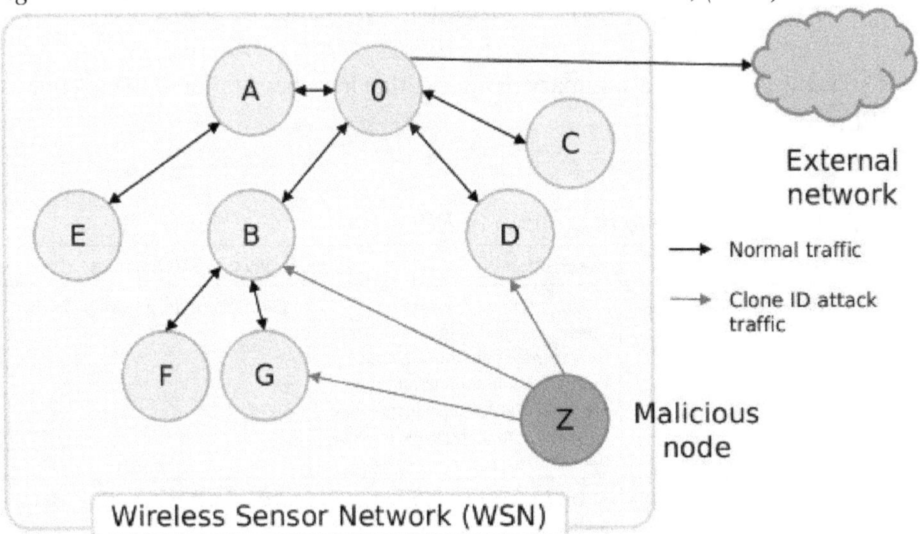

Denial of Service attacks: DoS and DDoS attacks aim to make nodes and/or the network unavailable. These attacks can be triggered against any layer of the IoT architecture. These attacks are simple to implement and very common because they have devastating consequences on the network Rghioui et al., (2014). In the point of view of the authors, all the aforementioned attacks may be cat- egorized as DoS or DDoS attacks since they overload the network with fake messages and exhaust resource which makes part of the whole network isolated and unavailable. Ioulianou et al., (2019), tackle Denial-of-Service (DoS) attacks in the RPL implementation of ContikiOS, specifically focusing on "Hello" flood- ing and version number modification attacks, which can lead to battery drain in IoT devices; they employ the Cooja simulator to illustrate the impact of these attacks on power consumption and suggest potential defenses, including a modular Intrusion Detection System (IDS) with distributed detection modules and a centralized detection module in the form of a border router. Belavagi & Muniyal., (2020), address Denial of Service (DoS) attacks in RPL-based net- works by proposing an IDS capable of identifying multiple intrusions, including DoS attacks, using machine learning algorithms and anomaly-based techniques, with the system monitoring the network for rank inconsistencies and other mali- cious activities based on the information collected by the DODAG's root and its members' behavior while accounting for different network sizes and percentages of malicious nodes.

The Table 2 provides a summary of attacks that have been tailored to the context of RPL.

Table 2. Attacks tailored to the context of RPL

Attack Type	Description	Protection Mechanisms
HELLO Flood Attacks	Utilizing favorable routing metrics, threat actors can impersonate neighboring nodes through the dissemination of DIO messages. Consequently, the response sent by legitimate nodes may be lost.	GRU-based deep learning model
Sinkhole attack	To route traffic through it, a malicious node presents itself as the optimal path	Rating and ranking the nodes in RPL protocol: SoS-RPL
Black-hole attacks	A malicious node that discards all data packets routed through it	Swarm-based Artificial Bee Colony (ABC) MK-Means algorithm

continued on following page

Table 2. Continued

Attack Type	Description	Protection Mechanisms
Selective-forwarding attacks	A malicious node may filter RPL control messages or drop data packets while forwarding only the traffic related to control messages.	Novel trust-based Routing Protocol for Low-Power and Lossy Networks
Wormhole attack	An attacker replays the network traffic in another part of the network, which create unoptimized routes	Lightweight multiclass Classification-based model
Sybil and CloneID attacks	An attacker copies the same logical identity on several physical nodes to take the control of the network	IDS using machine learning algorithms and anomaly-based techniques
Denial of Service attacks	All the previous attacks can be categorized as DoS attacks since they overload the network with fake messages and exhaust resource	Context-aware trust-based approach Gini index-based method

5.5. Application Layer

Referring now to the application layer, it embodies business solutions enabling end- users to remotely engage with the physical world, comprising various entities such as things, devices, and individuals. Additionally, it offers services for managing, an- alyzing, and visualizing measurements and outputs through user-specific interfaces. This layer outlines diverse applications for IoT deployment, including, for instance, smart health, smart cities, smart environment, and smart home solutions. Moreover, it allows to generate business models, graphs, and flowcharts from the received data. Because IoT devices have specific characteristics, instead of using HTTP, lightweight protocols like Constrained Application Protocol (CoAP) has been developed to sup- port application layer communications. As IoT network is directly connected to the unsecured Internet can undergo attacks from it (Internet) such as transactions replays, traffic congestion generation, and DoS and DDOS attacks. According to (Rghioui et al.,2013; Rghioui et al.,2014), attackers may trigger the Overwhelm attack to destroy the routing by generating huge traffic to the 6LBR, and the Path-based Dos attack to deplete resources by injecting false messages. For avoiding application layer attacks a firewall could be installed on the 6LBR to control the malicious packets from Internet. On the other end, Datagram Transport Layer Security (DTLS) may provide end-to-end security since it represents a solution to confidentiality, integrity, authentication and non-repudiation security problems for application layer communications using CoAP Raza et al., (2012a). Additionally, replay attacks may be mitigated with DTLS, using a different nonce value for each secured CoAP packet Granjal et al., (2015). The ex- posure to privacy breaches, stemming from known vulnerabilities, poses a significant risk, potentially leading to unauthorized access and data exfiltration. Furthermore, the looming threat of Denial of Service (DoS) attacks is a pressing concern, aiming to disrupt the operational availability of IoT applications, causing severe disruption.

Malicious code injection presents another critical challenge, enabling threat actors to exploit software vulnerabilities and gain unauthorized access to IoT systems. Lastly, social engineering tactics exploit interpersonal relationships within IoT networks, fa- cilitating information gathering that may serve as a foothold for various cyberattacks. These multifaceted application layer vulnerabilities underscore the imperative for the development and implementation of comprehensive and robust security measures to fortify IoT systems against these evolving threats, Nastase, L. (2017). Tampering with node-based applications, inability to receive security patches, hacking into the smart meter/grid, denial of service (DoS) attacks, unauthorized access, storage attacks, in- jecting fake information, spoofing attacks, sinkhole attacks, wormhole attacks, and man-in-the-middle attacks. These threats can significantly compromise the integrity and security of IoT systems. To mitigate these threats, various countermeasures have been proposed, including runtime type checking, physical security design, device au- thentication, network security protocols, IPsec, secure/multipurpose internet mail ex- tensions (S/MIME), and various authentication and encryption techniques to safe- guard the application, network, and perception layers of IoT systems,Masoodi et al., (2019).

The application layer attacks in the discussed IoT and messaging protocols encompass a range of vulnerabilities and risks. MQTT faces issues with authentication, authorization, message integrity, and encryption, potentially leading to unauthorized access and data compromise. CoAP is susceptible to threats like message parsing vulnerabilities, improper access control, and bootstrapping weaknesses, compromising device availability and integrity. AMQP, despite robust security mechanisms, experiences vulnerabilities related to broker configuration and access control, enabling priv- ilege escalation and Denial of Service attacks. DDS, while utilizing TLS and DTLS, suffers from vulnerabilities in security plugins and participant handshake protocols, potentially breaching confidentiality. XMPP, with SASL and TLS, remains exposed to threats due to the absence of end-to-end encryption, allowing for authentication and message validation flaws, as well as hardcoded account issues that risk compromising XMPP servers and sensitive data. These protocols necessitate careful implementation of security measures to counteract these vulnerabilities and safeguard IoT and messaging applications Nebbione & Calzarossa, (2020).

Figure 32. Access Control Vulnerability: temporal evolution of the MQTT interactions (Nebbione & Calzarossa, 2020)

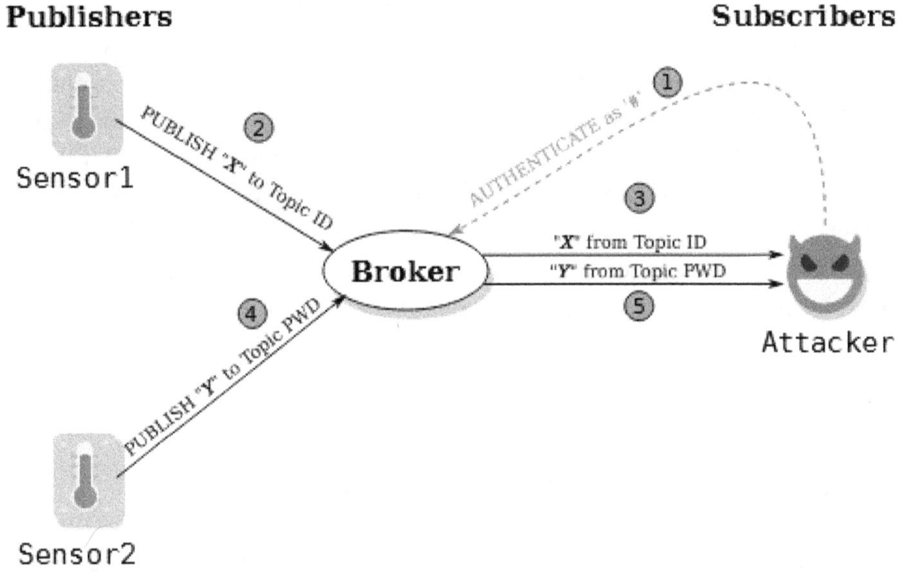

In the same context, application layer attacks include also Mirai Malware, which targets IoT devices by exploiting default Telnet or SSH credentials, IRCTelnet attacks that involve compromising IoT devices via Telnet ports, and Injection attacks that can lead to application vulnerabilities through untrusted data input. To mitigate these threats at the application layer, it is essential to employ lightweight encryption algorithms such as Secure MQTT (SMQTT) for data confidentiality and privacy. Furthermore, widely-used security protocols like SSL and Transport Layer Security (TLS) are crucial to ensure data integrity and safe- guard IoT systems from potential breaches and unauthorized access. Addressing these application layer attacks and implementing robust security measures is essential to protect IoT devices and networks effectively, Mrabet et al., (2020). As for Syed et al., (2020) the primary focus is on Application layer Denial of Service (DoS) attacks within the context of the MQTT protocol, which is commonly used in the Internet of Things (IoT) for Machine to Machine (M2M) communication. The authors highlight that as MQTT's popularity increases, it becomes more vulnerable to security threats, specifically protocol-based DoS attacks. These attacks have been known to disrupt services in various systems. To address this concern, the paper proposes a machine learning-based detection framework for Application layer DoS attacks on MQTT and conducts experiments to evaluate the effectiveness of this framework. The study un- derscores the severe impact of such attacks on MQTT brokers and demonstrates the capability of the proposed framework to detect them effectively. The findings show that attackers can overwhelm server resources, even when legitimate access is denied, mak- ing MQTT's security a critical issue for IoT deployments. The paper emphasizes the importance of identifying MQTT attack features for attack detection and the need to protect MQTT-based IoT infrastructures from these at- tacks. The IoT's rapid growth necessitates efficient communication among various applications and devices, which application layer protocols facilitate. Donta et al., (2022) notes that as the number of IoT applications increases, it becomes crucial to adapt or enhance these protocols to address specific IoT needs, such as dynamic adaptation to network conditions and interoperability. Various IoT application layer protocols are explored, and they in- troduces the idea of using machine learning to make these protocols more quick and dynamic. Highlighting open challenges to drive IoT application layer protocols in this direction.

6. COMPREHENSIVE SECURITY ANALYSIS

6.1. Security Considerations in RPL-Based IoT Networks

Advancing to examine security considerations within RPL-based IoT networks, the RPL specification integrates a dedicated security section within each RPL con- trol message to address multiple security facets. These encompass ensuring data confiden- tiality through the application of Message Integrity Codes, validating data authenticity via encryption, and implementing replay protection using the Consistency Check (CC) message. Still, in RPL network, some information can be monitored and used to mit- igate and/or minimize some attacks impacts. For instance the following numbers can be bounded within a given time in such a way the attacks cannot be triggered several time, and to quarantine neighbors having suspicious activities at unacceptable rates (Thubert et al.,2012):

The number of times a local repair procedure was triggered (Local repair at- tack) The number of times a global repair was triggered by the 6LBR (Version number attack)
The number of received malformed messages (Local repair attack) The number of times a node request to join a DODAG (DIS attacks)
The number of times routing tables are overflowed and the cause of overflow (DAO attacks)
The number of RPL control messages sent and received (Resource depleting attacks).

To fulfill the IoT requirements, as discussed above, several solutions have been pro- posed in the literature. Cryptography is considered the first line for solving the confi- dentiality, authentication and integrity. This mechanism, however, cannot solve other QoS securing requirements like availability, robustness and resiliency. It therefore needs to cooperate with Intrusion Detection System (IDS), which can monitor and detect malicious nodes from the early phase to eliminate further damage of the attacks. Per- azzo et al., (2017) implemented and evaluated different RPL security configurations: the unsecured mode, the preinstalled mode with light-security configuration, and the preinstalled mode with full-security configuration. The authors concluded that the net- work formation time increases with the network size, the overhead introduced by the replay protection increases with the network size, and the power consumption intro- duced by the security features increases with the network size too. However, globally the RPL security mechanisms have a negligible impact on the performances if there is no replay attack. Otherwise, if there is need to protect the network against replay attacks, the impact on performances is more pronounced.

As pointed out in (Alrababah et al.,2017; Griffin P.,2017), node authentication can solve most of the problems that may be caused by unauthorized uses such as Sybil and CloneID attacks. For example, DTLS may be used as a solution to confi-dentiality problems for the application layer by providing end-to-end security Raza et al., (2012a).

Figure 33. Rejoin procedure in the full-security configuration (Perazzo et al., 2017)

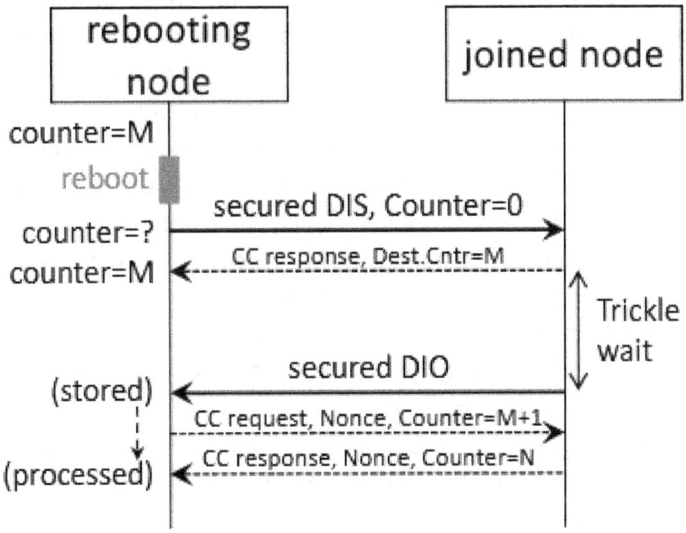

However, securely managing, processing, and storing cryptographic keys inside a re- source constrained and tamper-resistant embedded device deployed in an un- structured, distributed and untrusted environment is challenging problem. In this context, several works focus on secure key management mechanisms for IoT. For instance, authors in Raza et al., (2012b) proposed Internet Key Exchange (IKE) compression scheme to provide a lightweight automatic way to establish security associations for IPsec. Abdmeziem et al., (2017) proposed a compression scheme for MIKEY-Ticketkey ex- change protocol to provide light weighted and energy-aware version for the e-health application use.

Besides, existing cryptographic mechanism will fail in safeguarding the all network aspects since several attacks such as RPL-based attacks, selective-forwarding and black-hole attacks could not be prevented. In this context, tamper-resistant mod- ules and trusted computing technologies are required. Several works exist to secure RPL based on trust computation. For instance, authors in (Djedjig et al.,2015; Djed- jig et al.,2017) introduced a new trust-based metric to use when construction RPL topology. In this trust-based RPL, nodes cooperate to calculate trust metric of their respective neighbors based on nodes behaviors and some trust components (energy, honesty). If a node is detected as untrusted, it will be discarded from the list of parent and a local repair is triggered. Airehrour et al., (2016b) proposed a trust-aware RPL routing system to counter Blackhole attacks. This system uses the good forwarding behavior of neighbor nodes in the network to compute trust value for each node. The calculated trust value is dependent on the positive feedback awareness among the nodes and the trust evaluation analysis.

The classification of attacks directed at RPL into three principal categories, namely: attacks targeting network resources, those that manipulate network topology, and those that compromise network traffic provides a methodical framework for the anal- ysis and mitigation of potential threats within low-power and lossy network environ- ments. Attacks focused on network resources, such as the depletion of battery life or memory, necessitate the implementation of resource-efficient management and rate limiting mechanisms as countermeasures. In the context of attacks that tamper with network topology, which encompass issues like Sybil attacks, rank manipulation, and topology control attacks, it is imperative to institute robust authentication processes, trust systems, and continuous surveillance to safeguard the integrity of the network structure. Pertaining to attacks on network traffic, encompassing activities such as routing manipulation, data injection, and eavesdropping, the deployment of crypto- graphic safeguards, intrusion detection systems, and secure communication channels is paramount. This hierarchical taxonomy serves as an invaluable tool in the realm of risk assessment, enabling the customization of targeted mitigation strategies and bol- stering the efficacy of incident response plans, particularly within organizations reliant on RPL-based IoT and wireless sensor networks, Mayzaud et al., (2016).

Safeguarding IoT devices poses challenges, given their inherent resource constraints, including limited battery power and memory capacity. Notably, RPL, tailored for IoT environments, has unveiled vulnerabilities to a spectrum of security threats, with the Wormhole attack at the 6LoWPAN adaptation layer emerging as adversary. This kind of attack occurs when malevolent nodes collaborate to create a deceptive tunnel, lead- ing to potential delays and packet loss in network traffic. To counteract this peril, Deshmukh-Bhosale & Sonavane, (2019) unveils a comprehensive IDS system designed to identify and thwart Wormhole attacks. Leveraging the Received Signal Strength In- dicator (RSSI) for proximity assessment, the IDS operates through a hybrid approach, combining centralized and distributed modules, ensuring enhanced IoT network secu- rity and safeguarding the IoT landscape as it continues to burgeon.

(Muzammal et al.,, 2022), presents a trust-based security model, named SMTrust, and designed to address routing attacks within RPL-based Internet of Things (IoT) networks, particularly focusing on Rank and Blackhole attacks while considering static and mobile nodes. The model's phases encompass topology creation, trust metrics iden- tification, trust index calculation, attack detection, and the propagation of trustwor- thy nodes for routing. The trust parameters considered in SMTrust include historical observations, energy levels, direct trust, recommended trust, and mobility-based met- rics. The workflow is orchestrated to establish trustworthy nodes, thereby fortifying IoT network security. This research endeavor represents a significant contribution to the field, addressing critical routing attacks in RPL and setting a new standard in trust-based IoT security.

Rouissat et al., (2023) introduces a multifaceted contribution to the field of RPL- based IoT network security. Firstly, it identifies a novel security threat called the DIOF attack, which exploits vulnerabilities in RPL flooding attacks. Secondly, it proposes a sophisticated yet lightweight security framework called DSRPL to counter the DIOF attack, significantly enhancing network resilience. Finally, the research conducts an extensive evaluation of DSRPL's effectiveness by considering various attack scenarios. The DIOF attack's modus operandi involves circumventing RPL's restriction mech- anisms governing DIO message transmission frequency. It does so by broadcasting manipulated DIO messages with falsified trickle timing parameters, leading to the un- necessary proliferation of DIO messages and subsequent network congestion. DSRPL mitigates the DIOF attack by implementing distributed verification mechanisms for trickle timing updates in DIO messages. These mechanisms ensure that only legitimate updates are processed, minimizing the impact of the DIOF attack and preserving net- work efficiency. Overall, this research presents a comprehensive analysis of a new attack vector and an innovative security solution, offering valuable insights for safeguarding RPL-based IoT networks against emerging threats.

Figure 34. System Model of SMTrust (Muzammal et al., 2022)

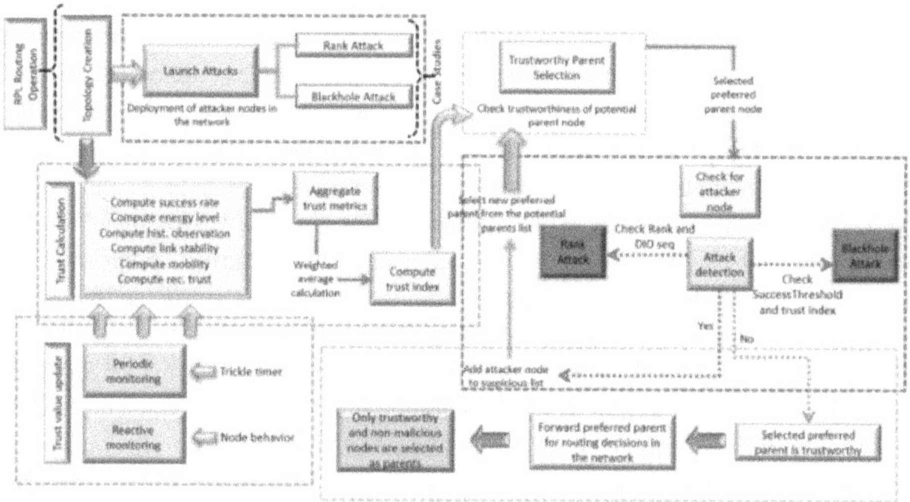

Alsukayti & Alreshoodi., (2023), delves into the vulnerabilities and potential threats posed by routing attacks within IoT networks employing the RPL protocol as their underlying framework. Notably, the study delves deeply into IoT devices, characterized by their inherent resource constraints, and meticulously situates them in various deployment configurations, such as uniform and random positioning, placing significant emphasis on the ramifications of attacks on network performance and stability. The investigative approach incorporates an array of attack types, namely Virtual Node (VN), rank, replay, and Witnessing Parent (WP) attacks, executed through judicious code alterations solely on attacker nodes. The research is exhaustive in its scrutiny, encompassing Quality of Service (QoS) parameters including throughput, Packet De- livery Ratio (PDR), delay, and Expected Transmission Count (ETX). It discerns a consistent pattern of substantial QoS deterioration across all these metrics, with VN attacks emerging as the most formidable threat, closely trailed by rank, replay, and WP attacks. In parallel, the investigation exposes the multifaceted impacts of these attacks on critical network stability indicators, namely beacon intervals and Preferred Parent Change (PPC) rates, revealing that network instability escalates under attack con- ditions. Additionally, network overhead metrics, as represented by the Dissemination Information Object (DIO) and Destination Advertisement Object (DAO) transmission rates, reflect a noteworthy amplification in control traffic, which is particularly evi- dent in composite and hybrid attack scenarios. Intriguingly, the research underscores a compelling facet related to energy efficiency, underscoring that these attacks could be especially detrimental to resource-constrained IoT devices, which rely heavily on energy conservation.

6.2. IDSs For IoT Networks

As stated in the previous sections, there are vulnerabilities from inside the network and from the Internet that go beyond the encryption and authentication first lines of defense for the IoT communications. In such cases Intrusion Detection Systems (IDS) are required as a second line of defense. Indeed, IDSs are primarily used to counter attacks against the network. They analyze activities and nodes behaviors in the network and trie to detect intruders that are trying to disrupt the network. IDSs can be classified depending on their location or on the method used for detection.

6.2.1. IDSs Over Networks

Intrusion Detection Systems (IDSs) in network security can be categorized into three main approaches: centralized (monolithic), distributed (cooperative/host-based), and hierarchical/hybrid. Centralized IDS involves all nodes sending their monitoring data to a central node for analysis, but this approach lacks scalability and is vulnerable to single points of failure. Distributed IDS, where each node hosts its own IDS and cooperates with others, is commonly used and effective but comes with challenges related to system maintenance and resource overhead. Hierarchical/hybrid IDS divides the network into hierarchical monitoring areas, allowing local analysis at each level and offering better scalability. However, changes in network topology may impact the hierarchy and the overall system's robustness

A. **Centralized (Monolithic):** In this approach, each node monitors the operation of all of its neighbors in the network and transfer collected information to a central intrusion detection nodes to be analyzed. It suffers from the following deficiencies:
    (1) Scalability: It is difficult to guarantee scalability as a network size grows. In ad- dition, because lots of information need to be transferred from monitoring nodes to a central node, this creates heavy overhead which causes severe degradation of the network performance.
    (2) Robustness: The central node represents a single point of failure making the overall IDS crippled in case of its failure or if it is attacked.
B. **Distributed (Cooperative)/Host-based:** In this approach, the IDS is placed on each node within the network. Each host-based IDS monitors only a small portion of the network (neighbors). The distributed host-based IDSs cooperate

to analyze and detect intruders. They can make a coherent inference and make a global decision. Monitoring host-based nodes are called watchdogs.

- * It is the most used approach in the literature because it resolves most of security breach in the network. Nevertheless, it is hard to repair and maintain the overall system in addition to the generated overhead on the monitored parts of the network. Furthermore, it requires a lot of memory and calculation resources.

C. **Hierarchical/Hybrid:** In this approach the network is divided on a number of hierarchical monitoring areas, where each IDS monitors a single area. Instead of transferring all the collected information from monitoring nodes to a central IDS, each single IDS at each level of monitoring area performs local analysis and sends its local analysis results up to the IDS at the next level in the hierarchy. Thus, IDS's at higher levels only need to analyze transferred local reports collectively.

(1) Scalability:Shows better scalability by allowing local analyses at distributed local monitoring areas.

(2) Robustness: If the topology of the network changed the network hierarchy changes as well and the whole mechanisms to aggregate monitored information must be changed. Furthermore, when a monitoring node residing at the high- est level is attacked or fails, the sub-network related to this node easily escape detection.

Figure 35. Classification of IDS for IoT (Khraisat et al.,2019)

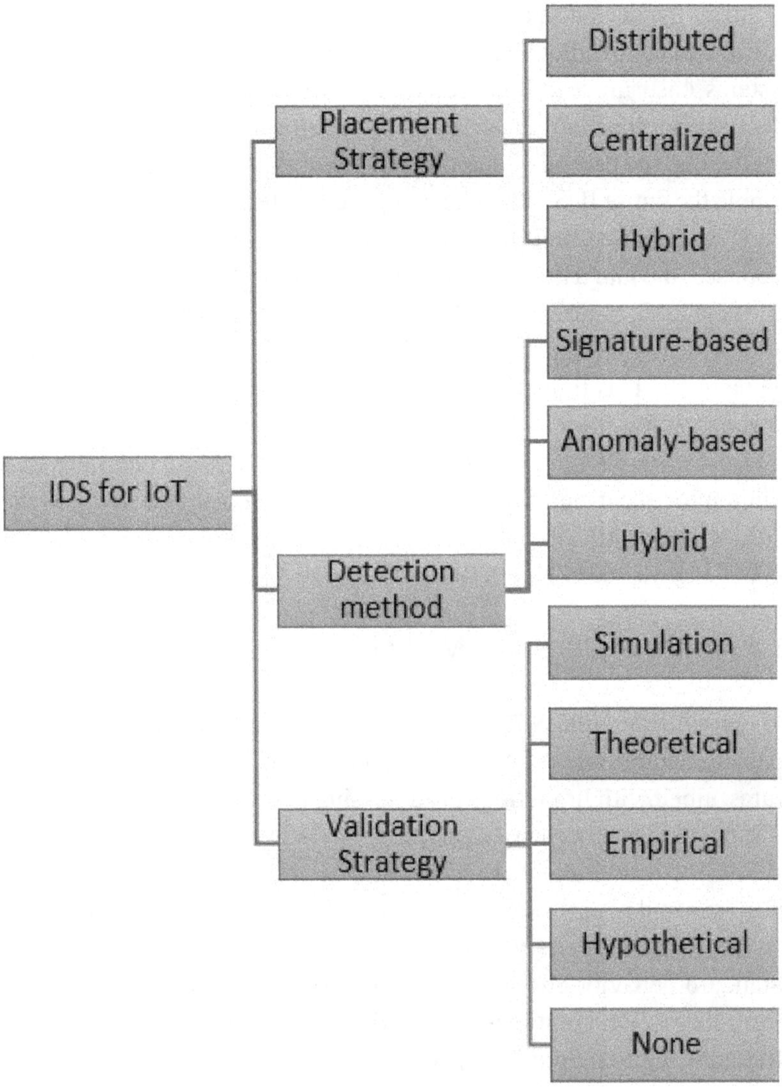

D. **Transitioning Towards Blockchain IDS for IoT:** Blockchain offers a decentralized framework wherein each node in the network maintains its copy of a dis- tributed ledger, thereby eliminating the need for a central authority. In the realm of IDS security for IoT, this distributed architecture can significantly enhance scalabil- ity and robustness. By employing a blockchain-based IDS, the network can achieve enhanced scalability as there's no central node burdened with processing information from all other nodes. Instead, each node can contribute to the detection and analysis of potential intrusions autonomously. This distributed nature alleviates the scalability concerns encountered in centralized systems, ensuring seamless operation even as the network size grows. Moreover, blockchain's inherent design ensures robustness against single points of failure. Unlike the centralized approach where the failure of the central node can cripple the entire IDS, a blockchain-based system distributes the workload and redundancy across multiple nodes. Thus, even if certain nodes fail or are compromised, the integrity and functionality of the IDS remain intact. Furthermore, the immutable nature of blockchain records enhances the security of the IDS by provid- ing a tamper-proof audit trail of all transactions and events within the network. This transparency and accountability are particularly valuable in detecting and mitigating intrusions in IoT environments, where data integrity and trust are paramount. While blockchain IDS offers a decentralized and robust solution for securing IoT environ- ments, it's crucial to delve into the various IDS methodologies to understand their effectiveness (Rekha H. & Siddappa M.,2023; Jayasree et al.,2020). .

6.2.2. IDSs Over Detection Method

IDSs can be differentiated by their detection methods, yielding five primary classifications. Signature-based IDS involves pattern matching against predefined attack signatures, albeit with limitations in addressing novel threats, storage requirements, and false positive and negative rates. Anomaly-based IDS establishes a normal behavior baseline, excelling at detecting new attack vectors but prone to increased false positives and negatives. Event-based IDS defines and stores malevolent event patterns and triggers alerts when such events occur. Specification-based IDS defines standard behavior and triggers alarms when deviations are detected. Lastly, the hybrid-based IDS combines various detection techniques to enhance precision and efficiency without unduly impacting network performance.

(1) Signature-based: In this IDS, signature of malicious activities or codes are stored in a database or a list. The signature patterns in packets are matched with the stored ones. If match founds then the IDS raises alarm for the attack. Hence, this IDS compares the current activities in a network or in a device against predefined and stored attack patterns (signatures). This approach cannot detect new attacks, needs specific knowledge of each attack, has a significant storage cost that grows with the number of attacks, and has a high false negative but low false positive rate.

(2) Anomaly-based: This type of IDS determines the ordinary behavior of a network or a device, uses it as a baseline, and detects anomalies when there are deviations from the baseline. This approach can detect new attacks but has comparatively high false positive and false negative rates because it may raise false alarms and/or cannot detect attack when attacks only show small deviations from the baseline.
(3) Event-based:In this IDS system, the malicious events patterns are defined, a priori, and stored in a database. Thus, the event based IDS captures the events triggered in the network and analyzes them. If an event is suspicious, the IDS raises alarm for attack detection.
(4) Specification-based:This type of IDS is also known as software engineering based IDS or Finite State Machine (FSM) based IDS. In this kind of IDS, a normal behavior is defined. If an abnormal behavior is detected, the IDS raises alarm.
(5) Hybrid-based: This type of IDS combines the aforementioned types to get better detection results with less negative impacts on the network performances.

6.2.3. IDSs in RPL-Based IoT Networks: Methods and Analysis

In the literature, several IDSs have been proposed to deal with RPL attacks. This section reviews some of the existing IDSs. Table 3 summarizes the different solutions. Le, A. et al., (2012) proposed a hybrid specification-based IDS idea for securing RPL against topology attacks. In this approach, nodes monitor routing information conveyed in control messages to detect attackers. As continuation to this work, Le, A. et al., (2016) implemented the proposed IDS using an Extended Finite State Machine (EFST) with statistic information about transitions and states for RPL. In this IDS, a cluster head requests its members to report its topology information periodically and process these information using EFST. Information used are: DIS sequence, number of DIS received, DIO sequence, number of DIO received, list of neighbors (Node ID, Rank, sequence of the DIO that provides this info, DIS sequence, number of DIS received, DAO sequence, number of DAO received, and a parent bit), and Preferred parent ID. The IDS aims to detect Rank, sinkhole, local repair, neighbor, and DIS attacks. This IDS is energy efficiency but showed less accurate when it works for a long time.

In Kasinathan et a. (2013a), the authors proposed a centralized signature-based IDS to detect DoS attacks in 6LoWPAN networks. The proposed IDS has been integrated into the network framework. In this IDS non-6LoWPAN monitoring nodes located in the network send periodically the 6LoWPANs sniffed traffic through wired connection to the IDS. Thus, if a DoS attack occurs and degrades the wireless transmission quality, IDS data transmission would not be affected. The IDS collaborates with a DoS protection manager to confirm the attack using jamming information. This solution targets only DoS attack and is not compatible to general network architecture.

Raza et al., (2013) introduced SVELTE, hybrid anomaly-based IDS for securing the RPL protocol. SVELTE modules were placed both in the 6LBR and in constrained nodes. These modules work on two stages: collecting and analyzing the IDS data. At the first stage the 6LBR requests the network nodes to send information about themselves and their neighbors. These information are: RPL Instance ID, the DODAG ID, the DODAG Version Number, Rank, parent ID, neighbors list and their corresponding Ranks, and a timestamp. At the second stage 6LBR analyzes the collected data and makes decisions. This IDS targets spoofed or altered information, sinkhole, and selective-forwarding attacks. Besides, authors proposed a distributed mini-firewall to protect the network against external attackers. Nevertheless, this IDS suffer from synchronization issue. For instance, the reported Rank information of a given node from the node itself and from its neighbors to the 6LBR are not the same because the recording time was not synchronized.

Zhang et al., (2015) proposed a stand-alone specification-based IDS with distributed Monitoring Nodes (MNs). In this IDS the detection data is network-based where in each MN is implemented a Finite-State-Machine (FSM) to detect RPL's abnormal behaviors. However, they relied their analysis on unrealistic assumptions; specifically the stable state of LLN environment.

Pongle & Chavan, (2015b) proposed a hybrid anomaly-based IDS to detect wormhole attack. Authors used Neighbor Discovery/Verification Based techniques for the detec- tion of wormhole attacks. In this solution four centralized modules are implemented on the 6LBR and four distributed modules on the monitoring in-network nodes. The monitoring nodes gather information about their respective neighbors and changes on the network (RSSI), and send them to the 6LBR. This later analyzes received data to detect intruders and makes decisions.

Cervantes et al., (2015) proposed an intrusion detection for Sinkhole attacks over 6LoWPANfor InterneT of ThIngs(INTI). INTI is a hybrid IDS that combines watchdogs, reputation and trust to detect sinkhole attackers. The INTI organizes the network on clusters where each node uses four modules to detect sinkhole attack: cluster configuration, routing monitoring, attack detection and attack isolation. After monitoring the traffic, if a node detects an attacker it alerts other nodes to isolate the attacker. The authors compared INTI with SVELTE regarding some metric, but they did not present INTI impact on energy consumption.

Thanigaivelan et al., (2016) presented a cross-layer anomaly-based detection system for IoT. The proposed IDS is composed of three subsystems located at the network and the link layers as follow: both the monitoring/grading subsystem (MGSS) and the reporting subsystem (RSS) operate at the network layer, whilst the isolation subsystem (ISS) operates at the link layer. If a node is detected to have abnormal behavior, the ISS is used to avoid packets from that node at the link layer level. When anomalies and network changes are detected, they are communicated from the node to the edge-router through subsequent parents. The edge-router analyses reports and makes decisions. Nevertheless, in this solution parents themselves can be compromised and anomalies notifications can be avoided.

Lai G.H;, (2016) introduced a distributed IDS aiming to identify wormhole attacks. The IDS uses nodes location. Each node uses the Rank information from RPL control messages to estimate the relative distance to the 6LBR and identify suspicious Rank values. Thus, the Rank value is compared with Ranks of the respective neighbors; if the discrepancy exceeds a threshold value, it signals that a wormhole might exist. In this IDS authors used the hop-count metric to calculate the Rank and detect the attack.

The question is: Is this solution effective when using different metric, such as ETX (Expected Transmission Count), to calculate the Rank? Is this solution effective in nodes mobility cases?

To identify mobility-based and specification-based attacks, Medjek et al., (2017a) proposed a hybrid, cooperative and hierarchical trust-based IDS: T-IDS. T-IDS integrates three cooperative modules: IdentityMod for identity management, MobilityMod for mobility management and IDSMod for intruder detection and isolation. IdentityMod uses the built-in Trusted Platform Module ID (TPM-ID) of each node to generate a unique identifier per in-network node. In order to authenticate and monitor a node at any stage of the network execution, the author propose to extend RPL control messages with the unique identifier of each node. The three actors of the IDS (Backbone Router (BR), 6LBR, and in-network nodes) maintain lists of mobile nodes within the network to monitor malicious behavior using Mobility Mod.

Gara et al., (2017) introduced a hybrid IDS to detect selective forwarding attack in IPv6-based Mobile IoT network. This IDS works using two modules: a centralized module on the sink node and a distributed one on routing nodes. Each monitoring node calculates periodically the number of packets received and the number of packets sent from each neighbor and sends the collected data to the sink node. These information are processed by the sink node to detect malicious behavior and decide wither a node is an attacker or not. If an attacker is detected, a global repair is triggered. Although this IDS offers good performance on attack detection, it generates high overhead.

Bostani & Sheikhan., (2017) proposed a new hybrid anomaly-specification-based IDS to detect sinkhole and selective-forwarding attacks in 6LoWPAN-RPL network. In this IDS, an anomaly-based intrusion detection module is located in the6LBR, and specification-based module are located in some router nodes. The 6LBR uses the information and analysis collected from the router nodes and itself to make a global decision. Alzubaidi et al., (2018), focuses on the intricate domain of Intrusion Detection Systems (IDS) deployed within network environments. It extensively explores the multifaceted aspects of IDS implementation, especially concentrating on two key dimensions: IDS deployment strategies across network infrastructures and the diversity of IDS detection methods employed to counteract various forms of cyberattacks.

In the first dimension, "IDS Over Network," the study delves into the overarching role of IDS within network architectures. It encompasses the critical placement of IDS sensors and analyzers throughout the network to effectively monitor traffic and security events. The work examines how IDSs are strategically positioned across networks, considering factors like network topology, the placement of sensors, and their scope in terms of the network's geographical and logical layout. This facet of the research seeks to optimize the network's security posture by ensuring comprehensive threat visibility and timely detection.

The second dimension, "IDSs Over Detection Method," delves into the diversity of detection methodologies employed by IDS systems. It delves into the technical intri- cacies of how IDSs leverage various detection techniques, including signature-based, anomaly-based, and heuristic-based methods, among others. The study explores how these detection mechanisms are implemented within IDS systems, highlighting their strengths and limitations. It examines the technical details of these detection methods, emphasizing their underlying algorithms, rule sets, and heuristics. Additionally, it investigates how these detection methods are combined and customized to create effective, multi-layered defense mechanisms.

In the third dimension, "Countered Attacks," the research investigates the real-world effectiveness of IDSs in identifying and mitigating cyber threats. It delves into specific categories of attacks that IDSs are designed to counter, such as malware, denial-of-service (DoS) attacks, intrusion attempts, and data breaches. The work takes a technical approach in dissecting the attack vectors and methodologies employed by cyber adversaries and scrutinizes how IDSs detect and respond to these threats.

The technical analysis extends to the countermeasures deployed by IDSs, including alerting, blocking, or initiating incident response actions to mitigate the impact of attacks on the network. Verma & Ranga, (2019), focuses on the development of Intrusion Detection Systems (IDS) using ensemble learning methods. The rationale for utilizing ensemble learning lies in its ability to improve classification outcomes, offering enhanced prediction accuracy and model stability in the context of IDS, a vital tool for detecting network-based attacks. Ensemble learning leverages multiple models to enhance classification results, with the aggregated output of the ensemble being less noisy compared to other machine learning approaches. Bagging methods are employed to mitigate overfitting issues in the models. The study utilizes the RPL-NIDDS17 dataset for training and testing classifiers, with a focus on evaluating their performance in terms of accuracy and the Area under ROC (Receiver Operating Characteristic) curve, which assesses the trade-off between true positive and false positive rates. The experimental flow design involves dataset preprocessing, encompassing cleaning (handling missing values), encoding (conversion of nominal data to numeric form), and scaling (feature normalization). The preprocessed dataset is then split into training and testing sets. The research employs ensemble classifiers, specifically Bagged Trees, Boosted Trees, Subspace Discriminant, and RUSBoosted Trees, which have been chosen based on their performance across various datasets, both balanced and imbalanced. The trained models are subsequently tested using the test set, with ensemble classifiers demonstrating their ability to classify network traffic as either normal or indicative of an attack. The proposed network intrusion detection system (NIDS), named EL-NIDS, adopts a signature-based approach and aims to detect routing attacks on RPL, including Sinkhole, Blackhole, Sybil, Clone ID, Selective Forwarding, Hello Flooding, and Local Repair attacks. The EL-NIDS architecture consists of components like the sniffer, sensor events/traffic repository, feature extraction module, analysis engine, signature database, user interface, and alarm/attack notification manager. The sniffer captures packet transmissions within the 6LoWPAN network, and the feature extraction module extracts useful features from the collected packet traces for traffic classification. The heart of EL-NIDS is the analysis engine, which leverages trained ensemble models to classify traffic and aggregates their predictions using a voting scheme. Results from the voting scheme trigger alarms if an attack is detected. The study concludes by demonstrating that ensemble classifiers, particularly Boosted Trees and RUSBoosted Trees, offer the best performance in terms of accuracy and Area under ROC, thus validating the effectiveness of EL-NIDS in securing RPL-based 6LoWPAN networks from routing attacks. Future work will focus on implementing and evaluating EL-NIDS on smart nodes and developing lightweight defense solutions for securing the Internet of Things. Aydogan et Al., (2019), proposes an Intrusion Detection System (IDS) for RPL-based Industrial Internet of Things (IIoT) networks, focusing on using genetic programming (GP) to detect various attacks against the RPL protocol. GP, inspired by natural evolution, generates conditional statements for attack detection through tree-like structures. The study uses a central intrusion detection architecture, with a root node responsible for monitoring and feature extraction. Results demonstrate effective attack detection, particularly with increased time intervals, and a central architecture outperforms a distributed one due to comprehensive data collection capabilities. The proposed GP-based IDS shows promise for enhancing IIoT network security.

Ambili & Jose., (2020), focuses on the development of a Trust-Based Intrusion Detection System (TN-IDS) designed for networks with limited resources, such as RPL-based Industrial Internet of Things (IIoT) systems. The primary objective is to detect and mitigate various types of attacks that can target these networks. In particular, the paper discusses attacks like Denial of Service (DoS), Sinkhole, and Version attacks, which can disrupt the normal operation of IIoT systems. The TN-IDS relies on a trust mechanism based on blockchain technology to assess the trustworthiness of network nodes. These trust values are stored in a blockchain and managed by a group of network administrators. This approach enables the system to identify and isolate malicious nodes, preventing them from causing harm to the network by transmitting unauthorized packets or interfering with network operations. One of the key advantages of the proposed TN-IDS is its efficiency in terms of power consumption and computational overhead. It can quickly identify and respond to threats while minimizing the impact on the network's performance. Additionally, the system is designed to handle unforeseen threats effectively, particularly when it comes to detecting and mitigating DoS attacks.

Canbalaban et al., (2020), presents a comprehensive Intrusion Detection System (IDS) tailored for RPL-based IoT networks, utilizing neural network-based methods for effective intrusion detection. The feature selection process encompasses data-related features, topology-related features, and novel link-layer features, which collectively provide valuable insights into network stability and the detection of various attacks. Simulated attacks, including Version Number Attacks (VNA), Worst Parent Attacks (WPA), and Hello Flood Attacks (HFA), are assessed across different network topologies and varying attacker densities. The IDS architecture employs fully-connected neural network layers with ReLU activation functions, dropout layers, and softmax outputs. Experimental results demonstrate high detection rates for each attack type, with performance variations influenced by network size and attacker density. Additionally, the IDS is capable of multiclass classification, albeit with initial challenges in distinguishing benign traffic from Version Number Attacks. Notably, the inclusion of link-layer features enhances the IDS's performance by reducing false positives and improving the detection rate for version number attacks.

Yadollahzadeh & Mataji, (2021), focuses on enhancing the security of Internet of Things (IoT) networks by developing an Intrusion Detection System (IDS) to detect sinkhole attacks. The proposed IDS involves several key components. It includes a lightweight data collection module within network nodes, which sends node information to the IDS agent on the border router. To ensure data security during transmission, mechanisms like IPsec or DTLS are assumed. The IDS agents are placed at both border routers and network nodes. The 6Mapper client module sends crucial network information to the 6Mapper module on the border router to reconstruct the network graph. The IDS agents at the border router consist of the 6Mapper module for graph reconstruction and a Detection module, responsible for the primary detection tasks, including preprocessing, classifier computation, and application of the classifier. The preprocessing phase involves data cleaning, feature selection, and the use of an evolutionary genetic algorithm to optimize feature selection. It employs linear regression for fitness assessment. The classifiers include Decision Tree, Support Vector Machine (SVM), and Na¨ıve Bayesian models, trained on the preprocessed data. In the detection phase, these models classify network data into normal or attack categories. Post-processing follows the initial detection, where alarms are produced based on the classification results. Alarms then undergo verification, and the degree of confidence is used to differentiate between true alarms and false positives. The verification process considers the impact of neighboring nodes in determining the final status of alarms.

Osman et al., (2021b) proposes an approach for detecting Version Number Attacks (VNA) within RPL-based networks. The proposed method, ML-LGBM, is structured into four crucial modules: data collection, data preprocessing, feature selection, and machine learning. These modules work synergistically to ensure high detection accuracy. The experimental setup is based on the Cooja simulator in conjunction with the Contiki operating system, simulating real hardware platforms. The dataset used for testing is generated by capturing radio traffic and saving it as a PCAP file through the 6Lowpan Package Analyzer tool. By simulating VNA scenarios, the study evaluates the model's performance, including the impact of the number of malicious nodes and client nodes. Raw data is collected and processed using Wireshark and then transformed into a dataset with essential features through Python feature extraction. Feature normalization is applied, and forward feature selection is used to minimize classification errors. The Light Gradient Boosting Machine (LightGBM), known for its efficiency, is employed as the core machine learning algorithm, offering fast and precise results. Various performance evaluation metrics, such as accuracy, precision, and detection rate, validate ML-LGBM's superiority over other machine learning models, making it a promising choice for intrusion detection in RPL-based networks and, notably, in Internet of Things (IoT) applications.

Figure 36. ML-LGBM, proposed model for detecting VNA (Osman et al., 2021b)

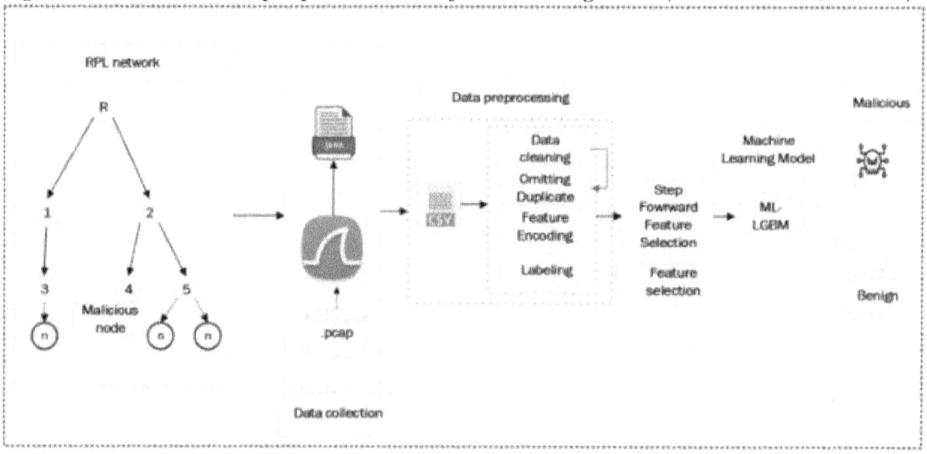

Arshad et al., (2022), present a novel and comprehensive approach to enhance the security of Internet of Things (IoT) networks, with a particular emphasis on countering Sybil attacks within RPL-based IoT communications. The proposed THC-RPL protocol introduces a distinctive hybrid trust mechanism, which incorporates both di- rect and indirect trust evaluations. This innovative approach serves a dual purpose: it efficiently detects Sybil nodes within the network while also optimizing energy usage, a critical concern for resource-constrained IoT devices. The research employs exten- sive simulations to showcase the protocol's superior performance across several critical metrics, including the detection of Sybil attacks, mitigation of packet loss, enhanced energy efficiency, and the minimization of computational and communication overhead. One of THC-RPL's key advantages is its suitability for mobile IoT environments, con- tributing to prolonged network longevity by conserving the energy resources of IoT devices.

Al Sawafi et al., (2023), proposed a hybrid Intrusion Detection System (IDS) for IoT networks, specifically targeting attacks on the Routing Protocol for Low-Power and Lossy Networks (RPL). They introduced a new dataset, called IoTR-DS, which con- tains simulated RPL attacks (DIS, Rank, and Wormhole) along with normal traffic data. The proposed IDS system is a hybrid model consisting of a supervised Deep Neu- ral Network (DNN) and a semi-supervised Deep Auto Encoder (DAE). This hybrid model is designed to detect both known and unknown attacks by comparing traffic reconstruction loss with normal traffic and known attack loss values. They evaluated the model's performance using standard machine learning metrics like accuracy, preci- sion, recall, and F1 score. The evaluation results demonstrated the model's detection accuracy rate for multi-class attacks. The hybrid IDS outperformed other classical and deep learning models like J48, KNN, SVM, and LSTM, showing its effectiveness in detecting and classifying RPL attacks.

Alkanhel et al., (2022), explores three foundational pillars: IDSs over Network, Detec- tion Methodology, and Counteracted Attacks, all converging to fortify the IoT security infrastructure. In the domain of IDSs over Network, the research meticulously crafts and optimizes IDSs, specifically tailored to the intricacies of IoT network infrastruc- tures. A pivotal focal point is data preprocessing, which involves cleaning of the diverse and often unruly IoT data. This cleansing process involves the removal of redundancies, addressing gaps through imputation, and the transformation of non-numeric data into machine-readable formats. With IoT data characterized by its inherent heterogeneity, encoding mechanisms are employed to translate non-numeric attributes into numeric values, rendering them comprehensible to the vigilant IDS. The Detection Methodol- ogy employed ventures into the realm of nature-inspired algorithms, specifically, Dip- per Throated Optimization (DTO) and Grey Wolf Optimization. DTO emulates the foraging patterns of birds through the tracking of their positions and velocities. Equa- tions underpin these bird-like movements with adaptive parameters that dynamically evolve during the optimization process. Meanwhile, Grey Wolf Optimization simulates the hierarchical structure within a pack of grey wolves, wherein alpha, beta, delta, and omega wolves undertake distinct roles in the optimization endeavor. The investi- gation also delves deep into the realm of Counteracted Attacks, addressing an array of menacing network threats, including Sybil, Blackhole, sinkhole, selective forwarding, local repair, and hello flooding attacks. The core battleground for this endeavor is to tackle the challenge of data imbalance head-on. In the realm of IDSs, datasets often exhibit significant imbalances, where benign network traffic dwarfs the instances of malicious attacks. To redress this imbalance, the proposed solution leverages the Syn- thetic Minority Over-sampling Technique (SMOTE), a technique that rectifies dataset disparities by oversampling the minority class, which, in this context, represents the attacks.

6.2.4. Mobility Related Security Issues

The dynamic nature of some IoT nodes increases link failures, collisions and packet loss. Obviously, the combination of mobility characteristic with a malicious behavior will aggravate the harm on the network performance. In fact, because RPL does not support mobility, lots of problems rise and have to be solved among them security. For instance, Aris et al., (2016) showed that the version number attacks performed by attackers far from the 6LBR and by mobile attackers have nearly the same destructive effects. Medjek et al., (2017b) demonstrated that SybM attack performed by mobile attackers generates more harm than DIS attack performed with static attackers. Thus, when attackers are mobile the performance the worst.

Most of the existing IDS solutions are based on static IoT environment. Nevertheless, because of the dynamic nature of such networks, an increasing number of works try to deal with security mobility issue in RPL-based networks. For instance, Gara et al., (2017) proposed an IDS to counter selective-forwarding attack in a mobile envi- ronment. Furthermore Medjek et al., (2017a) proposed a trust-based IDS to counter attacks in a mobile environment. Besides, Cervantes et al., (2015) considered mobil- ity in the IDS they proposed to detect sinkhole attack. Lai G.H;, (2016) introduced an IDS to detect wormhole attacks where nodes are mobile. Security in the context of mobility in IoT networks is a complex and multifaceted issue that requires thor- ough consideration. The dynamic nature of IoT networks, housing both mobile and stationary nodes, adds complexity to the security landscape. Mobility, particularly concerning mobile sink nodes, is often inadequately addressed in current security solu- tions, leading to security vulnerabilities. Trust-based security approaches hold promise for IoT networks due to their adaptability and scalability. These approaches evalu- ate node trustworthiness through metrics like successful transactions, energy levels, and historical behavior. Nevertheless, existing trust models often overlook mobility or provide insufficient support for mobile sink nodes, leaving security gaps.

Table 3. IDSs for RPL threats

Works	IDS Over Network	IDSs Over Detection Method	Countered Attacks
Kasinathan et al.(2013b)	Centralized	Signature-based	DOS attacks (IPv6 UDP flooding)
Raza et al.,(2013)	Hybrid	Anomaly-based	Rank, sinkhole, and selective-forwarding attacks
Zhang et al., (2015)	Distributed	Specification-based	Routing choice intrusion attack
Pongle & Chavan, (2015b)	Hybrid	Anomaly-based	Wormhole attacks
Cervantes et al.,(2015)	Distributed	Anomaly-based	Sinkhole attack
Thanigaivelan et al.,(2016)	Hybrid	Anomaly-based	-
Lai G.H;,(2016)	Distributed	Anomaly-based	Wormhole attacks
Medjek et al.,(2017b)	Hybrid	Specification-based	RPL-based attacks
Gara et al.,(2017)	Hybrid	Anomaly-based	Selective-forwarding attack

continued on following page

Table 3. Continued

Works	IDS Over Network	IDSs Over Detection Method	Countered Attacks
Bostani & Sheikhan.,(2017)	Hybrid	Anomaly-based	Sinkhole and selective forwarding attacks
Ioulianou et al.(2018)	Distributed	Signature-based IDS	DDOS Attacks, Device Spoofing
Jiang et al.,(2018)	Distributed	Packet loss detection algorithm: anomaly-based detection	Blackhole node is isolated by its children
Granjal & Pedroso,(2018)	Hybrid	specification-based	Resource exhaustion attacks in CoAP networks
Ali et al.,(2019)	Hybrid	Behavior Analysis	Data extraction, unauthorized access
Sherazi et al.,(2019)	Distributed	Fuzzy Logic and Q-Learning based	DDoS attack, specifically focusing on UDP flooding
Verma & Ranga,(2019)	Distributed	Anomaly-based	Distributed Denial of Service (DIS) flooding attacks
Luo et al., (2020)	Hybrid	Machine Learning-based IDS	Rogue Device Detection, Jamming
Djedjig et al., (2020)	Distributed	Specification-based	IPv6 ODP flooding, traffic profiling
Saba et al., (2022)	Centralized	Anomaly-based IDS	Insider Threats, IoT Botnets
Attique et al., (2022)	Hybrid	Anomaly-based IDS	Mirai-related attacks, distributed denial of service
Arshad et al., (2022)	Distributed	Anomaly-based	Sybil attacks
Kipongo et al.,(2023)	Distributed	Signature and anomaly-based IDS, Statistical, rule-based	Identifying intrusion attempts and malicious activities, DIS Flooding attacks,
Pasikhan et al.,(2023)	Hybrid	Machine learning-based methods	DIO Suppression, Blackhole, Grayhole, Wormhole, Increase Rank, Worst Parent
Ray et al.,(2023)	Centralized	DES-based IDS, active probing to detect rank and version attacks	Suboptimal routes, traffic redirection, and network partitioning

Mobility metrics, including node traceability, location, link stability, and movement frequency, are crucial for assessing node trustworthiness in mobile IoT environments. Ignoring these metrics could compromise security. With respect to various metrics,

Pradhan, R., Rakshit, S., & De, T. (2018), focuses on evaluating the performance of the RPL in the dynamic context of VANETs (Vehicular Ad Hoc Networks). RPL's effectiveness depends on the choice of the objective function (OF), which defines how RPL selects and optimizes data transmission routes. They considers three different objective func- tions: OF0, MRHOF (Minimum Rank with Hysteresis Objective Function), and the newly introduced MRHOFL, which incorporates latency as a routing parameter. In a VANET scenario with mobile vehicular nodes and a Roadside Unit (RSU) acting as a base station, vehicular nodes transmit data to the RSU, with varying speeds and reception ratios. In fact, vehicular networks face security challenges, such as data pri- vacy, message authenticity, and secure communication, due to their dynamic nature and the potential risks associated with vehicular communication.

Figure 37. Static, Mobile sender and Mobile sink scenarios: Simulation workflow (Muzammal et al.,2022)

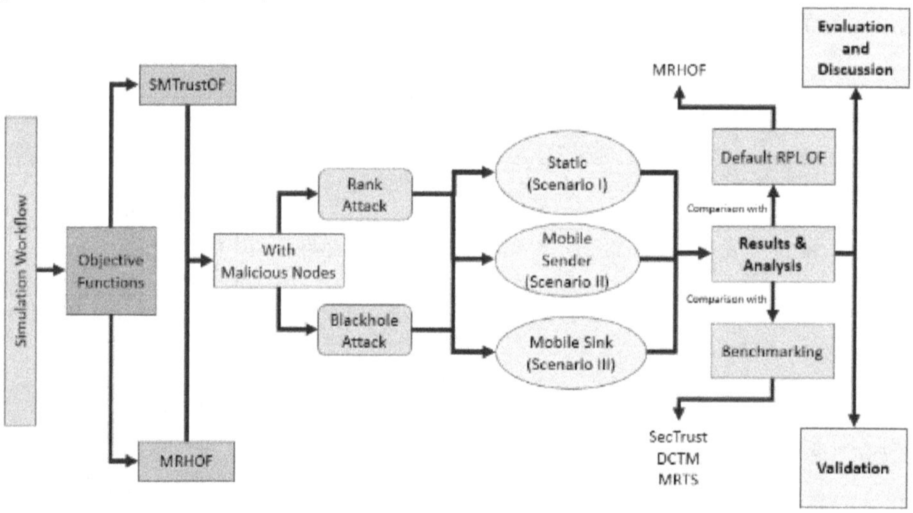

This was especially challenging in mobile IoT environments due to ever-changing network topologies and the potential for data interception and manipulation. Muzammal et al., (2022), focuses on incorporating mobility-based trust metrics into IoT routing protocols, enhancing security in IoT networks and providing superior protec- tion against RPL attacks in mobile scenarios. It aims to strike a balance between security, resource constraints, and energy efficiency, addressing the intricate demands of safeguarding data within dynamic and mobile IoT environments. In essence, security concerns in the mobility of IoT networks encompass a multifaceted landscape demand- ing innovative approaches to guarantee data confidentiality, integrity, and availability. Dhanvijay & Patil, (2021), emphases on security issues in healthcare applications, given the sensitive nature of patient data. Vulnerabilities in the mobility of Wireless Body Area Networks (WBAN) can lead to data interception, tampering, or unau- thorized access. These security concerns are addressed by emerging protocols like the CoAP-based Secure-aware Mobility Management Protocol (CoSMP), which fo- cuses on ensuring the privacy and authenticity of health-related data during mobility. CoSMP employs encryption and authentication techniques, like the Advanced Encryp- tion Standard Cipher Feedback Message Authentication Code (AES-CFMAC) and the Elliptic Curve Digital Signature Algorithm (ECDSA), to establish secure data trans- mission and minimize the risks associated with mobility. Overall, security remains a paramount requirement in healthcare IoT applications, especially concerning patient privacy, to safeguard sensitive medical information in an increasingly interconnected and mobile environment. Security concerns related to mobility, outlined in Muzam- mal et al., (2022), are critical and require more in-depth technical elaboration. Mobile IoT devices introduce unique challenges, particularly regarding trustworthiness, data transmission security, attack detection, dynamic key management, secure handovers, resource constraints, and resilience to mobility-related attacks. The work's mention of a trust-based approach is a positive step, but it lacks essential technical specifics. It is imperative to describe the specific cryptographic, trust establishment, and authentica- tion mechanisms employed to ensure the trustworthiness of mobile nodes. Secure data transmission, including encryption and authentication processes, should be discussed in detail, considering the dynamic nature of IoT environments. The work should also provide insights into how it detects and mitigates attacks, especially in the context of mobile IoT, and how it manages cryptographic keys for dynamic nodes. Secure handovers, a critical aspect of mobile IoT security, require special attention and ex- planation. Furthermore, the work should address how it balances the need for strong security with IoT device constraints. Lastly, the resilience of the system to mobility- related attacks, including handover vulnerabilities, should be thoroughly detailed.

7. CONCLUSION AND FUTURE WORKS

The Internet of Things (IoT), marked by extensive device interconnections across diverse sectors, has ushered in an era characterized by enhanced connectivity, innovation, and abundant opportunities. However, it is imperative not to underestimate the substantial security challenges that accompany this technological landscape. This study has identified vulnerabilities and proposed potential solutions to reinforce IoT security. Notably, Intrusion Detection Systems (IDSs) have emerged as pivotal tools for enhancing the defenses of IoT networks, with our discussion providing insights into their deployment within RPL-based IoT networks. Furthermore, our exploration of security concerns related to IoT mobility underscores the continuously evolving nature of this technology and the urgent necessity for robust security measures.

As the IoT continues to evolve and exert an expanding influence, the importance of ensuring its security cannot be overstated. The framework presented in this article, encompassing fundamental IoT concepts and security intricacies, equips readers with a deeper understanding of the complex IoT security landscape. It is evident that the success and widespread adoption of IoT technology heavily depend on the standardization and elaboration of security protocols. While no single security mechanism can comprehensively safeguard against all threats across every application area, this research lays the foundation for common security countermeasures and defense mechanisms applicable across various IoT applications.

Looking ahead, several key areas require further exploration to advance IoT security. Advanced IDSs driven by AI and machine learning represent a promising avenue for improving threat detection accuracy and adapting to the dynamic nature of IoT environments. Additionally, as IoT devices increasingly incorporate mobility, there is a pressing need for mobility-aware security protocols that can adapt to changing network topologies.

Moreover, the exploration of cross-layer security approaches that integrate measures across the physical, network, and application layers could provide a more comprehensive defense strategy. The standardization of IoT security protocols remains critical for ensuring interoperability and reducing risks across diverse platforms. With the advent of quantum computing, developing quantum-resistant cryptographic techniques tailored for IoT devices will be crucial to maintaining security in the future. Furthermore, as edge computing becomes more prevalent, understanding and addressing the security challenges specific to edge-based IoT environments will be essential.

Use Cases further illustrate the practical implications of these security measures:

1. Smart home security systems: IoT devices in smart homes, such as surveillance cameras and smart locks, are vulnerable to cyberattacks that could compromise home security. The integration of IDSs tailored for IoT can monitor and detect unusual device behavior, providing real-time alerts to homeowners in case of potential intrusions.
2. Healthcare IoT (IoMT): Connected medical devices, like insulin pumps and cardiac monitors, are critical for patient care. Securing these devices with standardized protocols and robust encryption mechanisms is essential to protect sensitive health data and ensure patient safety.

3. Smart cities and connected infrastructure: Urban environments rely on IoT devices to manage traffic lights, public lighting, and waste management systems. Implementing multi-layered security measures, including IDSs and anomaly detection systems, can safeguard these critical infrastructures from cyber threats.
4. Industrial IoT (IIoT): Industrial settings use IoT devices to automate and control essential production processes. Cross-layer security systems that protect both the network and individual devices are crucial for preventing disruptions, financial losses, and potential accidents.
5. Connected vehicles: As vehicles become more connected and autonomous, they face unique security challenges. Developing mobility-specific security solutions, such as secure communication protocols for moving vehicles and automotive IDSs, can help prevent accidents and ensure the safety of passengers.

Lastly, the ethical implications and privacy concerns associated with IoT deployments must be addressed, particularly as these devices handle vast amounts of personal data. Future research should focus on privacy-preserving technologies that balance data utility with robust privacy protections.

By exploring IoT characteristics, standard protocols, and threats within a three layer architecture, our goal is to empower the IoT community to proactively secure this transformative technology. The journey toward a truly secure IoT ecosystem remains ongoing, but this research represents a significant step in the right direction, offering valuable insights, use cases, and guidance to all stakeholders dedicated to building a safer and more interconnected future.

FUNDING

No funding was received for this research.

REFERENCES

Abbasi, M., Yaghmaee, M. H., & Rahnama, F. (2019, April). Internet of Things in agriculture: A survey. In 2019 3rd International Conference on Internet of Things and Applications (IoT) (pp. 1-12). IEEE.

Abdmeziem, M. R., Tandjaoui, D., &Romdhani, I. (2017), "Lightweighted and energy-aware MIKEY-Ticket for e-health applications in the context of internet of things.", International Journal of Sensor Networks.

Airehrour, D., Gutierrez, J., & Ray, S. K. (2016). Secure routing for internet of things: A survey. *Journal of Network and Computer Applications*, *66*, 198–213. DOI: 10.1016/j.jnca.2016.03.006

Airehrour, D., Gutierrez, J., & Ray, S. K. (2016). Securing RPL routing protocol from black- hole attacks using a trust-based mechanism. In *Telecommunication Networks and Appli- cations Conference (ITNAC), 2016 26th International* (pp. 115–120). IEEE.

Airehrour, D., Gutierrez, J., & Ray, S. K. (2017). A trust-aware RPL routing protocol to detect blackhole and selective forwarding attacks. *Journal ofTelecommunications and the Digital Economy*, *5*(1), 50–69. https://search.informit.org/doi/10.3316/informit.752286025338502. DOI: 10.18080/jtde.v5n1.88

Al-Amiedy, T. A., Anbar, M., Belaton, B., Kabla, A. H. H., Hasbullah, I. H., & Alashhab, Z. R. (2022). A Systematic Literature Review on Machine and Deep Learning Approaches for De- tecting Attacks in RPL-Based 6LoWPAN of Internet of Things. *Sensors (Basel)*, *22*(9), 3400. DOI: 10.3390/s22093400 PMID: 35591090

Al-Fuqaha, A., Guizani, M., Mohammadi, M., Aledhari, M., & Ayyash, M. (2015). Internet of things: A survey on enabling technologies, protocols, and applications. *IEEE Communications Surveys and Tutorials*, *17*(4), 2347–2376. DOI: 10.1109/COMST.2015.2444095

Al Sawafi, Y., Touzene, A., & Hedjam, R. (2023). Hybrid Deep Learning-Based Intrusion Detection System for RPL IoT Networks. *Journal of Sensor and Actuator Networks*, *12*(2), 21. DOI: 10.3390/jsan12020021

Ali, F., Yigang, H., & Yi, R. (2019). A novel security architecture of internet of things. *International Journal of Computer Theory and Engineering*, *11*(5), 89–96. DOI: 10.7763/IJCTE.2019.V11.1249

Alkanhel, R., El-kenawy, E., Abdelhamid, A., Ibrahim, A., Alohali, M., Abotaleb, M., & Khafaga, D. (2022). Network Intrusion Detection Based on Feature Selection and Hy- brid Metaheuristic Optimization. *Computers, Materials & Continua, 74*, 2677–2693. DOI: 10.32604/cmc.2023.033273

Alrababah, D., Al-Shammari, E., & Alsuht, A. (2017), "A Survey: Authentication Protocols for Wireless Sensor Network in the Internet of Things; Keys and Attacks.", In *The International Conference on New Trends in Computing Sciences (ICTCS).* DOI: 10.1109/ICTCS.2017.34

Alsukayti, I. S., & Alreshoodi, M. (2023). RPL-Based IoT Networks under Simple and Complex Routing Security Attacks: An Experimental Study. *Applied Sciences (Basel, Switzerland), 13*(8), 4878. DOI: 10.3390/app13084878

Alyami, S., Alharbi, R., & Azzedin, F. (2022, December 14). Fragmentation Attacks and Countermeasures on 6LoW- PAN Internet of Things Networks: Survey and Simulation. *Sensors (Basel), 22*(24), 9825. DOI: 10.3390/s22249825 PMID: 36560200

Alzubaidi, M., Anbar, M., Chong, Y. W., & Al-Sarawi, S. (2018). Hybrid monitoring technique for detecting abnormal behaviour in rpl-based network. *Journal of Communication, 13*(5), 198–208. DOI: 10.12720/jcm.13.5.198-208

Ambili, K. N., & Jose, J. (2020). TN-IDS for network layer attacks in RPL based IoT systems. Cryptology ePrint Archive.

Antao, L., Pinto, R., Reis, J., & Gonçalves, G. (2018, April). Requirements for testing and validating the industrial internet of things. In *2018 IEEE International Conference on Software Testing, Verification and Validation Workshops (ICSTW)* (pp. 110-115). IEEE.

Aris, A., Oktug, S. F., & Yalcin, S. B. O. (2016, April). "RPL version number attacks: In-depth study." In 2016 IEEE/IFIPNetwork Operations and Management Symposium (NOMS), (pp. 776-779). IEEE. DOI: 10.1109/NOMS.2016.7502897

Arshad, D., Asim, M., Tariq, N., Baker, T., Tawfik, H., & Al-Jumeily, O. B. E. (2022). THC- RPL: A lightweight Trust-enabled routing in RPL-based IoT networks against Sybil attack. *PLoS One, 17*(7), e0271277. DOI: 10.1371/journal.pone.0271277 PMID: 35901074

Attique, D., Wang, H., & Wang, P. (2022). Fog-Assisted Deep-Learning-Empowered Intrusion Detection System for RPL-Based Resource-Constrained Smart Industries. *Sensors (Basel), 22*(23), 9416. DOI: 10.3390/s22239416 PMID: 36502115

Aydogan, E., Yilmaz, S., Sen, S., Butun, I., Forsström, S., & Gidlund, M. (2019, May). A central intrusion detection system for rpl-based industrial internet of things. In 2019 15th IEEE International Workshop on Factory Communication Systems (WFCS) (pp. 1-5). IEEE.

Azzabi, T., Farhat, H., & Sahli, N. (2017, January). A survey on wireless sensor networks security issues and military specificities. In 2017 International conference on advanced systems and electric technologies (IC_ASET) (pp. 66-72). IEEE.

Badach, A. (2018). "A Literature Collection on Routing in the Internet of Things (IoT) with RPL - Part 1: RPL Framework." DOI: 10.13140/RG.2.2.14354.56008

Bang, A., & Rao, U. P. (2023). Impact analysis of rank attack on RPL-based 6LoWPAN net- works in Internet of Things and aftermaths. *Arabian Journal for Science and Engineering*, *48*(2), 2489–2505. DOI: 10.1007/s13369-022-07342-y

Barabanov, A., Markov, A., & Tsirlov, V. (2018). Information Security Controls against Cross- Site Request Forgery Attacks on Software Applications of Automated Systems. *Journal of Physics: Conference Series*, *1015*, 042034. DOI: 10.1088/1742-6596/1015/4/042034

Belavagi, M. C., & Muniyal, B. (2020). Multiple intrusion detection in RPL based net- works. *International Journal of Electrical and Computer Engineering*, *10*(1), 467–476. DOI: 10.11591/ijece.v10i1.pp467-476

Bostani, H., & Sheikhan, M. (2017). Hybrid of anomaly-based and specification-based IDS for Internet of Things using unsupervised OPF based on MapReduce approach. *Computer Communications*, *98*, 52–71. DOI: 10.1016/j.comcom.2016.12.001

Çorak, B. H., Okay, F. Y., Güzel, M., Murt, Ş., & Ozdemir, S. (2018, June). Comparative analysis of IoT communication protocols. In 2018 International symposium on networks, computers and communications (ISNCC) (pp. 1-6). IEEE.

Cakir, S., Toklu, S., & Yalcin, N. (2020). RPL Attack Detection and Prevention in the Internet of Things Networks Using a GRU Based Deep Learning. *IEEE Access : Practical Innovations, Open Solutions*, *8*, 183678–183689. DOI: 10.1109/ACCESS.2020.3029191

Canbalaban, E., & Sen, S. (2020). "A cross-layer intrusion detection system for RPL-based Internet of Things.", In Ad-Hoc, Mobile, and Wireless Networks: 19th International Confer- ence on Ad-Hoc Networks and Wireless, ADHOC-NOW 2020, Bari, Italy, October 19–21, 2020 [Springer International Publishing.]. *Proceedings*, *19*, 214–227.

Cervantes, C., Poplade, D., Nogueira, M., & Santos, A. (2015, May). "Detection of sink- hole attacks for supporting secure routing on 6LoWPAN for Internet of Things.", In In- tegrated Network Management (IM), 2015 IFIP/IEEE International Symposium on (pp. 606-611).IEEE.

Choudhary S. & Kesswani N.(2019). "Cluster-Based Intrusion Detection Method for Internet of Things.". 1-8. .DOI: 10.1109/AICCSA47632.2019.9035319

Chugh, K., Aboubaker, L., & Loo, J. (2012, August). "Case study of a black hole attack on LoWPAN-RPL." In Proceedingof the Sixth International Conference on Emerging Security Information, Systems and Technologies (SECURWARE), Rome, Italy (August 2012) (pp. 157-162).

Classen, J., Gringoli, F., Hermann, M., & Hollick, M. (2022, May). Attacks on wireless coexistence: Exploiting cross-technology performance features for inter-chip privilege escalation. In *2022 IEEE Symposium on Security and Privacy (SP)* (pp. 1229-1245). IEEE.

Deshmukh-Bhosale, S., & Sonavane, S. S. (2019), "A Real-Time Intrusion Detection System for Wormhole Attack in the RPL based Internet of Things", Procedia Manufacturing, Volume 32, 2019, Pages 840-847, ISSN 2351-9789, DOI: 10.1016/j.promfg.2019.02.292

Dhanaraj, R., Krishnasamy, L., Geman, O., & Izdrui, D. (2021). Black hole and sink hole attack detec- tion in wireless body area networks. *Computers, Materials & Continua*, *68*(2), 1949–1965. DOI: 10.32604/cmc.2021.015363

Dhanvijay, M. M., & Patil, S. C. (2021), "Internet of Things: A survey of enabling technologies in healthcare and its applications", Computer Networks, Volume 153, 2019, Pages 113-131, ISSN 1389-1286, DOI: 10.1016/j.comnet.2019.03.006

Djedjig, N., Tandjaoui, D., & Medjek, F. (2015, July). "Trust-based RPL for the Internet of Things.", In Computers and Communication (ISCC), 2015 IEEE Symposium on (pp. 962-967). IEEE. DOI: 10.1109/ISCC.2015.7405638

Djedjig, N., Tandjaoui, D., Medjek, F., & Romdhani, I. (2017)."New trust metric for the RPL routing protocol.", In Information and Communication Systems (ICICS), 2017 8th International Conference on (pp. 328-335). IEEE. DOI: 10.1109/IACS.2017.7921993

Djedjig, N., Tandjaoui, D., Medjek, F., & Romdhani, I. (2020, June). Trust-aware and cooperative routing protocol for IoT security. *Journal of Information Security and Applications*, *52*, 102467. DOI: 10.1016/j.jisa.2020.102467

Donta, P. K., Srirama, S. N., Amgoth, T., & Annavarapu, C. S. R. (2022). Survey on re- cent advances in IoT application layer protocols and machine learning scope for research directions. *Digital Communications and Networks*, *8*(5), 727–744. DOI: 10.1016/j.dcan.2021.10.004

Dvir, A., & Buttyan, L. (2011, October). "VeRA-version number and rank authentication in rpl.", In 2011 IEEE 8th International Conference onMobile Adhoc and Sensor Systems (MASS), (pp. 709-714). IEEE. DOI: 10.1109/MASS.2011.76

Farzaneh, B., Koosha, M., Boochanpour, E., & Alizadeh, E. (2020, April). A new method for intrusion detection on RPL routing protocol using fuzzy logic. In 2020 6th International Conference on Web Research (ICWR) (pp. 245-250). IEEE.

Gara, F., Saad, L. B., & Ayed, R. B. (2017, June). "An intrusion detection system for selec- tive forwarding attack in IPv6-based mobile WSNs.", In 2017 13th InternationalWireless Communications and Mobile Computing Conference (IWCMC), (pp. 276-281).IEEE.

Ghaleb, B., Al-Dubai, A., Hussain, A., Ahmad, J., Romdhani, I., & Jaroucheh, Z. (2023). "Resolving the Decreased Rank Attack in RPL's IoT Networks", arXiv e-prints,. /arXiv.2305.10025.DOI: 10.1109/DCOSS-IoT58021.2023.00018

Gnad, D. R. E., Krautter, J., & Tahoori, M. B. (2019). "Leaky Noise: New Side-Channel Attack Vectors in Mixed-Signal IoT Devices.", IACR Transactions on Cryptographic Hardware and Embedded Systems, 305–339. DOI:DOI: 10.13154/tches.v2019.i3.305-339

Gowtham, M., & Pramod, B. (2022, June). Semantic Query-Featured Ensemble Learning Model for SQL- Injection Attack Detection in IoT-Ecosystems. *IEEE Transactions on Reliability*, *71*(2), 1057–1074. DOI: 10.1109/TR.2021.3124331

Granjal, J., Monteiro, E., & Silva, J. S. (2015). Security for the internet of things: A survey of existing protocols and open research issues. *IEEE Communications Surveys and Tutorials*, *17*(3), 1294–1312. DOI: 10.1109/COMST.2015.2388550

Granjal, J., & Pedroso, A. (2018). An Intrusion Detection and Prevention Framework for Internet- Integrated CoAP WSN. *Security and Communication Networks*, *2018*, 1753897. DOI: 10.1155/2018/1753897

Griffin, P. H. (2017, March). "Secure authentication on the internet of things.", In Southeast- Con, 2017 (pp. 1-5). IEEE.

Groves, B., & Pu, C. (2019). *"A Gini Index-Based Countermeasure Against Sybil Attack in the Internet of Things," MILCOM 2019 - 2019 IEEE Military Communications Conference*. MILCOM., DOI: 10.1109/MILCOM47813.2019.9021050

Gupta, N., & Garg, U. (2022). A Proposed IoT Security Framework and Analysis of Network Layer Attacks in IoT. In Sharma, T. K., Ahn, C. W., Verma, O. P., & Panigrahi, B. K. (Eds.), *Soft Computing: Theories and Applications. Advances in Intelligent Systems and Comput- ing* (Vol. 1380). Springer., DOI: 10.1007/978-981-16-1740-9 9

Haas, Z. J., Yang, L., Liu, M. L., Li, Q., & Li, F. (2014). Current Challenges and Approaches in Securing Communications for Sensors and Actuators. In *The Art of Wireless Sensor Networks* (pp. 569–608). Springer Berlin Heidelberg. DOI: 10.1007/978-3-642-40009-4_17

Hanif, M., Ashraf, H., Jalil, Z., Jhanjhi, N. Z., Humayun, M., Saeed, S., & Almuhaideb, A. M. (2022). AI-Based Wormhole Attack Detection Techniques in Wireless Sensor Networks. *Electronics (Basel), 11*(15), 2324. DOI: 10.3390/electronics11152324

Hassan, M., Tariq, N., Alsirhani, A., Alomari, A., Khan, F. A., Alshahrani, M. M., & Humayun, M. (2023). Gitm: A gini index-based trust mechanism to mitigate and isolate sybil attack in rpl-enabled smart grid advanced metering infrastructures. *IEEE Access : Practical Innovations, Open Solutions, 11*, 62697–62720.

Hui, J. W. (2012). RFC6553. "Option for Carrying RPL Information in Data-plane Diagrams", 33, 3-8.Retrieved November 4th, 2017, from http://tools.ietf.org/html/rfc6553

Hummen, R., Heer, T., & Wehrle, K. (2011, March). "A security protocol adaptation layer for the IP-based internet of things", In Interconnecting smart objects with the Internet workshop (Vol. 3).

Ioulianou, P., Vasilakis, V., Moscholios, I., & Logothetis, M. (2018) (Accepted/In press). "A Signature-based Intrusion Detection System for the Internet of Things". Paper presented at Information and Communication Technology Form, Graz, Austria.

Ioulianou, P. P., Vassilakis, V. G., & Logothetis, M. D. (2019). Battery Drain Denial-of- Service Attacks and Defenses in the Internet of Things. *Journal of Telecommunications and Information Technology*, 2(2), 37–45. DOI: 10.26636/jtit.2019.131919

Ishaq, I., Carels, D., Teklemariam, G. K., Hoebeke, J., Abeele, F. V. D., Poorter, E. D., Moerman, I., & Demeester, P. (2013). IETF standardization in the field of the internet of things (IoT): A survey. *Journal of Sensor and Actuator Networks*, *2*(2), 235–287. DOI: 10.3390/jsan2020235

Jeridi, M. H., Elmadssia, S., & Ezzedine, T. (2021). Routing in IoT lossy net- works: Potential issues and proposed changes. *Journal of Interconnection Networks*, *21*(03), 2150018. DOI: 10.1142/S0219265921500183

Jiang, J., & Liu, Y. (2022). Secure IoT routing: selective forwarding attacks and trust-based defenses in RPL network. arXiv preprint arXiv:2201.06937.

Jiang, J., Liu, Y., & Dezfouli, B. (2018, November). A root-based defense mechanism against RPL blackhole attacks in internet of things networks. In 2018 Asia-Pacific Signal and Information Processing Association Annual Summit and Conference (APSIPA ASC) (pp. 1194-1199). IEEE.

Kasinathan, P., Costamagna, G., Khaleel, H., Pastrone, C., & Spirito, M. A. (2013, Novem- ber). An IDS framework for internet of things empowered by 6LoWPAN. In Proceedings of the 2013 ACM SIGSAC conference on Computer & communi- cations security (pp. 1337- 1340).ACM.

Kasinathan, P., Pastrone, C., Spirito, M. A., & Vinkovits, M. (2013, October). Denial-of- Service detection in 6LoWPAN based Internet of Things. In Wireless and Mobile Computing, Networking and Communications (WiMob), 2013 IEEE 9th International Conference on (pp. 600-607). IEEE.

Khan, F. I., Shon, T., Lee, T., & Kim, K. (2013, July). "Wormhole attack prevention mech- anism for RPL based LLN network." In Ubiquitous and Future Networks (ICUFN), 2013 Fifth International Conference on (pp. 149-154).IEEE.

Khan, R., Khan, S. U., Zaheer, R., & Khan, S. (2012). "Future internet: the internet of things archi- tecture,possible applications and key challenges." In: 2012 10th International Conference on Frontiersof Information Technology. pp 257-260

Khraisat, A., Gondal, I., Vamplew, P., Kamruzzaman, J., & Alazab, A. (2019). A Novel En- semble of Hybrid Intrusion Detection System for Detecting Internet of Things Attacks. *Electronics (Basel)*, *8*(11), 1210. DOI: 10.3390/electronics8111210

Kipongo, J., Swart, T. G., & Esenogho, E. (2023). Design and implementation of intrusion detection systems using RPL and AOVD protocols-based wireless sensor networks. International Journal of Electronics and Telecommunications, 309-318.

Kumar, J. S., & Patel, D. R. (2014). A survey on internet of things: Security and privacy issues. *International Journal of Computer Applications*, *90*(11).

Kumari, P., & Kumar, A. (2023, April). A comprehensive study of DDoS attacks over IoT network and their countermeasures. *Computers & Security*, *127*, C. DOI: 10.1016/j.cose.2023.103096

Kushalnagar, N., Montenegro, G., & Schumacher, C. (2007). "RFC: 4919. IPv6 over Low-Power Wireless Personal Area Networks (6LoWPANs): Overview, Assumptions, Problem State- ment, and Goals.", Retrieved November 4th, 2017, from http://tools.ietf.org/html/rfc4919

Kumar, A., Matam, R., & Shukla, S. (2016, December). "Impact of packet dropping attacks on RPL.", In *2016 Fourth International Conference on Parallel, Distributed and Grid Com-puting (PDGC)*, (pp. 694-698).IEEE.

Lai, G. H. (2016). Detection of wormhole attacks on IPv6 mobility-based wireless sensor network. *EURASIP Journal on Wireless Communications and Networking, 2016*(1), 274. DOI: 10.1186/s13638-016-0776-0

Le, A., Loo, J., Chai, K. K., & Aiash, M. (2016). A specification-based IDS for detecting attacks on RPL-based network topology. *Information (Basel), 7*(2), 25. DOI: 10.3390/info7020025

Le, A., Loo, J., Lasebae, A., Aiash, M., & Luo, Y. (2012). 6LoWPAN: A study on QoS secu- rity threats and countermeasures using intrusion detection system approach. *International Journal of Communication Systems, 25*(9), 1189–1212. DOI: 10.1002/dac.2356

Le, A., Loo, J., Luo, Y., &Lasebae, A. (2011, October). "Specification-based IDS for securing RPL from topology attacks.", In Wireless Days (WD), 2011 IFIP (pp. 1-3).IEEE.

Le, A., Loo, J., Luo, Y., & Lasebae, A. (2013, July). "The impacts of internal threats towards Routing Protocol for Low power and lossy network performance.", In *2013 IEEE Symposium on Computers and Communications (ISCC)*, (pp. 000789-000794). IEEE. DOI: 10.1109/ISCC.2013.6755045

Luo, Z., Zhao, S., Lu, Z., Sagduyu, Y. E., & Xu, J. "Adversarial machine learning based partial-model attack in IoT," in *ACM Workshop on Wireless Security and Machine Learning (WiseML)*, 2020. DOI: 10.1145/3395352.3402619

Hossain, M., Karim, Y., & Hasan, R. (2018, March). Secupan: A security scheme to mitigate fragmentation-based network attacks in 6lowpan. In *Proceedings of the EighthACM Conference on Data andApplication Security and Privacy* (pp. 307-318).

Mamdouh, M., Awad, A. I., Khalaf, A. A., & Hamed, H. F. (2021). Authentication and Identity Management of IoHT Devices: Achievements, Challenges, and Future Directions. *Computers & Security, 111*, 102491. DOI: 10.1016/j.cose.2021.102491

Masoodi, F., Alam, S., & Siddiqui, S. T. (2019, March). Security & Privacy Threats, Attacks and Countermeasures in Internet of Things [IJNSA]. *International Journal of Network Security & its Applications, 11*(2), 67–77. DOI: 10.5121/ijnsa.2019.11205

Mayzaud, A., Badonnel, R., & Chrisment, I. (2016). A Taxonomy of Attacks in RPL-based Internet of Things. *International Journal of Network Security, 18*(3), 459–473.

Mayzaud, A., Sehgal, A., Badonnel, R., Chrisment, I., & Sch¨onw¨alder, J. (2014, June). "A study of RPL DODAG version attacks.", In *IFIP International Conference on Autonomous Infrastructure, Management and Security* (pp. 92-104).Springer, Berlin, Heidelberg. DOI: 10.1007/978-3-662-43862-6_12

Medjek, F., Tandjaoui, D., Abdmeziem, M. R., & Djedjig, N. (2015, April). "Analytical eval- uation of the impacts of Sybil attacks against RPL under mobility.", In Programming and Systems (ISPS), 2015 12th International Symposium on (pp. 1-9). IEEE.

Medjek, F., Tandjaoui, D., Romdhani, I., & Djedjig, N. (2017a). "A Trust-based Intrusion Detection System for Mobile RPL Based Networks.", In 2017 IEEE 10th International Conference on Internet of Things (iThings-2017). DOI: 10.1109/iThings-GreenCom-CPSCom-SmartData.2017.113

Medjek, F., Tandjaoui, D., Romdhani, I., & Djedjig, N. (2017b). Performance Evaluation of RPL Protocol under Mobile Sybil Attacks. In *In 2017 IEEETrustcom/BigDataSE/ICESS* (pp. 1049–1055). IEEE. DOI: 10.1109/Trustcom/BigDataSE/ICESS.2017.351

Mendez, D. M., Papapanagiotou, I., & Yang, B. (2017). "Internet of things: Survey on security and privacy.", arXiv preprint arXiv:1707.01879.

Mishra, N., & Pandya, S. (2021). "Internet of Things Applications, Security Challenges, Attacks, Intrusion Detection, and Future Visions: A Systematic Review,". *IEEE Access : Practical Innovations, Open Solutions*, *9*, 59353–59377. DOI: 10.1109/ACCESS.2021.3073408

Morales-Molina, C. D., Hernandez-Suarez, A., Sanchez-Perez, G., Toscano-Medina, L. K., Perez-Meana, H., Olivares-Mercado, J., Portillo-Portillo, J., Sanchez, V., & Garcia-Villalba, L. J. (2021). Perez- Meana, H.; Olivares-Mercado, J.; Portillo-Portillo, J.; Sanchez, V.; Garcia-Villalba, L.J. A Dense Neural Network Approach for Detecting Clone ID Attacks on the RPL Protocol of the IoT. *Sensors (Basel)*, *21*(9), 3173. DOI: 10.3390/s21093173 PMID: 34063577

Mrabet, H., Belguith, S., Alhomoud, A., & Jemai, A. (2020). A Survey of IoT Security Based on a Layered Architecture of Sensing and Data Analysis. *Sensors (Basel)*, *20*(13), 3625. DOI: 10.3390/s20133625 PMID: 32605178

Muzammal, S. M., Murugesan, R. K., Jhanjhi, N. Z., Humayun, M., Ibrahim, A. O., & Abdelmaboud, A. (2022). A trust- based model for secure routing against RPL attacks in internet of things. *Sensors (Basel)*, *22*(18), 7052. DOI: 10.3390/s22187052 PMID: 36146400

Nandhini, P. S., Kuppuswami, S., & Malliga, S. (2021). Energy efficient thwarting rank attack from RPL based IoT networks: A review. *Materials Today: Proceedings*, *81*, 694–699. Advance online publication. DOI: 10.1016/j.matpr.2021.04.167

Nandy, T., Idris, M. Y. I. B., Noor, R. M., Kiah, L. M., Lun, L. S., & Juma'at, N. B. A. (2019). Review on Security of Internet of Things Authentication Mechanism. *IEEE Access : Practical Innovations, Open Solutions*, *7*, 1–36. DOI: 10.1109/ACCESS.2019.2947723

Nawir, M., Amir, A., Yaakob, N., & Lynn, O. B. (2016, August). "Internet of Things (IoT): Taxonomy of security attacks.", In 2016 3rd International Conference onElectronic Design (ICED), (pp. 321-326). IEEE. DOI: 10.1109/ICED.2016.7804660

Nazabal, J. A., Falcone, F., Fern'andez-Valdivielso, C., & Mat'ıas, I. R. (2013). Development of a Low Mobility IEEE 802.15.4 Compliant VANET System for Urban Environments. *Sensors (Basel)*, *13*(6), 7065–7078. DOI: 10.3390/s130607065 PMID: 23760089

Nebbione, G., & Calzarossa, M. C. (2020). Security of IoT Application Layer Protocols: Challenges and Findings. *Future Internet*, *12*(3), 55. DOI: 10.3390/fi12030055

Nixon, C., Sedky, M., & Hassan, M. (2019). Practical Application of Machine Learn- ing based Online Intrusion Detection to Internet of Things Networks. 1-5. .DOI: 10.1109/G- CIoT47977.2019.9058410

Pasikhan, A. M., Clark, J. A., & Gope, P. (2023). "Incremental hybrid intrusion detec- tion for 6LoWPAN", Computers & Security, Volume 135, 2023, 103447, ISSN 0167-4048, DOI: 10.1016/j.cose.2023.103447

Patel, K. K., & Patel, S. M. (2016). Internet of Things-IOT: Definition, Characteristics, Archi- tecture, Enabling Technologies, Application & Future Challenges. *International Journal of Engineering Science*, •••, 6122.

Perazzo, P., Vallati, C., Arena, A., Anastasi, G., & Dini, G. (2017, September). "An Imple- mentation and Evaluation of the Security Features of RPL." In *International Conference on Ad-Hoc Networks and Wireless* (pp. 63-76).Springer, Cham.

Perazzo, P., Vallati, C., Varano, D., Anastasi, G. P., & Dini, G. (2018). Implementation of a wormhole attack against a rpl network: Challenges and effects. 2018 14th Annual Conference on Wireless On-demand Network Systems and Services (WONS), 95-102.

Pongle, P., & Chavan, G. (2015). Real time intrusion and wormhole attack detection in internet of things. *International Journal of Computer Applications*, *121*(9), 1–9. DOI: 10.5120/21565-4589

Pongle, P., & Chavan, G. (2015a). "A survey: Attacks on RPL and 6LoWPAN in IoT." In 2015 International Conference onPervasive Computing (ICPC), (pp. 1-6). IEEE. DOI: 10.1109/PERVASIVE.2015.7087034

Pu, C. (2019). *"Energy Depletion Attack Against Routing Protocol in the Internet of Things," 2019 16th IEEE Annual Consumer Communications & Networking Conference.* CCNC., DOI: 10.1109/CCNC.2019.8651771

Raikar, M. M., & Meena, S. M. (2021, May). SSH brute force attack mitigation in Internet of Things (IoT) network: An edge device security measure. In 2021 2nd international conference on secure cyber computing and communications (ICSCCC) (pp. 72-77). IEEE.

Raizada, S., Matwani, L. G., & Singh, Y. (2023). Remote Code Execution: A Major Threat to Data Loss. In Misra, S., Jain, A., Kaushik, M., & Banerjee, C. (Eds.), *Futuristic Technology Perspectives on Entrepreneurship and Sustainable Innovation* (pp. 181–188). IGI Global., DOI: 10.4018/978-1-6684-5871-6.ch012

Rani, P., Verma, S., & Nguyen, G. N. (2020). Mitigation of black hole and gray-hole attack using swarm inspired algorithm with artificial neural network. *IEEE Access : Practical Innovations, Open Solutions, 8*, 121755–121764. DOI: 10.1109/ACCESS.2020.3004692

Ray, D., Bhale, P., Biswas, S., Mitra, P., & Nandi, S. (2023). A Novel Energy-Efficient Scheme for RPL Attacker Identification in IoT Networks Using Discrete Event Modeling. *IEEE Access : Practical Innovations, Open Solutions, 11*, 77267–77291. DOI: 10.1109/ACCESS.2023.3296558

Raza, S., Duquennoy, S., Chung, T., Yazar, D., Voigt, T., & Roedig, U. (2011, June). "Securing communication in 6LoWPAN with compressed IPsec.", In *2011 International Conference on Distributed Computing in Sensor Systems and Workshops (DCOSS)*, (pp. 1-8).IEEE. DOI: 10.1109/DCOSS.2011.5982177

Raza, S., Trabalza, D., & Voigt, T. (2012a, May). "6LoWPAN compressed DTLS for CoAP.", In Distributed Computing in Sensor Systems (DCOSS), 2012 IEEE 8th International Con- ference on (pp. 287-289).IEEE.

Raza, S., Voigt, T., & Jutvik, V. (2012b, March). "Lightweight IKEv2: a key management solution for both the compressed IPsec and the IEEE 802.15. 4 security.", In *Proceedings of the IETF workshop on smart object security* (Vol. 23).

Raza, S., Wallgren, L., & Voigt, T. (2013). SVELTE: Real-time intrusion detection in the Internet of Things. *Ad Hoc Networks, 11*(8), 2661–2674. DOI: 10.1016/j.adhoc.2013.04.014

Rehman, A., Khan, M. M., Lodhi, M. A., & Hussain, F. B. (2016, March). "Rank attack using objective function in RPL for low power and lossy networks.", In *2016 International Conference on Industrial Informatics and Computer Systems (CIICS)*, (pp. 1-5).IEEE. DOI: 10.1109/ICCSII.2016.7462418

Rekha, H., & Siddappa, M. (2023). Blockchain Mechanism-Based Attack Detection in IoT with Hybrid Classification and Proposed Feature Selection. *Cybernetics and Systems*. Advance online publication. DOI: 10.1080/01969722.2023.2166266

Rghioui, A., Bouhorma, M., & Benslimane, A. (2013, March). "Analytical study of security aspects in 6LoWPAN networks.", In Information and Communication Technology for the Muslim World (ICT4M), 2013 5th International Conference on (pp. 1-5). IEEE. DOI: 10.1109/ICT4M.2013.6518912

Rghioui, A., Khannous, A., &Bouhorma, M. (2014). "Denial-of-Service attacks on 6LoWPAN- RPL networks: Threats and an intrusion detection system proposition.", Journal of Ad- vanced Computer Science & Technology, 3(2), 143.

Rodríguez, G. E., Torres, J. G., Flores, P., & Benavides, D. E. (2020). Cross-site scripting (XSS) attacks and mitigation: A survey. *Computer Networks*, *166*, 106960.

Rouissat, M., Belkheir, M., Alsukayti, I. S., & Mokaddem, A. (2023). A Lightweight Mitigation Approach against a New Inundation Attack in RPL-Based IoT Networks. *Applied Sciences (Basel, Switzerland)*, *13*(18), 10366. DOI: 10.3390/app131810366

Saba, T., Rehman, A., Sadad, T., Kolivand, H., & Bahaj, S. A. (2022). Anomaly-based intrusion detection system for IoT networks through deep learning model. *Computers & Electrical Engineering*, *99*, 107810.

Sadiq, A., Anwar, M., Butt, R. A., Masud, F., Shahzad, M. K., Naseem, S., & Younas, M. (2021). A review of phishing attacks and countermeasures for internet of things-based smart business applications in industry 4.0. *Human Behavior and Emerging Technologies*, *3*(5), 854–864. DOI: 10.1002/hbe2.301

Sengupta, J., Ruj, S., & Das Bit, S. (2020). A Comprehensive Survey on At- tacks, Security Issues and Blockchain Solutions for IoT and IIoT. *Journal of Network and Computer Applications*, *149*, 102481. Advance online publication. DOI: 10.1016/j.jnca.2019.102481

Sharma, G., Grover, J., & Verma, A. (2023). Performance evaluation of mobile RPL-based IoT networks under version number attack. *Computer Communications*, *197*, 12–22. DOI: 10.1016/j.comcom.2022.10.014

Sheibani, M., Barekatain, B., & Arvan, E. (2022). A lightweight distributed detection algorithm for DDAO Attack on RPL routing protocol in Internet of Things. *Pervasive and Mobile Computing*, *80*, 101525. DOI: 10.1016/j.pmcj.2021.101525

Shelby, Z., & Bormann, C. (2011). *6LoWPAN: The wireless embedded Internet* (Vol. 43). John Wiley & Sons.

Shelby, Z., Hartke, K., Bormann, C., & Frank, B. (2014). "RFC 7252.Constrained Application Protocol (CoAP).", Retrieved November 4th, 2017, from http://tools .ietf.org/html/rfc7252

Sherazi, H. H. R., Iqbal, R., Ahmad, F., Khan, Z. A., & Chaudary, M. H. (2019). DDoS attack detection: A key enabler for sustainable communication in internet of vehicles. *Sustainable Computing : Informatics and Systems*, *23*, 13–20.

Shreenivas, D., Raza, S., & Voigt, T. (2017, April). Intrusion Detection in the RPL-connected 6LoWPAN Networks. In Proceedings of the 3rd ACM International Workshop on IoT Pri- vacy, Trust, and Security (pp. 31-38). ACM.

Syed, N. F., Baig, Z., Ibrahim, A., & Valli, C. (2020). Denial of service attack detection through machine learning for the IoT. *Journal of Information and Telecommunication*, *4*(4), 482–503. DOI: 10.1080/24751839.2020.1767484

Tandon, A., & Srivastava, P. (2019, August). Trust-based enhanced secure routing against rank and sybil attacks in IoT. In 2019 twelfth international conference on contemporary computing (IC3) (pp. 1-7). IEEE.

Tanveer, M., Abbas, G., Abbas, Z. H., Waqas, M., Muhammad, F., & Kim, S. (2020). S6AE: Se- curing 6LoWPAN Using Authenticated Encryption Scheme. *Sensors (Basel)*, *20*(9), 2707. DOI: 10.3390/s20092707 PMID: 32397469

Thanigaivelan, N. K., Nigussie, E., Kanth, R. K., Virtanen, S., & Isoaho, J. (2016, Jan- uary). "Distributed internal anomaly detection system for Internet-of-Things.", In 2016 13th IEEE Annual Consumer Communications & Networking Conference (CCNC), (pp. 319-320). IEEE.

Thubert, P. (Ed.). (2020). *RFC 8931: IPv6 over Low-Power Wireless Personal Area Network (6LoWPAN)*. Selective Fragment Recovery.

Thubert, P., Brandt, A., Hui, J., Kelsey, R., Levis, P., Pister, K., . . . Alexander, R. (2012). "RFC 6550.RPL: IPv6 routing protocol for low power and lossy networks, IETF. ", Retrieved November 4th, 2017, from https://tools.ietf.org/html/rfc6550

Tsao, T., Alexander, R., Dohler, M., Daza, V., Lozano, A., & Richardson, M. (2015). RFC 7416.A security threat analysis for the routing protocol for low-power and lossy networks (rpls).Retrieved November 4th, 2017, from https://tools.ietf.org/html/rfc7416

Valea, E., Da Silva, M., Flottes, M.-L., Di Natale, G., Dupuis, S., & Rouzeyre, B. (2019). "Providing Con- fidentiality and Integrity in Ultra Low Power IoT Devices," 2019 14th International Con- ference on Design & Technology of Integrated Systems. In *Nanoscale Era* (pp. 1–6). DTIS., DOI: 10.1109/DTIS.2019.8735090

Varadarajan, P., & Crosby, G. (2014, March). "Implementing IPsec in wireless sensor net- works.", In 2014 6th International Conference on New Technologies, Mobility and Security (NTMS), (pp. 1-5). IEEE.

Vasseur, J. P., Kim, M., Pister, K., Dejean, N., & Barthel, D. (2012). "RFC 6551. Routing Metrics Used for Path Calculation in Low-Power and Lossy Networks,I-ETF.", Retrieved November 4th, 2017, from https://tools.ietf.org/html/rfc6551

Verma, A., & Ranga, V. (2019, April). ELNIDS: Ensemble learning based network intrusion detection system for RPL based Internet of Things. In 2019 4th International conference on Internet of Things: Smart innovation and usages (IoT-SIU) (pp. 1-6). IEEE.

Verma, A., & Ranga, V. (2020). Security of RPL based 6LoWPAN Networks in the Internet of Things: A Review. *IEEE Sensors Journal*, *20*(11), 5666–5690.

Wallgren, L., Raza, S., & Voigt, T. (2013). Routing attacks and countermeasures in the rpl-based internet of things. *International Journal of Distributed Sensor Networks*, *9*(8), 794326. DOI: 10.1155/2013/794326

Weekly, K., & Pister, K. (2012, October). "Evaluating sinkhole defense techniques in RPL networks." In 2012 20th IEEE International Conference on Network Protocols (ICNP), (pp. 1-6).IEEE. DOI: 10.1109/ICNP.2012.6459948

Wei, M., Rong, C., Liang, E., & Zhuang, Y. (2022). An intrusion detection mechanism for IPv6-based wireless sensor networks. *International Journal of Distributed Sensor Networks*, *18*(3). Advance online publication. DOI: 10.1177/15501329221077922

Yadollahzadeh Tabari, M., & Mataji, Z. (2021). Detecting sinkhole attack in rpl-based internet of things routing protocol. *Journal of Artificial Intelligence and Data Mining*, *9*(1), 73–85.

Zahra, F., Jhanjhi, N., Brohi, S. N., Khan, N. A., Masud, M., & AlZain, M. A. (2022). Rank and Worm- hole Attack Detection Model for RPL-Based Internet of Things Using Machine Learning. *Sensors (Basel)*, *22*(18), 6765. DOI: 10.3390/s22186765 PMID: 36146111

Zaminkar, M., & Fotohi, R. (2020). SoS-RPL: Securing internet of things against sinkhole attack using RPL protocol-based node rating and ranking mechanism. *Wireless Personal Communications*, *114*(2), 1287–1312.

Zhang, L., Feng, G., & Qin, S. (2015, June). "Intrusion detection system for RPL from routing choice intrusion.", In *2015 IEEE International Conference on Communication Workshop (ICCW)*, (pp. 2652-2658). IEEE. DOI: 10.1109/ICCW.2015.7247579

Chapter 3
IoT Security and Privacy

Eshani Banik
https://orcid.org/0009-0005-5326-8523
Brainware University, India

Joyeeta Das
Brainware University, India

Piyal Roy
https://orcid.org/0009-0007-5378-2789
Brainware University, India

Sanchita Ghosh
https://orcid.org/0009-0007-9421-5246
Brainware University, India

ABSTRACT

The Internet is an innovative technology that is difficult to avoid since it is always evolving with new hardware and apps. Human-to-human and human-to-device communication is now the most popular forms of communication; however, the Internet of Things (IoT) foresees the near future when machine-to-machine (M2M) communication will become the standard. An item containing sensors, actuators, and CPUs may connect with other things to do useful tasks, according to the IoT concept. We have discussed the architecture of the Internet of Things in this post. The many IoT apps that users may use, along with their advantages and disadvantages, were also discussed. joined by a few IoT deployment tactics that are morally sound.

DOI: 10.4018/979-8-3693-5448-3.ch003

1. INTRODUCTION

The term "Internet of Things," or "IoT," refers to the aggregate network of interconnected gadgets. Computer scientist Kelvin Ashton originally used the term "Internet of Things" in 1999. IoT was originally introduced to phone calls, smart grids, and computers by Aston, a scientist. Additionally, Ashton suggested tagging things with radio frequency identification chips so that they could be tracked via a supply chain. It is embedded with software, sensors, and other technology. Its only function is to establish internet-based connections and data exchanges with other systems and devices. In the past several decades, the IoT has become a new technical technology that is enabling us to achieve new heights. In engineering, it integrates the devices using the internet in an active manner (Sadiku et al., 2016). This concept holds the key to a future full of promises and development. Along with development, IoT promises us many other facilities due to it being time-saving, cost effective and provides ease of life. One of its most impressing features lies in its automation. It provides faster output with minimum machine-man interaction needed. It had great potential to offer for making life and technology easier than ever. Nowadays, IoT is being applied to increase communication, field testing, ease of governing the machines and many other things. Being accessible and understandable to all, it has a bright future in the engineering. People have started to be more accepting towards such technology and it is truly remarkable because it reduces the work load by a lot. Figure 1 depicts the fundamental idea of the IoT, which is that objects and people may communicate with each other over a variety of wireless networks and channels at any time. The technologies that humans discover advance in tandem with human development. It's a balanced way of living. Furthermore, there is a lot of room for advancement for technologies like IoT in the future (Tiwary et al., 2018).

Figure 1. Basic Idea of the Internet of things (Alferidah et al., 2020)

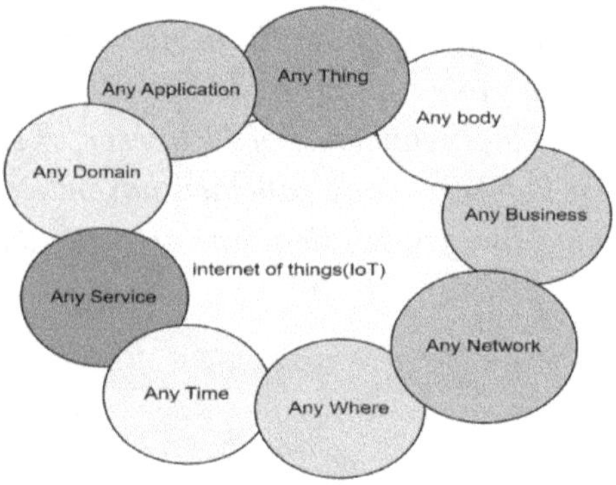

2. IOT TECHNOLOGIES AND PROTOCOLS

IoT primarily aims to create a self-talking environment where devices' data may be analyzed using machine learning to carry out specific activities. Large number of data produced by IoT devices are analyzed and worked on using sophisticated data analytics and machine learning (ML) to extract insightful information for decision-making. Over the last ten years, the IoT has enabled the automation of various systems, including transportation, medical technologies, smart household appliances, and electricity grids. This technology makes use of smart devices including actuators and sensors in addition to a number of communication methods like Bluetooth, WiFi, and ZigBee (2022) (Tiwary et al., 2018). In addition, the energy industry is utilizing IoT to integrate renewable energy sources, enhance system resilience, and manage smart grids. IoT may also be used for electricity generation, transmission, distribution, and consumption, which covers every aspect of the smart grid (Vaigandla et al., 2021). With the rise in popularity of edge and cloud computing, data processing may now take place next to the IoT devices themselves, which lowers latency and enhances real-time processing. Nonetheless, limited memory, low power, short battery life, and network constraints are typically associated with IoT sensors. Compute, store, access, and analyze IoT data in order to do this. And unified technology, like the cloud, which allows information to be accessed from anywhere, is needed to solve these problems (Shah & Yaqoob, 2016). A tree-like assembly control plane, part of the cloud-based architecture, comprises several layers of energy-saving strategies at various levels, including departments, buildings, and individual rooms. A few appliances include PCs or smartphones that come equipped with a variety of networking interfaces, including Bluetooth, 3G, Wi-Fi, and GPS (Vaigandla et al., 2021). We shall discuss the protocols and technologies in this area that will develop to satisfy the requirements of the future IoT application.

2.1 Key IoT Components

Among the most important features of the IoT are Artificial Intelligence (AI), sensors, connection, active participation, and the usage of tiny devices. A summary of these aspects is provided below:

AI: In essence, the IoT employs data collecting, networks, and AI algorithms to make nearly anything "smart," so enhancing all aspects of life.

Options for Connectivity: Networks are no longer only dependent on large carriers because to new networking enabling technologies, particularly IoT networking. Even at far lower and less expensive scales, networks are still valuable. IoT creates these small networks by connecting its system units.

Sensors: Without sensors, the IoT becomes less unique. They act as the defining instruments that transform the IoT from a conventional passive device network into a system with activity that can be incorporated into real systems.

Active Engagement: Most of the interactions people have with connected technology today are passive in nature. IoT creates a new model for active content, product, or service interaction.

Tiny technologies: It is to be expected that technologies have become more affordable, smaller, and capable throughout time.

IoT delivers its accuracy, scalability, and adaptability by utilizing specially designed tiny devices (Ren et al., 2018).

2.2 Protocols for Communication

Switches and Gateways Transferring and exchanging the data generated at the instrument level is necessary because the IoT is home to billions of smart objects, each with a unique set of functionalities. Nevertheless, it is projected that the IoT devices will be networked and able to communicate in a variety of ways. An IoT device should have the capabilities to communicate with other devices by determining the best path to take, understanding the messages it receives, and responding appropriately. Conventional practices so become necessary prerequisites for the IoT. This makes it easy to maintain the required level of network execution while accessing the full functionality of such restricted devices. A mobile gadget is one that is always on the go. Most of the time, it needs to be transferred from the existing attachment point to a new one. For most IoT devices, the communication protocols need to be aware of this. To ensure a smooth handover and minimize the delays imposed at various tiers, intelligent techniques are needed. We go over a few of the protocols taken into consideration for IoT devices in Section IV. To ensure a smooth handover and minimize the delays imposed at various tiers, intelligent techniques are needed. A few of the IoT protocols that are under consideration are covered in Section IV. In these large networks with many application systems, functions, and changeable data kinds, the middleware software layer is equally crucial. Interaction between the "Internet" and "things" is made possible by middleware. It serves as an interface that makes it possible for the different programs running on heterogeneous systems to interact with one another simply and effectively (Fuqaha et al., 2015). The middleware software layer plays a significant part in concealing the underlying information. Regardless of the underlying technologies, this facilitates the development of new software services and applications for distributed environments.

3. IOT APPLICATIONS IN ENGINEERING

If we look about, we can see that technologies have advanced significantly in the last few years, helping to advance IoT technology and solve problems in the real world. Applications for the IoT offer dependable communications and real-time message transmission, which boosts productivity, sustainability, and safety. They present all of the capabilities of the system to the end user through a myriad of networked devices. Networks and devices provide the physical connectivity, but IoT applications can provide strong device-to-device and human-to-device interactions (Vaigandla et al., 2021). IoT makes it possible that Google Car is currently able to provide a self-driving experience that includes real-time traffic, road conditions, weather conditions and other information. In future, there can be numerous possible applications that can promote controlling and monitoring equipment and processes, reducing the need for physical presence by improving response times (Shah & Yaqoob, 2016). with respect to application domain some of the prominent projects where implementation of IoT played impressive roles are Industrial automation and maintenance, Smart energy management, Smart agriculture in rural areas, Environmental monitoring, Healthcare and Biomedical Applications, Transportation and Traffic management, Smart City management etc. In this section we will talk about these applications more precisely (Costa et al., 2016).

3.1 Industrial Automation and Manufacturing

It is probable that the integration of IT in industrial automation is not necessary with this five-layer design. The objective is the direct transfer of measured data from field sensors to an Ethernet or cloud service. As a result, this is how future IT-based industrial automation—also referred to as cyber-physical systems (CPSs)—will look. Process control, asset & security management, intelligent manufacturing, machine-to-machine (M2M), service-oriented architectures, etc. automatically subscribe in real-time to the necessary data from cloud services. The integration of physical and computational processes is known as a CPS. It combines digital systems—usually computer-based systems—with physical ones, such as sensors and actuators. The physical processes, which are typically governed by feedback loops that influence calculations and vice versa, are managed by embedded computers and network monitors. The implementation of CPS is shown in Fig. 2 as moving from a classical automation system to an IT-based one. CPS is an autonomous network that combines the digital and real worlds. They can engage in lively interactions with their surroundings. Among other aspects, big data management offers real-time cloud services for smart manufacturing and process control, production rescheduling, asset & security management, automated remote monitoring, and service-oriented jobs. The incorporation of CPS into industrial systems offers excellent prospects for system adaptation and sustainability.

3.2 Smart Energy Management

Understanding the scope and characteristics of big data and data mining is essential for smart energy management to make the most of the big data idea. The first stage is to examine and comprehend the data attributes using the big data dimensions. After having a thorough knowledge of the data, researchers might select a useful data mining strategy to enhance its characteristics. For instance, there is a chance the data is not entirely accurate; therefore, the data mining strategy should include measures to make the data better. Several well-known data mining techniques are reviewed and their performances assessed by Garcia, Luengo, and Herrera. The classifier's performance on an unprocessed dataset was compared to multiple pre-processed datasets by the authors. There were noise and missing values in the raw dataset (Li et al., 2023).

3.3 Transportation and Logistics

Smart logistics may help with a variety of tasks, including acquiring new business models, taking advantage of opportunities given by rising technology, and managing the constraints posed by quickly shifting client expectations. The main technologies, uses, and industry and geographic distributions of IoT-based smart logistics research and applications are displayed in a bibliometric study based on publications from 2008 to 2019. The content analysis provides research needs for the progress of smart logistics and emphasizes the function and impact of IoT on smart logistics along with the challenges of IoT-based smart logistics. Among the current challenges are worries about IoT security and privacy, as well as limited IoT technical capacity and extension, problems with IoT standardization, difficulties acquiring and processing IoT data, and technical difficulties with radio frequency identification and wireless sensor networks (Ding et al., 2021). In order to solve the primary technical problems with the Internet of Things, advance different IoT applications in the logistics sector, and collaboratively create novel ICTs and management systems, research is very necessary.

3.4 Inspection of Structural Health

Assessing damage and conducting structural health monitoring (SHM) on civil engineering infrastructure are challenging tasks. The production stage of the materials, transportation, installation, workmanship, formwork removal, and concrete curing are only a few of the numerous variables that impact a structure's strength and structural integrity. The rising availability of Wi-Fi networks and technology improvements have led to the replacement of old wire-based technologies in SHM with IoT based real-time wireless sensors. It is possible to perform a thorough structural health evaluation using real-time test data from multiple IoT sensors that monitor multiple structural health indicators and store the data on cloud-based data storage systems. After that, there are a plethora of applications for the sensor data, including predicting the early-stage compressive strength of concrete, spotting potholes on roads, assessing construction quality, identifying different damage typologies, predicting when to remove formwork, managing vibration and curing quality, and assessing seismic vulnerability. The application of wireless Internet of Things technologies for infrastructure monitoring in civil engineering is explained in this review article. Furthermore, a number of case studies that track the structural integrity of civil engineering projects using actual structures and laboratory testing are included (Mishra et al., 2022).

3.5 Biomedical and Healthcare Applications

Recent developments in the IoT and big data analytics have created fascinating new opportunities for biomedical and healthcare technologies. Biomedical systems employ the IoT for telecare, diagnosis, prevention, treatment, and monitoring, among other purposes. Known as enabling technologies, radio frequency identification (RFID) systems and wireless body area networks (WBANs) are essential components of the IoT idea.

In healthcare systems, WBANs are a crucial part of the IoT. An access point and sensing nodes are components of a WBAN. An embedded device known as a sensing node is able to detect and transmit various vital physiological data from human bodies, such as blood pressure, SpO2, body temperature, and electrocardiogram (ECG), using a wireless component to transmit the data to a wireless access point. Afterwards, the information is received by the access point, which is also referred to as a network coordinator, and then sent to the healthcare monitoring canter following necessary preprocessing steps like extracting, filtering, and compressing. Quality of Service (QoS) represents a critical design concern for biomedical applications using WBANs, where two essential QoS metrics are data rate and transmission latency.

As well as, RFID systems also play a vital role in the IoT. RFID stands for radio frequency identification, and it consists of tags and a reader, sometimes referred to as an interrogator. A tag is made up of an antenna as well as a microchip that houses a special identification. The RFID tags have two modes of operation: active and passive. Which implies without a source of energy within. One way to access data is by the RFID reader and radio frequency data transmission band of 13.56 MHz/860

960 MHz/2.45 G, or 125–134.2 kHz Hz is applied to many different things. Usually, the reader charges a group of passive tags that are in its field of inquiry with the help of inductive coupling methods. In other words, the reader transmits an electromagnetic signal, which the tag's antenna subsequently picks up and puts the energy of the signal through an on-board capacitor (Zantalis et al., 2019). RFID technology is effectively used in the healthcare industry that lies within an IoT platform, alongside typical applications include identification of newborns and patients, tracking the locations of biological inventories, monitoring and validating medical treatments, managing procedures in a medical facility, and process management for surgery.

3.6 Agriculture and Environmental Monitoring

Various IoT applications influence IoT architecture. Figure 1 depicts a typical three- or four-layer IoT design. Take two examples, for instance. First scenario: Think about smart pollution monitors that include sensors that continuously communicate data to the central database about levels of sound pressure, carbon monoxide, nitrogen dioxide, and other pollutants. The data will be analyzed using these analytical methods, which will give specifics on the city's traffic police's degree of air pollution. This information makes it easier to exercise caution when it rises over the usual level. In this instance, the sensor layer indicates that sensors will be continually monitoring the air and sending data to the record through wired or wireless connection. The consumer's smartphone will receive the final, combined result through the air pollution control application after this data has been processed and reviewed. Four levels of architecture are therefore required. Scenario 2: Assume that a sensor is placed in the kitchen, next to a gas cylinder, or somewhere else to find a gas leak. Whenever this sensor notices a gas leak, it has to alert the user. vicinity as soon as possible, after which the owner must get the message. In this case, the sensor's internal analysis is necessary within the layer itself (Du et al., 2016).

3.7 Smart Cities and Urban Planning

People keep moving to cities because they offer better opportunities. This leads to an increase in population, and smart population management is required to deal with the issues that come with a growing city. Everybody has felt the annoyance we do. when we are stuck in traffic for hours on end. The stench of incorrect waste disposal in community, etc. IoT will provide solutions for all of these issues. illustrates a few of the possible uses for IoT in smart cities.

4. CHALLENGES AND FUTURE DIRECTIONS

IoT technologies are anticipated to operate as a catalyst for future technical knowledge, and their use is anticipated to increase rapidly from previous decades. There is still a worry about it being secure because of how much its utility is growing. Therefore, maintaining its security is turning into a constant issue. Indeed, current technical advancements driven by the IoT are modernizing a number of industries, including smart cities, energy efficiency, and health care, and stimulating innovation through inclusive industrialization. However, the attack surface is growing as well. as well as a bigger effect on privacy, which can due to the development of pervasive gadgets (such as smartphones or wearables), which produce a vast amount of data, this will ultimately impair people' safety (Karale, 2021). It is important to prevent unwanted identification and access in order to guarantee user security. In this section, we address security, privacy concerns in IoT systems, interoperability and scalability issues, energy efficiency, and data management while embracing edge computing, emerging technologies, sustainability, and the use of integrated AI and machine learning. We also briefly discuss potential approaches and future directions (Aledhari et al., 2022).

4.1 IOT System Security and Privacy Concerns

In addition to its benefits, IoT security is an important consideration. If the signal is hacked or stolen, it will have an instantaneous effect on the security of all IoT data. IoT offers a greater amount of information as it becomes more extensively used, which raises the possibility that such information may be exposed. IoT's growth will be severely limited if security concerns are not adequately addressed. Security issues are thus the most significant of all the IoT's difficulties. The proposed ITU-T Y.2002 design states that the three primary levels of an IoT architecture are the perception layer, the transportation layer, and the application layer. An IoT architecture in layers is shown in Figure 2.

Figure 2. Layers of IoT architecture

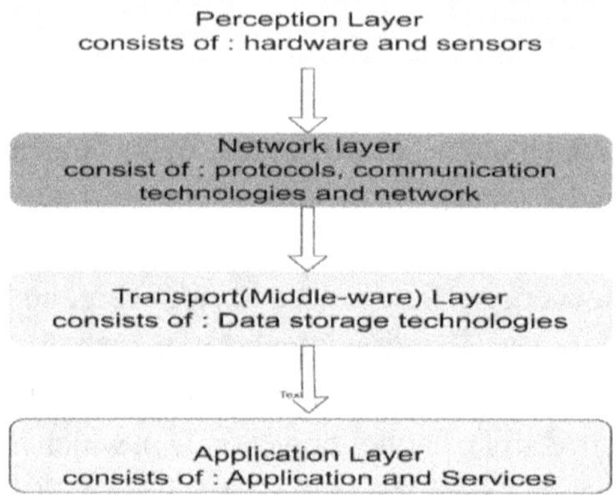

4.1.1 Perception Layer of Security

The primary functions of this layer include gathering data, seeing and controlling objects, and then sending that data to the transportation layer, which is the next layer below. A range of collecting and controlling modules, including as vibration, sound, and temperature sensors, are part of the perception layer. These sensors may collect information on many different subjects, including as position, motion, vibration, and changes in the air or surroundings. The perception node and the perception network, which communicates with the transportation network, are the two parts of the perception layer. While perception node is frequently used for data gathering and control, perception network sends control instructions to the controller or distributes collected data to the gateway (Jing et al., 2014). Because this layer has sensors, it includes technologies like RFID and WSNs that facilitate information gathering. Many types of attacks, including identity-based, replay, and node capture attacks, can target the perception layer. To prevent such attacks, a variety of techniques are used, such as hash-based encryption, lightweight cryptography, and other security-related techniques.

4.1.2 The Network Layer's Security

It is also known as a wireless sensor network. In the three tiers of IoT architecture, the network layer comes after the perception layer. The network layer is sometimes called the transmission layer because it acts as a crucial network between the application and perception layers. Additionally, it transmits and transfers the collected data. It is the primary responsibility of this layer to receive digital signals containing valuable information from the perception layer and forward those signals to processing systems in the application layer through a variety of protocols and transmission mediums (e.g., IPv4, IPv6, MQTT, DDS, WiFi, Bluetooth, WiMaX, Zigbee, GSM, 3G, etc.). Additionally, it is in charge of connecting network devices, smart objects, and networks. It is therefore quite susceptible to attacks from other parties. Network layer attacks such as Denial of Service (DoS) attacks and Main-in-the-Middle (MiTM) attacks can be thwarted by employing Intrusion Detection System (IDS), machine learning-based botnet traffic analyzers, and several other techniques (Meneghello et al., 2019).

Table 1. Pros and cons of different IoT protocols.

Protocol	Pros	Cons
MQTT (Message Queuing Telemetry Transport)	- Lightweight and ideal for low-bandwidth networks - Low power consumption - Supports Quality of Service (QoS) levels for message delivery - Widely supported across platforms and devices	- Limited security features - Relies on a central broker, which can be a single point of failure - Not designed for large-scale systems
CoAP (Constrained Application Protocol)	- Designed for constrained devices and networks - Low overhead with RESTful architecture - Can work with UDP for quick transmission - Supports multicast for efficient communication	- Lacks built-in security (needs DTLS) - Limited support for complex querying - May have issues in high-latency networks
HTTP (Hypertext Transfer Protocol)	- Ubiquitous and widely understood - Supports complex and structured data exchange - Works well in client-server models - Secure with HTTPS	- High overhead, not suitable for low-power or low-bandwidth devices - High latency compared to other protocols - No native support for continuous or real-time communication
AMQP (Advanced Message Queuing Protocol)	- High reliability with message delivery guarantees - Supports complex routing and queuing - Strong security features - Suitable for enterprise-level messaging systems	- Complex and resource-intensive - Overhead is high, making it unsuitable for constrained IoT devices - Less suitable for low-latency, real-time applications

continued on following page

Table 1. Continued

Protocol	Pros	Cons
DDS (Data Distribution Service)	- Real-time, low-latency communication - Peer-to-peer architecture - High scalability for distributed systems - Supports Quality of Service (QoS) policies	- High resource requirements, making it unsuitable for constrained devices - Complex to implement and configure - Not widely adopted compared to MQTT and CoAP
Zigbee	- Low power consumption, ideal for battery-operated devices - Supports mesh networking for extended range - Secure with built-in encryption - Low-cost and scalable	- Low data rate, not ideal for high-bandwidth applications - Limited range without mesh network - Interoperability issues with other protocols
LoRaWAN (Long Range Wide Area Network)	- Long-range communication with low power consumption - Suitable for rural and remote areas - Supports a large number of devices - Secure with AES encryption	- Low data rate - High latency in communication - Not ideal for real-time applications or high-bandwidth needs
Bluetooth Low Energy (BLE)	- Low energy consumption - Widely used for short-range IoT applications - Secure with built-in encryption - Low cost and widely supported	- Limited range - Low bandwidth, not suitable for large data transfers - Not ideal for applications requiring continuous data transmission
Z-Wave	- Low power consumption - Secure with AES encryption - Suitable for smart home applications - Supports mesh networking for extended coverage	- Limited interoperability with other protocols - Low data rate - More expensive than some other low-power protocols like Zigbee
6LoWPAN	- Integrates well with IPv6 - Low power consumption - Supports mesh networking - Scalable for large networks	- Limited bandwidth - Higher complexity in implementation - Limited range without mesh networking

4.1.3 Security Application Layer

It serves as an interface between IoT devices and the network they will be interacting with. The application is highly helpful in the large-scale growth of IoT as it encourages its development. This application layer's primary objective is to handle information and data formatting for all IoT applications. Smart cities, smart homes, smart transportation, smart health, smart animals, and smart agriculture are a few of the IoT applications. It is responsible for providing application-specific services to the users. The information gathered by the sensors from devices may be used to select the service. Attacks at the application layer are very different from those at the higher levels in that they target the device software directly instead of the communication infrastructure. Attackers may potentially target the integrity of machine learning algorithms, for example, by manipulating the algorithm's training procedure to cause undesirable behaviors (Kumar et al., 2020). Some of the most important security issues with IoT are shown in Figure 3.

Figure 3. Current security concerns

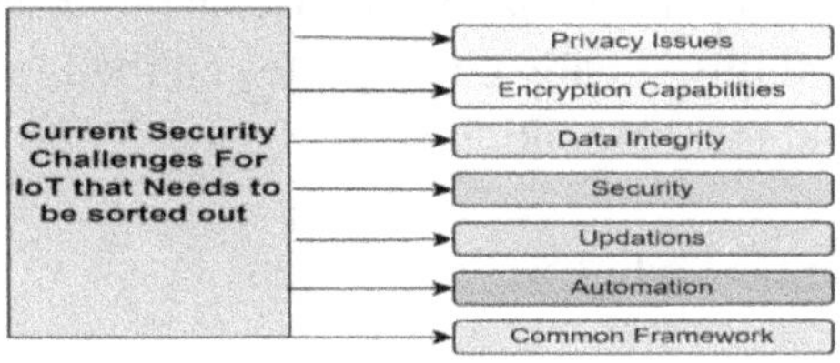

4.2 Scalability and interoperability challenges in IOT

Even while IoT applications are widely used across many industries, the technology is still in its infancy and, as a result, has several limitations and obstacles because of its large datasets. IoT services communicate with one another over the Internet in a variety of ways, much like traditional corporate services, to realize a global system workflow. IoT systems, on the other hand, necessitate the interaction of billions of services since the number of connected items is increasing at a faster rate than traditional business systems. There are already around 19 billion linked objects, and it is predicted that this number will continue to rise rapidly in the years to come. In 2020, Arellanes and Lau Scalability hence becomes quite important in this context. In order to satisfy the demands of services with an increasing performance or service quality of user, scalability is essential in IOT. However, the issue of information system interoperability has been around since 1988, if not earlier. Interoperability has been defined in a variety of ways in the literature. We shall, however, only cite those that are pertinent to our situation. Interoperability, according to the IEEE, is the capacity of two or more systems or components to share information and utilize that information. According to Čolaković & Hadžialić (2018), interoperability is the capacity of two systems to interact and exchange services with one another. IOT defines it as the capacity of different systems and devices to cooperate with one another. There are a lot of various protocols and technologies used in IoT development, along with a variety of vendor-specific solutions, which leads to a lot of heterogeneity and interoperability issues. Interoperability problems in the IoT are considered at all levels. To address this issue, a layered framework with a uniform architecture is required. Consequently, this method is used in ETSI White Paper No. 3, which proposes four layers of interoperability: technical, syntactical, and two types of interoperability: organizational and semantic.

4.2.1 Technical Interoperability

According to Hans van der Veer (2008), it is usually tied to platforms, systems, and hardware/software components that enable machine-to-machine communication. This kind of interoperability usually concentrates on the infrastructure necessary for (communication) protocol execution.

4.2.2 Syntax Interoperability

It usually has something to do with data formats. The messages exchanged via communication protocols need to have a clearly defined syntax and encoding, even if they are only bit-tables. High-level transfer syntaxes, such as ASN, XML, and HTML, can be used to describe the data or content that is transmitted by different protocols.1.

4.2.3 Semantic Interoperability

This level deals with the interpretation of the data and is mostly focused on how a human would understand it as opposed to a computer. Consequently, it shows that people have a common sense of the significance of the content (information) being conveyed at this degree of interoperability.

4.2.4 Organization-to-Organization Interoperability

The term pertains to the ability of an organization to transmit and convey relevant data or information across vastly divergent infrastructures, perhaps spanning several geographical and cultural contexts, even if it employs a variety of different information systems. Organizational interoperability actually depends on successful technical, syntactical, and semantic interoperability. Stabilizing the foundation of the actual world and ensuring technical connectivity across technologies to transfer massive volumes of information are the primary interoperability issues. Complementary difficulties include understanding and processing the information. Accordingly, the following are the primary problems for interoperability and our area of interest, as acknowledged by KUMAR et al. (2019):

(i) Integration of various data sources This illustrates the necessity of being interoperable at the information/occasion level to make joining/total information/ occasion originating from diverse information sources notably less difficult.
(ii) Unique ontological viewpoint on semantic interoperability It suggests possessing a remarkable philosophical viewpoint. An outsider tasked with interpreting the differences between the plans or by philosophical consolidation/mapping can comprehend this. There may also be customs for reaching agreement on a certain viewpoint. P2P Communication

(iii). The third one deals with the requirement for apps to transfer knowledge at higher levels by trading specific knowledge. It is possible to ignore interoperability at lower levels and to implement it at higher levels.

4.3 Integration of AI and ML With IoT

AI has been more important to IOT installations and applications in recent years. The Internet of Things is still in its infancy, mostly because of its processor's restricted capabilities. On the other hand, AI is evolving swiftly, and merging AI with IOT expands IOT's potential. We can now access a vast array of intelligent systems thanks to technology, such as smart homes, smart cities, smart grids, smart industries, smart cars, smart environmental monitoring, smart health, and smart healthcare. However, merely entering the IOT arena is not enough for businesses to fully realize its potential; this is only half the story. The Internet of Things and the quickly developing AI technologies must be combined. Combining different tactics will give rise to the creation of new and fascinating technology. Artificial Intelligence (AI) enables these sorts of intelligent applications, network solutions, and some useful methods to leverage the benefits of the Internet of Things. Combining AI with IOT allows us to have several systems interacting at the individual level through devices such as fitness trackers and Google Home, Amazon's Alexa, Apple's Siri, and Google Assistant (Katare et al., 2018). The effects of IoT and AI/ ML integration on smart facilities, such as smart healthcare, smart agricultural, smart transportation, smart grid, and smart industrial, will be particularly discussed in this section. It focuses on putting into practice efficient learning techniques to offer consumers beneficial services.

A. **Smart Healthcare:** The IoT offers applications in the healthcare industry that enhance patient care. Health-oriented smart systems monitor a patient's vital signs, including heart rate, temperature, weight, and so on. AI is also capable of recognizing health issues and distinguishing between a normal and ill individual. Additionally, it can review and analyze health device data, and if needed, it can notify patients to seek medical advice. Additionally, algorithms like as ANN, random forest, and gradient boosting aid in the monitoring of baby health, the prediction of glucose levels, and the incidence of strokes. Cameras may be used to capture and analyze data on calorie consumption, blood pressure, accidents caused by human activity, and sign languages. This data can then be analyzed using machine learning techniques to determine the best course of action. These objectives also involve the use of CNN-based algorithms, regression analysis, and Naive Bayes (NB) algorithms. Additionally, NB can be utilized to forecast any early-stage cardiac problems. In addition to these algorithms, lung cancer, chronic kidney disease (CKD), and heart-related disorders are also diagnosed using SVM, random forest, and ANN.

B. **Smart Agriculture:** As the need for smart agriculture grows, AI in the fields of crop assessment and prediction has grown significantly, enhancing agricultural yield. Crop age status analysis plays a major role in determining when to harvest, avoiding overfertilization, and reducing production costs. Estimating the crops' category age has shown to be a helpful use of computational intelligence-based picture analysis. Currently, the emphasis is on applying predictive computational intelligence methods to assess the nitrogen condition of wheat crops. The assessment is based on the examination of crop photos taken under various lighting conditions while in the field. In terms of the long-term impact of IoT technologies in agriculture, the main thing to note is that they will lead to major material and resource savings and, consequently, to cost optimization for agricultural businesses. New technologies will also increase yield, which will increase income. Future developments in this area will directly affect how marginal and competitive businesses are. Predictions about the expansion of the global food market by 2050, in addition to the projected 2.3 billion more people (almost 33%) and a rise in their standard of living on Earth, suggest that consumption will inevitably shift in favor of agricultural goods (Righetti et al., 2018). Furthermore, hyperspectral data processing may be carried out by ML algorithms to analyze information pertaining to irrigation. Data pertaining to crop and seed identification and classification, chlorophyll measurements in crop leaves, crop hydration state, and agricultural field preparation and classification may be handled with the aid of NNs, random forests, SVM, ELM, and MLP, as well as DL-based techniques. Nevertheless, a deficiency of high-quality photos and a lack of data amount may be the cause of ML algorithms' declining performance. This calls for the development of effective algorithms that can make accurate predictions with little or no data.

C. **Smart Transportation:** As information and communication technologies (ICTs) have advanced, there have been significant energy savings, a decrease in greenhouse gas (GHG) emissions, improvements in service quality, and several other advantages for both transportation authorities and commuters. Here, we presented a unique distributed multi-agent system-based AI-IoT integrated energy-efficient intelligent transport system for smart cities. An urban area is divided into many regions, each of which has a finite number of zones. Along with the edge computing devices, the ideal number of Road Side Units (RSUs) are deployed at each zone. Every RSU has a MAS installed, which collects massive volumes of data from the many sensors, devices, and infrastructures. The edge computing device processes, analyzes, and makes predictions using the raw data that was gathered from the MAS. MAS will use IoT to assist communicate the anticipated data with the cloud, automobiles, and neighborhood RSUs. By reducing traffic in metropolitan areas, freight vehicles may utilize the projected information to maintain a constant and smooth travel, so reducing energy and greenhouse gas emissions and eventually increasing the vehicles' mileage. With the use of GPS, Wi-Fi, and cellular connectivity, ML may leverage human trajectory data for transportation system modeling and human movement estimates. A case study has been examined in which an attempt has been made to anticipate the short-term traffic flow in both congested and non-packed conditions using regression and SVM models. In addition to estimating the time of arrival at the destination, DL models may be used to learn about human mobility and the many modes of transportation that are available. Identification of parking spaces based on visual information is a difficult problem. Traditional approaches often overlook the opportunity to pinpoint an exact parking spot since they take into account a limited number of unique scenarios. By using an image-based user interface, machine learning models such convolutional neural networks (CNNs), which are artificial neutral networks used for image identification and processing, can be utilized to locate open parking spots (Mahmood et al., 2022; Zantalis et al., 2019).

D. **Smart Grid:** Industry 4.0 refers to the current technological transformation. Any equipment or function that uses network sensors to connect to the internet generates a lot of data. ML is used to analyze data and information regarding smart manufacturing that has been created. Various ML models are applied to smart manufacturing within the context of the smart grid paradigm. Bayesian networks, k-nearest-neighbors, Support Vector Machine (SVM), instance-based learning, ensemble methods, artificial neural networks, multiple logistic regression, decision trees, gradient boosted, self-organizing maps, k-means, bag of words, random forests, locally weighted learning, and additive models are among the most well-known models. The evolution of smart manufacturing is correlated with changes in stakeholder and business model requirements as well as technological breakthroughs. Intelligent energy production systems gather data on energy usage in real time, enhance the precision of decision-making, improve plant performance and dependability, and raise total productivity. If defense mechanisms are absent from smart grid systems, intrusion becomes more susceptible. Therefore, low complexity based self-learning approaches are needed to protect the internet-based infrastructure of the smart grid system from security threats. Machine learning-based intrusion detection systems are capable of defending the smart grid infrastructure from online threats. An overview of a few machine learning algorithm-based intrusion detection methods can be found in. More specifically, these defense systems use rule learning. Among the automated learning techniques used to defend against DoS, over-current protection, and privilege escalation threats include Bayesian learning, decision trees, GAs, and PSOs (Mahmood et al., 2022). Thus, it can be concluded that the development of machine learning algorithms is unavoidable in guaranteeing the effectiveness of IoT systems.

E. **The Intelligent Sector:** A range of technologies, including smart homes, smart robots like the iRobot Roomba, smart systems, and more, are made possible by the integration of AI with IOT. On the other hand, real-world uses of AI and IOT in industries today. By incorporating wireless sensors, information technology, machinery control mechanisms, and computer software for production monitoring, Industrial IoT (IIoT) has embraced the idea of manufacturing. Big data analytics and automated learning approaches need to be integrated with the computer equipment in order to create even smarter industrial processes. Researchers are very interested in machine learning (ML) because it may improve data processing and analysis techniques for interpretation (Kumar et al., 2020). Particularly in the transportation sector, AI supports industrial robotics and automation systems in a variety of day-to-day applications with prominent industrial uses. The IoT is a network of physically linked things that uses software, sensor devices, and network connectivity to gather and share data (Figure 4). Numerous sectors have seen revolutionary changes because to the IoT. The phrase "INDUSTRY 4.0" has been used to describe the state of IoT in which machines and industrial equipment are networked and communicate to make choices using AI (M2M communication). Intelligent goods in the era of Industry 4.0 are capable of responding to specific customers under the ever-expanding customization paradigm. Nevertheless, individuals still frequently use conventional appliances despite their desire for smart items. For there to be a steady supply and demand, changes that are less detrimental to the existing manufacturing lines must be made. Smart appliances initiate more complex systems, such as those found in smart homes and buildings, to form smart cities. Since Industry 4.0, conventional homes are becoming into smart homes (El-Gendy, 2020).

Figure 4. Smart Industry

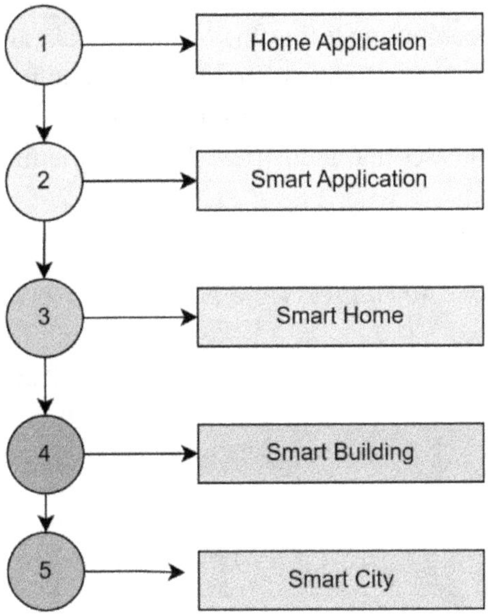

4.4 Edge Computing Advancements

Edge Computing is an emerging method for handling gathered data. The edge computing concept is applied to allow calculations to occur within the nodes in edge network systems. The number of smart devices connected to the Internet is rising as the Internet of Everything (IoE) undergoes a dramatic transformation, bringing with it vast amounts of information that have created problems. for instance, inadequate security and protection in the current computer paradigms. Different kinds of network features are needed for each category in terms of bandwidth, latency, dependability, mobility, data security, and policy management. when the need for Edge computing emerges. As shown in Figure 4, edge computing satisfies the aforementioned application needs by bringing computation closer to the network's edge. when a few benefits are demonstrated. Thus far, edge computing has been used to improve network speed during data transfer while lowering privacy and security threats. As a result, basic tasks in edge computing involve data sources (where data are stored and recorded from other data sources or clients), data processing steps (where the collected data are accessed and used to reveal insight patterns-based machine learning (ML) mechanisms), and showcasing the outcome from the previous phase in visualization tools. Edge computing tasks facilitate communication between IOT applications. the most recent innovations in edge computing technology that are fostering creativity and social and commercial progress. As a result, academic and corporate researchers invested a great deal of time and energy in studying the developments of edge computing technology (Lanka et al., 2021). Edge computing has expanded its uses across a range of industries, including gaming, retail, fleet management, healthcare, and more. It offers services to a number of different sectors, including businesses and industries. A few of the most important uses are listed in Figure 5.

Figure 5: Edge computer Advantages (Aheleroff et al., 2020)

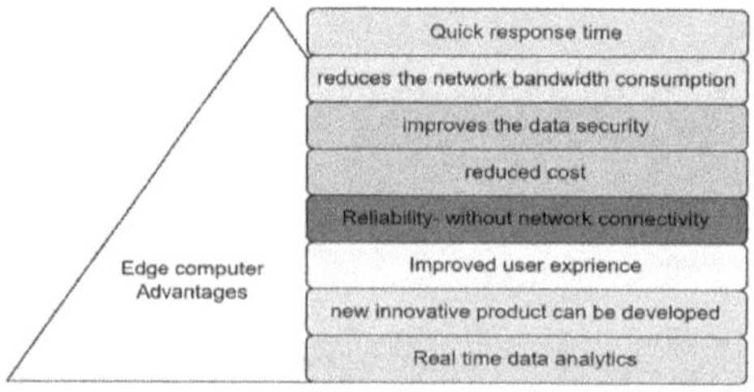

A. **Smart Healthcare:** The development of edge computing has led to the conclusion that real-time, affordable healthcare facilities are becoming ubiquitous in smart health care. Numerous applications are linked to it, including remote electrocardiography monitoring, smart homes, smart grid, smart hospitals, and smart health. The notion of intelligent healthcare has surfaced in several nations. Furthermore, it is extremely important since it is being developed for a variety of illnesses, including pneumonia, COVID-19, diabetes, cancer, and cardiac arrest. For this reason, in IoT-enabled healthcare organizations, the security of IoT devices and related data is crucial. One intriguing design that addresses these processing and computational issues is edge computing. By enhancing the transmission and processing speed of IoT devices inside a healthcare system, it may offer low latency data services. For instance, sensors gather and wirelessly transmit body temperature, heart rate, ECG signals, and other data to gateways for autonomous analysis and notification generation. In order to do this, a fog-based healthcare framework was proposed, which employed fog as an intermediary layer between the cloud and end devices in order to autonomously diagnose patients' heart rates. Privacy and security are significant considerations in healthcare applications. (Khan and others, 2020) The incurred delay between edge computing enabled infrastructure and traditional core-cloud infrastructure was recently analyzed by the authors, who found that edge computing resulted in reduced incurred delay, greater bandwidth, and faster reaction times.

B. **Fleet Administration:** Thus far, the use of cloud, fog, and edge computing approaches into an Intelligent Transportation System (ITS) has enabled the real-time and low-latency transmission and processing of terabytes of data. These data are provided by smart sensors that are networked with IoT devices. They are then transmitted to the cloud-based decision-making system unit, where various issues occur. Real-time tracking, vehicle health monitoring, and route optimization are difficult tasks in logistics and transportation. In essence, this puts a great deal of strain on server and cloud infrastructure. As a result, technologies like edge and fog computing were created in a variety of ways to lessen this overloading on the cloud (Lanka et al., 2021)(Khan et al., 2020).
C. **Smart Cities:** Smart traffic management, waste management, environmental monitoring, and public safety applications are also made possible by IoT and edge computing breakthroughs. Smart cities employ technology and data-driven solutions to boost productivity. The utilization of resources like electricity, water, and transportation systems was maximized in smart cities. Additionally, augmented reality and smart buildings were integrated, and edge computing was made possible. This led to increased operational efficiency and reduced latency with immediate computing and storage resources. A 5G-enabled software defined vehicular network (5G-SDVN) architecture, for instance, has been proposed. It consists of three planes: a data plane for data transmission, a social plane for networking among Vehicular Neighbour Groups (VNGs), and a control plane for using mobile edge computing (Khan et al., 2020).

Figure 6. Evolution of Edge computing (Khan et al., 2020)

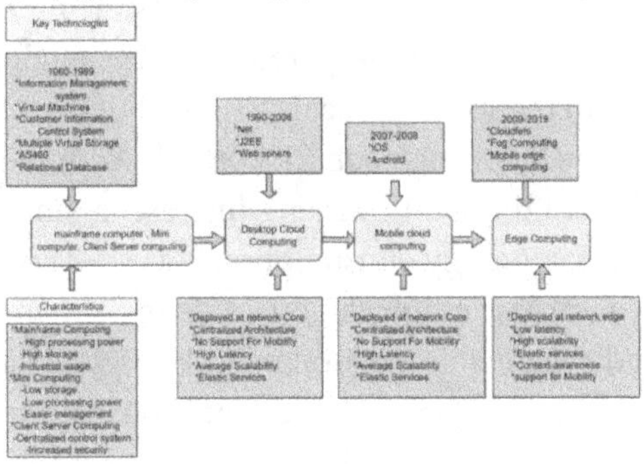

5. REGULATORY AND ETHICAL CONSIDERATIONS

Regulatory and ethical considerations in the context of IoT are referring to the laws, regulations and standards that control the deployment, usage, and operation of IoT devices and systems while keeping in mind to respect individual legal rights, privacy, and society values. As per government declaration of human rights, privacy is a basic right of every citizen. IoT technology has the potential to create many opportunities and to transform society. It is crucial to differentiate between technology's inventive and risky features. One real life instance is the "usage of 360-degree cameras" put in cars for ADAS, which can record from all angles in every second and endanger not only the owner's privacy but also for external users (Aledhari et al., 2022) (Pasricha, 2023). In this regard our main aim is the implementation of ethical considerations using IoT generated data which consists of Data privacy, Transparency, Data security, trust, Interoperability, environmental impact, operational Our primary performance is the integration of devices and communication protocols. Aside from governmental organizations, research and industry are also working on guaranteed initiatives and frameworks so that IoT applications can give ethical norms and that consumers will have trust in them, as they are prominent building blocks for success. Here, we have highlighted the above promising directions that may help facilitate the ethical design of computing systems (Karale, 2021).

5.1 Legal and Regulatory Aspects of IOT Deployment

Worldwide physical object connection is made possible via the IoT. People naturally cooperate with heterogeneous gadgets as science and technology progress, making the world a wiser place overall. The gadgets are transferring so much information that massive quantities of energy are required. Big data analytics and cloud computing are integrated with the IoT to forecast smart device behavior, offer practical business insights, and function as a feedback control. Additionally, the majority of businesses would adjust to the world's rising interest in smart technology, which would accelerate the rise in energy consumption.

Green IoT places a strong emphasis on lowering environmental issues and building an IoT-related sustainable environment. Energy-efficient technology use in the IoT has been shown to either mitigate greenhouse gas emissions or prevent the greenhouse effect in a variety of IoT applications. By introducing low-energy gadgets or electrical appliances, reducing greenhouse gas emissions, using carbon-free materials, and encouraging reusability, Green IoT (GIoT) helps to sustain the climate. Policies pertaining to energy-saving techniques for green data centers, cloud computing, and sensor networks are covered. Furthermore, the various Green IoT approaches are examined, and the adaptable concepts for achieving Green IoT are also covered. Additionally, a number of IoT apps enable users to operate smart home appliances, security cameras, locks, automatic doors, and electrical appliances via an app on a smartphone or other connected device. By integrating technology and offering essential services, smart cities benefit individuals by giving them a more intelligent perspective. IoT becomes greener when smart e-health is used to monitor and track patients in the current situation (Arshad et al., 2017).

5.2 Regulatory Implications of IoT Generated Data

The IoT paradigm is becoming more widely adopted in urban environments due to the ease with which any type of device may be connected to the Internet. The Smart City idea is becoming a reality thanks to numerous IoT applications that provide residents and local authorities with cutting-edge services. This evolution is dependent upon smooth information flow across various systems. Personal and/or sensitive information is included in exchanged data, so it must be handled carefully to prevent security and privacy problems (Righetti et al., 2018). Concurrently, advancements in robotics are facilitating the creation of self-governing entities (such as vehicles, buses, and unmanned aerial vehicles) that eliminate the need for human involvement. Autonomous agents will be essential to the provision of many services in the future metropolitan centers. In addition, autonomous agents—like robots—will grow emotionally intelligent and increasingly mimic humans. As so, they will replace individuals in a variety of municipal functions. Discussions have focused on important ethical and regulatory issues despite the advantages these technologies provide.

IoT, Cloud, and Edge Computing Technologies: Social, Legal, Regulatory, and Ethical Implications is a major source of knowledge on the non-technical impacts of adopting IoT technologies. Several legal, regulatory, and rights-related concerns are brought up by the IoT. In terms of legislation and policy, IoT devices both exacerbate many current problems and introduce brand-new ones. For instance, new challenges arise from the launching of innovative IoT device types in terms of ease of use for those with impairments, all the while maintaining compliance with existing accessible rules and standards. Laws and regulations relevant to consumer protection are among the many technology-neutral legal lenses that are being used to analyze IoT regulatory analysis. Information on privacy and safety, among other things, may be obtained by analyzing the legal ramifications of IoT devices from the perspective of eliminating unfair or misleading practices used against customers. Every now and again, IoT devices give rise to new legal and regulatory requirements as well as civil rights issues that were not there prior to these devices (Attia et al., 2019).

6. CONCLUSION AND FUTURE SCOPE

IoT has been gradually bringing a sea of technological advancements into our everyday lives through a range of technologies and applications, which therefore serves to make our lives easier and more pleasant. IoT finds applications in every sector of the economy, including government, healthcare, industry, transportation, education, mining, and housing. This study covers a wide range of IoT applications. With the help of IoT, human life will become more "smart" and "connected" in the future. There will be many more difficulties in the future, though. We must overcome the obstacles presented by the IoT in order to tackle these issues. Therefore, resolving these issues is necessary for future work.

REFERENCES

Aheleroff, S., Xu, X., Lu, Y., Aristizabal, M., Velásquez, J. P., Joa, B., & Valencia, Y. (2020). IoT-enabled smart appliances under industry 4.0: A case study. *Advanced Engineering Informatics*, *43*, 101043. DOI: 10.1016/j.aei.2020.101043

Al-Fuqaha, A., Guizani, M., Mohammadi, M., Aledhari, M., & Ayyash, M. (2015). Internet of things: A survey on enabling technologies, protocols, and applications. *IEEE Communications Surveys and Tutorials*, *17*(4), 2347–2376. DOI: 10.1109/COMST.2015.2444095

Aledhari, M., Razzak, R., Qolomany, B., Al-Fuqaha, A., & Saeed, F. (2022). Biomedical IoT: Enabling technologies, architectural elements, challenges, and future directions. *IEEE Access : Practical Innovations, Open Solutions*, *10*, 31306–31339. DOI: 10.1109/ACCESS.2022.3159235 PMID: 35441062

Alferidah, D. K., & Jhanjhi, N. Z. (2020). A review on security and privacy issues and challenges in internet of things. *International Journal of Computer Science and Network Security IJCSNS*, *20*(4), 263–286.

Arellanes, D., & Lau, K. K. (2020). Evaluating IoT service composition mechanisms for the scalability of IoT systems. *Future Generation Computer Systems*, *108*, 827–848. DOI: 10.1016/j.future.2020.02.073

Arshad, R., Zahoor, S., Shah, M. A., Wahid, A., & Yu, H. (2017). Green IoT: An investigation on energy saving practices for 2020 and beyond. *IEEE Access : Practical Innovations, Open Solutions*, *5*, 15667–15681. DOI: 10.1109/ACCESS.2017.2686092

Attia, T. M. (2019). Challenges and opportunities in the future applications of IoT technology.

Čolaković, A., & Hadžialić, M. (2018). IoT: A review of enabling technologies, challenges, and open research issues. *Computer Networks*, *144*, 17–39. DOI: 10.1016/j.comnet.2018.07.017

Costa, B., Pires, P. F., & Delicato, F. C. (2016, August). Modeling iot applications with sysml4iot. In 2016 42th Euromicro Conference on Software Engineering and Advanced Applications (SEAA) (pp. 157-164). IEEE. DOI: 10.1109/SEAA.2016.19

Ding, Y., Jin, M., Li, S., & Feng, D. (2021). Smart logistics based on the internet of things technology: An overview. *International Journal of Logistics*, *24*(4), 323–345. DOI: 10.1080/13675567.2020.1757053

Du, K., Chu, J., Sun, Z., Zheng, F., Xia, Y., & Yang, X. (2016). Design and implementation of monitoring system for agricultural environment based on WebGIS with Internet of Things. *Nongye Gongcheng Xuebao (Beijing)*, *32*(4), 171–178.

El-Gendy, S. (2020, December). IoT based AI and its implementations in industries. In 2020 15th International Conference on Computer Engineering and Systems (ICCES) (pp. 1-6). IEEE. DOI: 10.1109/ICCES51560.2020.9334627

Jing, Q., Vasilakos, A. V., Wan, J., Lu, J., & Qiu, D. (2014). Security of the Internet of Things: Perspectives and challenges. *Wireless Networks*, *20*(8), 2481–2501. DOI: 10.1007/s11276-014-0761-7

Karale, A. (2021). The challenges of IoT addressing security, ethics, privacy, and laws. *Internet of Things : Engineering Cyber Physical Human Systems*, *15*, 100420. DOI: 10.1016/j.iot.2021.100420

Katare, G., Padihar, G., & Qureshi, Z. (2018). Challenges in the integration of artificial intelligence and internet of things. *International Journal of System and Software Engineering*, *6*(2), 10–15.

Khan, L. U., Yaqoob, I., Tran, N. H., Kazmi, S. A., Dang, T. N., & Hong, C. S. (2020). Edge-computing-enabled smart cities: A comprehensive survey. *IEEE Internet of Things Journal*, *7*(10), 10200–10232. DOI: 10.1109/JIOT.2020.2987070

Suresh Kumar, V., Kuppusamy, P. G., & Karthikeyan, A. (2019). IOT-Interoperability and dynamic workflow composition–challenges and perspectives [articol].

Kumar, A., Saha, R., Alazab, M., & Kumar, G. (2020). A lightweight signcryption method for perception layer in Internet-of-Things. *Journal of Information Security and Applications*, *55*, 102662. DOI: 10.1016/j.jisa.2020.102662

Lanka, S., Win, T. A., & Eshan, S. (2021, December). A review on Edge computing and 5G in IOT: Architecture & Applications. In 2021 5th International Conference on Electronics, Communication and Aerospace Technology (ICECA) (pp. 532-536). IEEE.

Li, J., Herdem, M. S., Nathwani, J., & Wen, J. Z. (2023). Methods and applications for Artificial Intelligence, Big Data, Internet of Things, and Blockchain in smart energy management. *Energy and AI*, *11*, 100208. DOI: 10.1016/j.egyai.2022.100208

Mahmood, M. R., Matin, M. A., Sarigiannidis, P., & Goudos, S. K. (2022). A comprehensive review on artificial intelligence/machine learning algorithms for empowering the future IoT toward 6G era. *IEEE Access : Practical Innovations, Open Solutions*, *10*, 87535–87562. DOI: 10.1109/ACCESS.2022.3199689

Meneghello, F., Calore, M., Zucchetto, D., Polese, M., & Zanella, A. (2019). IoT: Internet of threats? A survey of practical security vulnerabilities in real IoT devices. *IEEE Internet of Things Journal*, *6*(5), 8182–8201. DOI: 10.1109/JIOT.2019.2935189

Mishra, M., Lourenço, P. B., & Ramana, G. V. (2022). Structural health monitoring of civil engineering structures by using the internet of things: A review. *Journal of Building Engineering*, *48*, 103954. DOI: 10.1016/j.jobe.2021.103954

Pasricha, S. (2023). Ethics for Digital Medicine: A Path for Ethical Emerging Medical IoT Design. *Computer*, *56*(7), 32–40. DOI: 10.1109/MC.2022.3216984

Ren, J., Pan, Y., Goscinski, A., & Beyah, R. A. (2018). Edge computing for the internet of things. *IEEE Network*, *32*(1), 6–7. DOI: 10.1109/MNET.2018.8270624

Righetti, F., Vallati, C., & Anastasi, G. (2018, June). IoT applications in smart cities: A perspective into social and ethical issues. In *2018 IEEE International Conference on Smart Computing (SMARTCOMP)* (pp. 387-392). IEEE. DOI: 10.1109/SMARTCOMP.2018.00034

Sadiku, M. N. O., Musa, S. M., & Nelatury, S. R. (2016). Internet of things: An introduction. *International Journal of Engineering Research and Advanced Technology*, *2*(3), 39–43.

Shah, S. H., & Yaqoob, I. (2016). A survey: IoT technologies, applications and challenges. 2016 IEEE Smart Energy Grid Engineering (SEGE), 381-385.

Tiwary, A., Mahato, M., Chidar, A., Chandrol, M. K., Shrivastava, M., & Tripathi, M. (2018). IoT: Research, architectures and applications. *International Journal on Future Revolution in Computer Science & Communication Engineering*, *4*(3), 23–27.

Vaigandla, K. K., Karne, R. K., & Rao, A. S. (2021). A Study on IoT Technologies, Standards and Protocols. IBMRD's. *Journal of Management Research*, *10*(2), 7–14.

Zantalis, F., Koulouras, G., Karabetsos, S., & Kandris, D. (2019). A review of machine learning and IoT in smart transportation. *Future Internet*, *11*(4), 94. DOI: 10.3390/fi11040094

Chapter 4

IoT Protocols for Healthcare for Enhancing Connectivity and Efficiency in Medical Systems

Agha Urfi Mirza
https://orcid.org/0009-0009-4897-801X
Department of Information technology, University of Technology and Applied Sciences, Al Mussanah, Oman

Rafath Samrin
https://orcid.org/0009-0002-0336-972X
Department of Computer Science, King Khalid University, Asir-Abha, Saudi Arabia

J. Venkatesh
https://orcid.org/0000-0002-4259-130X
Department of Computer Science and Engineering, Chennai Institute of Technology, India

R. Arulkumar
Department of Information Technology, Jeppiaar Institute of Technology, Chennai, India

K. Selvakumarasamy
Saveetha School of Engineering, Saveetha Institute of Medical and Technical Sciences, SIMATS, Chennai, India

P. S. V. Srinivasa Rao
Department of Computer Science and Engineering, Joginpally B.R. Engineering College, Hyderabad, India

ABSTRACT

The adoption of IoT protocols in health systems is a whole new way of increasing its interface and operability. This chapter highlights the potential of various IoT protocols concerning their roles in the medical system's reinvention, basing the

DOI: 10.4018/979-8-3693-5448-3.ch004

analysis on real-time monitoring of patients, data management, and interoperability. These include the main protocols, namely MQTT (Message Queuing Telemetry Transport), CoAP (Constrained Application Protocol), and HTTP (Hypertext Transfer Protocol), which illustrate the advantages of efficiency in data transmission, latency, and proper utilization of network resources. The next section looks at some of the challenges that this implementation poses in healthcare settings: data security, privacy concerns, and scaling. It provides insight, through case studies and current research, into how IoT protocols are enhancing the future of healthcare by adopting a holistic view in improving patient outcomes and streamlining medical processes.

INTRODUCTION

IoT is fundamentally changing many industries today, being one of the most promising fields in healthcare. IoT is based on a network of connected devices through communication with each other and centralized systems via the internet. In health, IoT describes a wide range of technologies and various applications that are targeted at patient care, efficiency in medical processes, and increasing overall efficiency of the system. IoT devices in healthcare are deployed to collect, send, and analyze real-time data over the patient. Generally, IoT sensors include wearables, smart medical equipment, and remote monitoring systems. Wearables like activity trackers and smartwatches track heart rate, blood pressure, and glucose levels for providing input to continuous health monitoring. Because smart medical equipment, including infusion pumps and imaging devices, can be monitored and controlled remotely, it ensures optimal performance and reduced downtown. IoT in health care provides great advantages (Ahad et al., 2020).

Advanced patient monitoring is one of the significant benefits. IoT devices enable continuous and real-time data and thus are helpful to health care providers in maintaining more complete and timely information regarding the health of the patients. This information, therefore, can be used to identify potential problems early and thus intervene earlier with much better outcomes for the patients. A good example is that the wearables in glucose monitoring will be able to alert abnormal levels of blood sugar through self-monitoring by the patient and physicians for timely adjustments in treatment. IoT also enables remote patient management, which is particularly useful in the management of chronic conditions. It facilitates remote monitoring systems, where patients are kept at home while continuing the monitoring process by health professionals. It reduces frequent visits to hospitals, reducing health costs and increasing patient comfort and convenience. Integrating IoT devices with telemedicine platforms even facilitates online consultation and remote diagnosis, thus extending healthcare services for which distance is a constraint (Ding et al., 2020).

Efficiency in medical processes is another critical benefit of IoT in healthcare. It can also free up nurses to perform less routine tasks, which include administering medication, maintaining equipment, and entering data into a computer. Smart infusion pumps, for example, may automatically adjust medication dosages based on real-time patient data, reducing the risk of human error. Besides, IoT devices can enhance inventory management by monitoring usage and supplies of medical equipment and pharmaceuticals to ensure such resources are available when needed. Interoperability, in this context, therefore, becomes inherently a typical character of IoT systems within healthcare. In IoT, it links various devices and systems to enable easy sharing and integration of data across heterogeneous platforms. Such interconnectedness brings forth an integrated approach toward patient care, where information sourced from various resources including EHRs, laboratory results, and wearable devices are aggregated together and analyzed to derive comprehensive insights into patient health (Pradhan et al., 2021).

Meanwhile, IoT application to healthcare is not devoid of challenges. First and foremost, data security and privacy represent major concerns, since health information is so sensitive that it needs to be well guarded from unauthorized access and breaches. Furthermore, integration of a variety of IoT devices and systems requires standardization and compatibility in order for them to operate seamlessly and allow for a smooth flow of data. This means scalability: the ability of healthcare organizations to scale up and handle a fast-growing number of connected devices and their data output. Even considering such challenges, the potential benefits of IoT for healthcare are enormous. By improving patient monitoring, enhancing remote care, and facilitating resource operational efficiency, IoT technologies have the potential to transform health care delivery. This means that even greater innovations, with IoT technologies continuing to evolve, advances in data analytics, and AI, are yet to come, thus enabling more personalized, efficient, and value-driven solutions for healthcare (Saba et al., 2020).

Connectivity and efficiency are becoming increasingly integral parts of the evolving healthcare landscape where better patient outcomes and optimized healthcare delivery are concerned. The integration of IoT technologies in medical systems underlines the importance of such factors since they can turn health care from reactive to proactive and from fragmented to cohesive. Connectivity may be defined as interoperability or non-interoperability that is characterized by seamless flow of data and information across devices, systems, and stakeholders within the healthcare ecosystem. In turn, all those devices, EHRs, and other systems would be connected, ensuring timely data exchange in IoT. Such interoperability is necessary for a number of reasons. First of all, interoperability allows much better patient care through timely observation and timely interference(Akkaş et al., 2020). For example, IoE-based wearables and sensors ensure real-time information regarding the patient's vital signs. Such information is conveyed to the caregivers just at the time it is captured. This allows for immediate analysis and action in the case of any anomaly, reducing the possibility of complications and improving outcomes. In patients with chronic conditions, like diabetes and heart-related diseases, continuous monitoring through IoT devices ensures that any deviation from normal health patterns is picked up early enough to allow medical intervention in time.

Second, increased interoperability enhances coordination among providers. Most traditional care settings have information compartmentalized into small silos, which results in fragmented care and may be subject to communication gaps. IoT technology thus bridges the gaps in integration of various data streams-from wearable devices to EHRs and diagnostic tools-into a single unified system (Qadri et al., 2020) . This holistic approach toward a patient's health provides more informed decision-making and collaborative care. For example, the data observed by a remote monitoring device of a patient can help a primary care physician coordinate with specialists for personalized treatment. Healthcare efficiency refers to the most effective use of resources, time, and processes that would lead to the best achievable outcomes.

The adoption of various IoT technologies contributes greatly to increasing this efficiency. One of the areas in which operational efficiencies hold great prominence. IoT devices can automate routine tasks through seamless integration into processes that earlier required a human interface and took a lot of time. For instance, smart infusion pumps can regulate dosages of medications automatically according to the real-time data inputs, thus reducing manual input and minimizing the possibility of human error. Similarly, the inventory management system enabled by IoT will track the consumption of medical consumables and notify the healthcare provider in advance when the level of inventory falls below the reorder point for them to have the consumables available at all times (Qadri et al., 2020).

Efficiency is also fueled with remote care and telemedicine. Through the employment of IoT, patients' monitoring and virtual consultation have been possible, hence reducing the numbers of frequent visits to the facilities, saving both patients and health care providers' time in the process. Besides convenience for the patient, it maximizes health care provision in that the practitioner is able to attend to more population. Remote care has been particularly useful in the management of chronic conditions and postoperative follow-ups where complications are prevented through constant monitoring, therefore reducing readmission to the hospital. Another aspect of efficiency is data management. IoT devices generate a large amount of data, and their analyses may provide much-valuable insight into patient health and system performance (Kelly et al., 2020). Advanced data analytics facilitate evidence-based decision-making to portray patterns and predict trends. This may involve analyzing various data from IoT devices to predict patient deterioration and thus optimize treatment plans in order to improve overall care quality and resource utilization.

Therefore, connectivity and efficiency in healthcare are very important. This, in turn, allows easy connectivity to enable seamless data exchange that enables real-time patient management and enhances coordination among the providers. Automation, remote care, and insights powered by data bring along efficiency, better resource utilization, and, ultimately, better outcomes for the patients. Continuing onward with IoT technologies in healthcare, the ruling factors of connectivity and efficiency will play an important role in driving transformational changes, yielding in an effective patient-centric healthcare system (Albahri et al., 2021).

Background

IoT protocols in healthcare provide a whole new dimension to the medical field through real-time monitoring, personalized treatment, and better patient outcomes. IoT protocols such as MQTT-Message Queuing Telemetry Transport, CoAP-Constrained Application Protocol, and BLE-Bluetooth Low Energy have been quite instrumental in allowing perfect communication across medical devices, sensors, and healthcare systems. These protocols guarantee secure data exchange with low latency, important in various applications: remote patient monitoring, wearables in health, smart medical equipment(Malathi et al., 2024; Upadhyaya et al., 2024; Venkateswaran et al., 2023).

IoT protocols in healthcare need to address issues such as data security, interoperability, and scalability. For instance, the security and privacy of data transferred across gadgets have to be ensured, since this gadget-related information is about health care and is very sensitive. Interoperability will also be an important factor because gadgets and systems provided from different manufacturers should work seamlessly together. Scalability will be needed to support an increasing number of IoT networks in a healthcare setting.

In this respect, the future possibility in this domain could also deal with the design of IoT devices powered with AI. Such devices can analyze all the important data locally at the 'edge' before a relay transmission to cloud systems for bigger analysis. It cuts latency to a minimum and speeds decision-making in critical healthcare situations. Practical Applications: The real-world examples of IoT in healthcare involve a remote monitoring system in chronic disease management where the devices track all the important parameters, such as vital signs, and send alerts to the health service providers in case an abnormality is detected. Examples include smart insulin pens that connect to apps for the management of diabetes, further illustrating the practical benefits brought about by the integration of IoT in real-life applications.

Scope

This chapter investigates the integration of IoT protocols in healthcare, focusing on enhancements in both connectivity and operational efficiency. The main IoT protocols considered for research in the review are MQTT, CoAP, and HTTP. These are analyzed in light of their respective roles regarding real-time patient monitoring, data management, and interoperability of a system. The paper has a scope that contemplates how such protocols can make improvements in patient outcomes and reduce burdens on the administrative aspect of clinical work. It also discusses challenges on data security, privacy, and scalability and therefore provides an overview of the present research and emerging trends in IoT applications for healthcare. Looking both at successful case studies and discussing future innovations, the aim of this chapter is to clearly provide insight into how IoT protocols can transform the general healthcare system and enable its evolution toward efficiency and effectiveness in patient care.

The objectives of this chapter are to:

- Study the main protocols in IoT, namely MQTT, CoAP, and HTTP, and their contribution to healthcare development in terms of connectivity and efficiency.
- Assess how these protocols improve real-time monitoring and data management while ensuring systems are interoperable.
- Indicate the various challenges arising in applying IoT in healthcare with respect to data security, privacy, and scalability; discuss the same.
- Also, present ongoing research work, case studies, and emerging trends in IoT for healthcare applications.
- Provide insight into how IoT can reshape healthcare delivery toward better patient care.

IOT PROTOCOLS FOR HEALTHCARE

Figure 1. Types of IOT Protocols for Healthcare

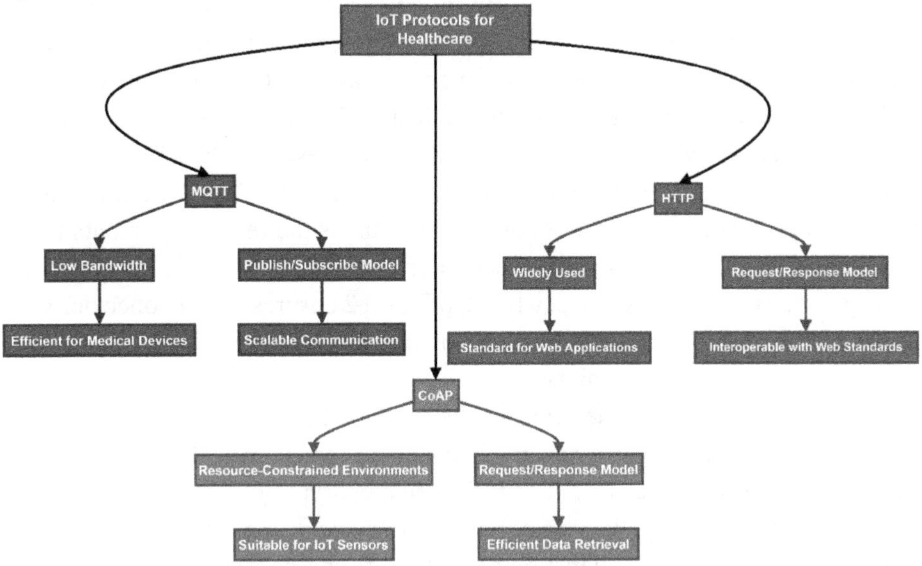

The flowchart in Figure 1 illustrates the relationships between various IoT protocols used in healthcare.

MQTT (Message Queuing Telemetry Transport)

The full form of MQTT is Message Queuing Telemetry Transport, a lightweight messaging protocol efficient in reliable communication with limited bandwidth and high latency. Quite essentially, it's appropriate for health-based IoT applications since it has an extremely low overhead, can manage a huge volume of data, and keeps power consumption extremely low. MQTT operates on a publish-subscribe model where two main components are concerned: publishers and subscribers. The Publishers send messages to a central broker. The broker then broadcasts these messages to all the subscribed clients. This decoupling of publishers and subscribers further enhances scalability and flexibility for the protocol. MQTT can be applied in medical environments to deliver, in real time, data from various medical devices-such as wearable sensors and remote monitoring devices-to healthcare providers(Bhuiyan et al., 2021).

Key advantages of MQTT rest in the efficiency that it has with regard to intermittent and unreliable network connectivity, which is particularly apt for healthcare environments where devices may be spread over several disseminated locations and whose connectivity can range from very variable. MQTT is lightweight and sends data efficiently to make sure critical health data is delivered promptly and accurately, even on challenging network conditions. Besides, MQTT encompasses QoS levels that are an assurance of delivering messages at different extents. This is the most important aspect when it comes to health, as the reliability in data transmission may be everything in patient care. For instance, QoS level 1 ensures that a message is delivered at least once, while QoS level 2 ensures exactly once delivery. This flexibility allows healthcare systems to make a good balance between message reliability and network constraints.

Another important consideration in the use of MQTT in healthcare involves security. The protocol supports several mechanisms for securing data transmission, which involve TLS, or Transport Layer Security, encryption. This is important for the security of sensitive patient data, but also to achieve some regulatory requirements, such as HIPAA (Health Insurance Portability and Accountability Act), in the United States. Finally, MQTT provides quite significant advantages to healthcare IoT applications: efficiency, scalability, and reliability. Its light structure and multiple QoS level support render it specifically suitable for real-time health data transport. Meanwhile, the security features of MQTT provide well-protected patient information. Through MQTT, health systems can facilitate better connectivity for enabling quality patient care by much more effective and efficient data communication.

CoAP (Constrained Application Protocol)

CoAP fills in the market for constrained protocol environments. In general, CoAP is much better suited to IoT health applications. Since CoAP was the result of addressing limitations on the part of constrained devices and networks, its contribution to efficient communication and interoperability in scenarios with reduced resources-such as bandwidth, processing power, and energy-actually is quite invaluable. CoAP is based on UDP, not TCP, which also makes it lighter when compared to other more complicated protocols. This makes CoAP even more effective within a context where the minimum amount of data is forwarded, and energy saving is crucial. In medical use cases, it is very essential that wearable devices or medical sensors or, in general, any IoT device be able to support lightweight and graceful communication methods(Bhuiyan et al., 2021).

CoAP allows the use of RESTful interactions, in a similar way to HTTP, dedicated to constrained devices. As such, CoAP makes use of methods like GET, POST, PUT, and DELETE for manipulating resources in a very familiar way. This can be utilized to request and update information from healthcare devices in an efficient manner. For example, a wearable health monitor could use CoAP to ask a server for instant data or publish patient health metrics to a central system. CoAP also includes resource discovery and registration mechanisms that allow devices to find and interact with each other in a dynamic manner. This becomes really useful in healthcare environments where new devices might join or leave the network relatively frequently. CoAP supports device capability registration and discovery from other resources, using resource directory service (Boopathi, 2023b; Boopathi & Khang, 2023; Karthik et al., 2023).

Thus, it is easy to integrate and communicate among heterogeneous medical IoT devices. Another important benefit of CoAP is the support for caching and asynchronous communication. CoAP uses response caching to reduce redundant data transfer and increase efficiency. It also enables devices to asynchronously subscribe to updates of particular resources, such as a patient's vital signs, without having to continuously poll for changes, thus getting notified of the same. This is very critical for an application that requires real-time health monitoring and timely updates. CoAP has been designed with security in mind, considering resource-constrained networks and devices, which is quite important in healthcare applications where sensitive patient data will be involved. CoAP natively supports DTLS, a protocol which adds encryption and authentication to CoAP messages. Thus, it provides CoAP with security during the transmission of data. DTLS will ensure that patient information remains confidential and is transmitted securely, addressing the concerns on privacy and regulatory matters.

CoAP has proven to be quite resourceful and adaptable in healthcare IoT applications regarding the management of limited resources, RESTful interactions, resource discovery, and asynchronous communication. Moreover, CoAP, being lightweight in nature with strong security features, makes it highly applicable in various environments for connecting and managing several medical devices in a resource-constrained environment, thereby enhancing the efficacy of the delivery of healthcare.

HTTP (Hypertext Transfer Protocol)

HTTP is one of the first protocols from the World Wide Web, establishing standard communication across clients and servers to exchange information with request and response messages over a network. While originally developed for use in web browsing, HTTP has become an important building block for many applications that support health and, for this reason, a major way to realize interoperability and health data exchange across different systems and devices (Reddy et al., 2023).

Functionality and Application in Healthcare: HTTP functions on top of TCP, which is Transmission Control Protocol that ensures reliability and ordered delivery of data packets. This, in fact, is incredibly essential in healthcare, as issues of data accuracy and total data set delivery are paramount. HTTP is finding applications in health care to permit information exchange among different systems such as Electronic Health Records, patient management systems, and medical devices. For example, the EHR system of a hospital will use HTTP to message the laboratory information system in order to retrieve test results with an intent in updating patient records. One important aspect of HTTP is the capability to work with RESTful or Representational State Transfer APIs, which is very common in the use of interactions on web services. RESTful APIs provide the ability for healthcare applications to read patient data, update medical records, and even submit diagnostic information, all with a simple use of standard HTTP methods such as GET, POST, PUT, and DELETE. This fact makes HTTP a very versatile protocol in integrating dissimilar healthcare systems, which paves the way for smooth data flow across platforms.

Benefits to Healthcare: The easy spread and compatibility of HTTP with other technologies make it a natural fit in healthcare applications that need robust and standardized protocols for communication. Its ease and ubiquity ensure that health systems and devices can easily interface with the existing web infrastructure. Besides, HTTP enables multiple data formats, including those in JSON (JavaScript Object Notation) and XML (Extensible Markup Language), which are commonly used in structuring healthcare data in order to make it possible to interoperate across systems. This protocol is stateless; it means that each individual request is independent of the previous one, which simplifies the interaction by minimizing the complexity of managing communications sessions. Due to such a benefit, statelessness finds application where the system has to conduct a large number of independent transactions - say, pull up patient information or make a record of a medical procedure (Kavitha et al., 2023; Reddy et al., 2023; Satav et al., 2023).

Challenges and Considerations: Despite its advantages, HTTP also brings forth some challenges when adopted within healthcare setups. Among the key concerns is that of data security. While HTTP has been designed in such a way that it does not provide encryption, it is most often practically deployed to be implemented in conjunction with HTTPS, which uses TLS to encrypt data in transit. HTTPS ensures the maintenance of both privacy and protection of sensitive patient information from unauthorized access during transmission in compliance with regulations such as HIPAA. Another aspect is the overhead associated with HTTP's request and response model. Although HTTP is efficient in most applications, the verbosity of the protocol is likely to cause increased utilization of bandwidth and increased latency, especially under large data volume conditions or when network resources are limited. This may highlight concern in health systems due to the expected promptness of deliveries; for example, in remote patient monitoring or telemedicine.

Integration with Internet of Things: HTTP also has its application fused with Internet of Things protocols for enhanced connectivity in healthcare. In such a care scenario, IoT devices might use HTTP to push the data to a cloud-based server or better delineate with the web-based application for analysis and visualization of data. This way, it leverages the already-installed infrastructure of HTTP and derives the real-time ability of IoT technologies. In other words, HTTP is a key protocol in health for communication, supporting the interoperability of information within and across different systems and applications. RESTful APIs and support for various data formats, along with wide compatibility with web technologies, makes it a handy tool for managing information. Although a few hurdles remain, notably data security and general overhead, adequate technology available today to leverage HTTPS and complementary technologies will handle these challenges, whereas HTTP will continue to play an enabling and critical role in achieving efficient and secure health information exchange.

IMPACT OF IOT PROTOCOLS

Real-Time Patient Monitoring

The integration of Internet of Things protocols in healthcare has revolutionized how real-time patient monitoring is carried out, enhanced quality of care, and improved patient outcomes. IoT protocols like MQTT, CoAP, and HTTP enable flawless communication of medical devices with healthcare systems for continuous and accurate tracking of patient health metrics (Boopathi, 2023b). The figure 2 illustrates the integration of IoT protocols into real-time patient monitoring.

Figure 2. integration of IoT protocols into real-time patient monitoring

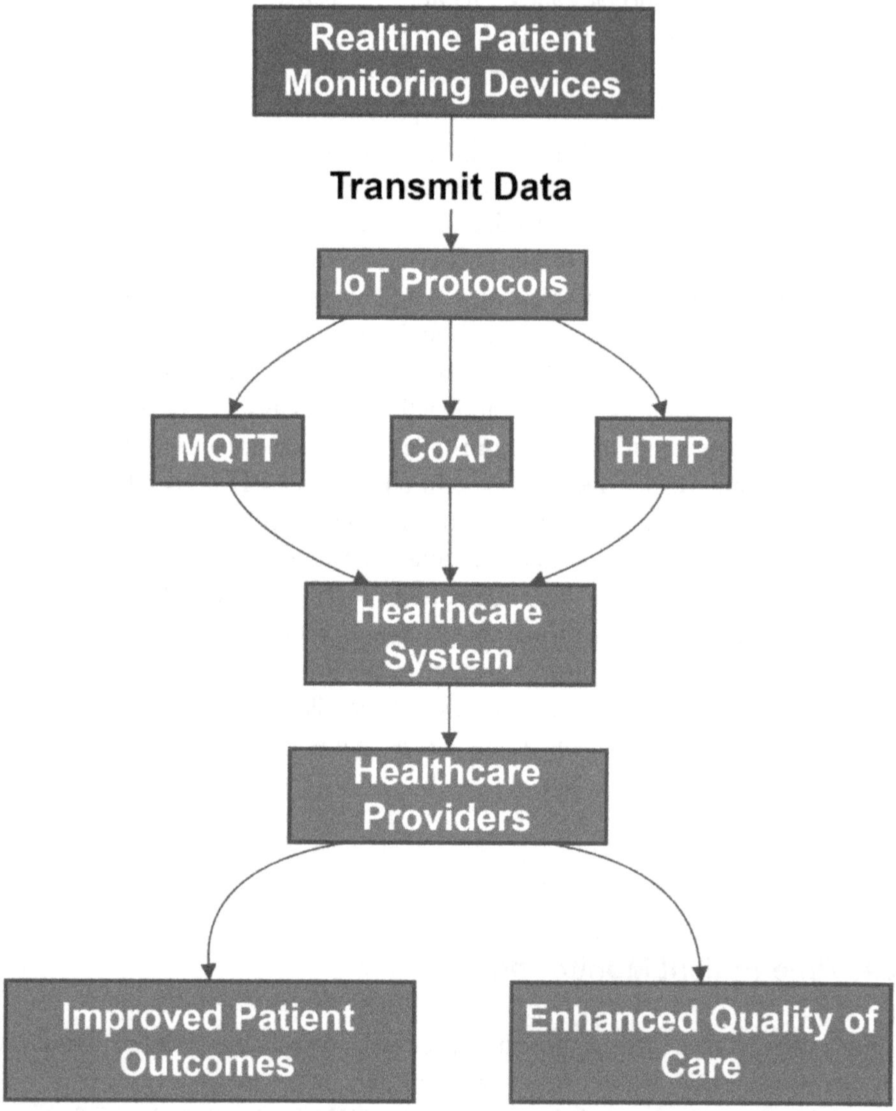

The major effect due to the IoT protocols on real-time patient monitoring will be improvement in data gathering and preciseness. Medical devices with IoT will always collect data regarding patients' vital signs, heart rate, blood pressure, glucose level with the help of wearables, glucose monitors, and cardiac sensors. MQTT and CoAP, as protocols, manage this data efficiently through lightweight, dependable methods of communication that reduce latency without sacrificing integrity. Such continuous collection of data will keep health professionals updated on the condition of a patient to make more accurate and timely interventions possible.

Early Detection and Preventive Care: IoT protocols are very important in early detection and preventive care since these enable real-time alerts and notifications. For instance, a patient with any chronic condition, like diabetes, can wear a monitor of glucose levels that transmit data to the central system using MQTT. In this case, in the event that the glucose levels go beyond the normal range, the system will immediately send alerts to the patient and the health professional for prompt action in changing the treatment or lifestyles. This proactive approach helps in managing conditions before they spiral into serious health concerns, thus reducing the number of emergencies and hospitalizations.

Improved Patient Participation and Adherence: IoT protocols enable real-time monitoring, hence offering better patient participation and compliance in treatment. Equipped with wearable devices, patients can monitor health metrics themselves and receive feedback on their performance regarding medication or exercise routines. For instance, protocols like CoAP support the devices in issuing notifications and reminders to the patients through asynchronous communication based on real-time data. In this manner, the patients are constantly informed about their health status and encouraged to be actively involved in managing their conditions.

Improved coordination and communication: The integration of protocols in IoT increases the coordination and communication within the providers themselves. For instance, aggregated data from multiple devices can be forwarded to a number of different platforms via HTTP-based RESTful APIs. Such interoperability provides various healthcare professionals with a general view of the patient's health condition, therefore allowing them to make more informed decisions and supporting collaborative care. In an emergency, real-time access to the data drives timely, quick, and coordinated responses that may result in positive outcomes based on improved patient outcomes and time to intervention.

IoT protocols for real-time patient monitoring do have their own challenges though there are certain advantages. Data security and confidentiality remain the major concerns because the sensitive health information needs protection from unauthorized access. MQTT and CoAP protocols bear securities like encryption and authentication; however, healthcare organizations should implement some extra measures that help to provide compliances related to HIPAA. Another challenge relates to how to manage and handle large volumes of data from IoT devices. Healthcare systems need to be more prepared to handle and analyze data in order to obtain actionable insights. Advanced analytics and machine learning algorithms will help process and interpret the data, though their implementation requires strong infrastructure and integration with existing healthcare systems.

In the future, IoT protocols and technologies will enhance the real-time monitoring of patients. Upcoming protocols, with enhanced security measures, could solve the challenges at present, while data analytics innovations will yield more sophisticated insights into patient health. The more the IoT technology evolves, the greater its impact on real-time monitoring and consequent improvement in patient care and healthcare delivery.

All in all, IoT protocols have transformed real-time patient monitoring by continuously collecting data, early detection of anomalies, and better engagement of patients. Though challenges are there, the development in IoT technologies promises to push further the effectiveness and efficiency of real-time monitoring in healthcare.

Data Transmission Efficiency

The implementation of IoT technology generally works effectively; however, the efficiency of data transfer is considered a vital factor in IoT technologies, particularly in the healthcare sector where appropriate and timely data transfer plays a significant role in patient care and system operations. Protocols in IoT ensure that effective optimization of data transmission can occur in order for the information to be delivered timely and reliably while managing constraints of network and device resources. IoT protocols are therefore instrumental in ensuring the delivery of services as required(Malathi et al., 2024).

Protocol-Specific Efficiency

a) MQTT is an abbreviation of Message Queuing Telemetry Transport. MQTT was developed against the backdrop of two important considerations: low-bandwidth, high-latency environments. In this regard, it is inherently very efficient in IoT applications since its lightweight message structure and the publish-subscribe model help reduce data transmission to a minimum. MQTT, with its three QoS levels, extends efficiency in transmission by providing options for different assurances on message delivery. QoS level 0 provides at most once delivery of messages; level 1 assures the delivery of messages at least once, whereas level 2 in the hierarchy delivers messages exactly once. In this manner, health services could use a range of reliability versus constraints on their network to enhance the validity of data transferred according to the criticality level of the information being exchanged.
b) CoAP: CoAP has, therefore, been optimized for these constrained environments and runs over UDP, which has less overhead as compared to TCP used by HTTP. The compact message format and binary encoding supported by CoAP

minimize the amount of data to be transmitted, hence making it more efficient. Apart from that, CoAP allows caching and may be used in asynchronous mode of communication. This reduces redundant data transfers and allows handling resource changes in an efficient way. Particularly for the use case of healthcare IoT devices dealing with real-time data, minimum latency, and energy loss will be important.

c) HTTP stands for Hypertext Transfer Protocol. Although it contains much more wordy messages and is also running atop TCP, which brings along even more overhead, it still maintains extreme support and integration with web services. However, the efficiency in data transmission for HTTP can be done in an even better manner by optimizations such as compression (for example, GZIP) and caching strategies. Such optimizations may reduce the overhead and enhance general efficiency in cases requiring big data exchanges for applications in healthcare, such as medical imaging and large health records. HTTPS adds encryption, which is quite necessary in protecting the data but results in additional overhead compared to HTTP. Sometimes, though, security trade-offs are greater than those in efficiency, mainly in health care where the protection of data is very significant.

Impact on Healthcare

IoT protocols, therefore, have very huge implications for efficiency in data transmission with regard to healthcare. Efficient data transmission directly impacts real-time health monitoring systems, remote patient management, and telemedicine applications. For example, efficient protocols ensure that vital signs, such as heart rate or glucose levels, are transmitted with minimum latency; this therefore enables health providers to respond swiftly to any abnormality. This becomes of the essence in conditions that have to do with follow-ups, such as chronic diseases or post-surgical care. Increased efficiency in data transmission also leads to efficiency in the general operation of health systems. The time utilized in transmitting data is reduced; bandwidth utilization is minimized with cost savings, hence improving system performance. Effective protocols can help healthcare organizations process a voluminous amount of data from several IoT devices without imposing significant overhead on the networks, thus going overboard in terms of cost(Satav et al., 2023; Upadhyaya et al., 2024).

Conclusion: IoT protocols are highly desirous in the establishment of efficiency in the transmission of data in healthcare. Overhead minimization, optimization of data formats, and the balancing of reliability against speed find a representation through protocols such as MQTT, CoAP, and HTTP. These protocols facilitate the establishment of real-time data communication in healthcare to enhance patient care and elevate operational efficiency.

Optimization of Network Resources

The integration of IoT protocols into health further enhances the optimization of network resources, ensuring the efficient use of network infrastructure for the better performance of medical systems. MQTT, CoAP, and HTTP are some of the protocols in IoT, which contribute differently toward the optimization of network resources by resolving bandwidth, latency, and device management issues.

Efficient Bandwidth Utilization: This is one of the major impacts that the protocols from IoT have on network resource optimization. For instance, MQTT is efficiently designed with least overhead; therefore, small-sized and compact-sized messages are sent that reduce the amount of data transmitted over the network. Its event-publish-subscribe model ensures that only relevant data is transmitted to interested subscribers to minimize irrelevant traffic. This is particularly useful in healthcare, where most devices, ranging from wearable sensors to remote monitoring devices, generate a great amount of data. MQTT minimizes network congestion and bandwidth consumption by reducing redundant data that travels across the network. Similarly, CoAP targets constrained networks and devices, and it runs over UDP instead of TCP. Since UDP has less overhead than TCP, it means that CoAP will be able to carry out data transmission with much more efficiency, which is really important for these devices, usually used in IoT, operating in bandwidth-constrained environments. Resource discovery and caching in CoAP enhance efficiency in bandwidth even more. In this respect, for example, caching frequently used data reduces the need for CoAP to transfer data over and over again, thus saving bandwidth and generally improving network performance.

Latency: Generally speaking, latency or delay in transmitting packets has been considered one of the most critical factors in network resource optimization. Thus, most IoT protocols such as CoAP and MQTT are designed in such a way that most of this latency is minimized. Minimizing latency becomes crucial in many healthcare-based applications wherein real-time data involvement is concerned. Its lightweight design and native support for asynchronous communication enable CoAP devices to receive updates in time without the need to continuously poll. This cuts down on data transmission time and ensures that healthcare providers get the information on time, which may be critical for certain activities like remote patient monitoring or emergency response. Also, MQTT helps in reducing latency due to its mechanism of message delivery. Its publish-subscribe model guarantees fast message delivery to subscribed clients. This real-time data delivery becomes critical in cases where timely information may create the difference between life and death, either through vital sign monitoring or management of medical alarms.

Device Management and Scalability: Scalability and efficient device management are the most important areas in which large-scale healthcare IoT deployments improve network resources. MQTT is a lightweight protocol, supports all three levels of QoS, and therefore can easily handle a large number of devices over a network. Due to its efficient message handling and meager resource requirements, CoAP easily supports an increasing number of connected devices without degradation. CoAP efficiently offers device management using its Resource Directory service, which allows the devices to dynamically discover and interact with other devices. This, in turn, is one of the most necessary capabilities needed for managing large networks of medical devices with seamless communication and integration during network growth.

Data Security and Compliance: While optimizing network resources, IoT protocols also address data security and compliance issues. Both MQTT and CoAP support encryption and secure data transmission, including TLS for MQTT and DTLS for CoAP. Therefore, this provides a guarantee that sensitive information of patients will not be disclosed during transmission and hence protects the integrity of the network, besides meeting the requirements of regulatory laws.

The key conclusion is that IoT protocols take the lead in the optimization of network resources through providing better bandwidth usage, lower latency, and in terms of device management and scalability. In health, the listed ones provide for more effective and efficient network operation: MQTT by its lightweight design and the publish-subscribe model, CoAP through efficient use of UDP and supporting asynchronous communication, and HTTP due to wide compatibility. These protocols will, in turn, enable the smooth integration of IoT devices by performing the main challenges related to network resources, hence improving performance and the reliability of healthcare systems.

CHALLENGES AND CONSIDERATIONS

Since IoT protocols are escalating healthcare systems with better connectivity and operational efficiency, a few challenges and considerations have to be made for their proper implementation (Figure 3). The main issues remain concerns over data security and privacy, scalability issues, and interoperability challenges (Karthik et al., 2023; Kavitha et al., 2023; Satav et al., 2023; Venkateswaran et al., 2023).

Figure 3. Several critical challenges- IoT protocols

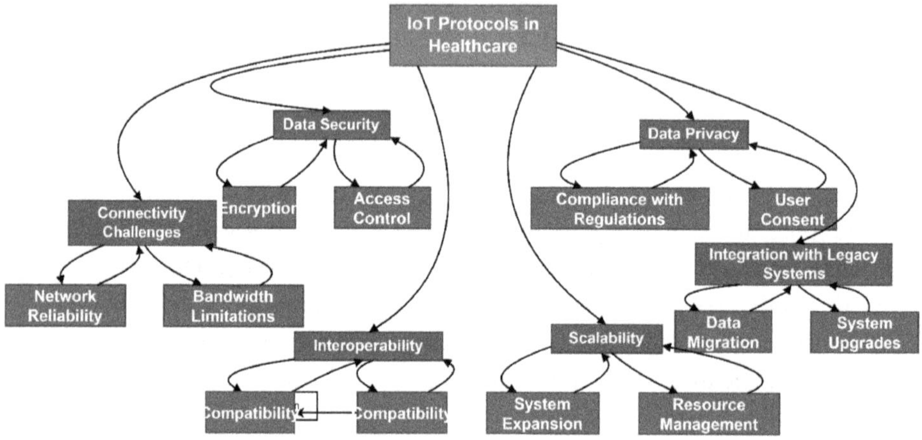

Data Security and Privacy

Data Security: Data security is the primary concern of a healthcare IoT system because the information regarding patients is highly sensitive. This is because IoT devices, such as wearable sensors and remote monitoring tools, collect a great deal of data about a person's health, thus making it a hotbed for malicious actors. Protocols such as MQTT and CoAP have instituted mechanisms for securing the transmission of data through security protocols like TLS for MQTT and DTLS for CoAP. Of course, making all this work seamlessly with the device is anything but straightforward.

Privacy: It is a matter that comes under the ambit of laws, such as those within the United States, concerning Health Insurance Portability and Accountability Act. The regulation of similar nature requires making sure that certain measures, which guarantee controlled use and handling, are in place and exercised for personal health information. Indeed, adherence to such laws necessitates secure encryption, access controls, and various data anonymization techniques. Even when these protocols are watertight, data breaches or unauthorized accesses are not entirely ruled out. For this reason, care providers must adopt holistic data protection strategies that include regular security audits, threat detection systems, and employee training in the best practices for protecting patient privacy.

IoT Challenges: Some IoT challenges include data encryption, access control, and regulatory compliance. Ensuring that all the data transmitted by IoT devices is always encrypted end-to-end can be very cumbersome and may affect system performance. Similarly, for strict access control, advanced identity and access management systems will be required.

Scalability Issues

IoT Device Management: As the IoT networks of healthcare expand, connected devices are growing massively, which are a huge management challenge. Networking infrastructure scalability issues may occur, along with protocol performance. Several protocols like MQTT and CoAP have been designed considering a huge number of devices; still, in practical deployment, this limitation problem can be faced. For instance, in MQTT, the Publish-Subscribe Model will allow talking effectively to multiple devices. However, without a good system architecture, it will be cumbersome to maintain a massive number of subscriptions.

Network Performance: Scalability has an effect on network performance. The number of connected devices can lead to network congestion, latency, and therefore reduced data throughput. For avoidance of these issues, healthcare systems need to consider robust strategies for network management that are widely based on load balancing and optimization of traffic.

IoT challenges revolve around infrastructure capacity, protocol efficiency, and data storage. To be prepared for the oncoming levels and masses of devices on the network infrastructure, careful planning and investment in scalable hardware and software solutions have to be made. Optimizing protocols and incorporating advanced network management tools will help mitigate these issues. Scaling of data storage solutions also involves physical storage considerations and database management challenges.

Interoperability Concerns Protocol compatibility: A system or device integration from multiple manufacturers generates issues related to interoperability. Furthermore, different protocols and data formatting might raise a number of barriers for the smooth communication of devices. For instance, MQTT and CoAP are different. Thus, systems using different protocols may need translation layers or gateways for communication. Compatibility and standardization become essential so that all devices and systems will work harmoniously together.

Standardization: Absence of general standards in the field of IoT in healthcare may lead to its fragmentation and inability to cope with diverse systems. While there are some standards, like HL7 and FHIR, which permit a framework to enable the exchange of health data, these are still not good enough to respond to the demands of all IoT applications. The establishment and enforcement of common standards within the industry will help in working out interoperability and ensure that devices and systems are managed together.

The key to successful implementation of IoT protocols in healthcare therefore pertains to solving challenges imposed by data security and privacy, scalability, and interoperability. Strong protection of data, scaling-up operations in large deployments, and encouraging standardization are issues that would go a long way toward full exploitation of the IoT technologies. Each of these considerations offers a way through which healthcare organizations can improve their IoT systems to be more effective without diminishing the security and interoperability standards.

CASE STUDY: IMPLEMENTATION OF IOT IN CHRONIC DISEASE MANAGEMENT

Chronic diseases such as diabetes and cardiovascular conditions pose a big challenge to most health systems the world over. Most of these chronic diseases require continuous monitoring, timely intervention, and coordinated care-all of which is extremely costly and complex. In recent years, new opportunities for the betterment of chronic disease management have been afforded with the introduction of IoT technologies through real-time monitoring, personalized care, and better engagement of patients. Precisely, the case study focuses on the application of IoT to chronic disease management and narrates an experience of a hypothetical healthcare organization in integrating IoT technologies in diabetes management (Ramudu et al., 2023) .

HealthCare Innovations Inc., or HCI in short, is a mid-sized healthcare organization operating several clinics and one hospital. It attends to a large population suffering from different chronic conditions. Diabetes features as one of the major health concerns in its patient population. Given the need to improve care for patients with diabetes and improve the outcomes of patients, it was decided that HCI will adopt an IoT-based solution for continuous glucose monitoring, providing personalized insights for pro-active care.

Improve glucose monitoring by providing patients with real-time glucose monitoring devices, improving patient engagement through actionable insight, enabling timely interventions by allowing providers to receive alerts when their patients have abnormal glucose, and optimizing care coordination by integrating diverse data on one platform for personalized patient care.

Implementation

HCI chose as a device package continuous glucose monitor devices combined with wearable fitness trackers. CGMs permanently measure the glucose level in the blood and send the information to a cloud-based service. Wearable fitness trackers collect data about physical activities, sleep patterns, and heart rate. Using IoT protocols, they are connected with MQTT for efficient data transfer and HTTP to integrate it into the existing health systems.

In order to manage the volume of data emanating from CGMs and wearable devices, HCI created an integrated data management platform. Of course, that data is aggregated from these various streams, processed through machine learning algorithms, and then presented in an easy-to-understand form. The platform provides real-time alerts to healthcare providers and patients if glucose levels are outside the target range.

HCI developed a mobile application that connects with IoT devices to provide the patient with their glucose level, personalized recommendation about medicines and their precautions, and progress monitoring. The application also comprises a library of educational resources: how to manage the disease, dietary tips, and recommendations of exercise. Push notifications and reminders allow patients to follow treatment plans and make healthier lifestyle choices.

The web-based dashboard of HCI supports healthcare providers with access to aggregated patient data. It will provide the provider with a comprehensive view of the patient's trend in glucose, medication adherence, and various aspects of lifestyle. Providers can set up alerts for specified conditions such as when the glucose levels are too low or high. Historical trends may be reviewed to determine the possible decisions of any adjustments in treatment.

Results

a) Continuous glucose monitoring significantly improved glucose control among patients. In fact, real-time data paved the way for sufferers to make timely adjustments in dosage and life-style habits that improved the stability of blood sugar levels and reduced incidents of hyperglycemia or hypoglycemia.
b) Patients felt more involved in their diabetes management. The interactivity of the mobile app, the educational content, and the real-time feedback called for activity on the part of the patients in the care of themselves. As a result, treatment plans and healthy lifestyle choices were better followed.
c) Real-time notifications for abnormal glucose levels among the patients enabled health professionals to act swiftly and promptly. For example, one of the providers was notified about a patient whose glucose level was running predominantly high. This immediately called for an urgent consultation, after which some adjustments in the treatment plan were effected. Such proactive measures went a long way in preventing complications and generally improving outcomes for the patients.

d) Data from CGMs and wearable devices were integrated into the central management platform for coordinated care among HCI's multidisciplinary teams. The providers could work in better collaboration, thereby sharing their insights and making more data-driven decisions.

The holistic approach to the patient guaranteed comprehensive and customized care. Implementation of various IoT technologies across HealthCare Innovations Inc. showcased multiple possibilities regarding continuous monitoring, personalized insight, and proactive care in improving the management of a chronic disease. Integration of CGMs and wearables with a centralized data management platform ensured better glucose control, improved patient adherence, and much more efficient coordination of care. There were challenges that needed to be overcome, like those associated with data security and system integration, but overall, the results underpin the transformative power of IoT in the management of chronic diseases and set a high bar for future innovation in healthcare.

INTEGRATION OF IOT PROTOCOLS IN HEALTHCARE- IMPLEMENTATION

Figure 4. Integration of IoT Protocols in Healthcare: Implementation

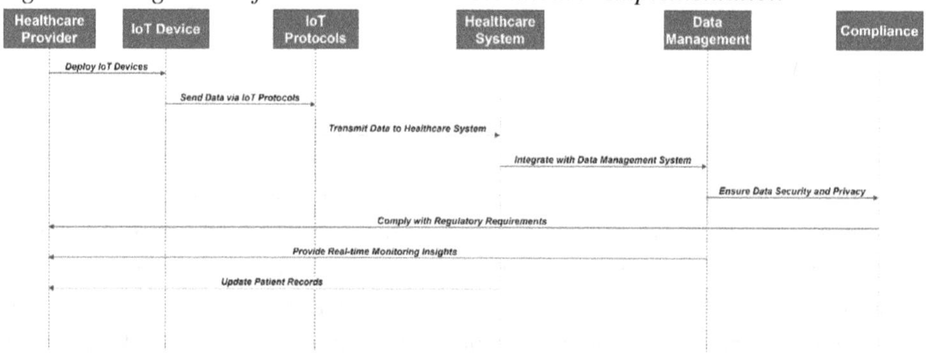

Integration of IoT protocols in healthcare is acting as a game-changing phenomenon for patient care, integrating various aspects: enhanced connectivity, real-time monitoring with data management. Some key components and factors to be kept in mind are there while implementing IoT protocols in healthcare. Figure 4 depicts the application of IoT protocols in healthcare(Boopathi & Khang, 2023; Karthik et al., 2023; Kavitha et al., 2023; Upadhyaya et al., 2024).

a. Internet of Things Protocol Selection: MQTT finds its perfect implementation in the low-bandwidth networks that usually have high latency; hence, MQTT will efficiently carry out the transmission of data from medical devices to healthcare. CoAP is designed for simple devices and constrained environments; hence, CoAP enables communication in such resource-constrained scenarios. HTTP is usually used for web-based applications, which is good enough for general data exchange requirements.
b. Integration with Devices: Ensuring that different devices can connect reliably to the networks using chosen protocols. Their integration with existing healthcare systems should provide a seamless flow and integration of data.
c. Data Management: IoT protocols shall be used in the continuous monitoring of patient health metrics such as heart rate, glucose level, and blood pressure. Sensitivity regarding patients' information must be ensured through encryption and proper access controls during transmission and storage.
d. System Integration: Integration of IoT data with EHR and other management systems for seamless end-to-end knowledge of the patient. Solving the incompatibility problems of the available infrastructure to allow for smooth integration and data migration.
e. Regulatory Compliance: Adapting to regulations such as GDPR, HIPAA, on patient data privacy and consent. Industry standards while implementing IoT to maintain integrity of data and reliability in the system.
f. Future Considerations: System design to handle an ever-increasing amount of devices and volume from expansions in healthcare IoT. The integration of healthcare IoT with emerging developments in AI, machine learning, and wearable technologies will continue to extend the capabilities and benefits of IoT systems.

The use of IoT protocols in healthcare can ensure huge benefits but needs adequate planning and execution to handle the challenges to realize the fullest potentials.

CURRENT RESEARCH AND FUTURE DIRECTIONS

Integration of IoT protocols in healthcare is an evolving field that moves with wide research and advancement, promising bright future opportunities. The current research and future directions have been represented in Figure 5. Shifting trends and potential innovation thus mark the future of IoT healthcare, with its improvement happening at a rapid rate. This section explores these trends and innovations, highlighting their impact on improving healthcare delivery and patient outcomes (Boopathi, 2023a; Malathi et al., 2024).

Figure 5. Current Research and Future Directions

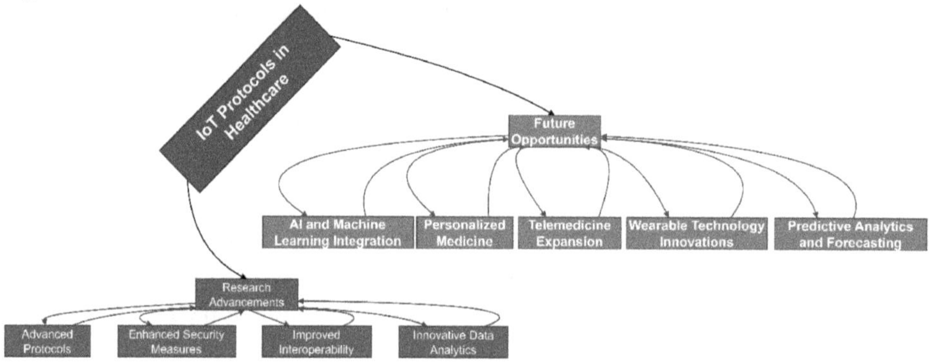

Emerging Trends

Integration of Advanced Analytics with AI: This is one of the major trends in IoT in healthcare: integrating advanced analytics and artificial intelligence or AI. AI algorithms, such as machine learning or ML, and deep learning have begun to be increasingly used to voluminous data emanating from IoT devices. These technologies make predictive analytics possible by which potential health issues can be foreseen much before they become critical. The data recorded by such wearable devices can be analyzed by AI to forecast events like cardiac problems among others.

Edge Computing: Edge computing is another uprising trend in healthcare IoT. This involves data being processed at the device or edge of the network rather than transferring all the data to centralized cloud servers. This saves time due to reduced latency and bandwidth use, hence allowing real-time data analysis and decision-making. In health, edge computing may enable immediate responses to critical health data, including alerting medical personnel of wearables detecting abnormal vital signs.

Improved Security Protocols: Many IoT devices are being integrated into the healthcare industry daily, making improvement of the security protocols all the more urgent. Among several varying proposals being considered by researchers for protection to be accorded to sensitive health data are advanced encryption methods, secure communication protocols, and decentralized security models. Innovations include the use of blockchain technology to provide immutable and secure records of patient data and transactions.

Interoperability Standards: Interoperability continues to be a key area of interest in ongoing research. In this direction, establishing and implementing global standards that can facilitate the smooth interaction of various devices and systems that constitute IoT is being addressed. Other approaches include the FHIR standard and integration of the IoT protocols with the health information exchange system in order to address the challenges related to interoperability and share data between systems of different platforms.

Potential Innovations Smart Health Environments: The future developments in health IoT might be integrated into the sphere of smart health; these would include IoT device-based, sensor-based, AI-powered continuous monitoring and management of patient health in either a smart home or hospital environment. For example, smart environments could adjust the lighting, temperature, and medication schedule based on real-time health data for personalized care with a view to improving patient comfort(Karthik et al., 2023; Kavitha et al., 2023; Malathi et al., 2024).

Advanced Wearable Technologies: A new generation of advanced wearable technologies may completely change the perspective on healthcare monitoring. The wearables of the future may well include sophisticated sensors that are able to provide a wider range of physiological parameters, such as the level of blood glucose, hydration status, and respiratory function. Such wearables would offer continuous, non-invasive monitoring and real-time feedback to the patients and healthcare providers (Boopathi et al., 2022; Jeyakymar et al., 2022; Myilsamy Sureshkumar, 2018).

Integration of IoT with Personalized Medicine: This is an exciting and rapidly emerging field of research. Integration of data from IoT devices to incorporate genetic, environmental, and lifestyle information enables treatment and intervention to be offered in a more personalized manner. For example, IoT devices may monitor the response of a patient to a specific medication and adjust dosages to achieve optimal therapeutic outcomes and minimize adverse side effects.

Improved Management and Visualization of Data: Innovations in data management and visualization tools are likely to make health data more usable and accessible. Advanced dashboards and real-time visualization of data may allow healthcare providers to interpret complex datasets more intuitively. These devices may integrate various sources of data, like the IoT, EHR, and clinical databases, for integrated insights on patient health.

Improvement and Growth of Remote Care Solutions: Improvement and growth in remote care solutions development are further probable innovations driven through the adoption of IoT technologies. Indeed, remote patient monitoring and telemedicine platforms have increasingly become sophisticated, therefore enabling healthcare providers to offer high-value, distanced care. Innovation around this space may be in the form of next-generation telehealth platforms, remote diagnostic tools, and more user-friendly patient engagement solutions that provide easier access to care and health management for patients.

Current research and emerging trends of IoT in healthcare underpin the transformational use of advanced technologies that improve patient care and operational efficiencies. Analytics, edge computing, enhanced security protocols, and interoperability standards will continue to be major factors in driving innovative solutions shaping the future of healthcare. As these technologies continue to evolve even further, they will promise to bring new opportunities for personalized medicine, smart health environments, and more effective remote care. By keeping pace with these evolving developments, healthcare organizations have the opportunity to apply IoT innovations and serve as drivers of the future of healthcare delivery.

CONCLUSION

Basically, the integration of IoT protocols in healthcare represents a quantum leap toward a better-connected, efficient, and patient-friendly medical system. Through this chapter, we have taken a journey into some of the most important IoT protocols-MQTT, CoAP, and HTTP-and their critical roles with respect to enhancing healthcare delivery. Each one of these protocols has different advantages targeted at different needs within the healthcare environment-from the lightweight low-bandwidth efficiency of CoAP to the strong scalable communication capability of MQTT and the wide applicability of HTTP.

In healthcare, IoT means more than just connectivity; it's about improvement in data accuracy, engagement of patients, and efficiency in care. Enabling timely monitoring and personalized intervention, IoT technologies transform chronic disease management with better patient outcomes and reduced load on healthcare professionals. This is well reflected in the case study of HealthCare Innovations Inc., where IoT solutions lead to better glucose control among patients with heightened engagement and timelier interventions.

However, the way to full utilization of IoT in healthcare is anything but smooth. The assurances on data security and privacy concerns seriously need strict measures to ensure that confidential information can be kept without tampering. Scalability issues require high caution since the systems have to be done in such a way that they will hold volumes of data and huge numbers of devices. Interoperability concerns indeed demand adherence to universal standards to guarantee ease of integration and communication between diverse systems and devices.

Future innovation may, therefore, be informed by the emerging trends in edge computing, advanced analytics, and improved security protocols. From the current adoption of the IoT technologies within healthcare organizations, further research and development will be very important in addressing challenges that remain today while creating new opportunities. Ongoing IoT evolution in healthcare could lead to huge gains being achieved in patient care, optimization of resources used, and advancements attained in medical systems in the future.

In a nutshell, the introduction of IoT protocols in healthcare will mark a revolution for the industry, opening up new avenues for connectivity, efficiency, and better outcomes. As technological changes continue to take place, the key to unlocking the full transformative potential of IoT in healthcare lies with addressing the challenges associated with it.

REFERENCES

Ahad, A., Tahir, M., Aman Sheikh, M., Ahmed, K. I., Mughees, A., & Numani, A. (2020). Technologies trend towards 5G network for smart health-care using IoT: A review. *Sensors (Basel)*, *20*(14), 4047. DOI: 10.3390/s20144047 PMID: 32708139

Akkaş, M. A., Sokullu, R., & Çetin, H. E. (2020). Healthcare and patient monitoring using IoT. *Internet of Things : Engineering Cyber Physical Human Systems*, *11*, 100173. DOI: 10.1016/j.iot.2020.100173

Albahri, A. S., Alwan, J. K., Taha, Z. K., Ismail, S. F., Hamid, R. A., Zaidan, A., Albahri, O. S., Zaidan, B., Alamoodi, A. H., & Alsalem, M. (2021). IoT-based telemedicine for disease prevention and health promotion: State-of-the-Art. *Journal of Network and Computer Applications*, *173*, 102873. DOI: 10.1016/j.jnca.2020.102873

Bhuiyan, M. N., Rahman, M. M., Billah, M. M., & Saha, D. (2021). Internet of things (IoT): A review of its enabling technologies in healthcare applications, standards protocols, security, and market opportunities. *IEEE Internet of Things Journal*, *8*(13), 10474–10498. DOI: 10.1109/JIOT.2021.3062630

Boopathi, S., Hari balaji, V., Mageswari, M., & Asif, M. M. (2022). Influences of Boron Carbide Particles on the Wear Rate and Tensile Strength of AA2014 Surface Composite Fabricated by Friction-Stir Processing. *Mater-Tehnol.Si. Materials Technology*, *56*(3), 263–270.

Boopathi, S. (2023a). Internet of Things-Integrated Remote Patient Monitoring System: Healthcare Application. In *Dynamics of Swarm Intelligence Health Analysis for the Next Generation* (pp. 137–161). IGI Global. DOI: 10.4018/978-1-6684-6894-4.ch008

Boopathi, S. (2023b). Securing Healthcare Systems Integrated With IoT: Fundamentals, Applications, and Future Trends. In *Dynamics of Swarm Intelligence Health Analysis for the Next Generation* (pp. 186–209). IGI Global.

Boopathi, S., & Khang, A. (2023). AI-Integrated Technology for a Secure and Ethical Healthcare Ecosystem. In *AI and IoT-Based Technologies for Precision Medicine* (pp. 36–59). IGI Global. DOI: 10.4018/979-8-3693-0876-9.ch003

Ding, J., Nemati, M., Ranaweera, C., & Choi, J. (2020). IoT connectivity technologies and applications: A survey. *IEEE Access : Practical Innovations, Open Solutions*, *8*, 67646–67673. DOI: 10.1109/ACCESS.2020.2985932

Jeyakymar, M., Munusami, V., Shanmugam, P., Boopalan, N., Boopathi, S., & Sureshkumar, M. (2022). An investigation on wear loss and hardness of nitro-carburizing coated stainless-steel grade-316. *Materials Today: Proceedings*, *66*(3), 1398–1404. DOI: 10.1016/j.matpr.2022.05.223

Karthik, S., Hemalatha, R., Aruna, R., Deivakani, M., Reddy, R. V. K., & Boopathi, S. (2023). Study on Healthcare Security System-Integrated Internet of Things (IoT). In *Perspectives and Considerations on the Evolution of Smart Systems* (pp. 342–362). IGI Global.

Kavitha, C., Varalatchoumy, M., Mithuna, H., Bharathi, K., Geethalakshmi, N., & Boopathi, S. (2023). Energy Monitoring and Control in the Smart Grid: Integrated Intelligent IoT and ANFIS. In *Applications of Synthetic Biology in Health, Energy, and Environment* (pp. 290–316). IGI Global.

Kelly, J. T., Campbell, K. L., Gong, E., & Scuffham, P. (2020). The Internet of Things: Impact and implications for health care delivery. *Journal of Medical Internet Research*, *22*(11), e20135. DOI: 10.2196/20135 PMID: 33170132

Malathi, J., Kusha, K., Isaac, S., Ramesh, A., Rajendiran, M., & Boopathi, S. (2024). IoT-Enabled Remote Patient Monitoring for Chronic Disease Management and Cost Savings: Transforming Healthcare. In *Advances in Explainable AI Applications for Smart Cities* (pp. 371–388). IGI Global.

Pradhan, B., Bhattacharyya, S., & Pal, K. (2021). IoT-based applications in healthcare devices. *Journal of Healthcare Engineering*, *2021*(1), 6632599. PMID: 33791084

Qadri, Y. A., Nauman, A., Zikria, Y. B., Vasilakos, A. V., & Kim, S. W. (2020). The future of healthcare internet of things: A survey of emerging technologies. *IEEE Communications Surveys and Tutorials*, *22*(2), 1121–1167. DOI: 10.1109/COMST.2020.2973314

Ramudu, K., Mohan, V. M., Jyothirmai, D., Prasad, D., Agrawal, R., & Boopathi, S. (2023). Machine Learning and Artificial Intelligence in Disease Prediction: Applications, Challenges, Limitations, Case Studies, and Future Directions. In *Contemporary Applications of Data Fusion for Advanced Healthcare Informatics* (pp. 297–318). IGI Global.

Reddy, M. A., Reddy, B. M., Mukund, C., Venneti, K., Preethi, D., & Boopathi, S. (2023). Social Health Protection During the COVID-Pandemic Using IoT. In *The COVID-19 Pandemic and the Digitalization of Diplomacy* (pp. 204–235). IGI Global. DOI: 10.4018/978-1-7998-8394-4.ch009

Saba, T., Haseeb, K., Ahmed, I., & Rehman, A. (2020). Secure and energy-efficient framework using Internet of Medical Things for e-healthcare. *Journal of Infection and Public Health*, *13*(10), 1567–1575. DOI: 10.1016/j.jiph.2020.06.027 PMID: 32682657

Satav, S. D., Hasan, D. S., Pitchai, R., Mohanaprakash, T., Sultanuddin, S., & Boopathi, S. (2023). Next generation of internet of things (ngiot) in healthcare systems. In *Sustainable Science and Intelligent Technologies for Societal Development* (pp. 307–330). IGI Global.

Sureshkumar, M., & Boopathi, S. (2019). The improving wear resistance properties of molybdenum alloy. *Int. J. Mech. Prod. Eng. Res. Dev*, *8*, 724–728.

Upadhyaya, A. N., Saqib, A., Devi, J. V., Rallapalli, S., Sudha, S., & Boopathi, S. (2024). Implementation of the Internet of Things (IoT) in Remote Healthcare. In *Advances in Medical Technologies and Clinical Practice* (pp. 104–124). IGI Global. DOI: 10.4018/979-8-3693-1934-5.ch006

Venkateswaran, N., Kumar, S. S., Diwakar, G., Gnanasangeetha, D., & Boopathi, S. (2023). Synthetic Biology for Waste Water to Energy Conversion: IoT and AI Approaches. *Applications of Synthetic Biology in Health. Energy & Environment*, •••, 360–384.

Chapter 5
AI and IoT in Edge Computing for Climate Analysis and Sustainability

Vidya M. J.
https://orcid.org/0000-0001-6858-6165
School of Computer Science and Engineering, RV University, Bengaluru, India

Veena Divya Krishnappa
Department of Electronics and Instrumentation Engineering, RV College of Engineering, Bengaluru, India

Rajasree P. M.
https://orcid.org/0009-0002-2415-2685
Department of Electronics and Instrumentation Engineering, RV College of Engineering, Bengaluru, India

ABSTRACT

AI, IoT, and edge computing are integrated to perform sustainability and climate analyses. AI has been employed in modeling and weather prediction, while IoT performs the gathering of real-time environmental data. The approach guarantees sensitive data processing locally with less bandwidth and delay. It provides real-time decision making, enhanced accuracy of climate data, and faster processing. However, issues such as data privacy, interoperability, and technical complexity have also been observed. Standard protocols, scalable infrastructure, and best practices have been done for data management.

DOI: 10.4018/979-8-3693-5448-3.ch005

INTRODUCTION

Innovative technical solutions like edge computing, IoT integration, and AI were developed to address climate change and achieve sustainability urgently. These technologies remain key factors in sustainability since they greatly improve environmental monitoring and climate analysis. A general overview of these technologies and their functions in climate analysis is provided in this introduction. Because it maintains intricate algorithms that model climate change and forecast weather patterns with unprecedented accuracy, AI is a serious tool in climate science (Bibri et al., 2024).

AI makes sense of large datasets by identifying patterns and trends that often evade conventional methods using machine learning techniques. This forecasting capability is important in predicting severities of weather events and their impact, thus, helping in the finding of long-term strategies for climate resilience. The so-called Internet of Things, or integrated sensor device network, provides data on the study of climate. Such devices send real-time information from different locations by observing environmental conditions related to temperature, humidity, air quality, and soil moisture (Bourechak et al., 2023). Frequency and granularity of data allow for deep analytics, enabling a deeper understanding of local and global changes. Edge computing technology reduces bandwidth usage and latency while it improves privacy by processing and analyzing data closer to the source. Especially useful in real-time applications, such as in climate analysis, data processing on-site with a natural catastrophe can speed up decision-making and response times to save lives and reduce damages caused by it all (Slama, 2022).

The combination of edge computing, IoT, and AI provides a strong foundation for sustainability and climate studies. IoT sensors pick up a great amount of environmental data, which are then subject to processing with AI algorithms right on the edge for real-time analytics and immediate insights. This methodology makes for faster reactions to situations or changes in the environment. Commonly, the data from IoT devices streams to the architecture edge computing nodes, which generate actionable insights for both short- and long-term climate policies (Iftikhar et al., 2023). These two synergy factors will enhance the real-time analysis and improvement of data quality. Disaster preparedness and effective resource management with well-informed policy decisions are the consequences of these impacts.

With minimum quantity of chemical inputs and more efficient usage of water, precision agriculture makes the most of real-time soil and meteorological data. Healthy cities may be a result of continuously monitored traffic patterns and air quality to enhance urban design. In particular, exponential growth in environmental data due to edge computing saves energy in the transportation of data and in centralized processing, diminishing its carbon footprint (Chang et al., 2021). Some of the issues that arise relate to interoperability, technological challenges, and data protection vulnerabilities in the integration of edge computing, IoT, and AI. Standard standards should be drafted and created, and stakeholders need to ensure participation and innovation. They are going to be even more effective in the future years; therefore, they are going to strengthen the contribution they make toward sustainability and climate analysis (Kalyani & Collier, 2021a).

The incorporation of edge computing, IoT, and AI marks a significant milestone for the future of sustainability and climate analysis. The technologies enable predictive analytics, immediate response, and real-time monitoring of the environment. These inventions will be crucial in the implementation of a sustainable future inspired by challenges observed within aspects of climate change. Climate analysis presents a gateway to understanding Earth's climate system dynamics, forecasting the future, and preparing for extreme weather events. Such accurate analysis of climate allows the forecasting of hurricanes, droughts, flooding, and other natural disasters while providing, in parallel, governments and communities with the information they need to activate emergency response plans to protect people and property. In sum, climate data plays a critical role in the development of evidence-based policy that will help both enterprises and governments manage resources effectively, cut greenhouse gas emissions, and transition toward renewable energy sources (Sittón-Candanedo, Alonso, García, et al., 2019).

The application of sustainable practices allows resources to be set aside for future generations through the balance of economic growth and environmental preservation. This is possible with reliable data concerning the climate. Climate analysis in agriculture offers insight into the selection of crops, the best planting, and harvesting schedules, and how to use the water resource correctly. Knowledge of the pattern of climate decreases risks while enhancing productivity to achieve food security. Precision farming, with the use of climate information, manages to utilize inputs like water, fertilizers, and pesticides more effectively, but with lesser negative environmental consequences than ever, along with maintaining high levels of agricultural production (Ferrández-Pastor et al., 2018).

Urban planners also carry out the climate analysis in the backdrop of responding to climate change factors such as heat waves, rising sea levels, and air pollution to develop resilient and sustainable cities. The latter includes development of green spaces, increasing energy efficiency, and extension of transit infrastructure by reducing carbon emissions and improving livability and community sustainability in return. Climate analysis involves comprehension of how climate change affects biodiversity and ecosystems. Changes in temperature and rainfall, among other factors, force the animals to migrate, destroy their habitats, or simply become extinct (Ting et al., 2022). It is also crucial for conservationists to be able to take this climate data and use it in creating plans for preserving vital services for human well-being, rehabilitating ecosystems, and safeguarding endangered species.

Climate analysis enables the elaboration of specific data related to weather conditions, solar radiation, wind speeds, and the movement of water that help in the establishment of the best places to install renewable energy sources. Therefore, this view contributes to the better installation of renewable energy infrastructures, increasing less dependence on fossil fuels and carbon emissions. Reliable climatic data also promote the use of energy storage and smart grids to further improve the reliability and efficiency of systems based on renewable energy sources (Fraga-Lamas et al., 2021).

Background

These include some of the major socio-economic impacts of climate change: poor health, reduced quality of life, and livelihood impacts-particularly among more vulnerable populations from developing countries. Climate analysis data can, therefore, be used to concentrate such acts early in advance warning systems, resilient infrastructure, and equitable distribution of resources. Appreciating how climate change has affected economic and social conditions enhances social justice as well as community adaptability. Climate analysis is one of the major factors to reach sustainability and address challenges such as biodiversity conservation, renewable energy development, and agricultural productivity. These will help build a future where human society and nature can survive together.

Objectives

- To discuss the adoption of AI, IoT, and edge computing in climate analysis and sustainability platforms regarding near real-time environmental monitoring and processing.
- AI, IoT, and edge computing are increasingly applied for the betterment of climate analysis, enhanced weather predictions, real-time decision-making, assurance of data privacy, and energy efficiency that will drive sustainability efforts much higher.
- The presentation will discuss the integration of AI, IoT, and edge computing in climate sustainability, discussing technical and operational challenges, and future trends and innovations.

ARTIFICIAL INTELLIGENCE IN CLIMATE ANALYSIS

Artificial intelligence rejuvenates climate analysis, making state-of-the-art tools and techniques in weather forecasting and environmental data analysis available. This shows how strong AI is on the ground in the field of climate science, especially the application of machine learning techniques. The atmosphere is turbulent; hence, forecasting the weather is a difficult task. Traditional techniques use numerical weather prediction models that are very computationally intensive and raise problems of handling huge volumes of data. It is here and now that AI algorithms, specifically the ones based on machine learning, play their role as a strong alternative (Nishant et al., 2020).

- Deep Learning Models: Deep learning is a subcategory of machine learning concerned with knowledge learning from vast datasets using neural networks that contain multiple layers. The major applications to weather forecasting with deep learning are two: Convolutional Neural Networks and Recurrent Neural Networks, including Long Short-Term Memory. CNNs are more suitable for time series forecasting because it processes data spatially, whereas RNNs, especially the LSTMs, process data sequentially.
- Ensemble Learning: This is an approach that involves the combination of several AI models to improve predictive performance. It can reduce error and improve accuracy. They may be applied in weather predictions to incorporate uncertainties in their forecasts, hence giving more reliable outputs.
- Generative Adversarial Networks: GANs generate realistic data in view of the training samples. A GAN is comprised of a discriminator network and a generator network.

Through learning from historical data, GANs can be further applied to simulate future weather situations for the prediction of weather. In fact, this skill really pays off when high-resolution climate forecasts are needed to cope with in-depth regional climate studies and planning.

Machine Learning Techniques Applied to Environmental Data

Large-scale environmental data of temperature, humidity, wind speed, and level of pollutants will be analyzed through machine learning approaches to identify trends and insights that will further assist in climate analysis and decision-making (Jha et al., 2019).

- Supervised Learning: The system is trained by examples of known outcomes in supervised learning. The most widely used techniques include random forests, gradient boosting, and support vector machines. For example, Random Forests can predict air quality indices based on past experiences, while Support Vector Machines classify the weather patterns. These models will be able to train themselves in predicting specific climatic variables-say, temperature or rainfall-from historical and current inputs.
- Unsupervised Learning: Unsupervised learning finds hidden patterns in data without any defined labels. Clustering algorithms, such as K-means and Hierarchical Clustering, may be conducted to identify patterns and anomalies by forming groups of similar weather events or climate conditions. Dimensionality reduction techniques, including Principal Component Analysis, which simplify complex datasets, make the visualization and analysis of environmental data more readable.
- Reinforcement learning can be defined as the training of models on decision-making based on rewards and penalties. In climate analysis, reinforcement learning can be used to optimize the assignment of resources to methods that could be used to mitigate climate change. This may include, but is not limited to, supporting the most effective use of renewable sources of energy based on forecasts of the weather, or the efficient allocation of water resources during droughts.
- Time Series Analysis: Understanding and forecasting the climate pattern is extremely pertinent to time series analysis. As mentioned earlier, machine learning models are especially good at processing time series data, thus really enabling one to make very accurate projections about future climate conditions. Conventional methods such as ARIMA models and STL can be improved with machine learning in order to increase the accuracy of the forecast.

Artificial Intelligence and machine learning bring in effective ways of analyzing and forecasting weather with high accuracy; hence, they are revolutionizing the study of climate. This in return enables more advanced planning and decision-making for climate resilience and sustainability. The evolution of AI will integrate into climate research-sophisticated approaches toward understanding and mitigating climate change.

INTERNET OF THINGS (IOT) IN ENVIRONMENTAL MONITORING

By utilizing networked sensors and gadgets to gather data on environmental characteristics in real time, the Internet of Things (IoT) is completely changing environmental monitoring. The many kinds of IoT sensors and devices, together with the protocols and techniques for gathering data, are examined in this section(Xu et al., 2019). The IoT process is being utilized to monitor the environment, as depicted in Figure 1.

Figure 1. Process of Internet of Things (IoT) in Environmental Monitoring

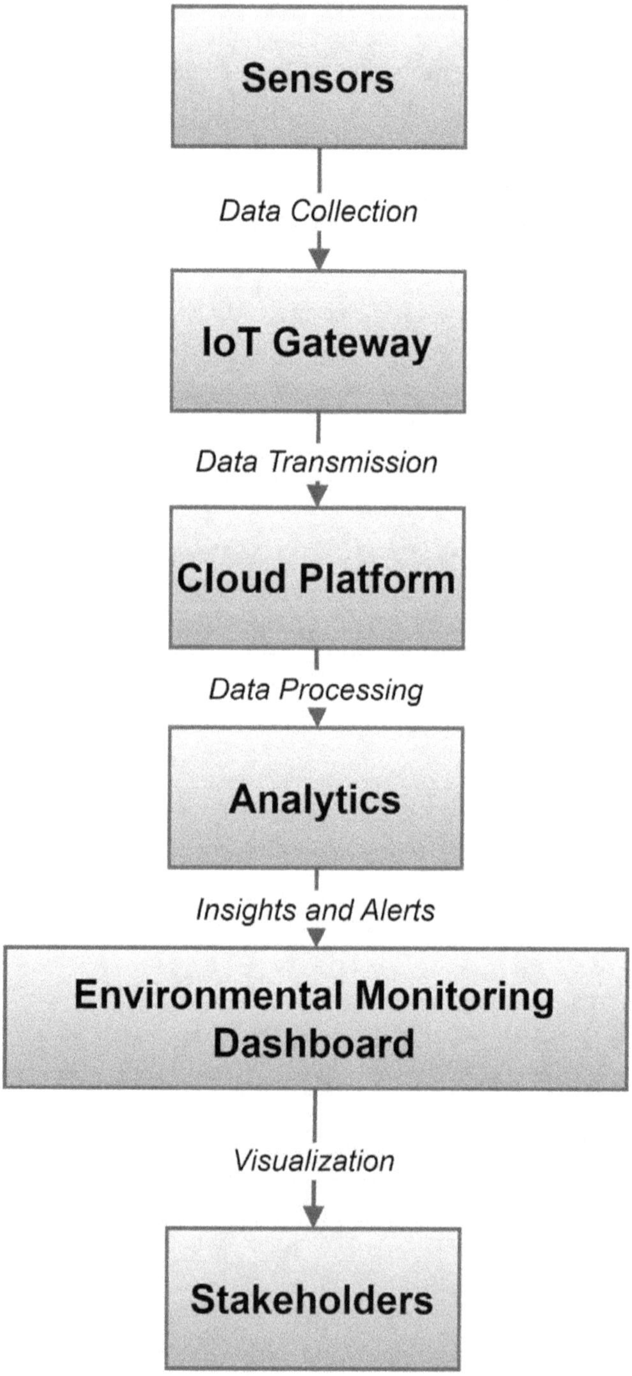

Types of IoT Sensors and Devices

IoT sensors are the crucial components to monitor the environment, as their capability for gathering data with high accuracy in several situations can be traced (Xu et al., 2019).

Temperature sensors measure the ambient temperature of a range of environments and monitor climatic trends to note any variations like heat waves. Humidity sensors measure the moisture of air to predict precipitations, understand weather patterns, or assess the risks regarding mold growth. Air quality sensors monitor urban pollution, health concerns, and regulatory enforcement by detecting $PM_{2.5}$, NO_2, SO_2, CO, and VOCs. These tools are critical in formulating the weather trends, predicting precipitation, and ascertaining the chances that mold could develop in structures.

In maintaining water safety and controlling aquatic ecosystems, water quality sensors detect pH, turbidity, dissolved oxygen, and pollutants in rivers, lakes, and coastal areas. Soil moisture sensors measure the water content in the soil for forestry, land management, and agriculture purposes. Light sensors are important in understanding plant photosynthesis processes and in the optimal placement of solar panels, as they measure the intensity and quality of sunlight. These sensors find applications in many fields for controlling aquatic ecosystems and ensuring water safety.

Sensors for wind direction and speed deliver very important information to help prevent wind erosion of landscapes and for renewable energy applications, including weather forecasting. Noise levels and patterns detected by acoustic sensors can be used to monitor industrial processes, wildlife activities, and road noise.

Data Collection Methods and Protocols

Full utilization of the Internet of Things in the case of environmental monitoring, in turn can be done by the employment of various techniques and protocols with regards to accurate as well as efficient data collection, transmission, and analysis (Al Mamun & Yuce, 2019).

The large scale and adaptable wireless sensor networks are employed for environmental monitoring in areas which are either inaccessible or quite remote. Low-power wide-area networks, such as Sigfox and LoRaWAN, allow for communication with low power and over longer ranges. Thereby, they can be applied to a wide variety of IoT devices requiring longer lifetimes for batteries. As traditional network infrastructure is not available, current and future cellular networks with high reliability and wide coverage, such as 4G and 5G, are suitable for applications needing mobile or remote environmental monitoring. Thereby, these networks can serve IoT devices operating in a remote or mobile context.

The satellite-based communication is very reliable for environmental monitoring over remote areas, like the ocean or polar regions. IoT devices can transmit data from almost anywhere on Earth. With edge computing, latency and bandwidth usage are minimized, and allow real-time observation and prompt reaction over crucial events. It also frequently happens that the data aggregation regularly takes place and is then sent to the cloud platforms for processing, analytics, and storage. The cloud platforms also scale the resources and provide advanced analytics tools that enable one to analyze a great amount of datasets and their integration into other data sources.

IoT systems use standardized protocols like MQTT, CoAP, and HTTP/HTTPS that are used in relaying data. These guarantee that communication between gateways, sensors, and cloud platforms will be effective and secure. Data integrity and unauthorized access are prevented by encryption, authentication, and protocols that ensure securities in communications, hence making the security and privacy of data very important. All these technologies taken together provide a comprehensive real-time environmental monitoring system that is informative about the status of the environment and helps in decision-making and involving sustainability in practice. The IoT technology will be more engaged in environmental monitoring as it develops, presenting increased width and accuracy in the control and protection of the environment.

EDGE COMPUTING: BRINGING COMPUTATION CLOSER TO DATA SOURCES

Edge Computing

Instead of sending all the gathered data to centralized cloud servers, edge computing is a form of distributed computing strategy where it focuses on storing and processing data closer to the source. Some of the benefits of this method include reduced latency, which is important for real-time applications found in industrial automation, autonomous vehicles, and environmental monitoring. By processing the data locally, edge computing reduces the time it takes to send data to and from a remote cloud server, guaranteeing both best practices and security precautions (Chen & Ran, 2019).

Some advantages of edge computing are increased data privacy and security, scalability, and bandwidth savings. This would reduce the requirement for constant data transfers, saving bandwidth and operating costs by processing and filtering the data locally. Ensuring data privacy through local handling of private information reduces vulnerabilities to security breaches and builds trust in users concerned about privacy. Besides sharing the processing load, edge computing ensures system scalability will be effective and efficient to handle additional devices and data without causing bottlenecks. Taking everything into consideration, edge computing therefore provides an IoT device solution that is far more cost-effective and faster.

Comparison to Cloud Computing

Where there are a number of advantages to edge computing, it needs to be distinguished from traditional cloud computing, which is based on the central processing and storage of data. Thus, cloud computing, because of the enormous processing and storage capacity it allows, is ideal for large-scale data archiving, training of machine learning models, and big data analytics. Scalable infrastructure allows big applications, which require central control. Each paradigm has advantages, and they are often applied in combination with others (Jiang et al., 2019).

With the emphasis on low latency and real-time processing, edge computing technology shrinks the time that is required to analyze data and make decisions. Climate analysis is one such application that, with its high demand for rapid processing, could greatly benefit from this concept regarding environmental responsiveness and sustainability initiatives. However, the limitation in processing, memory, and storage over an edge device may render the execution of complex calculations or local storage of large datasets challenging. Where edge computing turns to for increasingly complex tasks quite often is the cloud. It's due to this fact that cloud computing and edge computing coexist and are often combined to make sure things are efficient, scalable, and showing the best performance. Where long-term storage, sophisticated analytics, and data integration are dealt with by cloud computing, edge computing handles a vast number of dispersed devices.

However, unlike the cloud with its centralized infrastructure, edge computing requires more complex management to demand powerful frameworks and processes for performing reliable upgrades, security fixes, and device compatibility. In the case of environmental monitoring, IoT sensors perform data collection about temperature, humidity, and air quality at the edge. Using real-time processing allows it to detect issues in weather patterns or surges of pollutants. Data and alerts flow to the cloud for detailed analysis and historical storage, and sharing insights across platforms.

By moving compute closer to sources, edge computing has advantages like scalability, bandwidth efficiency, lower latency, and data privacy. It's a complementary architecture to the centralized methodology of cloud computing and offers stable and adaptive infrastructure for diverse applications. Integration of edge and cloud computing in the future will drive even more efficiencies and innovations in industries that include industrial automation and environmental monitoring, among others, as technology evolves.

INTEGRATION OF AI AND IOT WITH EDGE COMPUTING

Architecture of an Edge Computing System for Climate Analysis

Real-time data processing and analytics made possible by edge computing's integration of AI and IoT is transforming the field of climate analysis(Hartmann et al., 2022a; Jiang et al., 2019). The edge computing system for climate analysis is examined in this section, showing how AI and IoT technology support sustainability initiatives. The system's data flow is depicted in a figure 2.

Figure 2. Architecture of an Edge Computing System for Climate Analysis

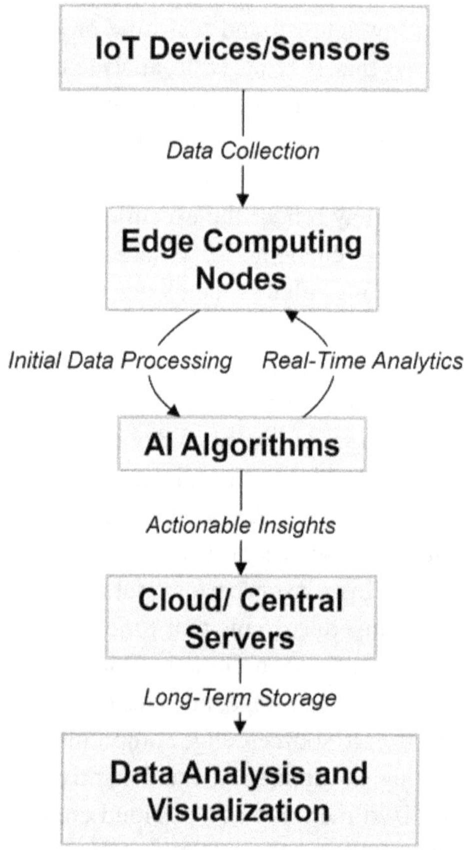

Architecture of an Edge Computing System for Climate Analysis

IoT devices, edge computing nodes, and AI algorithms are the constituents of an edge computing system for climate study, each indispensable in gathering, processing, and analyzing data in real time (Hartmann et al., 2022a).

- IoT Devices and Sensors: IoT devices are the key collectors in this respect, and they continuously collect data about the environment from various sources, including satellites, weather stations, and sensors on the ground. These devices integrate a host of sensors measuring temperature, humidity, air quality, and precipitation amongst others. The collected data is then sent further for processing to the nearest edge computing nodes.
- Edge Computing Node: These are basically small intermediary processing units, often kept in proximity with IoT devices, that is, near a device. The nodes do the initial processing, filtering, and analyzing of data and then forward the relevant data to the central servers or the cloud. This therefore means edge computing reduces latency by doing processing locally for faster reaction time while putting less load on centralized systems.
- AI Algorithms: Advanced data analytics and decision-making by AI algorithms executed at the edge. The algorithms, utilizing machine learning techniques, analyze the patterns, trends, and anomalies of the environmental data gathered from the IoT devices. Artificial intelligence models can predict weather patterns and detect environmental changes that would give valuable information to sustainability and climate studies.

The climate analysis system, on the other hand, is a hierarchical distributed structure of processing and analysis on different levels. IoT devices and edge computing nodes collect data and process it in real time, sending it for further analysis and storage on central servers or the cloud.

Advantages of AI and IoT Integration with Edge Computing

With the integration of edge computing with AI and IoT in climate studies, many benefits arise for sustainability projects (Maguluri, Ananth, et al., 2023; Samikannu et al., 2022).

- Real-time Data Processing: With AI and IoT technologies on, the data gathered would be processed right on the spot, that is, near the very source of data gathering. Real-time insight into environmental situations thus enables immediate action with regard to weather patterns, natural disasters, and other environmental events.

- Low Latency: By reducing the time taken for data to move from Internet of Things devices to central servers and back, edge computing reduces latency. Applications such as emergency management, precision agriculture, and disaster response that rely on fast decisions depend on this low latency.
- Bandwidth Optimization: With edge computing, processing and filtering of data are done locally before sending the relevant data to the cloud. This will ensure that bandwidths are utilized to their maximum extent. This kind of selectivity in transmission reduces the operational expenditures associated with data transmission and saves network bandwidth resources.
- Improved Security and Privacy: Because most of the data in the edge computing architecture is processed in a local environment, this minimizes the chance of sensitive data exposure during the transmission process. This, in turn, improves its security and privacy. The aspect of security and privacy becomes even more crucial in working out climate data, given that private or confidential information about specific parties may be at risk.
- Flexibility and Scalability: The architecture for edge computing is intrinsically flexible and scalable. It is easy to deploy and manage edge computing nodes for any environment because of its scalability. That's one guarantee that the system will be able to adapt to changing environmental conditions and can cope with the ever-growing data and IoT devices.
- Energy Efficiency: Because edge computing reduces the demand to transfer data to a central server, energy use will be lowered. Since energy efficiency equates to the overall reduction of carbon footprint associated with data processing and storage, it is key to any undertaking on sustainability.

A combination of edge computing, AI, and the Internet of Things together forms a powerful backbone for any climate analysis and sustainability-related activity. This architecture facilitates informed decision-making and climate mitigation with real-time processing, reducing latency, optimizing bandwidth usage, improving privacy and security, and aiding energy conservation.

PERFORMANCE Of EDGE COMPUTING IN CLIMATE ANALYSIS

By enhancing data processing speed, accuracy, and timeliness and providing advantages for sustainability and energy efficiency, edge computing is transforming climate analysis. Its performance benefits and implications for climate science and sustainability initiatives are examined in this section(Maguluri, Arularasan, et al., 2023a; Sharma et al., 2024).

Figure 3. Performance of Edge Computing in Climate Analysis

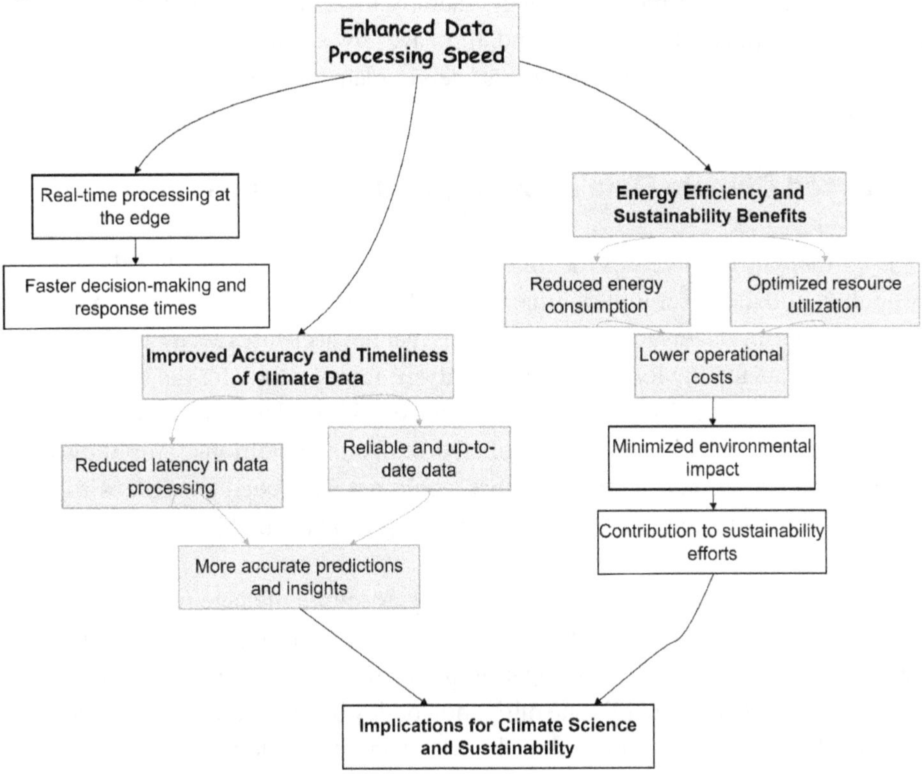

The topic of edge computing's performance benefits, implications for energy efficiency, sustainability, and the timeliness and quality of climate data, as well as its wider implications for climate science and sustainability initiatives, are depicted in the figure 3.

Enhanced Data Processing Speed

It gives edge computing, since data is processed right at its source with diminished latency, an unmistakable advantage in climate analysis. Traditional cloud-based methods are used, which call for transferring data to computers situated far away and creates congestion of networks, therefore raising latency. With edge computing, processing closer to the source diminishes the time taken in analysis.

It accelerates edge computing for real-time applications such as environmental monitoring, disaster management, and weather prediction. In a scenario where there is a natural disaster, it reduces latency, thereby increasing the speed of decision-making and response times. At the end, this saves lives and reduces overall damage to the need to issue evacuation orders, deployment of emergency services, and assessment of the impact of disasters.

Enhanced Climate Data Accuracy and Timeliness

Edge computing increases the timeliness and quality of climate data through the evaluation of data closer to its source, thus reducing the chances of contamination, loss, or tampering during transmission. In this manner, it ensures accurate, consistent, and current data for any form of climate analysis. It also allows for the integration of real-time data streams from satellites, Internet of Things devices, among others for thorough and continuous monitoring of environmental conditions. This allows it to catch minute variations and patterns that would not have been possible with more traditional, sporadic methods of data capture (Agrawal et al., 2023; Venkateswaran et al., 2024).

This is very important in climate modeling, resource management, and policy-making. The information provided by such data ranges from planting schedules on the farm to flood and drought mitigation plans by the water resource manager, to the successful execution of climate mitigation policy by the policymakers. These are further enhanced in their applications by edge computing along with sustainability and resilience. Energy Efficiency and Sustainability Benefits

It has sustainable and energy-efficient advantages since edge computing processes data locally at the edge, thereby removing the need for centralized cloud servers and carbon emissions. This will lessen the energy consumed in the transfer of data, further reducing the carbon footprint in climate analysis systems. This also cuts down computational overhead and power consumption by filtering out redundant data, prioritizing key operations, and optimizing resource use. Such edge computing designs can also use renewable sources of energy generated through solar or wind for further sustainability.

Since edge computing is energy-efficient, it finds its perfect application in isolated or off-grid areas that have only a few power sources. By minimizing energy consumption and optimally using the resources, it enables the installation of sustainable climate analysis systems in a wide range of settings.

Climate analysis on edge computing offers some considerable performance advantages, including accelerated data processing, increased accuracy, and real-time data. In addition to ensuring data integrity and reducing latency, it enables real-time environmental status monitoring. These advantages allow faster weather forecasting, efficient catastrophe management, and wiser decision-making with great implication in climate research and sustainability projects. By decreasing carbon emissions and the carbon footprint of climate analysis systems, edge computing also contributes to a green environment. As it continues to evolve, edge computing will no doubt play an increasingly important role in climate analysis, driving efficiency and innovation for projects related to sustainability and climate science.

CHALLENGES AND SOLUTIONS

Challenges and Solutions in Edge Computing for Climate Analysis

Climate analysis presents edge computing with a number of operational and technical issues, including resource limitations, data management, scalability, and security. On the other hand, best practices and solutions can guarantee the efficient implementation of edge computing systems(Hartmann et al., 2022b).

Figure 4. Technological and operational challenges in edge computing for climate analysis

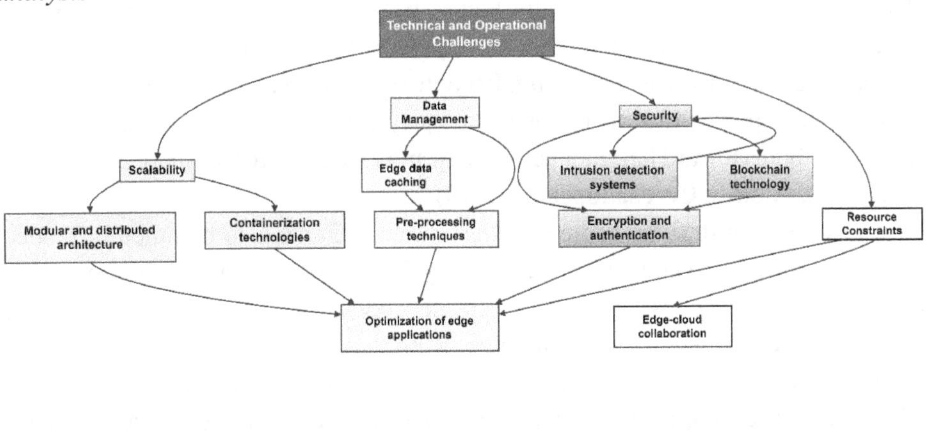

Figure 4 presents many approaches and best practices to address the technological and operational challenges in edge computing for climate analysis. The importance of overcoming these obstacles for the effective use of edge computing systems in climate analysis and sustainability initiatives is emphasized in the conclusion.

Technical and Operational Challenges

In addition, edge computing systems will need to be scalable, from managing a growing number of IoT devices to handling data being produced. However, it is difficult to ensure performance and dependability in dynamic contexts. Efficient storage, processing, and analytics skills will be required to manage data for consistency and accessibility across the dispersed edge nodes. As edge computing technologies are generally susceptible to certain security threats like virus attacks, data breaches, and unauthorized access, they are of paramount importance in terms of security regarding devices, networks, and data flows in protecting sensitive climate data from violation of privacy or confidentiality.

These edge computing settings present challenges in fulfilling performance requirements, optimizing resource usage, and reducing energy consumption because of their small size of memory, storage, and processing capacity.

Solutions and Best Practices

The various benefits that seem to be provided by the edge computing systems include data management, security, and scalability. Horizontal expansion is done by adding or removing edge nodes, respectively. Besides containerization technologies such as Docker and Kubernetes, easy deployment and orchestration are catered for. Techniques related to the caching and preprocessing of edge data reduce the overhead at the centralized cloud servers while improving responsiveness. While pre-processing reduces costs in data transfer and efficiently utilizes bandwidth, edge data caching saves data locally that is frequently accessed. Access control, authentication, encryption, and other strong security protocols guard systems from all types of attacks in edge computing. In contexts with edge computing, blockchain technology, employed in verification of integrity and decentralized management of trust, contributes toward heightened security and brings more transparency (Sittón-Candanedo, Alonso, Corchado, et al., 2019).

In resource-constrained contexts, edge computing can also be tuned to reduce resource consumption and increase efficiency. Energy consumption and processing overhead on edge devices can also be decreased, considering strategies such as edge analytics, edge caching, and AI inference. With the usage of collaboration models between edges and cloud layers, the computing load and resource limits can be further balanced, outsourcing resource-intensive workloads to centralized servers. These challenges associated with climate analysis can be overcome by strategies and best practices for the successful implementation of edge computing systems regarding sustainability initiatives and environmental monitoring. Ensuring scalability, data management, security, and resource restrictions will result in the enhancement of processing real-time data, decision-making, climate resilience, and sustainability.

FUTURE TRENDS AND INNOVATIONS

Recent technologies such as Artificial Intelligence, the Internet of Things, and Edge computing will keep bringing in radical change on environmental issues by redefining the landscape of sustainability (Kalyani & Collier, 2021b; Roman et al., 2018).

Emerging Technologies and Methodologies

- AI-driven sustainability solutions, as AI is capable of predictive modeling, advanced analytics, and decision-making, sustainability initiatives can be expected to get a complete makeover. Machine learning algorithms analyzing large-scale data sets will be able to find patterns in them, optimally allocate resources, and make unprecedentedly accurate predictions regarding environmental occurrences. AI will improve resource efficiency and reduce environmental impact in predictive maintenance, precision agriculture, and smart energy management systems.
- IoT-Enabled Environmental Monitoring: Most IoT devices are embodied with sensors and actuators that could be employed in monitoring and controlling environmental conditions in real time. The IoT network monitors air quality, water consumption, energy utilization, and other parameters in industrial areas and smart cities for proactive environmental management and mitigation. IoT innovations enhance the scalability, reliability, and coverage of environmental monitoring systems. Examples include edge computing, wireless networking, and low-power sensors.
- Edge Computing for Decentralized Sustainability: Edge computing enables real-time analysis and decision-making closer to the edge of the network. It facilitates localized environmental monitoring, autonomous systems, and distributed energy management within sustainability. Such computing can reduce dependency on centralized infrastructure, shrink data in transit, and enhance efficiency, robustness, and sustainability across many fields.

Potential Betterment of AI, IoT, and Edge Computing for Sustainability

The integration of Artificial Intelligence, the Internet of Things, and Edge Computing has the potential to disrupt all sustainability efforts in terms of efficient resource management, stronger structure, and a greener future for forthcoming generations through emerging trends and innovations (Bourechak et al., 2023; Fraga-Lamas et al., 2021; Ting et al., 2022).

- Resource optimization will be so complicated due to the fact that AI algorithms like deep learning, reinforcement learning, and natural language processing are at a high development stage. The AI systems can adapt to volatile environmental conditions and optimize energy use while minimizing waste in transportation, metropolitan areas, and industrial procedures.
- IoT-driven Ecosystem Monitoring-as offered by sensors, unmanned aerial vehicles, satellite imaging devices-geared IoT networks will avail deep insights into biodiversity and ecological health. In addition, real-time monitoring of pollution and deforestation with wild life habitat protection safeguards threatened ecosystems and species through timely responses and conservation.
- Edge for Resilient Infrastructure: It will be further developed to support decentralized and robust solutions for sustainability. In the event of natural disasters, for example, AI-enabled edge nodes might work out the best traffic flow, manage energy networks autonomously, and coordinate the response effort. It helps edge computing to let localized decision-making be performed along with adaptive control that makes critical infrastructure more resilient against such disruptively changing phenomena like climate change.

STRATEGIC INSIGHTS AND RECOMMENDATIONS

In order to maximize environmental monitoring and management initiatives, this section offers strategic insights for sustainable edge computing. It places a strong emphasis on data accuracy, real-time decision-making, energy efficiency, scalable infrastructure design, collaboration, and standardization(Ting et al., 2022).

Figure 5. Strategic to maximize environmental monitoring and management initiatives

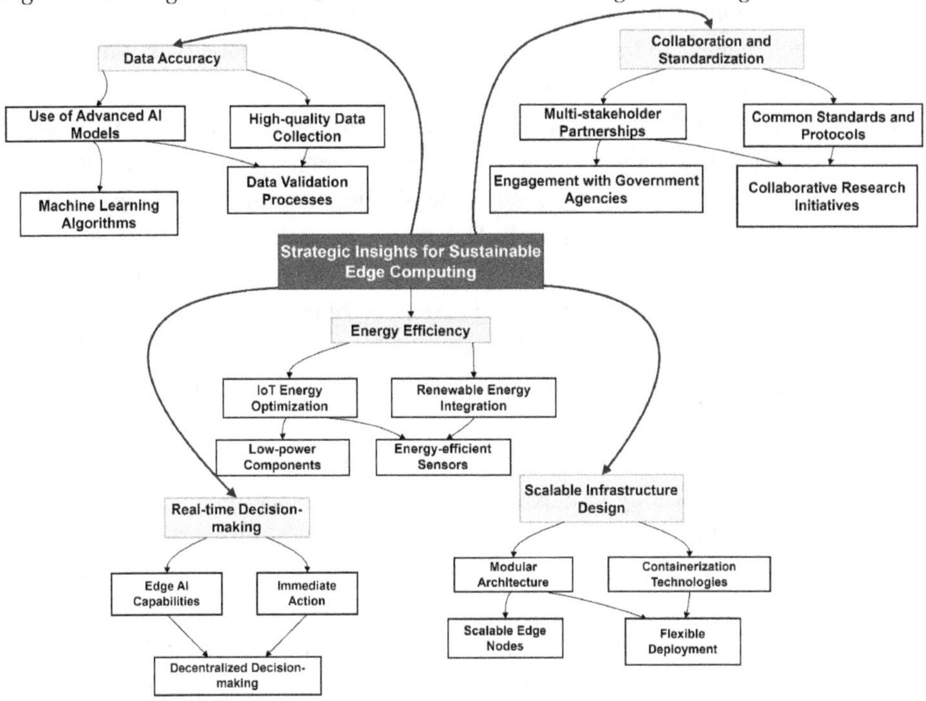

A flowchart providing strategic insights for edge computing that is sustainable is shown in Figure 5. In addition to offering particular recommendations within each area, it highlights important elements including data accuracy, real-time decision-making, energy efficiency, scalable infrastructure design, collaboration, and standardization.

Increasingly, AI models are a prerequisite for the accuracy of data from environmental monitoring. Algorithms of such a type allow us to highlight quite precariously patterns in complex datasets and even outliers. It is improved by the exposure of valid and reliable data from different sources, including satellite photography and IoT devices. The procedures in place for its continuous validation and improvements have, in turn, helped increase the accuracy of data collected over a period of time (Kumar et al., 2023; Ramudu et al., 2023; Sampath et al., 2022).

Real-Time Decision Making and Edge AI

Edge AI will be critical for environmental management and reaction decisions in real time. It reduces latency, hence allowing for quick responses by processing and analyzing data at the edge of the network. AI algorithms on IoT devices or edge servers empower organizations to infuse more agility and resilience into environmental monitoring and management operations. These algorithms can perform analytics on data, detect changes in the environment, and trigger automatic reactions (Das et al., 2024; Maguluri, Arularasan, et al., 2023b).

Energy-Efficient IoT Devices as IoT devices go, they have to be energy-efficient as well. This is because they are supposed to have a very small negative impact on the environment and would last for a long time. Without compromising the functionality, energy consumption can be reduced by designing such devices with low-power components, energy-efficient sensors, and better communication protocols. Key advantages that could be reaped with the integration of renewable sources of energy are decreasing reliance on traditional sources of power and raising sustainability.

The infrastructures for sustainable edge computing design should be focused on scalability and flexibility. The infrastructures should have modular architecture, support growing counts of IoT devices and data without being affected in performance. Containerization technology like Docker and Kubernetes facilitates application orchestration and deployments at the edge. It's the creation of infrastructures that are adaptive, allowing interoperability and integration with the existing infrastructures, which essentially favor flexibility and agility in front of the changing environmental conditions and technological discontinuities. Collaboration and Standardization

It necessarily requires several related initiatives on sustainable edge computing through the active cooperation and standardization among a wide range of non-profits, government agencies, educational institutions, and business partners. It initiates how one can collaborate and share information and resources. Common standards, protocols, and best practices allow for interoperability and scalability across ecosystems. Multi-stakeholder partnerships for valuable environmental outcomes allow collaborative research, innovation, and implementation of sustainable solutions.

With strategic implications and recommendations, edge computing activities can be furthered to make a positive impact on emerging environmental issues arising globally. By leveraging AI models for enhanced accuracy of data capture, enabling real-time decision-making, promoting energy efficiency in the IoT devices, designing scalable infrastructure, and collaborating with others, an organization can ensure optimal environmental monitoring and management for the transformational change, sustainability, and resiliency that future generations deserve.

CONCLUSION

This chapter looks at some of the possibilities for sustainability and climate analysis offered by edge computing. The chapter considers that AI, IoT, and edge computing technologies do empower real-time data processing and decision-making, including optimization of resources. Key takeaways are expected from the state-of-the-art AI models used to advance data precision, energy conservation promotion within IoT devices, building scalable infrastructure, and collaboration of several stakeholders. In fact, only those organizations that can implement strategic advice will exploit the promise of sustainable edge computing in optimize resource use, improve climate resilience, environmental monitoring, and management in various attempts to help state environmental concerns and further sustainability programs globally. Innovation and ever-changing technologies are seeing an increased use of edge computing in sustainability and climate analysis. By being at the helm of trends and embracing ways in which collaboration shall thrive, organizations are capable of setting a good example for environmentally positive impact and helping to create a sustainable world for future generations.

REFERENCES

Agrawal, A. V., Magulur, L. P., Priya, S. G., Kaur, A., Singh, G., & Boopathi, S. (2023). Smart Precision Agriculture Using IoT and WSN. In *Handbook of Research on Data Science and Cybersecurity Innovations in Industry 4.0 Technologies* (pp. 524–541). IGI Global. DOI: 10.4018/978-1-6684-8145-5.ch026

Al Mamun, M. A., & Yuce, M. R. (2019). Sensors and systems for wearable environmental monitoring toward IoT-enabled applications: A review. *IEEE Sensors Journal*, *19*(18), 7771–7788. DOI: 10.1109/JSEN.2019.2919352

Bibri, S. E., Krogstie, J., Kaboli, A., & Alahi, A. (2024). Smarter eco-cities and their leading-edge artificial intelligence of things solutions for environmental sustainability: A comprehensive systematic review. *Environmental Science and Ecotechnology*, *19*, 100330. DOI: 10.1016/j.ese.2023.100330 PMID: 38021367

Bourechak, A., Zedadra, O., Kouahla, M. N., Guerrieri, A., Seridi, H., & Fortino, G. (2023). At the confluence of artificial intelligence and edge computing in iot-based applications: A review and new perspectives. *Sensors (Basel)*, *23*(3), 1639. DOI: 10.3390/s23031639 PMID: 36772680

Chang, Z., Liu, S., Xiong, X., Cai, Z., & Tu, G. (2021). A survey of recent advances in edge-computing-powered artificial intelligence of things. *IEEE Internet of Things Journal*, *8*(18), 13849–13875. DOI: 10.1109/JIOT.2021.3088875

Chen, J., & Ran, X. (2019). Deep learning with edge computing: A review. *Proceedings of the IEEE*, *107*(8), 1655–1674. DOI: 10.1109/JPROC.2019.2921977

Das, S., Lekhya, G., Shreya, K., Shekinah, K. L., Babu, K. K., & Boopathi, S. (2024). Fostering Sustainability Education Through Cross-Disciplinary Collaborations and Research Partnerships: Interdisciplinary Synergy. In *Facilitating Global Collaboration and Knowledge Sharing in Higher Education With Generative AI* (pp. 60–88). IGI Global.

Ferrández-Pastor, F.-J., Mora, H., Jimeno-Morenilla, A., & Volckaert, B. (2018). Deployment of IoT edge and fog computing technologies to develop smart building services. *Sustainability (Basel)*, *10*(11), 3832. DOI: 10.3390/su10113832

Fraga-Lamas, P., Lopes, S. I., & Fernández-Caramés, T. M. (2021). Green IoT and edge AI as key technological enablers for a sustainable digital transition towards a smart circular economy: An industry 5.0 use case. *Sensors (Basel)*, *21*(17), 5745. DOI: 10.3390/s21175745 PMID: 34502637

Hartmann, M., Hashmi, U. S., & Imran, A. (2022a). Edge computing in smart health care systems: Review, challenges, and research directions. *Transactions on Emerging Telecommunications Technologies*, *33*(3), e3710. DOI: 10.1002/ett.3710

Hartmann, M., Hashmi, U. S., & Imran, A. (2022b). Edge computing in smart health care systems: Review, challenges, and research directions. *Transactions on Emerging Telecommunications Technologies*, *33*(3), e3710. DOI: 10.1002/ett.3710

Iftikhar, S., Gill, S. S., Song, C., Xu, M., Aslanpour, M. S., Toosi, A. N., Du, J., Wu, H., Ghosh, S., Chowdhury, D., Golec, M., Kumar, M., Abdelmoniem, A. M., Cuadrado, F., Varghese, B., Rana, O., Dustdar, S., & Uhlig, S. (2023). AI-based fog and edge computing: A systematic review, taxonomy and future directions. *Internet of Things : Engineering Cyber Physical Human Systems*, *21*, 100674. DOI: 10.1016/j.iot.2022.100674

Jha, K., Doshi, A., Patel, P., & Shah, M. (2019). A comprehensive review on automation in agriculture using artificial intelligence. *Artificial Intelligence in Agriculture*, *2*, 1–12. DOI: 10.1016/j.aiia.2019.05.004

Jiang, C., Cheng, X., Gao, H., Zhou, X., & Wan, J. (2019). Toward computation offloading in edge computing: A survey. *IEEE Access : Practical Innovations, Open Solutions*, *7*, 131543–131558. DOI: 10.1109/ACCESS.2019.2938660

Kalyani, Y., & Collier, R. (2021a). A systematic survey on the role of cloud, fog, and edge computing combination in smart agriculture. *Sensors (Basel)*, *21*(17), 5922. DOI: 10.3390/s21175922 PMID: 34502813

Kumar, P. R., Meenakshi, S., Shalini, S., Devi, S. R., & Boopathi, S. (2023). Soil Quality Prediction in Context Learning Approaches Using Deep Learning and Blockchain for Smart Agriculture. In *Effective AI, Blockchain, and E-Governance Applications for Knowledge Discovery and Management* (pp. 1–26). IGI Global. DOI: 10.4018/978-1-6684-9151-5.ch001

Maguluri, L. P., Ananth, J., Hariram, S., Geetha, C., Bhaskar, A., & Boopathi, S. (2023). Smart Vehicle-Emissions Monitoring System Using Internet of Things (IoT). In *Handbook of Research on Safe Disposal Methods of Municipal Solid Wastes for a Sustainable Environment* (pp. 191–211). IGI Global.

Maguluri, L. P., Arularasan, A., & Boopathi, S. (2023a). Assessing Security Concerns for AI-Based Drones in Smart Cities. In *Effective AI, Blockchain, and E-Governance Applications for Knowledge Discovery and Management* (pp. 27–47). IGI Global. DOI: 10.4018/978-1-6684-9151-5.ch002

Nishant, R., Kennedy, M., & Corbett, J. (2020). Artificial intelligence for sustainability: Challenges, opportunities, and a research agenda. *International Journal of Information Management*, *53*, 102104. DOI: 10.1016/j.ijinfomgt.2020.102104

Ramudu, K., Mohan, V. M., Jyothirmai, D., Prasad, D., Agrawal, R., & Boopathi, S. (2023). Machine Learning and Artificial Intelligence in Disease Prediction: Applications, Challenges, Limitations, Case Studies, and Future Directions. In *Contemporary Applications of Data Fusion for Advanced Healthcare Informatics* (pp. 297–318). IGI Global.

Roman, R., Lopez, J., & Mambo, M. (2018). Mobile edge computing, Fog et al.: A survey and analysis of security threats and challenges. *Future Generation Computer Systems*, *78*, 680–698. DOI: 10.1016/j.future.2016.11.009

Samikannu, R., Koshariya, A. K., Poornima, E., Ramesh, S., Kumar, A., & Boopathi, S. (2022). Sustainable Development in Modern Aquaponics Cultivation Systems Using IoT Technologies. In *Human Agro-Energy Optimization for Business and Industry* (pp. 105–127). IGI Global.

Sampath, B., Pandian, M., Deepa, D., & Subbiah, R. (2022). Operating parameters prediction of liquefied petroleum gas refrigerator using simulated annealing algorithm. *AIP Conference Proceedings*, *2460*, 070003. DOI: 10.1063/5.0095601

Sharma, D. M., Ramana, K. V., Jothilakshmi, R., Verma, R., Maheswari, B. U., & Boopathi, S. (2024). Integrating Generative AI Into K-12 Curriculums and Pedagogies in India: Opportunities and Challenges. *Facilitating Global Collaboration and Knowledge Sharing in Higher Education With Generative AI*, 133–161.

Sittón-Candanedo, I., Alonso, R. S., Corchado, J. M., Rodríguez-González, S., & Casado-Vara, R. (2019). A review of edge computing reference architectures and a new global edge proposal. *Future Generation Computer Systems*, *99*, 278–294. DOI: 10.1016/j.future.2019.04.016

Sittón-Candanedo, I., Alonso, R. S., García, Ó., Gil, A. B., & Rodríguez-González, S. (2019). A review on edge computing in smart energy by means of a systematic mapping study. *Electronics (Basel)*, *9*(1), 48. DOI: 10.3390/electronics9010048

Slama, S. B. (2022). Prosumer in smart grids based on intelligent edge computing: A review on Artificial Intelligence Scheduling Techniques. *Ain Shams Engineering Journal*, *13*(1), 101504. DOI: 10.1016/j.asej.2021.05.018

Ting, L., Khan, M., Sharma, A., & Ansari, M. D. (2022). A secure framework for IoT-based smart climate agriculture system: Toward blockchain and edge computing. *Journal of Intelligent Systems*, *31*(1), 221–236. DOI: 10.1515/jisys-2022-0012

Venkateswaran, N., Kiran Kumar, K., Maheswari, K., Kumar Reddy, R. V., & Boopathi, S. (2024). Optimizing IoT Data Aggregation: Hybrid Firefly-Artificial Bee Colony Algorithm for Enhanced Efficiency in Agriculture. *AGRIS On-Line Papers in Economics and Informatics*, *16*(1), 117–130. DOI: 10.7160/aol.2024.160110

Xu, G., Shi, Y., Sun, X., & Shen, W. (2019). Internet of things in marine environment monitoring: A review. *Sensors (Basel)*, *19*(7), 1711. DOI: 10.3390/s19071711 PMID: 30974791

Chapter 6
5G and Cloud Computing Synergy in Smart City Networks:
A Review

Mukul Misra
School of Computer Engineering, KIIT University (Deemed), Bhubaneswar, India

Abhishek Pradhan
https://orcid.org/0009-0006-0125-0097
School of Computer Engineering, KIIT University (Deemed), Bhubaneswar, India

Pratham Mishra
School of Computer Engineering, KIIT University (Deemed), Bhubaneswar, India

Satyam Kumar Singh
https://orcid.org/0009-0007-9558-7797
School of Computer Engineering, KIIT University (Deemed), Bhubaneswar, India

Ankit Kumar
https://orcid.org/0009-0000-6796-0979
School of Computer Engineering, KIIT University (Deemed), Bhubaneswar, India

Hitesh Mohapatra
https://orcid.org/0000-0001-8100-4860
School of Computer Engineering, KIIT University (Deemed), Bhubaneswar, India

ABSTRACT

As urban populations continue to grow, the demand for efficient, interconnected, and sustainable smart city solutions becomes increasingly imperative. In this context, the convergence of 5G wireless technology and cloud computing presents a promising opportunity to transform urban environments into intelligent and interconnected ecosystems. This paper explores the synergy between 5G and cloud computing within the context of smart city networks, aiming to identify key research challenges,

DOI: 10.4018/979-8-3693-5448-3.ch006

opportunities, and implications for future development. In this paper we examine the fundamental characteristics and capabilities of 5G technology, highlighting its ultra-fast data speeds, low latency, and massive device connectivity. We then delve into the foundational principles of cloud computing, emphasizing its scalability, flexibility, and on-demand resource provisioning. By integrating these two transformative technologies, we then look at the emerging cloud paradigms which utilize 5G and advances that result from them to solve the emerging challenges of smart city.

1. INTRODUCTION

The integration of 5G technology and cloud computing has unlocked unprecedented potential for the development of smart city networks. As cities strive to become more connected, efficient, and sustainable, this synergy offers a robust foundation for innovation and transformation. 5G, the fifth-generation wireless technology, promises ultra-fast speeds, low latency, and massive connectivity, while cloud computing provides scalable, on-demand access to computing resources and data storage. Together, they form a dynamic duo driving the evolution of smart cities, enhancing various aspects of urban life such as transportation, healthcare, energy management, and public safety. We explore the synergies between 5G and cloud computing within smart city networks, highlighting their transformative impact and the opportunities they present for creating more intelligent, responsive, and resilient urban environments.

1.1 Smart City and why it is needed

The world is experiencing a rapid urbanization process, with the urbanization level increasing from 39% to 52% in the last 3 decades (Chen et al., 2014), with 60% of the population forecasted to live in an urbanized environment by 2030 (Petrolo et al., 2014). This has led to traditional models of city planning to come up short as the complexity and sizes of cities keeps increasing. Poor urban infrastructure - such as unreliable power systems, congested roads, poor public transport systems and inefficient allocation of public resources to list a few can reduce a cities' competitiveness and economic prospects, according to (Joshi et al., 2016). To be able to fully capitalize on the increasing population and urbanization, the concept of smart cities was proposed, which aims to utilize the advancements made in the field of Information Communication Technologies (ICTs) (Parida et al., 2024) through smart mobile devices and sensors to optimize the use of public assets, improve coordination & manageability, and enhanced Quality of Experience (QoE).

Figure 1. Services provided by smart city

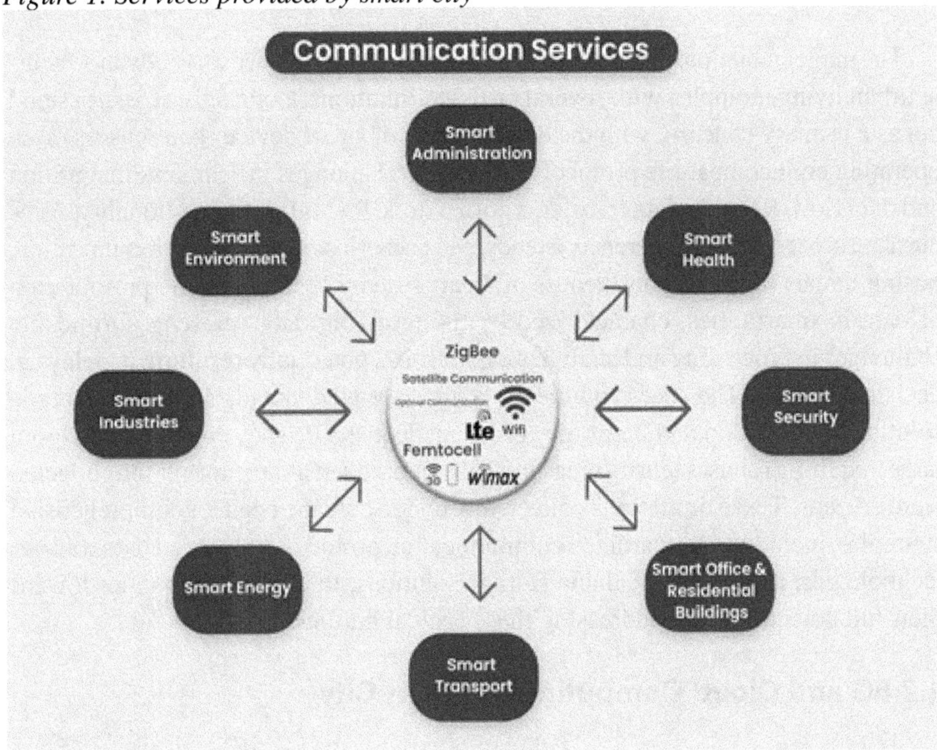

In the vision of a smart city, the proliferation of networking devices is set to undergo an unprecedented surge with the declining costs of sensory and mobile devices and increasing demand for data driven management systems. The implementation of a smart city relies heavily on an intricate network of interconnected devices and sensors, collectively forming the Internet of Things (IoT) as shown in Fig. 1. These devices, ranging from smart traffic lights and environmental sensors to connected vehicles and surveillance systems create a complex web of communication, offering the capacity to generate fresh insights and useful information from vast quantities of real-time data, as well as to remotely monitor, manage, and control equipment (Anthopoulos, 2015). For instance, meteorological data sets are growing at a very quick speed. In terms of agricultural development, identifying and extracting useful information from massive amounts of weather data can be quite helpful. Additionally, weather data analytics can assist in warning people ahead of time of potentially dangerous situations (Kim et al., 2017).

1.2 Limitations in implementation of Smart Cities

The implementation of smart cities, while promising transformative advancements in urban living, grapples with several critical limitations. Connectivity issues stand out as a primary concern, with the diverse array of smart devices and sensors often operating on incompatible protocols, leading to challenges in seamless integration and data flow. Reliable data transfer, a cornerstone for real-time functionality, faces hurdles such as signal interference, latency, and packet loss in wireless communication, posing threats to the responsiveness of smart systems. Moreover, the proliferation of data in smart cities, characterized by the term 'big data,' presents formidable challenges in processing and analytics capabilities, potentially resulting in delays in decision-making. The sheer volume of data generated demands innovative storage solutions, as traditional systems may prove inadequate. Privacy concerns also loom large, requiring robust security measures to safeguard sensitive information collected from citizens. These limitations collectively underscore the need for comprehensive strategies, including standardized communication protocols, advanced data transfer technologies, and secure, scalable storage solutions, to propel smart cities toward their full potential while addressing these critical hurdles.

1.2 5G and Cloud Computing in Smart City

The 5th generation of wireless networks (5G), having key characteristics as high bandwidth and data rates, massive connectivity, broad coverage, and low latency is the chosen moniker for the next generation of wireless communication, providing inter-networking of information between the different systems with overlapping purposes. The 3 emerging application scenarios of 5G are stated to be enhanced Mobile Broadband (eMBB), massive Machine Type Communications (mMTC), and ultra-Reliable and Low-Latency Communications (uRLLC) (Hashem et al., 2016). The sheer number of connected devices envisioned in a smart city presents a significant challenge for existing networks. 5G's ability to handle massive machine-type communication (mMTC) empowers the seamless integration of billions of sensors, actuators, and smart devices, forming the intricate nervous system of the city. It's ultra-reliable and low-latency communication (uRLLC), crucial for many services proposed under Smart City and eMBB, which ensures swift, reliable connectivity for IoT devices. Its high data transfer rates and low latency enhance real-time applications, such as traffic management and security systems.

In conjunction with 5G, cloud computing emerges as a linchpin in the realization of the Smart City vision (Morgado et al., 2018). The immense influx of data generated by the multitude of sensors and devices deployed throughout the urban landscape requires robust storage, processing, and analytical capabilities (Parida et al., 2022). Cloud computing provides the necessary infrastructure to efficiently manage and harness this vast amount of data. By leveraging cloud-based solutions, cities can create scalable and flexible platforms capable of handling the dynamic nature of the smart city ecosystem. Cloud computing also plays a critical role in fostering collaboration and innovation within the smart city framework. It facilitates the development and deployment of diverse applications and services that enhance urban living, ranging from intelligent transportation systems to healthcare monitoring and environmental sustainability initiatives. The cloud acts as a catalyst for the creation of a vibrant ecosystem of developers, city planners, and businesses working collaboratively to address the complex challenges of urbanization. Cloud-based analytics enable Smart Cities to derive meaningful insights from the vast datasets generated by sensors and devices (Shafi et al., 2017). With real-time data processing and analysis, city officials gain immediate insights into traffic patterns, energy consumption, air quality, and other critical parameters. These insights empower informed decision-making, allowing for swift responses to dynamic urban challenges and the optimization of city services, thus proving to be a critical component of Smart Cities. Furthermore, the scalability of cloud computing aligns seamlessly with the dynamic requirements of smart city deployments. As the number of connected devices grows, cloud infrastructure can easily scale to accommodate the increased data volume and computational demands. This ensures that the smart city ecosystem remains responsive and adaptable to evolving urban challenges and technological advancements.

2. LITERATURE REVIEW

A considerable body of literature includes case studies and best practices from Smart Cities around the world, providing insights into successful implementations, lessons learned, and the impact on urban living conditions. These case studies contribute valuable empirical evidence, offering practical guidance for city planners, policymakers, and stakeholders seeking to harness the potential of cloud computing and 5G in their respective urban contexts. Research on 5G in smart cities has focused on improving connectivity, data transfer, and urban services. Studies explore the integration of 5G for low-latency communication, efficient IoT device connectivity, and enhanced real-time data analytics. Intelligent Transportation System (ITS): According (Skouby & Lynggaard, 2014), The automotive sector is progressing towards enhancing the awareness of vehicles about their environment. Consequently, vehicles are evolving into real time sensor platforms that both transmit and receive data from their surroundings. This real time data needs to be transmitted with ultra-low latency, high mobility, high reliability, and scalability. Thus, 5G is vital for enabling the establishment of services like autonomous self-driving cars, vehicle platooning, and remote driving, along with helping reduce traffic accidents, increase traffic efficiency and improve access to emergency vehicles. The drawbacks being the security concerns relating to reliability of the sensor data, danger of malicious hacking and ethical concerns regarding autonomous vehicles.

2.1. High-Performance Data Analytics:

As discussed in the paper (Nowicka, 2014), in the dynamic landscape of Smart Cities, the application of high-performance data analytics goes beyond mere data processing; it serves as the cornerstone for unlocking actionable insights that drive real-time decision-making across diverse domains. Research in (Mohapatra, 2021) underscores the transformative potential of advanced Artificial Intelligence (AI) algorithms and Machine Learning (ML) techniques in this context. Cloud computing, with its capacity for parallel processing and scalable infrastructure, empowers Smart Cities to harness the power of these algorithms for the analysis of complex, multidimensional datasets. Furthermore, highlight the use of high-performance data analytics in optimizing resource allocation. Through real-time analysis of data streams from IoT devices, smart sensors, and urban infrastructure, cities can dynamically adjust services, manage energy consumption, and respond proactively to events such as traffic incidents or environmental changes. The integration of cloud computing and 5G ensures the necessary computational power and connectivity, fostering a sophisticated analytics ecosystem that enhances the overall efficiency and resilience of Smart Cities.

2.2. Predictive Analysis:

Predictive analysis emerges as a critical tool in the Smart City toolkit, allowing cities to move beyond reactive approaches to proactive, anticipatory strategies. Research emphasizes the role of predictive analysis in forecasting future states of systems and facilitating multiple scenario exploration. Cloud computing, with its ability to store and process vast datasets, provides the infrastructure needed for predictive modeling and analysis. The integration of 5G networks further enhances predictive analysis by providing low-latency communication and reliable connectivity (Dalla Cia et al., 2018). This is instrumental in scenarios such as traffic management, where predictive models can anticipate congestion and dynamically adjust traffic signals to optimize flow. The collaborative synergy between cloud computing and 5G technologies empowers Smart Cities to predict and prevent issues, contributing to improved efficiency, sustainability, and resilience. 5G Technologies: The advent of 5G technologies plays a pivotal role in the successful implementation of Smart

Cities, offering a comprehensive set of features that cater to the unique demands of contemporary urban environments. The multifaceted role of 5G, providing high reliability, low latency, and support for a high density of Internet of Things (IoT) devices. The high data rates and low-latency communication capabilities of 5G networks lay the foundation for real-time data exchange among interconnected devices. Furthermore, the reliability and capacity of 5G networks contribute to the seamless integration of various smart city services. This ranges from connected vehicles and smart grids to healthcare applications and public safety systems. As cities evolve into intelligent, interconnected hubs, the deployment of 5G technologies becomes instrumental in creating a robust and adaptive urban fabric that enhances the overall quality of life for residents (Xue & Qin, 2023).

2.3. ICT Components:

The Information and Communication Technology (ICT) components form the backbone of a Smart City, encompassing a diverse array of technologies essential for the digital transformation of urban environments. Research delves into these components, which include sensor webs, communication networking technologies, and various networking types like Machine-to-Machine (M2M), Home Area Networking (HAN), Personal Area Networking (PAN), Local Area Networking (LAN), Metropolitan Area Networking (MAN), and Wide Area Networking (WAN). According to the paper (Wieclaw et al., 2017), the modern municipality faces a complex balancing act: optimizing budget allocations, enhancing service delivery, and adapting to ever-evolving technological landscapes. Cloud services have emerged as a powerful tool to address these challenges, offering scalable, flexible, and cost-effective solutions. Among these offerings, three stand out for their potential impact on municipal operations: Network-as-a-Service (NaaS), Storage-as-a-Service (STaaS), and Sensor-as-a-Service (SSaaS).

2.4. Network-as-a-Service (NaaS):

Imagine a city able to instantly adapt its network capacity to accommodate fluctuating demands, be it a bustling festival or a critical emergency response. NaaS makes this a reality. By eliminating the need for dedicated, physical infrastructure, municipalities gain access to virtual networks built upon diverse underlying infrastructures managed by service providers. This translates to enhanced scalability, cost-effectiveness, and disaster resilience. Municipalities can scale network resources up or down as needed, avoiding hefty upfront investments and ongoing maintenance costs. Additionally, geographically distributed infrastructure ensures network availability even during disruptions (Hogan et al., 2011).

2.5. Storage-as-a-Service (STaaS):

Secure Data Repositories in the Cloud - Managing and securing vast amounts of data can be a daunting task for municipalities. STaaS eases this burden by offering secure, scalable cloud storage solutions. Gone are the days of maintaining physical servers and worrying about data breaches. Instead, municipalities can leverage provider-managed security measures and industry-standard compliance certifications, ensuring data integrity and adherence to regulations. Furthermore, pay-as-you-go pricing models align with fluctuating storage needs, optimizing costs and freeing internal IT resources for core competencies (Sabt et al., 2015).

2.6. Sensor-as-a-Service (SSaaS):

Unlocking the Power of Sensor Data - A city pulsing with sensors, collecting real-time data on traffic flow, air quality, and resource utilization. SSaaS makes this sensor-driven vision a reality by providing access to sensor data through cloud services. Municipalities can collect, analyze, and integrate this data into various applications, leading to informed decision-making, improved operational efficiency, and reduced infrastructure costs. For example, real-time traffic data can optimize traffic light management, while environmental sensor data can inform air quality initiatives. The strategy creation process inevitably includes the use of cloud-based technologies in the architecture of smart city information systems for the implementation of context-sensitive services.

2.7. Infrastructure-as-a-Service (IaaS):

Imagine a vast pool of virtual building blocks - servers, storage, and networks - accessible on-demand through the cloud. IaaS offers precisely that, allowing you to construct your personalized digital city infrastructure. Choose the specific components you need, scaling them up or down as your demands fluctuate. Think of it as having a virtual data center, minus the upfront costs and ongoing maintenance of physical hardware. This flexibility empowers cities to adapt to diverse needs with ease. Manage traffic data on scalable servers, store crucial citizen data in secure cloud storage, or deploy real-time traffic management systems – all within your own customized environment. You control the foundation, laying the groundwork for innovative smart city initiatives.

2.8. Platform-as-a-Service (PaaS):

It is a pre-built development platform in the cloud. Instead of setting up complex infrastructure, you get a ready-to-use environment complete with programming languages, tools, databases, and more. It's like having a dedicated developer playground readily available, allowing you to focus on crafting your applications, not infrastructure management. This translates to faster development cycles. Leverage pre-built tools and automation to streamline the process, focusing your energy on application logic and innovation. Deployment becomes effortless - use built-in tools to launch your applications on the cloud quickly and seamlessly. Imagine developing smart city apps like citizen engagement platforms or environmental monitoring systems faster than ever before, with automated deployment ensuring scalability for future growth.

2.9. Software-as-a-Service (SaaS):

It offers the ultimate in convenience - no installation, no maintenance, just instant access to powerful applications through the cloud. Think of it as subscribing to software instead of owning it. The provider handles everything - server upkeep, updates, security patches - while you simply log in and start using the application. This translates to immediate benefits. Eliminate installation delays and start using applications like data analytics platforms for citizen insights or communication tools for public engagement instantly. No technical expertise is required - simply log in and utilize the software for immediate benefit. Benefit from continuous improvement and new features without manual updates or downtime, ensuring continuous improvements always having the latest version.

2.10. AI, Blockchain and ML integrated cloud:

In the rapidly evolving landscape of smart cities, the integration of AI, machine learning, and blockchain technologies is revolutionizing urban living. AI and machine learning provide advanced data analytics and automation, enabling efficient resource management and enhanced public services. Blockchain ensures secure, transparent transactions and data integrity, critical for trust in public systems. Cloud computing serves as the backbone, offering scalable infrastructure and seamless connectivity, facilitating real-time data processing and storage. Together, these technologies are transforming smart cities into intelligent, secure, and adaptive urban environments (Ahmad et al., 2018).

2.11. Deep Belief Networks (DBNs)

Recent advancements in Deep Belief Networks (DBNs) and sensor networks are revolutionizing various fields, particularly in intelligent sensing and data analysis. In DBNs, improved training algorithms, such as layer-wise greedy learning combined with advanced optimization methods like gradient clipping and adaptive learning rates, have enhanced their performance in deep learning tasks. Integration of DBNs with reinforcement learning and generative models is allowing more complex decision-making and pattern recognition in real-time applications [46]. On the sensor networks side, developments in edge computing, energy-efficient communication protocols, and advanced data fusion techniques are improving network reliability and extending operational lifetimes. When combined, these innovations enable more intelligent, scalable, and responsive sensor systems that can analyze vast amounts of data locally, reducing latency and enhancing decision-making for applications like smart cities, environmental monitoring, and healthcare (Du et al., 2020).

3. PROPOSED STUDY

The rapid evolution of smart city networks, powered by advancements in 5G and cloud computing, stands at the forefront of modern urban development. These technologies promise to revolutionize the way cities operate, manage resources, and serve their citizens. However, despite their potential, the integration of 5G and cloud computing within smart city infrastructures exposes a complex landscape of research gaps and challenges that must be addressed to fully realize their capabilities. As urban areas continue to expand and demand for smarter, more efficient services grows, the synergy between 5G and cloud computing becomes increasingly critical. This introduction aims to shed light on the existing research gaps in the field of 5G and cloud computing synergy within smart city networks, highlighting the need for innovative solutions to pave the way for the cities of tomorrow (Mohapatra et al., 2023).

Security Risks: Attacks including denial-of-service (DDoS), side-channel, malware insertion, and authentication and authorization are still possible. Regretfully, compared to the cloud, the edge environment is far more complex. The following four issues affect cloud computing security (Khan et al., 2019).

- Weak Computation Power: Cloud computing units, pivotal in processing vast amounts of data for smart cities, are often constrained by the need to meet real-time processing requirements. This limitation significantly impacts their ability to execute complex security algorithms, which are crucial for protecting the data integrity and privacy of smart city networks. The dilemma here is the balancing act between processing speed and the computational depth required for robust security measures. Enhancing computation power without sacrificing real-time performance requires innovative approaches, such as distributed computing models that leverage the computational capabilities of edge devices or the development of more efficient algorithms that reduce computational overhead while maintaining high security standards (Hu et al., 2017).

- Attack Unawareness: The pervasive nature of smart devices, many of which operate without a traditional user interface, poses a unique challenge in cybersecurity management within smart city frameworks. Users and even system administrators may remain oblivious to cyberattacks until significant damage has occurred (Bonomi et al., 2014). This scenario underscores the need for advanced, automated monitoring systems capable of detecting and responding to threats in real-time. Solutions might include the deployment of AI-driven anomaly detection systems that can identify and mitigate unusual activities without human intervention, thereby safeguarding the network even in the absence of direct user monitoring.
- OS and Protocol Heterogeneities: The diversity in operating systems and protocols across devices accessing edge platforms in smart cities adds a layer of complexity to network management and security. This heterogeneity complicates efforts to ensure seamless communication and interoperability among devices, leading to inefficiencies and potential security vulnerabilities. Addressing this challenge requires the development of universal or adaptive communication protocols that can automatically adjust to different device specifications and operating systems. Moreover, there is a pressing need for comprehensive management platforms that can oversee the operation of such diverse devices in a unified manner, ensuring consistent performance and security across the network (Zhang & Zhou, 2007).
- Coarse-Grained Access Control: The prevalent models of access control within edge computing environments often lack the granularity necessary for effective data protection and privacy management. Most systems employ coarse-grained access controls that do not accommodate the nuanced needs of smart city applications, such as differentiated permissions (e.g., read-only, write-only, no read/write, and full access). This limitation can lead to either excessive access restrictions, hampering the efficiency of city operations, or overly lenient access, exposing the system to data breaches (Zhang et al., 2017). The solution lies in the adoption of fine-grained access control (FGAC) models that allow for precise specification of access rights based on the user, context, and the type of data being accessed. Implementing FGAC

in smart city networks would significantly enhance security and operational flexibility, enabling more precise and dynamic access management tailored to the specific needs of each application (Ren et al., 2017).

5G's new technology will expose users to greater threats and provide more opportunities for hackers to enter the system (Ahmed & Ahmed, 2016). As an illustration, network slicing technology raises security concerns about interslice communications and network slice isolation (Bonomi et al., 2014). The growing reliance of the 5G network on software will raise the risks associated with software development and updates. Those dangerous backdoors that threat actors have introduced are more difficult to find. As a result, security issues surrounding the integration are more complicated and include issues with end-device security, virtualization security, edge server security, wireless network security, and core network security (Mao et al., 2017). There aren't many specific security designs for the edge computing paradigm, but it doesn't mean new security mechanisms have to be created from scratch. Since cloud and grid computing, for example, share some of the same security mechanism presumptions as edge computing, their security solutions may eventually adapt to the edge environment (Wang et al., 2014). Furthermore, specific security technologies such as hardware-assisted trusted execution environments [16] and machine learning-based security schemes (Yi et al., 2015) are required for edge computing and 5G networks.

Heterogeneity Issues: The heterogeneity within edge computing, particularly in the context of smart city networks utilizing 5G and cloud computing synergy, introduces a multifaceted set of challenges that stem from the diverse array of infrastructures, architectures, platforms, and communication technologies (Vaquero & Rodero-Merino, 2014). This diversity affects each layer of the typical three-tier architecture—end devices, edge nodes, and cloud platforms—along with the communication networks that interlink these tiers. Below, we further expand on these points to highlight the intricacies of managing such heterogeneity.

- End Devices Heterogeneity: End devices, serving as the primary data sources in the outermost layer, exhibit a wide range of hardware types and data characteristics. This variety spans simple sensors collecting environmental data to complex systems like smartphones and IoT devices with multifunctional capabilities. The challenge here extends beyond just the physical hardware differences; it encompasses the diverse nature of the data generated, including variations in format, volume, velocity, and veracity. Addressing this heterogeneity requires adaptable and resilient data processing frameworks that can normalize and interpret the disparate data types efficiently. Moreover, ensuring seamless integration and communication between these varied devices necessitates sophisticated protocols that can dynamically adjust to the device-specific requirements (Vaquero & Rodero-Merino, 2014).

- Edge Nodes Heterogeneity: Edge nodes, deployed across different environmental contexts, display significant variation in their form factors, ranging from routers and gateways to dedicated servers and base stations (Satyanarayanan et al., 2011). This diversity is not limited to physical hardware but extends to differences in computational and storage capabilities, operating systems, and the software applications they run. Such heterogeneity complicates the management and coordination of edge computing resources, especially in terms of workload distribution, resource allocation, and maintenance. Developing standardized yet flexible management systems that can cater to the unique needs of each edge node, while ensuring efficient operation and security, is crucial to overcoming these challenges.
- Cloud Platforms Heterogeneity: While individual cloud platforms may present a unified hardware and software environment, heterogeneity arises in scenarios involving the collaboration of multiple cloud services. This can lead to compatibility issues, data silos, and increased complexity in data management and application deployment across platforms. Tackling this form of heterogeneity demands interoperable cloud services and APIs that facilitate seamless data exchange and function orchestration. Furthermore, adopting multi-cloud management tools and frameworks that provide a unified interface for managing diverse cloud resources can greatly alleviate these issues.
- Communication Networks Heterogeneity: The communication networks linking the three tiers are characterized by their own form of heterogeneity, comprising both high-speed connections to the cloud and various wireless links to edge devices, such as Wi-Fi, Bluetooth, 3G, and 4G (Chen et al., 2014). This diversity affects network reliability, latency, bandwidth, and security, presenting a significant hurdle in ensuring consistent and efficient data transmission across the smart city network. The solution lies in the development of advanced network management systems capable of dynamically optimizing connections based on the requirements of each application and the current network conditions. Additionally, the impending widespread adoption of 5G technology offers the promise of addressing many of these challenges through its enhanced speed, lower latency, and greater capacity, which could unify communication standards across the network (Tufail et al., 2021).

Addressing the heterogeneity inherent in edge computing for smart city networks requires a concerted effort toward developing adaptable, resilient, and standardized technologies and protocols. By focusing on these challenges, it is possible to harness the full potential of 5G and cloud computing synergy, paving the way for more efficient, secure, and responsive smart city infrastructures.

Self-management and configuration: In addition to increasing configuration and management expenses, the vast array of linked sensors in IoT networks may make it impossible for users to keep track of every connected sensor. In addition to having the essential components for intelligent task performance, a sensor should also be capable of self-management, which includes the ability to adapt to changes in its surroundings, switch between networking protocols, perform predictive maintenance, optimize its own performance, and self-heal from errors or faults while sensing (Tufail et al., 2021).

Computation offloading: Sensors with limited hardware capacity and non-scalable resources, like computational power, storage capacity, and battery life, are ideal for a wide range of applications. In addition, edge, fog, and cloud computing are only a few of the platforms and paradigms that IoT and 5G networks offer to improve its offerings. Remote platforms may handle heavy computing activities locally, which would improve data processing, boost sensor application performance, and conserve energy (Zhang et al., 2017).

3.1. Critical Analysis of existing models

In the realm of cloud computing, several optimization techniques have been proposed in order to enhance the performance and to provide better solutions to problems raised by Smart City and IoT networks. Some important variants are listed below:

A. Transparent Computing

In order to separate software, including operating systems (OS), from the diverse hardware of mobile devices and the Internet of Things, transparent computing was proposed (Zhang et al., 2017). It completely "transparently" serves users while hiding the specifics of service provisioning (Wang et al., 2014). In order to achieve this, TC gives devices the ability to select services via networks as needed, ignoring the specifics of service supply, like software updates and maintenance. In general, the following attributes can be used to describe TC:

1. By operating in a client-server model, TC conceptually unifies the dispersed devices on the network into a single system. Based on the capabilities of the devices and the state of the networks, the system intelligently offers services. In order to deliver elastic services to clients connected to a network, centralized resource management is assigned to the server side.

2. TC creates an on-demand service loading and execution architecture that allows heterogeneous devices to logically separate their software and hardware. This architecture gives the client devices the capacity to dynamically execute cross-platform services from remote servers through high-speed networks.
3. TC allows client devices to perform local block-streaming execution of distant services by fetching them on-demand from the server side, hence making full use of the computation resources present in the client devices. Block-streaming execution allows clients to load and run only the relevant portions of software when customers want a particular service, rather than the entire programme. Thus, it is possible to greatly enhance both the energy efficiency and the latency of service providing (Sharma et al., 2021).

Figure 2. Layers of a TC System

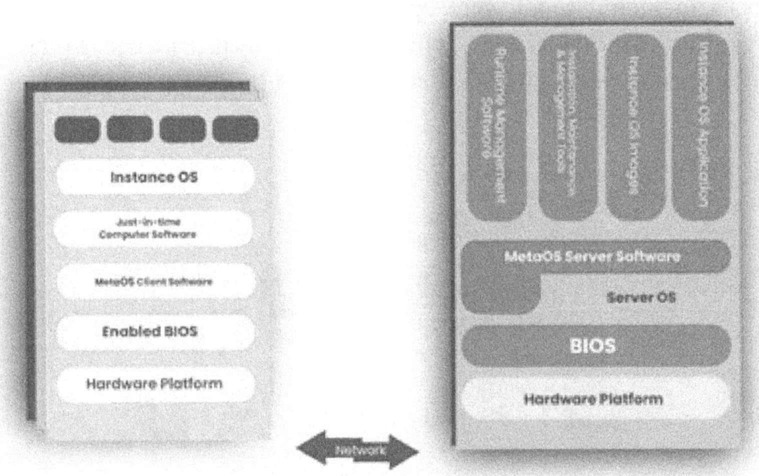

Figure 2 shows an example of a typical TC implementation. In both the spatial and temporal realms, it expands upon the traditional von Neumann design. A single computer is expanded to include network-connected computers and devices in the TC architecture. The system/service software that can be dynamically loaded and executed on client devices using the Meta OS platform is stored on the servers. The Meta OS, which is based on an underlying operating system, is intended to hide the heterogeneity of hardware and standardize the interfaces to higher layers. The Meta OS platform can thus initialize and manage several commodity OS and applications.

The benefits of TC have recently come to light in the context of the Internet of Things (IoT), which highlights the need for a well-thought-out solution to manage the massive volume of software for extremely diverse hardware infrastructures, ranging from expensive servers, laptops, and smartphones to inexpensive sensors. A scalable TC-based Internet of Things architecture was put forth in (Ahmed et al., 2022) in order to provide prompt and adaptable services at the network's edge. In their suggested architecture, the cloud serves as the storage and administration center, providing centralized control for the entire system, while nearby edge servers function as the TC server, storing some frequently used Apps and data and achieving rapid response for the service requests of IoT devices. Developed as the MetaOS for Internet of Things devices, LiteTCOS is in charge of block-streaming service execution and loading commodity operating systems. In addition, they suggested Block-stream as a Service (BaaS), a novel TC-based service paradigm, to enable ambient service computing for Internet of Things devices. As seen in Fig. 2, they clearly outline their BaaS architecture, which divides all software into a series of code blocks, including middleware, apps, libraries, and operating systems (Badidi, 2022). A portion of the code blocks will be dynamically provided to IoT devices upon receipt of a service request by the server, enabling prompt and effective response.

Figure 3. Architecture of a Mobile Edge System

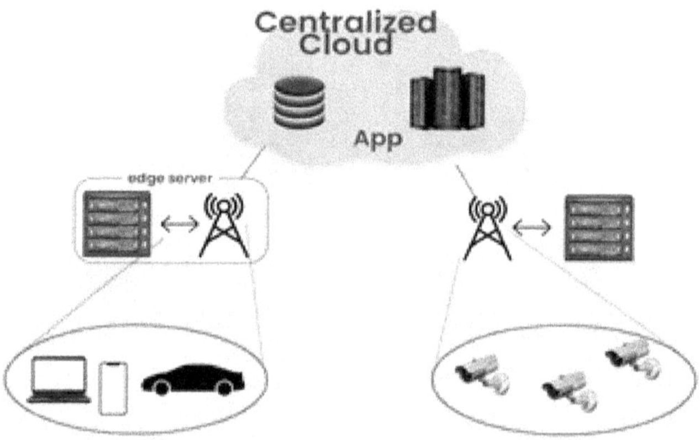

B. Mobile Edge Computing

The European Telecommunication Standards Institute (ETSI) launched MEC to make cloud computing services available to mobile subscribers in close proximity (Badidi, 2022). As seen in Fig.3, MEC can process user requests at the network edge with lower latency and location awareness thus enhancing user experience, as well as reducing the load over the core network, by placing MEC servers at the macro or micro base stations. In addition to software-defined networks and network function visualization, MEC has been identified as a critical enabling technology for the 5G future.

In order to promote the smooth participation of suppliers, service providers, and other external parties in MEC, an industry standardization committee has been formed inside ETSI to create guidelines for a standardized and transparent MEC environment. Huawei, Intel, Nokia, Vodafone, NTT DOCOMO, and other companies are among the members of the ETSI MEC ISG. 2014 saw the release of the first technical white paper that explained MEC and the MEC platform's standard architecture (Kiruthika & Ponnuswamy, 2021). It also covers the main MEC enabling strategies and difficulties. The group has documented a number of MEC specifications between early 2017 and 2015. These include the technical requirements of MECs as well as service scenarios and nomenclature. The moniker MEC has been modified to multi-access edge computing, which may reflect the increased interest from non-cellular operators, despite the group's assertion that it primarily focused on the early integration of cloud computing technology into cellular networks. Both academics and industry have put forth a number of strategies in recent years to improve MEC performance, including system implementation, multiuser resource allocation, modeling, and so forth. Numerous investigations have examined MEC's development from various angles. Additionally, they go over the taxonomy of MEC from a variety of angles, including its features, actors, applications, applications' access technologies, and important enablers. Reference presents an overview of the core enabling technologies for MEC. System performance and MEC platforms are taken into account when discussing MEC orchestration, illuminating the various orchestration setups. The writers also provide an introduction to the MEC architectures and common deployment situations. The authors of (Xibilia et al., 2020) examine the most recent MEC research, with an emphasis on the cooperative management of processing and communication resources. Computation offloading, the primary use case of MEC, is covered in the survey. The primary facilitators of MEC, such as cloud computing, SDN/NFV, and smart devices, are surveyed by the writers of Reference (Hwang & Huang, 2017). Additionally, they go over a few important technologies for mobile edge networks, including SDN/NFV, cloud computing, and smart devices.

C. Fog Computing

In (Chen et al., 2021), Fog computing is defined as “a scenario where a huge number of heterogeneous (wireless and sometimes autonomous) ubiquitous and decentralized devices communicate and potentially cooperate among them and with the network to perform storage and processing tasks without the intervention of third parties. These tasks can be for supporting basic network functions or new services and applications that run in a sandboxed environment. Users leasing part of their devices to host these services get incentives for doing so”. The Internet of Things (IoT) necessitates location awareness and prompt response in addition to wireless access and mobility assistance, which is why fog computing was first proposed. Furthermore, fog computing emphasizes that every network device along the data routing channel can offer data processing and storage services for end devices by utilizing an n-tier architecture to deliver more flexible services. To facilitate computing, storage, and networking resource pooling, a fog tier is physically positioned between the cloud and the Internet of Things devices, as seen in Fig. 4, The fog tier is made up of a huge number of heterogeneous nano servers, which can be temporary devices like cell phones, sophisticated sensors, and cars, or dedicated equipment like edge routers and set-up boxes.

Figure 4. Architecture of a Fog Computing System

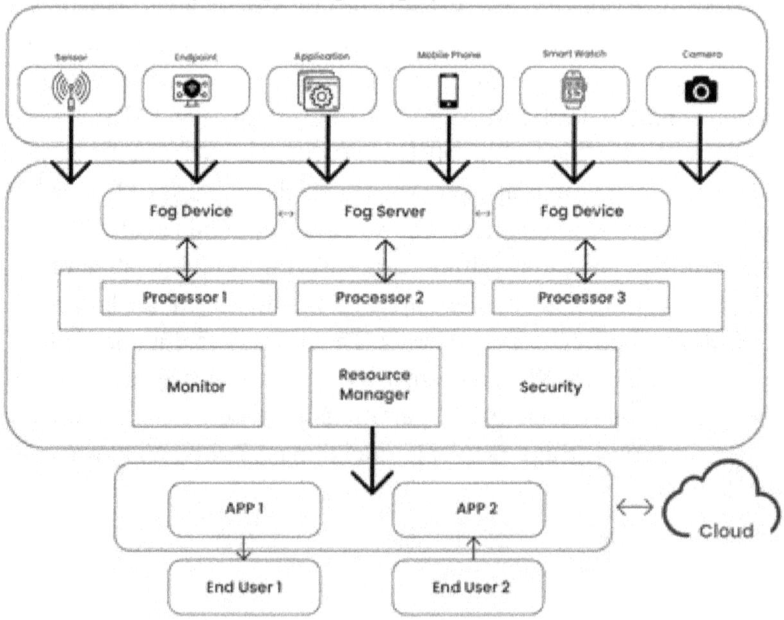

Applications for fog computing include those that need to work with cloud servers in order to provide real-time control to improve system performance instantly and batch data analytics for long-term strategic insights. In order to reduce service latency, real-time control uses fog servers that are widely spread, whereas batch data analytics makes use of cloud servers' massive processing and storage capacities. Fog computing is being advanced, as seen by the formation of the Openfog Consortium in 2015 by prominent IT businesses including Cisco, Intel, and Microsoft, among others. Fog computing's adaptability is demonstrated by its many applications in fields such as wind farms, smart grids, and smart traffic systems. A substantial body of literature has explored fog computing, each with its unique focus. Paper (Hwang & Huang, 2017) provides an overview of enabling technologies and future prospects in fog computing. Another survey addresses fog computing's definition, application scenarios, and system design challenges. Some researchers anticipate fog computing's potential as a platform for challenging IoT scenarios. Others highlight its advantages for IoT and outline suitable application scenarios. Additionally, a set of evaluation criteria has been proposed to assess fog computing's effectiveness compared to other computing paradigms. These contributions collectively enrich the understanding and development of fog computing.

D. Cloudlet

The year 2009 saw the introduction of the concept of cloudlet by a Carnegie Mellon University research team. A cloudlet is a little data center that is placed strategically close to mobile users; these can be found in places like classrooms or coffee shops, as seen in Fig. 5, A cloudlet's main goal is to make mobile applications more responsive, especially those with strict end-to-end latency and jitter requirements. These applications require millisecond-scale reaction times, which are difficult to attain over the Internet. Closely located cloudlets allow servers to provide extremely responsive cloud services by bridging this gap, adding to the traditional three-tier cloud hierarchy of mobile users, cloudlets, and cloud infrastructure. Furthermore, this architecture anonymizes personal data prior to sending it to the cloud, emphasizing consumer privacy. Cloudlets have been shown to have a number of applications, such as training assistance for ports and assisted driving help.

Figure 5. Simplified Architecture of a Cloudlet

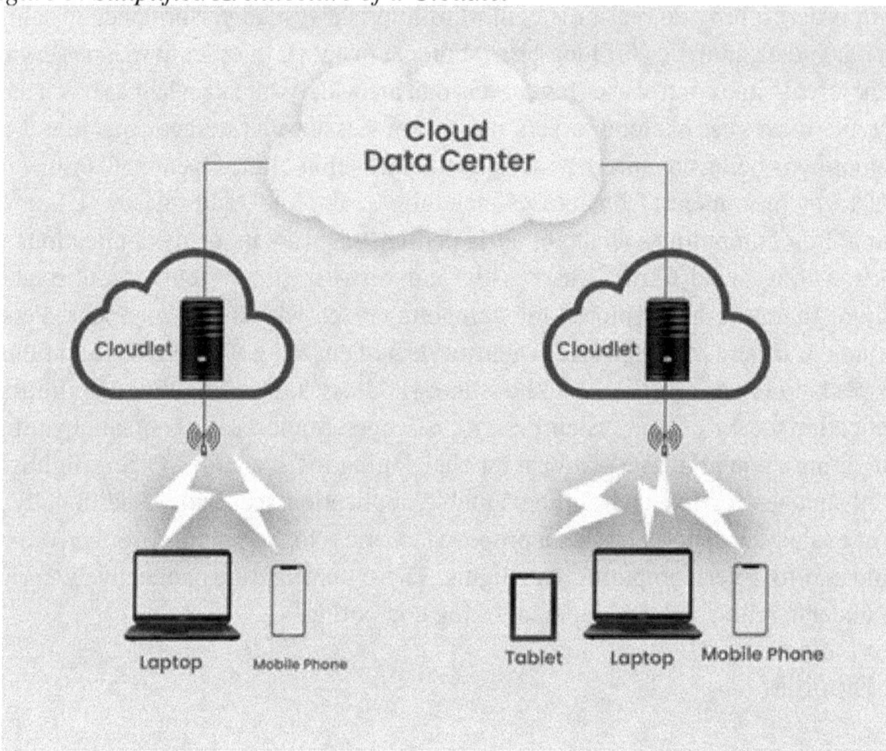

Since cloudlets were first introduced, the CMU research group has published a great deal, mostly concentrating on two areas: finding useful uses for cloudlets and creating virtual machines that facilitate user mobility. According to the authors of (Meng et al., 2016), real-time cognitive support—like speech and facial recognition—might be the "killer app" for cloudlets, necessitating fast reactions in order to gently direct user attention. Additionally, work in (Joshi et al., 2016) suggests a virtual machine overlay architecture for cloud computing that is built on the OpenStack ecosystem. This design minimizes data flow across cloudlets while simplifying cloudlet identification and hand-off in mobile scenarios. Acknowledging the commercial possibilities of cloudlets, CMU launched the open edge computing effort in collaboration with major industry partners such as Intel, Vodafone, Nokia, etc. The objective of this programme is to assess user acceptability and create important technologies related to cloudlets. Rather than placing specialized computing facilities close to customers, several current works utilize mobile devices' idle processing power to offer computing services; this idea is known as "Ad-hoc cloudlet". Ad-hoc cloudlets are IoT devices with low processing power, such RFID sensors, where mobile devices work together on data analytics using short-range radio communication protocols. Although this method reduces the amount of work needed to create and maintain the facility, the mobility and dynamic local load of mobile devices pose major obstacles to effective computing and service delivery in an ad-hoc cloudlet.

E. Moisture Computing

The word "moisture" comes from the field of meteorology, where it refers to the vapor that is found in the atmosphere, which is often found between the earth's surface and the clouds. As a result, the computing infrastructure will be situated above the network's edge but beneath the cloud infrastructure. The benefits of the two most popular paradigms—cloud computing, which is characterized by centralization, and edge computing, which is characterized by high distribution are combined in the moisture computing paradigm. Notably, fog computing suggests that in order to improve performance in various ways, the processing, storage, and associated infrastructure be moved to the edge of the network and near the user equipment (UE). Such computing paradigms have been dubbed MEC, cloudlet, and mist computing, depending on how close the infrastructure is to the UE. Unlike other fog-based paradigms like MEC, which is one hop away, and mist computing, where computation takes place within the UE, the proposed paradigm, called MC, advocates for placing the infrastructure a few hops away from the UE. This is in line with the fog computing principle of pushing computation towards the UE.

4. COMPARATIVE ANALYSIS

In order to develop a TC paradigm, the distributed clients must be able to perform computations over the network by accessing the operating system and software images that are kept on the centralized transparent servers. Congestion in the servers' access is caused by the large number of queries distributed clients make to obtain operating systems and application software. Therefore, in order to improve overall system performance, the latency in data transfer between clients and servers within TC systems needs to be carefully examined. To address this problem, block caches from both servers and clients are combined into a two-tier cache architecture. In order to minimize input/output processing time, client cache is used to store service requests from clients and the related data received from remote servers. Expanding the Von-Neumann architecture in a network situation is the goal of transparent computing [38]. Reliability and efficiency in computing are dependent on a robust communication network. Mobile computer devices are becoming more and more common as a result of the development of mobile communication technologies. Investigating cutting-edge caching techniques appropriate for transparent computing in mobile network contexts is therefore necessary. It was suggested to improve the block-level caching approach by incorporating an access technology which is stored locally with a block level caching method based on the temporal characteristics of server access, while also accounting for unreliable connection of wireless networks and limited bandwidth. Within the domain of mobile networks based on the all-IP cellular infrastructure, traditional centralized mobile network's capacity is insufficient to satisfy the growing need for an abundance of a variety of multimedia content. Cache placement issues for the user devices, radio access network (RAN), and core network have been the focus of recent efforts. A significant amount of traffic from RAN and user devices that communicate data between and within ISPs (Internet service providers) overloads the core network. As a result, the central network faces significant difficulties when handling high volumes of traffic. Consequently, the core network becomes an essential location for the broad implementation of content caching.

Research projects aimed at caching popular content in the mobile core network have shown promise in reducing content traffic by a large percentage, between one and two thirds [39]. These projects also look into the locations where content delivery technologies are being used for caching. MEC has the potential to reduce transmission energy consumption by taking advantage of consumers near proximity to MEC servers. Utilizing cutting-edge wireless communication techniques like adaptive power management and interference cancellation can help reduce the energy used to offload jobs. In order to reduce computation-related delay and energy consumption on mobile devices as well as MEC servers, dynamic adjustment of CPU-cycle frequency is essential. To be more precise, increasing the frequency of CPU cycles shortens processing times but increases energy consumption. An ideal balance between processing lag and energy usage requires optimal control of CPU-cycle frequency. When the MEC server is overworked or has limited processing capability, compute scheduling delays significantly in addition to task-execution latency. The overall compute latency can be effectively reduced by implementing intelligent scheduling principles and load-balancing mechanisms.

Table 1. Comparison of Various Paradigms Discussed

Paradigms	Year Introduced	Supporting Organization	Computing Infrastructure location	Research Focuses
Transparent Computing	2004	None	End devices and nearby devices	Cross-platforming and providing services on demand
Mobile Edge Computing	2014	ETSI	Edge of the Network (Base Station)	Offloading uplink compute from downlink caching
Fog Computing	2012	OpenFog Consortium	Anywhere Between Cloud and End Device	Offloading uplink compute from downlink caching
Cloudlet	2009	Open Edge Computing Initiative	Edge of the Network (Typically One Hop away from the End Device)	Offloading computation, managing virtual machines
Moisture Computing	2021	None	Above the Edge of the Network (Typically two hops away from the End Device)	Uplink computation offloading, reducing latency for real time tasks

The end device may be several kilometers away from the cloud computing infrastructure, which causes increased latency and slower reaction times. Table 1 provides a basic comparison of closeness. On the other hand, proximity to mobile edge computing infrastructure might only reach a few meters from the end device, resulting in significant expenses because numerous computing infrastructure points are required to cover a given geographic area. The incapacity of mobile edge computing infrastructure to adapt to changes in traffic density is a clear indication of its limits. When the computing infrastructure is overloaded, latency increases significantly. For Internet of Vehicles applications, SDN/NFV and fog computing-based solutions provide improved Quality of Service standards; nevertheless, when duties are offloaded to Mobile Edge Computing (MEC) servers, latency may be compromised, especially in densely populated networks. Moisture computing, which applies a semi-distributed computing paradigm to lower end-to-end latency while enhancing the dependability of the data processed through computations and analytics closer to the user equipment, was offered as a solution to these drawbacks [40].

The establishment of Fog Computing Nodes (FCNs) at different points between end devices and cloud Data Centre Networks (DCNs) is recommended for the Fog Computing implementation. A greater degree of freedom in terms of choosing devices to serve as FCNs is offered by fog computing. However, FCNs adds more storage and processing power to already-existing devices, but at the expense of less computational and storage capacity than Cloudlets and Mobile Edge servers. On the other hand, compared to Fog Computing, the adoption rate of Mobile Edge Computing and Cloudlet Computing is hampered by the need for specialized hardware.

Still, these devices can serve as servers for Mobile Edge Computing (MEC) and as Cloudlets. When it comes to close proximity to the periphery in the context of fog computing, the FCN may not function as the end device's principal access point because old devices are used as FCNs. For example, the first router that is connected to the end device could not have enough capacity to run an FCN framework, so the closest FCN is located several hops away. This makes it easier to provide Fog Computing with inter-node communication capability [41].

On the other hand, while using Cloudlet and Mobile Edge Computing, the devices connect directly to the node using Wi-Fi and the base station's mobile network, respectively. By using gateways as FCN devices, fog computing makes it possible to support non-IP based protocols like BLE and Zigbee. As a result, Fog Computing can connect to a wider range of endpoints thanks to this feature, which also makes protocol translation possible. It is clear from looking at each of these implementations' request processing system and services that they all follow a similar hierarchical structure. With this method, a supervisory entity monitors the underlying nodes and communicates with them to get information about resource availability and status. Because Fog devices are diverse and heterogeneous, an abstraction layer is required, in contrast to conventional systems where nodes are specialized devices. MEC, on the other hand, provides the advantage of comprehensive data regarding the location and network load of the end user, resulting in improved context awareness [42]. When it comes to Cloudlets, the data that is handled and processed on these entities already has a backup copy on the cloud and is updated after processing is finished, thus it is in a soft state. As a result, another benefit of Cloudlets is that end devices can use a new Virtual Machine on the Cloudlet for every session, while the Cloudlet manages post-use cleanup by backing up processed data to the cloud and pre-use customized resource allocation.

In the comparative analysis of edge computing paradigms within the context of 5G and cloud computing synergy in smart city networks, several key findings emerged, along with considerations of their limitations. Cloudlets demonstrated their capability to reduce latency by bringing cloud resources closer to end-users, offering scalability and flexibility despite potential challenges in deployment complexity. However, their reliance on local infrastructure may lead to higher costs and management overhead. Fog computing architectures showed promise in facilitating distributed processing and real-time analytics at network edges, particularly beneficial for applications such as traffic management and environmental monitoring. Yet, maintaining a large number of fog nodes across diverse locations poses challenges in resource allocation and network management [43]. Mobile Edge Computing (MEC) showcased its ability to enhance mobile application performance by offloading computational tasks to the edge of cellular networks, thereby reducing latency and improving user experience. Nevertheless, issues such as limited computational capacity and potential congestion in the radio access network need to be addressed [44]. Transparent computing emphasized seamless integration of cloud resources with existing infrastructure, highlighting benefits in resource utilization optimization and service continuity. However, ensuring interoperability and security across heterogeneous systems remains a significant challenge. While still in its nascent stages, moisture computing exhibited potential for energy-efficient environmental sensing and monitoring applications in smart cities. Yet, its reliance on moisture-based devices and limited scalability may hinder widespread adoption. These findings underscore the importance of considering both advantages and disadvantages when selecting the most suitable edge computing paradigm to maximize the benefits of 5G and cloud computing synergy in shaping the future of smart cities [45].

4.1. Ethical and privacy concern of data

In environmental monitoring, the collection of data through various sensors and devices raises significant ethical and privacy concerns. While gathering information on air quality, water pollution, or wildlife is essential for preserving ecosystems and public health, it often involves monitoring areas where people live and work. This can inadvertently lead to the collection of personal data, such as location or behavioral patterns, raising issues around consent and data ownership [48]. It's crucial to establish clear guidelines on what data is collected, how it is used, and who has access to it. Transparency with communities and ensuring that data collection is aligned with the public interest can help build trust. Additionally, implementing data anonymization techniques and stringent security measures can mitigate risks, ensuring that environmental monitoring efforts remain focused on sustainability while respecting individual rights and privacy [49].

5. CONCLUSION

In conclusion, the comparative analysis of different proposed paradigms within the framework of 5G and cloud computing synergy in smart city networks reveals a nuanced landscape of opportunities and challenges. Cloudlets, fog computing, Mobile Edge Computing (MEC), transparent computing, and moisture computing each offer distinct advantages in reducing latency, improving scalability, and enhancing user experience in smart city environments. However, they also come with their respective limitations, including deployment complexity, resource management challenges, interoperability issues, and scalability constraints. Despite these challenges, the findings underscore the significance of leveraging the strengths of each paradigm to address specific application requirements and infrastructure constraints effectively. Moreover, the study highlights the need for further research and innovation to overcome existing limitations and unlock the full potential of 5G and cloud computing synergy in shaping the future of smart cities. By addressing these challenges and capitalizing on the opportunities presented by diverse edge computing paradigms, stakeholders can pave the way for more resilient, sustainable, and interconnected urban environments. Ultimately, the successful integration of 5G and cloud computing in smart city networks holds the promise of revolutionizing urban infrastructure and improving the quality of life for residents worldwide.

REFERENCES

Ahmad, I., Kumar, T., Liyanage, M., Okwuibe, J., Ylianttila, M., & Gurtov, A. (2018). Overview of 5G security challenges and solutions. *IEEE Communications Standards Magazine*, *2*(1), 36–43. DOI: 10.1109/MCOMSTD.2018.1700063

Ahmed, A., & Ahmed, E. (2016). A survey on mobile edge computing. *2016 10th International Conference on Intelligent Systems and Control (ISCO)*. https://doi.org/DOI: 10.1109/ISCO.2016.7727082

Ahmed, I., Zhang, Y., Jeon, G., Lin, W., Khosravi, M. R., & Qi, L. (2022). A blockchain- and artificial intelligence-enabled smart IoT framework for sustainable city. *International Journal of Intelligent Systems*, *37*(9), 6493–6507. DOI: 10.1002/int.22852

Anthopoulos, L. G. (2015). Understanding the smart city domain: A literature review. *Public Administration and Information Technology*, 9-21. https://doi.org/DOI: 10.1007/978-3-319-03167-5_2

Badidi, E. (2022). Edge AI and blockchain for smart sustainable cities: Promise and potential. *Sustainability (Basel)*, *14*(13), 7609. DOI: 10.3390/su14137609

Bonomi, F., Milito, R., Natarajan, P., & Zhu, J. (2014). Fog computing: A platform for Internet of things and analytics. *Studies in Computational Intelligence*, 169-186. https://doi.org/DOI: 10.1007/978-3-319-05029-4_7

Bonomi, F., Milito, R., Natarajan, P., & Zhu, J. (2014). Fog computing: A platform for Internet of things and analytics. *Studies in Computational Intelligence*, 169-186. https://doi.org/DOI: 10.1007/978-3-319-05029-4_7

Chen, M., Hao, Y., Li, Y., & Lai, C.-F. (2015). On the computation offloading at ad hoc cloudlet: Architecture and service modes. *IEEE Communications Magazine*, *53*(6), 18–24. DOI: 10.1109/MCOM.2015.7120041

Chen, M., Zhang, H., Liu, W., & Zhang, W. (2014). The global pattern of urbanization and economic growth: Evidence from the last three decades. *PLoS One*, *9*(8), e103799. DOI: 10.1371/journal.pone.0103799 PMID: 25099392

Chen, Q., Pan, G., Chen, W., & Wu, P. (2021). A novel explainable deep belief network framework and its application for feature importance analysis. *IEEE Sensors Journal*, *21*(22), 25001–25009. DOI: 10.1109/JSEN.2021.3084846

Dalla Cia, M., Mason, F., Peron, D., Chiariotti, F., Polese, M., Mahmoodi, T., Zorzi, M., & Zanella, A. (2018). Using smart city data in 5G self-organizing networks. *IEEE Internet of Things Journal*, *5*(2), 645–654. DOI: 10.1109/JIOT.2017.2752761

Du, M., Wang, K., Xia, Z., & Zhang, Y. (2020). Differential privacy preserving of training model in wireless big data with edge computing. *IEEE Transactions on Big Data*, *6*(2), 283–295. DOI: 10.1109/TBDATA.2018.2829886

Hashem, I. A., Chang, V., Anuar, N. B., Adewole, K., Yaqoob, I., Gani, A., Ahmed, E., & Chiroma, H. (2016). The role of big data in smart city. *International Journal of Information Management*, *36*(5), 748–758. DOI: 10.1016/j.ijinfomgt.2016.05.002

Heylighen, A., & Martin, G. (2004). That elusive concept of concept in architecture. *Design Computing and Cognition*, *04*, 57–76. DOI: 10.1007/978-1-4020-2393-4_4

Hogan, M., Liu, F., Sokol, A., & Tong, J. (2011). NIST cloud computing standards roadmap:. https://doi.org/DOI: 10.6028/NIST.SP.500-291v1

Hu, P., Dhelim, S., Ning, H., & Qiu, T. (2017). Survey on fog computing: Architecture, key technologies, applications and open issues. *Journal of Network and Computer Applications*, *98*, 27–42. DOI: 10.1016/j.jnca.2017.09.002

Hwang, R. J., & Huang, Y. Z. (2017). Secure data collection scheme for wireless sensor networks. *2017 31st International Conference on Advanced Information Networking and Applications Workshops (WAINA)*, 553-558. https://doi.org/DOI: 10.1109/WAINA.2017.2

Joshi, S., Saxena, S., Godbole, T., & Shreya, . (2016). Developing smart cities: An integrated framework. *Procedia Computer Science*, *93*, 902–909. DOI: 10.1016/j.procs.2016.07.258

Khan, W. Z., Ahmed, E., Hakak, S., Yaqoob, I., & Ahmed, A. (2019). Edge computing: A survey. *Future Generation Computer Systems*, *97*, 219–235. DOI: 10.1016/j.future.2019.02.050

Kim, T., Ramos, C., & Mohammed, S. (2017). Smart city and IoT. *Future Generation Computer Systems*, *76*, 159–162. DOI: 10.1016/j.future.2017.03.034

Kiruthika, M., & Ponnuswamy, P. P. (2021). Fusion of IoT, blockchain and artificial intelligence for developing smart cities. *Blockchain, Internet of Things, and Artificial Intelligence*, 155-177. https://doi.org/DOI: 10.1201/9780429352898-9

Mao, Y., You, C., Zhang, J., Huang, K., & Letaief, K. B. (2017). A survey on mobile edge computing: The communication perspective. *IEEE Communications Surveys and Tutorials*, *19*(4), 2322–2358. DOI: 10.1109/COMST.2017.2745201

Meng, T., Li, X., Zhang, S., & Zhao, Y. (2016). A hybrid secure scheme for wireless sensor networks against timing attacks using continuous-time Markov chain and queueing model. *Sensors (Basel), 16*(10), 1606. DOI: 10.3390/s16101606 PMID: 27690042

Mohapatra, H. (2021). Socio-technical challenges in the implementation of smart city. *2021 International Conference on Innovation and Intelligence for Informatics, Computing, and Technologies (3ICT)*, 57-62. https://doi.org/DOI: 10.1109/3ICT53449.2021.9581905

Mohapatra, H., Mohanta, B. K., Nikoo, M. R., Daneshmand, M., & Gandomi, A. H. (2023). MCDM-based routing for IoT-enabled smart water distribution network. *IEEE Internet of Things Journal, 10*(5), 4271–4280. DOI: 10.1109/JIOT.2022.3216402

Morgado, A., Huq, K. M., Mumtaz, S., & Rodriguez, J. (2018). A survey of 5G technologies: Regulatory, standardization and industrial perspectives. *Digital Communications and Networks, 4*(2), 87–97. DOI: 10.1016/j.dcan.2017.09.010

Mouradian, C., Naboulsi, D., Yangui, S., Glitho, R. H., Morrow, M. J., & Polakos, P. A. (2018). A comprehensive survey on fog computing: State-of-the-Art and research challenges. *IEEE Communications Surveys and Tutorials, 20*(1), 416–464. DOI: 10.1109/COMST.2017.2771153

Nowicka, K. (2014). Smart city logistics on cloud computing model. *Procedia: Social and Behavioral Sciences, 151*, 266–281. DOI: 10.1016/j.sbspro.2014.10.025

Parida, B., Rath, A., Pati, B., Panigrahi, C., Mohapatra, H., Tien-Hsiungweng, T., & Buyya, R. (2024). SSEPC cloud: Carbon footprint aware power efficient virtual machine placement in cloud milieu. *Computer Science and Information Systems, 21*(3), 759–780. DOI: 10.2298/CSIS230923018P

Parida, B. R., Rath, A. K., & Mohapatra, H. (2022). Binary self-adaptive salp swarm optimization-based dynamic load balancing in cloud computing. *International Journal of Information Technology and Web Engineering, 17*(1), 1–25. DOI: 10.4018/IJITWE.295964

Petrolo, R., Loscrí, V., & Mitton, N. (2014). Towards a smart city based on cloud of things. *Proceedings of the 2014 ACM international workshop on Wireless and mobile technologies for smart cities*. https://doi.org/DOI: 10.1145/2633661.2633667

Ren, J., Guo, H., Xu, C., & Zhang, Y. (2017). Serving at the edge: A scalable IoT architecture based on transparent computing. *IEEE Network, 31*(5), 96–105. DOI: 10.1109/MNET.2017.1700030

Sabt, M., Achemlal, M., & Bouabdallah, A. (2015). Trusted execution environment: What it is, and what it is not. *2015 IEEE Trustcom/BigDataSE/ISPA*. https://doi.org/ DOI: 10.1109/Trustcom.2015.357

Satyanarayanan, M., Bahl, V., Caceres, R., & Davies, N. (2011). The case for VM-based cloudlets in mobile computing. *IEEE Pervasive Computing*. Advance online publication. DOI: 10.1109/MPRV.2009.64

Serrano Mamolar, A., Pervez, Z., Alcaraz Calero, J. M., & Khattak, A. M. (2018). Towards the transversal detection of DDoS network attacks in 5G multi-Tenant overlay networks. *Computers & Security*, *79*, 132–147. DOI: 10.1016/j.cose.2018.07.017

Shafi, M., Molisch, A. F., Smith, P. J., Haustein, T., Zhu, P., De Silva, P., Tufvesson, F., Benjebbour, A., & Wunder, G. (2017). 5G: A tutorial overview of standards, trials, challenges, deployment, and practice. *IEEE Journal on Selected Areas in Communications*, *35*(6), 1201–1221. DOI: 10.1109/JSAC.2017.2692307

Sharma, A., Podoplelova, E., Shapovalov, G., Tselykh, A., & Tselykh, A. (2021). Sustainable smart cities: Convergence of artificial intelligence and blockchain. *Sustainability (Basel)*, *13*(23), 13076. DOI: 10.3390/su132313076

Skouby, K. E., & Lynggaard, P. (2014). Smart home and smart city solutions enabled by 5G, IoT, AAI and cot services. *2014 International Conference on Contemporary Computing and Informatics (IC3I)*, *2*, 874-878. https://doi.org/DOI: 10.1109/IC3I.2014.7019822

Tufail, A., Namoun, A., Abi Sen, A. A., Kim, K., Alrehaili, A., & Ali, A. (2021). Moisture computing-based internet of vehicles (IoV) architecture for smart cities. *Sensors (Basel)*, *21*(11), 3785. DOI: 10.3390/s21113785 PMID: 34070719

Vaquero, L. M., & Rodero-Merino, L. (2014). Finding your way in the fog. *Computer Communication Review*, *44*(5), 27–32. DOI: 10.1145/2677046.2677052

Wang, S., Zhang, X., Zhang, Y., Wang, L., Yang, J., & Wang, W. (2017). A survey on mobile edge networks: Convergence of computing, caching and communications. *IEEE Access : Practical Innovations, Open Solutions*, *5*, 6757–6779. DOI: 10.1109/ACCESS.2017.2685434

Wang, X., Chen, M., Taleb, T., Ksentini, A., & Leung, V. (2014). Cache in the air: Exploiting content caching and delivery techniques for 5G systems. *IEEE Communications Magazine*, *52*(2), 131–139. DOI: 10.1109/MCOM.2014.6736753

Wieclaw, L., Pasichnyk, V., Kunanets, N., Duda, O., Matsiuk, O., & Falat, P. (2017). Cloud computing technologies in "smart city" projects. *2017 9th IEEE International Conference on Intelligent Data Acquisition and Advanced Computing Systems: Technology and Applications (IDAACS)*. https://doi.org/DOI: 10.1109/IDAACS.2017.8095101

Xibilia, M. G., Latino, M., Marinkovic, Z., Atanaskovic, A., & Donato, N. (2020). Soft sensors based on deep neural networks for applications in security and safety. *IEEE Transactions on Instrumentation and Measurement*, *69*(10), 7869–7876. DOI: 10.1109/TIM.2020.2984465

Xue, Y., & Qin, J. (2023). Partial connection based on channel attention for differentiable neural architecture search. *IEEE Transactions on Industrial Informatics*, *19*(5), 6804–6813. DOI: 10.1109/TII.2022.3184700

Yi, S., Li, C., & Li, Q. (2015). A survey of fog computing. *Proceedings of the 2015 Workshop on Mobile Big Data*. https://doi.org/DOI: 10.1145/2757384.2757397

Zhang, Y., Guo, K., Ren, J., Zhou, Y., Wang, J., & Chen, J. (2017). Transparent computing: A promising network computing paradigm. *Computing in Science & Engineering*, *19*(1), 7–20. DOI: 10.1109/MCSE.2017.17

Zhang, Y., & Zhou, Y. (2007). 4VP: A novel Meta OS approach for streaming programs in ubiquitous computing. *21st International Conference on Advanced Networking and Applications (AINA '07)*, 394-403. https://doi.org/DOI: 10.1109/AINA.2007.6

Chapter 7

A Hybrid Approach for Smart Home Activity Recognition Using Sensor Data and Deep Learning Techniques

K. Kavitha
Vellore Institute of Technology, Chennai, India

Shashank Pandey
Vellore Institute of Technology, Chennai, India

Piyali Saha
https://orcid.org/0009-0001-1794-3334
Vellore Institute of Technology, Chennai, India

Atul R. Patel
Vellore Institute of Technology, Chennai, India

ABSTRACT

Recognizing human activities in smart homes poses challenges due to environmental variability, sensor systems, and sparse signals. Deep learning models struggle to extract meaningful features autonomously, necessitating additional context. This study proposes a novel hybrid approach merging natural language processing and time series classification techniques to address feature extraction for activity recognition. Sensor events are encoded into frequency-based terms, generating embeddings capturing semantic relationships. Evaluation on two smart home data-

DOI: 10.4018/979-8-3693-5448-3.ch007

sets demonstrates the effectiveness of encoding-based embeddings for improving automatic feature learning. Comparisons with Cascade LSTM and other models show the superiority of the proposed approach. The hybrid technique, centered around Cascade LSTM, effectively leverages contextual information from sensor event sequences for recognizing complex human activities in smart homes.

I. INTRODUCTION

Over the years, the methodology for recognizing human activities from sensor data in smart homes has evolved significantly. Initially, conventional machine learning approaches relying on handcrafted features were prevalent. However, these methods often struggled to capture the rich temporal dependencies present in sensor sequences. With the advent of deep learning, particularly recurrent neural networks (RNNs) like Long Short-Term Memory (LSTM) and Gated Recurrent Units (GRU), there was a paradigm shift towards models better suited for temporal data analysis. Additionally, convolutional neural networks (CNNs) were successfully applied by treating sensor sequences as images, paving the way for novel approaches in activity recognition. Despite these advancements, certain gaps remained prevalent in the field. Challenges persisted due to the variability in sensor readings, overlapping activities, and the lack of well-defined boundaries between activities. Moreover, traditional deep learning models often struggled with noise, sparsity, and irregular sampling of sensor data. These limitations underscored the need for innovative methodologies capable of addressing these challenges effectively. In response to these identified gaps, this paper introduces a novel hybrid approach that combines natural language processing and time series classification techniques to tackle the feature extraction problem in activity recognition. Central to our methodology is the encoding of sensor events into terms based on frequency, akin to text representations. This encoding process generates embeddings that capture semantic relationships between events, thus facilitating more effective feature learning. Furthermore, we identify and address the gap in leveraging domain knowledge within the model architecture. By incorporating a Cascade LSTM architecture, we capitalize on the temporal dependencies present in sensor sequences, enhancing the model's ability to capture complex activity patterns.

Figure 1. Overview of Activity Recognition

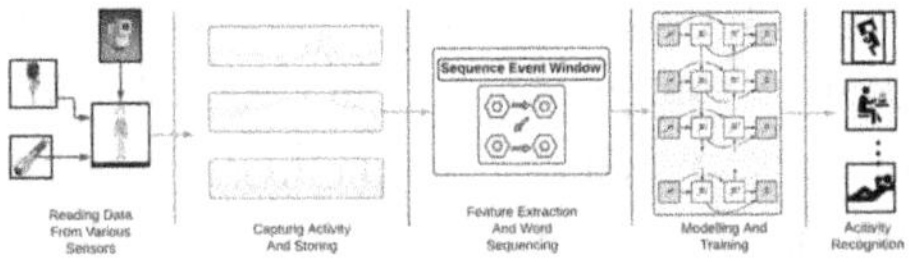

Our key contributions include the development of a segmentation algorithm for extracting discrete activity instances from continuous sensor streams, a novel encoding scheme to transform variable-length sensor sequences into fixed-length word sequences, and the utilization of a Cascade LSTM model incorporating word embeddings for activity recognition from pre-processed sensor data. Through rigorous evaluation on real-world smart home datasets, we demonstrate the effectiveness of our proposed approach. Notably, our method consistently outperforms traditional LSTM models, showcasing its capability to accurately classify activities from sensor data while leveraging contextual information effectively.

In summary, our hybrid technique presents a promising solution to the challenges inherent in human activity recognition in smart homes. By addressing key gaps in existing methodologies and leveraging domain knowledge effectively, our approach contributes to advancing the field of home automation and enhancing resident well-being.

II. LITERATURE SURVEY

Triboan et al. (2019): Existing smart home activity recognition focuses on coarse-grained activities using simple sensors without confirming object interactions. While some integrate multiple sensor data, they only identify high-level activities. Fuzzy logic has been used for activity modelling but not fine-grained recognition at object-level. Prior ambient sensing does not combine off-the-shelf and custom object sensors. Evaluation also only considers high-level rather than detailed object interactions. This paper aims to address this by proposing a fuzzy ontology approach to recognize fine-grained activities from multimodal object data. It develops a microservice architecture and non-intrusive heterogeneous sensing method at object-level for real-world testing of fine-grained recognition.

Guo and Fang (2016): In this study, real-time recognition of human activities is addressed by employing binary tree support vector machines (BT-SVM) in conjunction with a sliding sensor window. The construction of an optimal binary tree is facilitated through the utilization of genetic algorithms (GA).

Training of SVM models and assessment of recognition algorithm accuracy are conducted using real data obtained from participants engaged in various activities. Comparative analysis with one-against-one SVM and BT-SVM incorporating human knowledge reveals that the proposed BT-SVM with GA yields superior performance in activity recognition within smart home environments, as evidenced by the experimental results.

Hamad et al. (2019): The proposed method improves upon existing activity recognition approaches for smart homes which only consider preceding sensor data when making predictions. To provide more contextual information, especially for applications like dementia diagnosis, the method delays recognition to also include some subsequent sensor activations. It extracts features using multiple incremental fuzzy temporal windows encompassing data before and after the prediction time. Evaluated on a sensor dataset with convolutional and recurrent neural networks, the method significantly outperforms a real-time baseline by leveraging following sensor activations for enhanced activity recognition performance through delayed predictions with fuzzy temporal windows.

Jethanandani et al. (2019): HAR is important for health monitoring in smart homes. Traditional HAR methods focus on single resident recognition.This paper proposes a Binary Relevance method for multi-resident HAR. The method decomposes the multi-label problem into binary classification tasks. Evaluation on a real-world dataset shows competitive performance. The results demonstrate the effectiveness of the Binary Relevance method for multi-label, multi-resident HAR.

Calzada et al. (2014): RIMER+ is a generic decision model for handling different types of information and uncertainty in real environments. This paper adapts RIMER+ for human activity recognition using binary sensors in smart homes. IMER+ is evaluated against other state-of-the-art classifiers in terms of accuracy, efficiency, and applicability. IMER+ performs well, especially in situations with incomplete input data. The results demonstrate the potential of RIMER+ for sensor-based activity recognition in smart homes.

Nafea et al. (2021): Focusing on Human Activity Recognition (HAR), a significant advancement involves a two-stream deep learning architecture utilizing Convolutional Neural Networks (CNN) and Bi-directional Long Short-Term Memory (BiLSTM). This model adeptly captures spatial and temporal features, leading to enhanced activity recognition. A notable innovation is observed in the spatial stream's architecture, incorporating multiple convolution layers with varying kernel dimensions, allowing feature capture at diverse resolutions. Comparative analyses across prominent datasets, such as Wireless Sensor Data Mining (WISDM) and UCI-HAR, indicate superior performance of the proposed method, achieving higher accuracy on the WISDM dataset (98.53%)

compared to the UCI dataset (97.05%). Additionally, research explores ensemble approaches, combining diverse HAR models, and transfer learning techniques, leveraging pre-trained models for improved performance in scenarios with limited labeled data. These studies collectively highlight the efficacy of innovative model architectures and their positive impact on HAR accuracy, reinforcing the importance of result-driven advancements in this field.

Akter et al. (2023): In this study on Human Activity Recognition (HAR), a novel methodology employing a 2D CNN-based Deep Learning (DL) model with convolutional block attention modules (CBAM) is proposed. The model utilizes spectrograms constructed from raw sensor data (accelerometer and gyroscope) for feature extraction, avoiding hand-crafted feature engineering. The proposed model, named AM-DLFC, exhibits a flexible architecture with six convolutional blocks, each incorporating a CBAM module to refine features. The model achieved high classification accuracies of 96.86%, 93.48%, and 93.89% on the KU-HAR, UCI-HAR, and WISDM datasets, respectively. The approach demonstrates superior performance compared to previous works, emphasizing the effectiveness of the CBAM-enhanced 2D CNN in HAR. The study provides insights into the integration of spectrograms, CBAM modules, and the overall architecture, showcasing its competence in human activity classification across diverse datasets.

Bouchabou et al. (2021): This study focuses on enhancing the efficiency of Long Short-Term Memory (LSTM)-based structures for recognizing daily living activities in smart homes. While LSTM captures temporal dependencies in sensor activations, it often falls short in understanding sensor semantics and context. The proposed approach incorporates two Natural Language Processing embedding methods, Word2Vec and ELMo, to augment LSTM-based structures in activity-sequences classification tasks. By transforming sensor activations into semantic representations, the model gains insights into the context and meaning of sensor data. The method involves segmenting event sequences, encoding them into categorical values, and utilizing pre-trained embeddings for semantic enrichment. The results, based on real smart home datasets, demonstrate improved classification performance, reduced confusion between activity classes, and the ability for transfer learning. The approach offers valuable information, such as a sensor organization map, and addresses challenges posed by competing activities from other residents or pets in smart home environments. The study underscores the significance of considering sensor context and semantics for enhanced human activity recognition.

Chung et al. (2019): This paper presents a systematic study on the on-body sensor positioning and data acquisition for Human Activity Recognition (HAR) systems, employing a testbed with eight body-worn Inertial Measurement Units

(IMU) sensors and an Android mobile device. The study utilizes a Long Short-Term Memory (LSTM) network framework for training a deep learning model on human activity data collected in real-world and controlled environments. Results indicate that a sampling rate as low as 10 Hz from four sensors (both wrists, right ankle, and waist) is sufficient for recognizing Activities of Daily Living (ADLs), including eating and driving activities. The paper introduces a two-level ensemble model for sensor fusion, demonstrating that classifier-level fusion techniques can improve classification performance. Custom weights for multimodal sensor fusion are derived based on the accuracy of individual sensors on different types of activities. The study provides insights into optimal sensor positioning and effective sensor fusion strategies for HAR applications, offering practical guidance for further research in the field.

Alani et al. (2020): This paper addresses the challenge of classifying activities in smart homes using real-time sensor data, considering the complexities of the data, its noisiness, and imbalances in class occurrences. Focusing on Human Activity Recognition (HAR), the study explores the effectiveness of fusing imbalanced multi-modal sensor data in comparison to using unimodal data. The experiments employ deep learning models, namely Convolutional Neural Network (CNN), Long Short-term Memory (LSTM), and a hybrid CNN-LSTM architecture, on a large multi-modal sensor dataset from the Sensor Platform for Healthcare in a Residential Environment (SPHERE). The results emphasize the difficulties posed by imbalanced multi-modal data and highlight the importance of addressing class imbalance for training and testing deep learning models effectively. The study reveals that, when fusing data and applying the Synthetic Minority Oversampling Technique (SMOTE) to correct class imbalance, the CNN-LSTM model achieves the highest classification accuracy of 93.67%, surpassing CNN and LSTM models. The paper discusses the architectures of the models, the dataset characteristics, and the use of resampling methods to handle imbalanced data, providing insights for addressing challenges in activity recognition within smart home environments.

Sarkar et al. (2023): The paper discusses the proposed model for Human Activity Recognition (HAR) based on sensor data. It utilizes a Spatial Attention-aided CNN as the feature extractor and a novel feature selection technique. The model is tested on five public datasets and outperforms state-of-the-art methods. Additionally, the study evaluates the impact of feature selection on model performance, highlighting its effectiveness for different datasets. The use of Genetic Algorithms (GA) for feature selection in HAR is also discussed, along with the application of Continuous Wavelet Transform (CWT) for encoding sensor data into image form. The paper emphasizes the importance of efficient feature selection and the application of advanced algorithms for accurate human activity recognition. Furthermore, it provides insights into the challenges and future improvements for HAR models, and the source code of the paper is publicly available. The text also covers various approaches and techniques for human activity recognition using deep learning, convolutional neural networks, genetic algorithms, and feature selection methods, as well as their applications in other domains such as image processing, stock market forecasting, cardiovascular disease prediction, and cancer classification.

Mihoub (2021): The paper provides a comprehensive overview of the field, covering various aspects such as data-driven models, wearable devices, sensor data analytics, machine learning algorithms, and deep learning approaches. The survey includes references to related research and datasets, highlighting the use of deep learning algorithms for activity recognition using mobile and wearable sensor networks. Additionally, the survey discusses the development and evaluation of deep learning-based frameworks for activity recognition in smart homes, particularly focusing on the "Orange4Home" dataset. The frameworks include data preprocessing, feature selection, and model building using neural network architectures, demonstrating their effectiveness in classifying activities in the smart home environment. The survey also addresses the implications of the findings for smart home technology and future research directions, providing valuable insights for researchers and practitioners in the field of human activity recognition.

Tan et al. (2021): The paper encompasses a range of studies and research papers related to monitoring the activities of daily living (ADLs) of elderly individuals, with a focus on technologies such as 3D key human postures, action recognition, deep learning, and sensor-based data. The survey also addresses privacy concerns, unobtrusive monitoring, and the use of binary sensors for activity recognition. Notably, the survey includes a study proposing a privacy-preserving activity recognition model that outperformed existing state-of-the-art models, achieving a high F1 score of 0.917 and a classification latency of less than 30 ms. Additionally, the references cover topics such as population aging, living alone among older adults, and the use of Internet of Things (IoT) and wearable sensors for elderly healthcare and well-being. Another article discussed a new method for recognizing the activities of daily living (ADLs) of elderly people living alone using binary motion sensors and a recurrent neural network (RNN), achieving a relatively high F1 score of 0.917. These studies collectively aim to improve the quality of care for elderly individuals living alone or in assisted living environments using advanced technologies and innovative models for activity recognition and monitoring.

Bulbul et al. (2018): The paper discusses the utilization of smartphone sensors for human activity recognition, aligning with previous research that has explored the potential of sensor data for activity recognition. Prior studies have demonstrated the significance of accelerometer and gyroscope signals in capturing human movement patterns and activities. Additionally, the use of machine learning classification approaches for activity recognition has been widely investigated in the literature. Various models, including Support Vector Machines (SVM), have been applied and compared for their efficiency and precision in recognizing human activities using sensor data. However, there is a gap in the literature regarding the incorporation of data from other sensors commonly found in smartphones for activity recognition. Therefore, this paper contributes to the existing body of knowledge by providing a detailed analysis of different classification approaches and their success rates, while also highlighting the potential for further research to incorporate additional smartphone sensor data for enhanced activity recognition models.

Kumar and Suresh (2022): The paper introduced FLAAP dataset to the field of human activity recognition (HAR) using smartphone sensors, addressing the need for high-quality, publicly available datasets for developing and evaluating HAR models. The dataset, collected from eight subjects performing ten activities at a sampling rate of 100Hz, provides raw and trimmed sensor data, along with detailed information about the experimental setup and data collection procedure. In comparison to other publicly available HAR datasets, the FLAAP dataset stands out for its comprehensive nature and potential for advancing learning models. The article also presents experimental results using the Random Forest learning algorithm, achieving a recognition accuracy of approximately 77.22%. This dataset and the associated experimental findings add to the existing body of literature on HAR, offering valuable insights for future research and model development. Additionally, the paper references related work in the field, further contextualizing the significance of the FLAAP dataset and its potential applications.

Table 1. Summarized overview of literature survey

Title	Model used	Finding or Gap	Dataset	Results
Fuzzy-based Fine-grained Human Activity Recognition within Smart Environments (Triboan et al., 2019)	ontology-based approach for Human Activity Recognition (HAR), extending it to incorporate fuzzy ontology for reasoning with imprecise non-binary sensor data measurements.	The system demonstrates potential applications in detecting casual mishaps or dangerous incidents related to everyday object usage. Future work is outlined to evaluate the approach with real-world SH environment data and enhance scalability for complex activity scenarios.	real-world Smart Home (SH) environment data.	proposed approach enhances accuracy and detection of fine-grained actions based on object usage.
Real-time human activity recognition in smart home with binary tree SVM (Guo & Fang, 2016)	Binary tree support vector machines (BT-SVM) with genetic algorithms (GA)	BT-SVM with GA is employed for real-time activity recognition. GA is used to construct an optimal binary tree. The proposed method achieves higher accuracy compared to other methods. Experimental results demonstrate the effectiveness of the proposed system.	CASAS smart home dataset	The proposed BT-SVM with GA achieved a recognition accuracy of 74.25%, outperforming other methods.
Binary Relevance Model for Activity Recognition in Home Environment using Ambient Sensors (Jethanandani et al., 2019)	Binary Relevance, Multi-Label Classification, Random Forest	The study focuses on Human Activity Recognition (HAR) in smart homes, employing the Binary Relevance Method with a Random Forest classifier on the ARAS dataset. They utilize 10-fold cross-validation for result prediction.	ARAS dataset (Triboan et al., 2019)	average accuracy of 83.7% and 88.2% for House A and House B, respectively. Precision values are 69.2% and 80.4%, with a hamming loss of 14.4% for House A and 7.9% for House B.

continued on following page

Table 1. Continued

Title	Model used	Finding or Gap	Dataset	Results
Sensor-Based Activity Recognition Using Extended Belief Rule Based Inference Methodology (Calzada et al., 2014)	Extended Belief Rule-Based Inference Methodology (RIMER+), specifically enhanced with a Decision Rule Aggregation (DRA) mechanism using the Hamming distance.	RIMER+ is a generic decision model for handling different types of information and uncertainty in real environments. This paper adapts RIMER+ for human activity recognition using binary sensors in smart homes. IMER+ is evaluated against other state-of-the-art classifiers in terms of accuracy, efficiency, and applicability. IMER+ performs well, especially in situations with incomplete input data	Activity Recognition Dataset with varying levels of sensor failure, including scenarios with 4 and 5 deactivated sensors.	R+DRAH accuracy in 4 sensors deactivated scenario: 95.92% R+DRAH accuracy in 6 sensors deactivated scenario (50% train): 95.10%

Table 2. Summarized overview of literature survey

Title	Model used	Finding or Gap	Dataset	Results
Sensor-Based Human Activity Recognition with Spatio-Temporal Deep Learning (Nafea et al., 2021)	Two-stream deep learning architecture	Enhanced activity recognition through CNN and BiLSTM; Spatial stream innovation with multiple convolution layers; Ensemble approaches and transfer learning explored for limited labelled data.	Wireless Sensor Data Mining (WISDM), UCI-HAR	WISDM: 98.53%, UCI: 97.05% accuracy
Human Activity Recognition Using Attention-Mechanism-Based Deep Learning Feature Combination (Akter et al., 2023)	2D CNN-based DL model with CBAM modules	Novel AM-DLFC model using spectrograms from raw sensor data; Achieved high accuracies on KU-HAR, UCI-HAR, and WISDM datasets; Outperformed previous works with CBAM-enhanced 2D CNN.	KU-HAR, UCI-HAR, WISDM	KU-HAR: 96.86%, UCI-HAR: 93.48%, WISDM: 93.89% accuracy
Using Language Model to Bootstrap Human Activity Recognition Ambient Sensors (Bouchabou et al., 2021)	LSTM-based structures with Word2Vec and ELMo embeddings	Improved efficiency by incorporating language models; Addressed shortcomings of LSTM in understanding sensor semantics and context; Real smart home datasets used; Transfer learning demonstrated.	Smart home datasets	Improved classification performance, reduced confusion, valuable sensor information

continued on following page

Table 2. Continued

Title	Model used	Finding or Gap	Dataset	Results
Sensor Data Acquisition and Multimodal Sensor Fusion for Human Activity Recognition (Chung et al., 2019)	LSTM network framework, ensemble model for sensor fusion	Optimal on-body sensor positioning; Effective sensor fusion strategies using a two-level ensemble model; Insights for real-world and controlled environments.	On-body IMU sensors, Android mobile device	Sampling rate as low as 10 Hz sufficient; Classifier-level fusion improves performance
Classifying Imbalanced Multi-modal Sensor Data for Human Activity Recognition (Alani et al., 2020)	CNN, LSTM, CNN-LSTM hybrid architecture	Addressed challenges of classifying activities in smart homes with imbalanced multi-modal sensor data; Demonstrated the importance of addressing class imbalance for effective deep learning models.	Sensor Platform for Healthcare in a Residential Environment	CNN-LSTM achieved 93.67% accuracy, surpassing CNN and LSTM models

III. PROPOSED APPROACH

In response to the challenges inherent in recognizing human activities from sensor data in smart homes, this section outlines a novel methodology aimed at enhancing feature extraction for activity recognition. Traditional approaches in this domain have encountered limitations stemming from the complexity of home environments, diverse sensor systems, variability in user behaviors, and the presence of sparse or redundant signals. To address these challenges and advance the methodology in activity recognition, a new technique is proposed, offering both novelty and efficacy. The proposed methodology draws inspiration from two distinct fields: natural language processing (NLP) and time series classification. By leveraging concepts from NLP, particularly in the encoding and representation of sensor events, and combining them with established time series classification techniques, the methodology aims to capture and extract meaningful patterns from sensor data. The integration of these methodologies offers a holistic approach to feature extraction, enabling a more comprehensive understanding of human activities within smart home environments.

The need for this new technique is underscored by the limitations of existing approaches, which often struggle to effectively extract features from heterogeneous and noisy sensor data. In contrast, the proposed methodology fills critical gaps by harnessing the semantic relationships between sensor events, thereby providing a more nuanced and contextually rich representation of activity data. Furthermore, the incorporation of a Cascade LSTM model enhances the methodology's ability to capture complex temporal dependencies, offering improved accuracy and robustness in activity recognition.

Overall, the proposed methodology represents a significant advancement in the field of activity recognition in smart homes. By combining insights from NLP and time series classification, it offers a novel approach to feature extraction that addresses key challenges and holds promise for enhancing the accuracy and reliability of activity recognition systems in real-world settings. Figure 2 depicts the architecture of proposed approach for Activity Recognition in Smart Homes utilizing Cascade LSTM model for contextual analysis of sensor event sequences.

Figure 2. Architecture of proposed approach

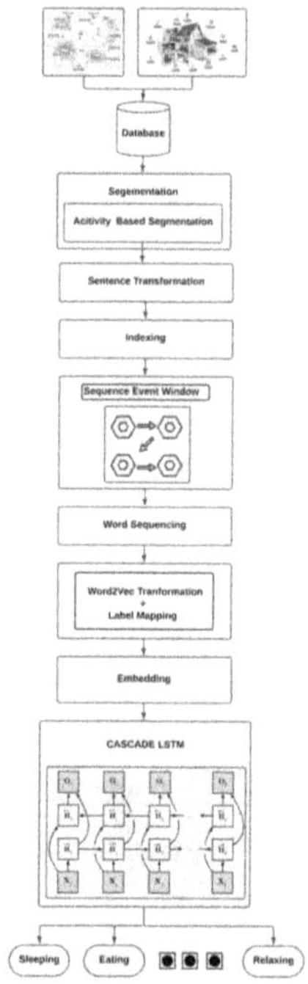

3.1 Basic Structure of CASCADE LSTM

In the proposed methodology for activity recognition in smart homes, the CASCADE LSTM model stands as a pivotal component, enabling the contextual analysis of sensor event sequences. This section delves into the intricate architecture and functioning of the CASCADE LSTM, elucidating its role in capturing temporal dependencies and facilitating accurate activity identification. In CASCADE LSTM's architecture, each layer consists of Long Short-Term Memory (LSTM) units organized hierarchically, facilitating a cascading approach to information processing. Unlike traditional LSTM architectures where information flows linearly through layers, CASCADE LSTM allows outputs from one layer to serve as inputs to subsequent layers. This cascading mechanism enables the model to capture temporal dependencies across multiple scales, enhancing its ability to discern nuanced patterns within sensor event sequences. By iteratively processing and refining information across layers, CASCADE LSTM performs a more comprehensive analysis of temporal relationships, ultimately improving the accuracy and robustness of activity recognition in smart home environments.

3.1.1 Cascading Mechanism and Contextual Analysis

Figure 3. Cascade LSTM for extracting temporal dependencies

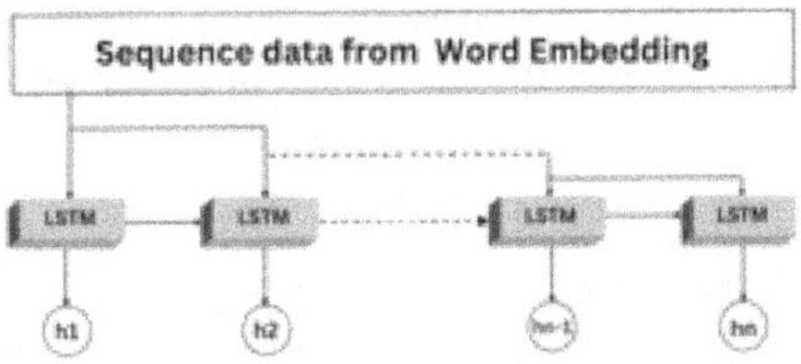

CASCADE LSTM's cascading mechanism facilitates the propagation of information across multiple temporal scales, enabling the model to capture both short-term and long-term dependencies within sensor event sequences. At each layer, LSTM units process sequential input data, preserving relevant information while filtering out noise and irrelevant signals. Leveraging this cascading approach, CASCADE LSTM conducts contextual analysis of sensor event sequences, taking into account the temporal context surrounding each event. This contextual understanding enables the model to infer underlying activities with higher accuracy, even in the presence of noise or irregularities in the data. Furthermore, CASCADE LSTM serves as an effective feature extractor, automatically learning discriminative features from the input sensor data. The hierarchical structure of the model allows for the extraction of hierarchical features, capturing both low-level sensor patterns and high-level activity semantics. Through its cascading mechanism and contextual analysis capabilities, CASCADE LSTM enhances the accuracy and robustness of activity recognition in smart home environments.

IV. IMPLEMENTATION

4.1 Data Collection

In the initial phase of the implementation, a comprehensive sensor deployment strategy is implemented within smart home environments. Various types of sensors, including motion sensors, temperature sensors, and door sensors, are strategically positioned throughout the home to capture a diverse array of data. These sensors continuously monitor and record pertinent environmental and activity-related information, such as motion patterns, room temperatures, and door openings/closures. Each sensor generates data points consisting of timestamps and corresponding sensor readings, ensuring temporal accuracy and enabling precise tracking and analysis of events over time. The collected data is meticulously organized and stored in a dedicated database warehouse, facilitating easy accessibility and efficient management for subsequent processing stages. This robust sensor infrastructure provides researchers with comprehensive insights into the activities occurring within the smart home environment, laying the foundation for subsequent data processing and analysis tasks.

4.2 Activity Segmentation

Activity segmentation constitutes a pivotal stage following the data collection phase, serving as a crucial preprocessing step before activity recognition. In this phase, sophisticated algorithms are deployed to partition the collected sensor data into discrete activity segments. These segments are delineated based on predefined criteria derived from sensor readings and patterns, aiming to identify distinct activities within the dataset accurately. The segmentation process involves analysing temporal patterns and sensor readings to identify transitions between different activities. Various techniques, such as sliding window analysis or change-point detection algorithms, may be employed to identify boundaries between activities. Additionally, context-aware segmentation approaches may leverage contextual information, such as time of day or user behavior patterns, to refine the segmentation process further. The primary goal of activity segmentation is to break down the continuous stream of sensor data into meaningful segments corresponding to different activities, thus facilitating more focused and effective analysis in subsequent stages of activity recognition. A robust activity segmentation framework lays the groundwork for precise activity recognition algorithms, enabling the system to accurately identify and classify various activities occurring within the smart home environment.

4.3 Sentence Transformation and Indexing

Following the segmentation of sensor data into activity segments, the segmented data undergoes a transformative process wherein each segment is converted into a sequence of sentences representing related activities. This transformation aims to provide a more granular understanding of the sequential nature of activities within the smart home environment. Each activity segment is translated into a structured sentence format, capturing essential information such as the type of activity, duration, and any associated contextual details. This transformation enhances the interpretability and contextuality of the data, facilitating more accurate analysis and classification in subsequent stages of the methodology. Moreover, to streamline retrieval and processing, the transformed activity sequences are meticulously indexed. Indexing involves assigning unique identifiers to each activity sequence, enabling efficient storage and retrieval of data during subsequent analysis tasks. By indexing the activity sequences, the methodology ensures rapid access to relevant data points and facilitates seamless navigation through the dataset. This indexed approach enhances the efficiency of data processing, enabling swift retrieval and analysis of activity sequences, ultimately contributing to the effectiveness of the activity recognition system within the smart home environment.

4.4 Windowing Technique for Activity Sequencing

A windowing technique is utilized to partition indexed activity sequences into smaller, temporally coherent windows, aiming for a finer granularity in capturing activity patterns. Carefully calibrated parameters, including window size and overlap, ensure comprehensive coverage of activities while preserving temporal relationships. This approach deepens the analysis of activity dynamics within the smart home environment. Each window encapsulates a subset of activities within a specific timeframe, facilitating focused analysis. By retaining temporal coherence between adjacent windows, the methodology accurately represents activity sequences. Overall, the windowing technique enhances the effectiveness of activity sequencing, providing researchers with detailed insights into activity patterns for robust activity recognition systems.

4.5 Word Sequencing and Word2Vec Transformation

In this stage, the methodology converts segmented and windowed activity sequences into word representations, where each word denotes a specific activity or event. Subsequently, the Word2Vec transformation is employed to generate high-dimensional vector representations of these activity words. This transformation captures semantic similarities between activities, enriching the representation of activity sequences. By embedding activities into a high-dimensional vector space, the methodology enhances the subsequent classification process, enabling the model to better discern complex activity patterns. Overall, word sequencing and Word2Vec transformation contribute to a more nuanced understanding of activity sequences, facilitating more accurate activity recognition within the smart home environment.

4.6 Label Mapping and Embedding Creation

In this stage, activity labels, such as eating, sleeping, and relaxing, are mapped to numerical values to facilitate classification. Concurrently, the Word2Vec representations of activity sequences are created. These mapped labels and Word2Vec representations are then utilized to generate embedded vectors. These embedded vectors, which encapsulate both semantic information and activity labels, are stored as vector NPX files for subsequent utilization in the methodology. By combining numerical mappings of activity labels with Word2Vec representations, the methodology establishes a comprehensive framework for activity classification, enabling accurate recognition of diverse activities within the smart home environment.

4.7 CASCADE LSTM Model for Activity Classification

At the heart of the methodology lies the CASCADE LSTM model, meticulously designed for activity classification within the smart home environment. Configured to accept embedded vectors as input, the CASCADE LSTM model leverages the rich semantic information encoded within these vectors to accurately classify activity sequences into different labels corresponding to various activities within the smart home. By harnessing the hierarchical structure and temporal dependencies captured by LSTM units, the model excels in discerning complex activity patterns. Finally, the classified activity sequences undergo rigorous evaluation using established performance metrics, ensuring the reliability and efficacy of the classification process within the smart home environment.

4.8 Training and Optimization

Training CASCADE LSTM involves a meticulous optimization process aimed at minimizing the disparity between predicted and ground truth activity labels. This optimization is crucial for enhancing the model's predictive accuracy and robustness. Leveraging techniques such as backpropagation through time (BPTT) and gradient descent, CASCADE LSTM iteratively adjusts its parameters to iteratively improve its predictions based on the available training data. During training, the model learns to capture complex temporal dependencies and semantic relationships within the data, refining its ability to accurately classify activity sequences. By continuously updating its parameters through the optimization process, CASCADE LSTM strives to achieve optimal performance in activity recognition tasks within the smart home environment.

V. EXPERIMENTAL RESULTS AND DISCUSSION

After completing the model training process, an impressive accuracy of 0.93 was achieved on the test set, indicating robust performance. However, an imbalance issue persists within the dataset, where the majority of instances are classified as "other," with "eat," "leave home," and "sleep" following as prominent categories. Conversely, the "work" activity exhibits a notably minor presence in the dataset. Figure 4 shows the graph that illustrates the training and validation accuracy of a CASCADE LSTM model over multiple epochs during the training process. The blue line represents the training accuracy, showing an increasing trend over epochs, indicating the model's ability to learn from the training data. The orange line depicts the validation accuracy, which fluctuates around a similar trend as the training accuracy, demonstrating the model's generalization performance on unseen validation data. Figure 5 depicts the ROC curve where the area under the curve (AUC) indicates the model's ability to distinguish between positive and negative instances of each activity class, where higher AUC values suggest better classification performance. Figure 6 depicts a correlation matrix showing the relationship between predicted and actual activity labels. Each cell represents the number of instances where a predicted activity matches the actual activity.

Figure 4. Accuracy curve

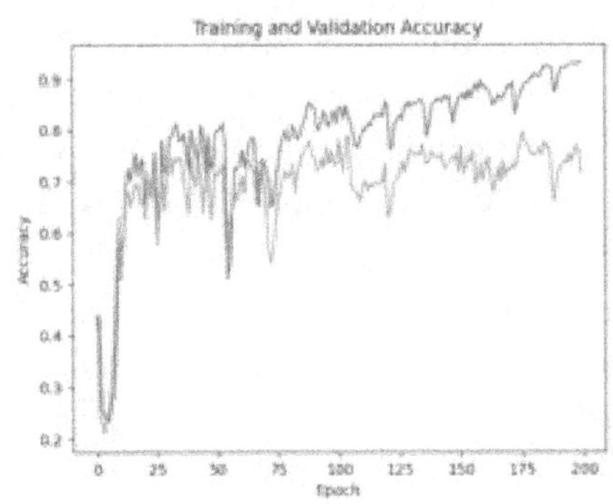

Figure 5. ROC curve

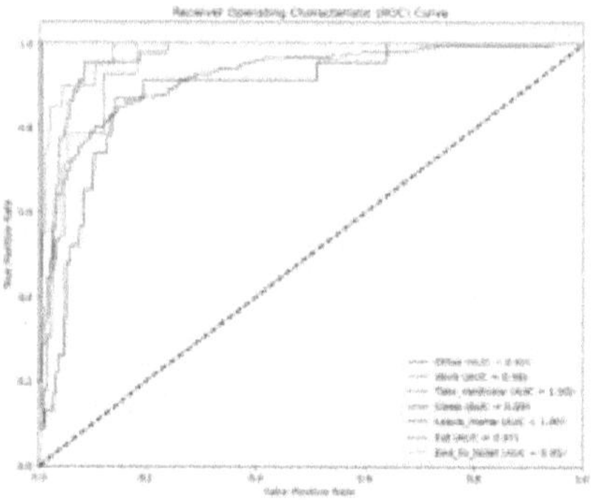

Figure 6.Correlation Matrix

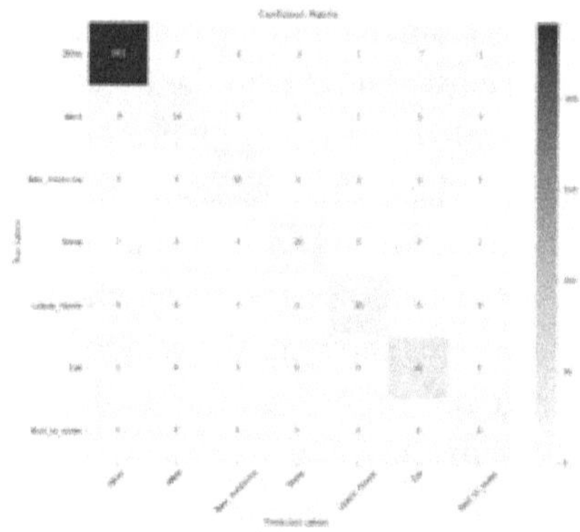

In the correlation matrix it is observed that the "Work" class appears to be perfectly classified, as indicated by the dark blue diagonal line along its corresponding row and column. This suggests that the model accurately predicted instances of the "Work" activity without any misclassifications. The "Take medicine" and "Leave_HOME" classes also show high correlation scores, suggesting moderate to strong classification performance with minimal misclassifications. However, the "Other" and "Eat_to_toilet" classes exhibit some misclassifications, indicating instances where the model predicted these activities incorrectly. However, the "Sleep" activity stands out with a higher rate of misclassification compared to other activities, indicating a challenge for the model in accurately identifying this particular activity.

VI. CONCLUSION AND FUTURE SCOPE

In conclusion, the analysis of the confusion matrix and ROC curve highlights the robust performance of the activity classification model within the smart home environment, particularly in accurately predicting activities such as "Take medicine" and "Leave home." However, challenges persist, notably in accurately classifying the "Sleep" activity, suggesting the need for further refinement. Future research should focus on addressing class imbalance issues, exploring advanced machine learning algorithms, incorporating domain-specific knowledge, considering contextual information, and deploying the model in real-world settings to enhance its effectiveness and usability. By addressing these areas of improvement, the research aims to advance activity recognition systems in smart homes, ultimately contributing to improved support for daily living activities.

REFERENCES

Akter, M., Ansary, S., Khan, M. A. M., & Kim, D. (2023). Human activity recognition using attention-mechanism-based deep learning feature combination. *Sensors (Basel), 23*(12), 5715. DOI: 10.3390/s23125715 PMID: 37420881

Alani, A. A., Cosma, G., & Taherkhani, A. (2020). Classifying imbalanced multimodal sensor data for human activity recognition in a smart home using deep learning. In *2020 International Joint Conference on Neural Networks (IJCNN)* (pp. 1-8). IEEE. DOI: 10.1109/IJCNN48605.2020.9207697

Bouchabou, D., Nguyen, S. M., Lohr, C., LeDuc, B., & Kanellos, I. (2021). Using language model to bootstrap human activity recognition ambient sensors based in smart homes. *Electronics (Basel), 10*(20), 2498. DOI: 10.3390/electronics10202498

Bulbul, E., Cetin, A., & Dogru, I. A. (2018). Human activity recognition using smartphones. In *2018 2nd International Symposium on Multidisciplinary Studies and Innovative Technologies (ISMSIT)* (pp. 1-6). IEEE. DOI: 10.1109/ISMSIT.2018.8567275

Calzada, A., Liu, J., Nugent, C. D., Wang, H., & Martinez, L. (2014). Sensor-based activity recognition using extended belief rule-based inference methodology. In *2014 36th Annual International Conference of the IEEE Engineering in Medicine and Biology Society* (pp. 2694-2697). IEEE. DOI: 10.1109/EMBC.2014.6944178

Chung, S., Lim, J., Noh, K. J., Kim, G., & Jeong, H. (2019). Sensor data acquisition and multimodal sensor fusion for human activity recognition using deep learning. *Sensors (Basel), 19*(7), 1716. DOI: 10.3390/s19071716 PMID: 30974845

Guo, L., & Fang, H. (2016). Real-time human activity recognition in smart home with binary tree SVM. In *2016 35th Chinese Control Conference (CCC)* (pp. 6977-6981). IEEE. DOI: 10.1109/ChiCC.2016.7554456

Hamad, R. A., Hidalgo, A. S., Bouguelia, M. R., Estevez, M. E., & Quero, J. M. (2019). Efficient activity recognition in smart homes using delayed fuzzy temporal windows on binary sensors. *IEEE Journal of Biomedical and Health Informatics, 24*(2), 387–395. DOI: 10.1109/JBHI.2019.2918412 PMID: 31135373

Jethanandani, M., Perumal, T., Liaw, Y. C., Chang, J. R., Sharma, A., & Bao, Y. (2019). Binary relevance model for activity recognition in home environment using ambient sensors. In *2019 IEEE International Conference on Consumer Electronics-Taiwan (ICCE-TW)* (pp. 1-2). IEEE. DOI: 10.1109/ICCE-TW46550.2019.8991837

Kumar, P., & Suresh, S. (2022). Flaap: An open human activity recognition (har) dataset for learning and finding the associated activity patterns. *Procedia Computer Science*, *212*, 64–73. DOI: 10.1016/j.procs.2022.10.208

Mihoub, A. (2021). A deep learning-based framework for human activity recognition in smart homes. *Mobile Information Systems*, *2021*, 1–11. DOI: 10.1155/2021/6961343

Nafea, O., Abdul, W., Muhammad, G., & Alsulaiman, M. (2021). Sensor-based human activity recognition with spatio-temporal deep learning. *Sensors (Basel)*, *21*(6), 2141. DOI: 10.3390/s21062141 PMID: 33803891

Sarkar, A., Hossain, S. S., & Sarkar, R. (2023). Human activity recognition from sensor data using spatial attention-aided CNN with genetic algorithm. *Neural Computing & Applications*, *35*(7), 5165–5191. DOI: 10.1007/s00521-022-07911-0 PMID: 36311167

Tan, T. H., Badarch, L., Zeng, W. X., Gochoo, M., Alnajjar, F. S., & Hsieh, J. W. (2021). Binary sensors-based privacy-preserved activity recognition of elderly living alone using an RNN. *Sensors (Basel)*, *21*(16), 5371. DOI: 10.3390/s21165371 PMID: 34450809

Triboan, D., Chen, L., & Chen, F. (2019). Fuzzy-based fine-grained human activity recognition within smart environments. In *2019 IEEE SmartWorld, Ubiquitous Intelligence & Computing, Advanced & Trusted Computing, Scalable Computing & Communications, Cloud & Big Data Computing, Internet of People and Smart City Innovation (SmartWorld/SCALCOM/UIC/ATC/CBDCom/IOP/SCI)* (pp. 94-101). IEEE. DOI: 10.1109/SmartWorld-UIC-ATC-SCALCOM-IOP-SCI.2019.00059

Chapter 8
Achieving Smart Connectivity cum Automation Through Data-Driven Applications of IoT in Electronic:
Automation With Smart Electronics Connectivity

Gunjan Mukherjee
https://orcid.org/0000-0002-3959-3718
Brainware University, India

Arpita Shome
Abacus Institute of Engineering and Management, India

ABSTRACT

The IOT application has become multifaceted and diversified now days in its applications in the versatile fields including electronics. The Internet of things (IOT) has also revolutionized and rejuvenated the ways and strategies how the electronics interact with the surroundings possibly with the exchange of data. The IOT provides the smart solution for seamless connectivity with devices in some real time mode to make them much adaptable to the changing environments as per the user needs in some autonomous fashion. Some application areas include the predictive maintenance, data forecasting as a preventive measure of any sort of failure. Optimizing the scheduling time and downsizing some time specific measurement. Another application of the IOT is also to monitor and control the energy consumption, promoting the resources

DOI: 10.4018/979-8-3693-5448-3.ch008

in optimal fashion and reduces cost of operation. The IOT technology contributes to the personalizing interactions with other electronic devices, customizing their functional operations and interpretations of different connected behavior patterns.

INTRODUCTION

The integration of Internet of Things (IoT) technology into data-driven electronics has revolutionized various industries by enabling smarter and more connected devices. IoTor Internet ofthings refers to the network of interconnected devices that communicate and exchange data over the internet, allowing for seamless data collection, analysis, and decision-making. The IOT infrastructure enables all other devices dispersed in the whole big network to seamlessly communicate with the other remote device prompting the delivery of the exact and accurate information in very small amount of time. The electronic devices are working on the switch mode operation leading to the path of automation of task . When applied to electronics, IoT enhances the functionality, efficiency, and intelligence of electronic devices by leveraging real-time data insights. The electronics approach to the IOT for data preparation and transmission has enhanced the speed of transmission and the carriage of the bulk of data in short time and to the distant devices. The ensuing research works in this path enables the processing of data with all possible security adherence in more easy way .The IOT application has been proved to exist in the versatile domains such as health sector, social networking, Agriculture, remote monitoring, traffic management, transportation and many more. One case study has been represented in this chapter to elucidate the concept of IOT based application in more deeper sense of operations.

Here are several ways IoT enhances data-driven electronics Many different works have been done and the continuous research forces related to the security of data and interoperability of the data has come to the limelight of active research trends. The literature survey on the pertinent papers have been presented in the Table 1.

Review of Different IOT Based Applications in The Electronic Fields

Table 1. Related works with methodologies

Reference	Paper title	Methodology	Accuracy
(Rubi et al., 2021)	IoT-based platform for environment data sharing in smart cities	the research suggests an Internet-based platform that facilitates interoperability by gathering information through information extraction and visualization via the use of a semantic framework for environmental indicators.	The study examined REST, CoAP, and MQTT protocol latency and utilization of resources measures. In comparison, MQTT and REST used 3259 kband 2965 kb, correspondingly, while the CoAP adapter performed the best, using 2751 kb.
(Petrasch et al., 2022)	Data Integration and Interoperability: Towards a Model-Driven and Pattern-Oriented Approach	The present study describes a technique that uses structures and patterns to design data integration. The utilization of these models eliminates the requirement for human implementation by using specified data integration patterns to produce simple and straightforward descriptions of dataconversions	This method's benefits include the ability to reuse pre-existing data models, quick outcomes because data integration patterns are standardized, and simple upkeep and expansion of data integration requirements.
(Condry et al., 2016)	Using Smart Edge IoT Devices for Safer, Rapid Response with Industry IoT Control Operations	The model described in this research uses cryptography, image analysis, and connectivity features for authentication and authorization, and it links smart Internet of Things appliances with electronic control interfaces for safely monitoring operations.	Improved safety, adaptability, and real-time efficiency are provided by the model, which also highlights Intel's Internet of Things (IoT) framework for automated system controllers across multiple sectors.
(Qian et al., 2019)	HybridIoT: Integration of Hierarchical Multiple Access and Computation Offloading for IoT-Based Smart Cities	A scalable and robust Internet of Things infrastructure that combines offloading of computation between MEC and the cloud with UDN-based hierarchical access underpins the SC vision.	When considering standard IoT systems, it can significantly decrease both the total energy usage and the end-to-end analyzing delay.
(Subramaniam et al., 2023)	Interoperable IoMT Approach for Remote Diagnosis with Privacy-Preservation Perspective in Edge Systems	Using methods including energy-efficient clustering, data validation, tracking the environment, device verification, and TwineLiteNet-based encryption, the proposed interoperable IoMT implementation maximizes the performance of IoT networks.	The achieved throughput outperforms both the EIR-CIoT and BDCSIoMT approaches by 35% and 23%, accordingly with 20% more throughput, 15% less PDR, and 35% longer network lifetime

continued on following page

Table 1. Continued

Reference	Paper title	Methodology	Accuracy
(Li et al., 2022)	Big data analysis of the Internet of Things in the digital twins of smart city based on deep learning	The study combines convolutional neural networks for massive data processing in smart cities. It intends to enhance authority and data handling in smart cities by using multi-hop transmission and digital twin technologies for safe and effective data processing.	The study shows that data transmission in IoT-BDA systems becomes more energy efficient when the least amount of energy is gathered, and that signal transmission efficiency depends on a proper power diversion factor. The prediction accuracy of the model is 97.80%.
(Pazos et al., 2015)	ConnectOpen - automatic integration of IoT devices	This research presents a software-enabled approach that uses adaptable communication agents installed on a gateway to address fragmentation. These agents link sensors to a centralized infrastructure for consolidating data and REST API exposure, adapting to various data formats and protocols.	As a step toward plug-and-play design, the modular architecture speeds up device connections. Protocol, parser, and push modules are implemented as OSGi bundles by an eclipse plug-in. In contrast to the protocol bundle, which is extremely reusable, the push bundle can be used on any device.
(Sheng et al., 2015)	Lightweight Management of Resource-Constrained Sensor Devices in Internet of Things	The present research provides an extremely light RESTful web application that allows easy maintenance as well as access to wireless sensor devices by implementing IPv6 across a low-power connections and CoAP protocols.	The proposed approach for the effective and successful control of wireless sensor equipment was shown by the prototype cloud system.
(Daniele et al., 2016)	Interoperability for Smart Appliances in the IoT World	EEBus and Energy@house industry groups collaborated to develop SAREF4EE, a customized version of SAREF that enables easy communication between intelligent appliances from various suppliers via any house or cloud-based energy management system.	SAREF and SAREF4EE's network should be utilized more methodically in order to increase interoperability in the segmented industry, as they satisfy the requirements for IoT standardization.

continued on following page

Table 1. Continued

Reference	Paper title	Methodology	Accuracy
(Nasution et al., 2019)	Designing an IoT-based air quality monitoring system	The study created an Internet of Things (IoT)-based air purity monitoring system that measures CO, O3, SO2, and particle levels with sensors. An Arduino microcontroller is used to read data, which is then transmitted to the ThingSpeak system in the cloud.	A system trial that was established on campus measured PM10 particles, SO2, CO2, and ozone to monitor the state of the air quality. Every minute, ThingSpeak receives data that can be viewed on the channel sites that are supplied.
(Hossain et al., 2919)	A smart IoT based system for monitoring and controlling the sub-station equipment	The research paper presents a network approach based on the Internet of Things (IoT) for managing and monitoring sub-station equipment, improving resource management and efficiency.	The suggested system continuously detects the secondary current of two CTs, transmits the results to an internet server, and presents the information on a web page. It is an automated, independent verification oil quality and level system.

IOT BASED APPLICATIONS IN VERATILE FILEDS

Real-Time Monitoring and Control IoT-Enabled Electronics

This ideas can continuously monitor various parameters and conditions in real-time. For example, IoT sensors embedded in industrial machinery can monitor temperature, pressure, vibration, and other critical factors to detect anomalies and prevent breakdowns. This real-time monitoring allows for proactive maintenance, reducing downtime and optimizing performance. The data collection has been an important step for implementation of the IOT based monitoring of the system. Data is being gathered from various sources related to different machineries, environments, equipment etc. The sensor can also be employed to measure several values of related parameters like temperature, pressure, humidifies . The connectivity of the system to the local network has some vital impacts as collected data are to be transmitted over the network to either centralised system or the cloud based platform. Many device oriented connections like Bluetooth, Wifi, Cellular networks and other specialised IOT protocols like LoRaWAN can be used. The analysis of data provides trends, correlations and the main underlying pattern of the data to get the in-depth insight of flow of it leading to the better tendency analysis of data. The assessment of critical parameters of the data is also posing some significant values towards tracking the trends and visualisation of the same in future perspective. The alert management has become the integral part of it and can be conveyed through the sms, email notification or the mobile application. The remote control operation alongwith the automation and other trigger actions can be helpful for the overall automation of the system. The realtime IOT enabled system also performs the diagnostic scan for the potential failures or any sign of the same. The predictive analysis could provide an edge for tracking the unplanned approach and the unoptimized asset for catching any sort of flaws in the system. The efficiency of the monitoring systems has been effective enough to optimise the energy optimisation for improvement of opportunities for efficiency up gradation. Hussain et al.,(2021) reported the real time load monitoring and the control system based on LoRa. The quantitative measurement in terms of current, voltage and the qualitative measurement in terms of the power-factor have been reported. The IOT based system efficiently measure the different power measurement parameters. Another work in this regard WU et al., (2023) reported about the elastic net algorithm dealing with the high dimensional data stream collected by the IOT device to search out the faulty variables . The finding of the root cause alongwith necessary optimisation has been reported in the paper with good accuracy value.The work cited as Dui et al., (2024) .Reported about the IOT enabled real time traffic monitoring and the control information related to the traffic congestion process. Control flow mechanism over the traffichas been stated with the potential strategies for the optimum route choice toward the improvement of the transportation system.

Data-Driven Insights IoT devices

The unique concept of generate vast amounts of data that can be analysed to extract valuable insights. By collecting and analysing data from sensors, electronics manufacturers can gain deeper insights into product usage patterns, customer behaviour, and performance metrics. These insights can inform product design improvements, enhance user experiences, and drive strategic decision-making. The concept of data driven IOT devices can lead to many valuable opportunities. Alert management has become one of the potential outcomes of such data driven strategies of the IOT device. The sudden rise of any sort of flaw in the manufacturingenvironment, equipment failures needs the immediate alert generation to bring it to the notice so that the prompt actions can be taken for the immediate repairment or restoration. Supply chain management and the corresponding optimisation of business strategies and logistics along withthe improvement of the inventory management and reduction of the transportation cost helps in minimizing the fuel consumption and optimisation of the delivery routes. The energyproduction, consumption and distribution are the strategies of energy management techniques. The occupancy pattern and energy usage can enable the building automation systems to adapt to the proper lighting, heating and cooling conditions for the optimal energy efficiency. The data availability with IOT devices can enable the remote monitoring of the patient health and the corresponding decision makingand the requisite analysis of data. The setting of personalizedtreatment goal, heath recommendations and the choice for the informed lifestyles has become the feasible mean as the outcome of such research. The traffic and pollution management could also be done by the presence of IOT devices leading to the state of least amountof pollution. The tracking of such dynamic pattern selection can be recognized promptly to tackle the entire traffic management, optimize the traffic flow and reduce the overall commute time.Some of the communications in this line are Tripathi et al., (2019) reported about the achievement of IOT enabled data applications and its communications, optimisations and overall security. The role of the communication bandwidth, cloud storage and energy saving and its utilisations reserving the properservice qualities have been well reported in this paper. The publication Dave et al., (2022) stated the multifaceted challenges for communication among the heterogenous devices and its architecture. The paper also states the standard protocols, comparative approaches with the other emerging technologies like the block chain, edge computing etc. The concept of data integration and the data interoperability has also been studies and analysed in the broad review presented. Chataut, et al., (2023) reported the real connectivity and other issues related to device-oriented functions in relations to the real-world objects. The impacts of the wireless technologies and its immenseimpacts in other technological areas. This paper also providesthe deeper understanding and the future growth prospect of the device oriented approach . Maurya et al., (2023) reported about the spatial distribution of the devices and othersensors.The paper also has addressed some critical issues related to the security,intolerabilities, energy optimisation to greater extents. The exploration of the emerging trends of IOT and its interactions with the AI, ML and the deeplearinig concept has been elaborately outlined.

Predictive Analytics

This unique concept of IOT Leveraging machine learning algorithms and predictive analytics, IoT-enabled electronics can anticipate future events and trends based on historical data patterns. For instance, IoT sensors in smart home appliances can analyse usage patterns to predict when maintenance is required or when consumables need replenishment. Predictive analytics enables proactive decision-making, cost savings, and improved user experiences. The more engagement of various statistical functions can enable the interpretation of behaviour of data and analysis of its future trends. The predictive analysis of the IOTbased modeldepends on the initial volumes of data and it inter relevance.The proper data collections excepting the false data can be the eligible batch for the good prediction. The pre-processing of data step the process further towards the good accuracy in the analytical process. The contribution of modeltowards the good score of prediction is also significant.The IOT enabled electronic devices like the sensors and have been made to be the best suited as monitoring devices. The remote monitoring can add also to the detection and corresponding analytics. The health monitoring and environment monitoring can be served as the applicationof predictive benefitsof the IOT based model. Kumar et al., (2018) reported the sore based analysis for prediction. The exploitation of the historical and transactional data pattern for multifaceted problems related to the business and other science problems. The detection of myriad risks and associated opportunities are the main aspects to be addressed. Taneja, et al., (2017) has stated the main concept of the IOT generated data and the utilisation of the same for predictive analysis. The paper also highlights the different prediction model for getting some precise result leading to the perfect prediction in future data. Shorfuzzaman et al., (2021) reported the predictive model based on the temperature and humidity sensor based data and their analysis along with the total energy consumption of the devices using the LSTM approach. The main salient feature of such paper is the use of the optimisation.

Enhanced Connectivity

The connectivity of IoT enables seamless connectivity between electronic devices, enabling them to communicate and collaborate intelligently. For example, in a smart home ecosystem, IoT-enabled devices such as thermostats, lighting systems, and security cameras can communicate with each other to optimize energy usage, enhance security, and improve overall convenience for users.The main objective of the enhanced connectivity of the IOT devices are implied in the collection, exchange and analysis of data of different scenarios leading to the improved functionalities and efficiency of operations of devices. The cobbectiveity of the IOT system could be enhanced more by deploying the wireless communications. Some of the wireless applications like Wi-Fi, Bluetooth, Zigbee, Zwave has impacted the and enhanced the technological aspects for reliable The low power connectivity can be ensured by means of soe innovative technologies . some of the concerned technologies like Bluetooth Low energy, LoRaWAN and Sigfox based low-power wide area connectivity. The enhancement of connectivity across the wide distance can be maintained by means of the low power . The ultimate reliabilityofthe IOT systems can be backed up by the enhanced technological progress and other security concerns of the edge computing approach. The proper security is being maintained by the authentication, integrity, confidentiality and availibity of the IOT systems. The bandwidth intensification as is followed in the 5 G network implied in the high throughputs and communications approach. The concept o f mesh networking systems can also add to the overall capability of the dynamic environment in addition to offering of the resiliency of IOT. This also leads to also the reliability of the IOT devices as well. Some of the publications in this regard can be listed as López, O. L., Rosabal, O. M., Ruiz-Guirola et al., (2023) refers to the sustainable energy utilisation by the IOT devices. How has the concept of self sustainability be maintained, has become the subject of this paper. This paer also elaborated about the recent research and advancement in the IOT based energy utilisation . Shafique et al., (2020) reported about the higher bandwidth, larger channel capacity and enhanced data rate providing the ease of operations. This paper also provides some highlights on the MIM and MMIMO and other device to device communications . The concept of deep learning and its connectivity has been expressed in this paper. The new concept of the software defined sensor technology and its effects on the mobile computing paradigm has been expressed in the reported paper. Rahmani, et al., (2023) reported the next generation IOT system for the sustainable development in the industry automation. The rapid and largescale innovation has led to the increasing number of IOT system based paradigm shifts . The eco friendlyand non biodegradable approach along with the sustainable powering by exploiting the energy harvesting with the good presence of the wireless technology has been highlighted in the concerned paper leading to the battery less operation. Nižetić et al., (2020) has reported about the IOT based technology dshift towards the sustainability of human population and the associated improvement of the quality of life with the smart utilisaton of limited global resources. The IOT technology application towards the upkeeping of the better environment in eco-friendly manner has been reported in this paper.

Remote Monitoring and Management

The very concept of remote monitoring of the IoT allows for remote monitoring and management of electronic devices from anywhere with an internet connection. Manufacturers can remotely monitor device performance, troubleshoot issues, and deploy software updates or patches in real-time. Remote management capabilities improve operational efficiency, reduce maintenance costs, and enhance scalability.

Remote Monitoring and Management (RMM) utilizing the Internet of Things (IoT) is a cutting-edge approach that revolutionizes how we oversee and control various systems and equipment remotely. Through the integration of IoT devices embedded with sensors, this concept enables the continuous collection of data pertaining to different environmental and operational parameters. These sensors, ranging from temperature and humidity to motion and vibration detectors, facilitate the acquisition of real-time insights into the performance and condition of monitored assets. The data gathered by these devices is transmitted via wired or wireless networks to centralized cloud platforms, where it undergoes comprehensive processing and analysis. Leveraging cloud-based infrastructure, users gain access to intuitive web or mobile applications that provide comprehensive dashboards and alerts, allowing for immediate action when anomalies or maintenance requirements are detected. The secured challenged communication based on the IOT device plays a crucial role in tendering the information in prompt manner without creating any sort of hampering to the other component of the systems.

Furthermore, IoT-based RMM systems empower users with the capability to remotely manage and adjust device settings, initiate maintenance procedures, and even implement automated responses based on predefined rules and algorithms. The highly ideated approach towards the development of such rules can produce the good and prompt outcomes.

By seamlessly integrating with other enterprise systems, such as ERP and CRM platforms, these solutions facilitate enhanced decision-making, operational efficiency, and proactive maintenance strategies. The concept of the interoperability in the software and other hardware components for smooth rung of the system with ulterior security is of great interest to the recent trends of innovations with sufficient automation harnessing the power of IOT technology. In summary, Remote Monitoring and Management utilizing IoT technologies offer a paradigm shift in how organizations monitor, manage, and optimize their assets, driving efficiency gains and cost savings across a wide array of industries. The flawless management of the whole of such component integration in seamless manner maintaining the requite security can lead to the maximum production line of in the business paradigm. Totade et al., (2022) expounded the concept of the IOT in remote sensing. The micro controller and relay sensor based approach and its connectivity to the clouds has become an important point discussed in this paper. Iranpak, et al., (2021) reported a about the remote patient management using the connectivity with the cloud and the LSTM. The accuracy of the method achieved has been estimated to 94%. Islam et al.,(2023) Stated about the deep learning based IOT system in emote monitoring of the health status and detection of the health issues depending on several parametric studies and their analysis. Some of the health hazards like assessment of the cardiac conditions, heart beat rate and maturity assessment of the heart beats etc. The hints of some critical abnormalities are also prominent from the paper.

Personalized Experiences

This concept has garnered by analysing user data collected from IoT devices, electronics manufacturers can deliver personalized experiences tailored to individual preferences and behaviours. For example, wearable devices equipped with IoT sensors can track user activity levels, sleep patterns, and health metrics to provide personalized recommendations for fitness routines, diet plans, and lifestyle adjustments.

Wearable devices integrated with IoT sensors represent a transformative technology that enables comprehensive tracking and monitoring of various activities in real-time. The wearable devices are generally fitted with multiple sensors and can be capable of doing many efficient tasks like, physical movements, biometric indicators, and environmental factors. By leveraging the Internet of Things (IoT), this captured data (audio, video or textual data) is seamlessly transmitted to centralized platforms for analysis and interpretation, unlocking valuable insights into individual activity patterns, health metrics, and performance levels. The maintenance of the overall security for this purpose is highly solicited and is also assured enough by its implementation. Whether it's a fitness tracker, smartwatch, or health monitoring device, wearables with IoT sensors offer unparalleled capabilities in monitoring activities such as exercise routines, sleep patterns, heart rate variability, and calorie expenditure. Advanced sensors embedded within these devices accurately capture and record data, providing users with real-time feedback and personalized insights into their overall well-being and fitness progress. The efficient sensors will also be capable of capturing the data from some considerable distance. The commonly attachable devices are also concerned to the continuous monitoring of respective physical parameters are those for measuring the blood pressure or monitoring of the blood sugar of n the human body . Moreover, the integration of IoT technology enables wearables to connect with smartphones, tablets, or cloud-based platforms, allowing users to access and analyze their activity data from anywhere, at any time. In eldercare, these devices provide caregivers with real-time updates on the activities and health status of elderly individuals, enabling timely interventions and support. Overall, wearable devices equipped with IoT sensors represent a powerful tool for tracking and monitoring activities across diverse domains, offering individuals, organizations, and healthcare providers unparalleled visibility into performance, health, and well-being. The research scopes of such notion can be broadened further by judging the critical concepts of myriad physical parameters and tracking their values in real time mode. Maghsudi, et al., (2021) reported about the personalized education and its upgradation using the core concept of the AI and ML approach. The future direction of the studies in this respect has stated indepth in this paper. Murtaza et al., (2022) stated the original requirement, challenges and concerned methodologies in the AI based learning systems. The future education process based on the AI concept has also ben illustrated in this paper.

The personalized marketing has become the mile stone in the achievement of AI based research work. The work Alqurashi et al., (2023) has referred to the core concept of the personalized content marketing and the dynamic platform by means of the strategic integration of different AI channels.

Efficient Resource Utilization IoT-enabled electronics

This concept facilitate more efficient resource utilization by optimizing energy consumption, minimizing waste, and improving resource allocation. For example, smart energy management systems can dynamically adjust power usage based on demand patterns, leading to energy savings and reduced environmental impact. Overall, IoT technology enhances data-driven electronics by enabling real-time monitoring, data-driven insights, predictive analytics, enhanced connectivity, remote management, personalized experiences, and efficient resource utilization. As IoT continues to evolve and proliferate, its integration into electronics will drive innovation, efficiency, and transformation across industries. Data-oriented electronics applications through IoT encompass a wide range of use cases where electronic devices leverage IoT technology to collect, analyze, and utilize data for various purposes. Here are some examples of data-oriented electronics applications through IoT. There are number of application of such data driven IOT enabled electronics devices specially in the fields of automation. Prasad et al,, (2023) stated about the versatile agreement platform for the adaptions of the ecommerce based applications to the IOT and cloud computing environment. The alignment of the service level agreement has also been exclusively done. Mishra et al., (2023) stated about the integration of the IOT and the deep reinforcement learning in terms of the smart grid application. The concept of efficient energy resource utilisation and its adaptability concept hasbeen stated in this paper.

Smart Home Automation

The typical automation IoT-enabled electronic devices in smart homes collect data from sensors embedded in appliances, thermostats, lighting systems, security cameras, and other devices. This data is used to automate home functions, such as adjusting room temperatures based on occupancy, optimizing energy usage, remotely controlling appliances, and enhancing home security through motion detection and surveillance. Automation lies at the heart of the transformative potential of IoT devices in a smart home ecosystem. With a plethora of interconnected devices, that can be interconnected with each others to achieve the maximum gain being in the place of comfort. The easy and hassle free approach of such echnology concerns the IOT and other allied technologies . When setting up automation routines, it's essential to consider the unique rhythms of day to day life . A for instance the automated thermostat can easily offer solutions to control and regulate the room temperature depending on the weather parameters undergoing some changes depending on the allied and external influences. The smart light technology cn come to the aid of automated lighting of the room light keeping the ambient temperature low and soothing. The automatic turn off and on of the switch depending on the ambient temperature fluctuation has become also another concerned point to be noticed. Another important task to be addressed towards the automation of task is the initiation of the smart irrigation system. The regulation of thamount of water to be prayed depending onthe soil moisture condition has become another active way of soilhealth maintenance. The most important of all activities is the creation, notification and modification of the schedule of automation. The easy control on the concerned factors by using mobile control has become another step towards the automation. Singh et al., (2019)

Industrial IoT (IIoT)

In industrial settings, IoT-enabled electronic devices gather data from sensors installed in manufacturing equipment, machinery, and infrastructure. This data is utilized for predictive maintenance, quality control, process optimization, and supply chain management. IIoT applications improve operational efficiency, minimize downtime, and enable real-time monitoring and control of industrial processes. Theconcept represents a revolutionary shift in the way industries operate, integrating advanced digital technologies with traditional industrial processes to unlock unprecedented levels of efficiency, productivity, and connectivity. At its essence, IIoT encompasses a vast network of interconnected sensors, actuators, devices, and machinery embedded within industrial infrastructure. These components communicate and exchange data in real-time, forming a cohesive ecosystem that enables seamless monitoring, control, and optimization of industrial operations. The interoperability among the respective devices are also importantas it caters to significant contribution to the versatile applicationsby bringingthem under the single umbrella. One of the key pillars of IIoT is data acquisition and analysis. Through a multitude of sensors strategically placed throughout the industrial environment, IIoT systems continuously collect a wealth of data pertaining to equipment performance, environmental conditions, energy consumption, and more. This influx of data serves as the foundation for advanced analytics, where sophisticated algorithms and machine learning techniques are employed to derive actionable insights. By leveraging predictive analytics, organizations can anticipate equipment failures before they occur, thereby minimizing unplanned downtime and reducing maintenance costs. Through continuous monitoring of equipment health metrics such as temperature, vibration, and energy consumption, IIoT-enabled predictive maintenance systems can detect anomalies and issue alerts to maintenance personnel, allowing for timely intervention and preventive action. The real time mode forthe generation of alert signal can be effective enough to track the situation and which also enables the prompt action initiation. Furthermore, IIoT enables the concept of the smart factory, where interconnected machines, robots, and production systems collaborate autonomously to optimize manufacturing processes. Through real-time data exchange and communication, IIoT systems can orchestrate production workflows, adjust parameters on the fly, and dynamically allocate resources to maximize throughput and minimize waste. This level of agility and flexibility not only enhances operational efficiency but also enables manufacturers to respond swiftly to changing market demands and customer preferences. The trends analysis of the IOT based device becomes helpful to get through the crucial environment posing hindrances to the smooth collection of data sometimes.Beyond the factory floor, IIoT extends its reach into supply chain management, logistics, and asset tracking. By embedding sensors and RFID tags into products, containers, and vehicles, organizations can gain real-time visibility into the movement and status of goods throughout the supply chainBy offering value-added services such as equipment-as-a-service, predictive maintenance subscriptions, or performance-based contracts, industrial organizations can create new sources of recurring revenue while delivering greater value to customers. The profit escalation of the market connected to transactions in business and commercial sector can provide the wide scope of customer trends analysis and their retentions. Deshpande et al.(2020) carried an extensive survey works on the industrial things for industry purpose showing the actual need of such technology for the industry based production.

Healthcare Monitoring and Telemedicine

Wearable electronic devices equipped with IoT sensors collect health-related data, such as heart rate, blood pressure, blood glucose levels, and activity levels. This data is transmitted to healthcare providers or caregivers in real-time, enabling remote patient monitoring, early detection of health issues, and personalized healthcare interventions. Telemedicine platforms leverage IoT to facilitate virtual consultations, remote diagnostics, and treatment monitoring. The advancement of such technology helps in reducing the overall mortality rate. Healthcare monitoring and telemedicine represent transformative advancements in healthcare delivery, catering to the evolving needs of patients and providers alike. Healthcare monitoring encompasses a multifaceted approach that leverages a plethora of technologies and devices, each designed to track and manage various aspects of an individual's health remotely. Wearable devices, such as smartwatches and fitness trackers, equipped with sensors capable of measuring vital signs like heart rate, blood pressure, and activity levels, offer real-time insights into an individual's well-being. These devices, coupled with mobile applications and cloud-based platforms, enable continuous monitoring and data collection, empowering individuals to proactively manage chronic conditions, track fitness goals, and make informed lifestyle choices. So many difficult and chronic diseases can be controlled easily by means of such influence of technologies to achieve such coveted target potential. Moreover, healthcare monitoring extends beyond personal devices to include remote patient monitoring (RPM) systems utilized by healthcare providers to track patients' health status outside traditional clinical settings. RPM systems integrate various medical devices and sensors to monitor patients with chronic conditions, postoperative care needs, or those at risk of complications.

In tandem with healthcare monitoring, telemedicine has emerged as a cornerstone of modern healthcare delivery, particularly in the wake of the COVID-19 pandemic, which highlighted the importance of remote healthcare services. Telemedicine harnesses telecommunications technology to facilitate virtual consultations between patients and healthcare providers, transcending geographical barriers and improving access to medical expertise. Through secure video conferencing platforms, patients can consult with healthcare professionals, receive diagnoses, discuss treatment options, and even obtain prescriptions from the comfort of their homes.The approach to some invincible disease and the patients can be controlled and monitored with the good success rate for mitigating the overall risk and abatement of the risk explosions. The benefits of telemedicine extend beyond convenience, offering cost-effective solutions for both patients and healthcare systems. By reducing the need for in-person visits, telemedicine minimizes travel time and associated expenses for patients while optimizing the utilization of healthcare resources. Furthermore, telemedicine enables healthcare providers to extend their reach to underserved populations, rural communities, and individuals with mobility limitations, ensuring equitable access to quality care. The remote technology also plays a pivotal role in controlling the overall things with new touch of the various technologies. Artificial intelligence (AI) and machine learning (ML) algorithms further enhance the capabilities of healthcare monitoring and telemedicine platforms. These technologies analyse vast amounts of patient data, identify patterns, and generate actionable insights to support clinical decision-making. AI-driven diagnostic tools can assist healthcare providers in interpreting medical images, detecting anomalies, and predicting disease progression with high accuracy, thereby improving diagnostic efficiency and reducing errors.The overall technological aspect has become the big shadow behind the advancement of teleporting and telemedicine issues. Kelly et al., (2020) proposed the the IOT device design and its applications in the health sector in keeping the security aspect and confidentiality of the data flow through the device.

Smart Agriculture

IoT-enabled electronic devices in agriculture gather data on soil moisture levels, temperature, humidity, weather conditions, crop growth, and pest activity. This data is used to optimize irrigation schedules, monitor crop health, automate pesticide and fertilizer application, and forecast weather patterns. Smart agriculture applications improve crop yields, conserve water resources, and reduce environmental impact. Smart agriculture, termed as precision agriculture or digital farming, encompasses a suite of advanced technologies and data-driven approaches aimed at optimizing agricultural practices and increasing productivity while minimizing resource inputs and environmental impact. At its core, smart agriculture leverages interconnected systems comprising sensors, drones, satellite imagery, artificial intelligence (AI), and Internet of Things (IoT) devices to monitor and manage various aspects of crop cultivation, livestock farming, and agricultural logistics. These technologies enable farmers to collect real-time data on soil moisture levels, nutrient content, weather conditions, pest infestations, and crop health, allowing for precise decision-making throughout the agricultural cycle. The device centric software application has become the trust worthy component for realizing the system to tackle all real time scenarios. One of the key components of smart agriculture is precision farming, which involves the targeted application of inputs such as water, fertilizers, and pesticides based on site-specific data and analysis. By deploying sensors and IoT devices in fields, farmers can gather detailed information about soil composition and crop performance, enabling them to tailor irrigation and fertilization schedules to meet the specific needs of each crop and minimize waste. This not only conserves water and reduces environmental pollution but also improves crop yields and quality. The thermal and hyperspectral images play an important role in determining so many difficult and challenging scenarios like the freshness of different fruits and vegetables to the ultimate level to get the good yield in some timely manner. In livestock farming, smart agriculture technologies enhance animal welfare, productivity, and management efficiency. Wearable devices equipped with sensors can monitor the health, behaviour, and movement patterns of livestock, enabling early detection of diseases, estrus cycles, and calving events. By continuously tracking individual animals' vital signs and activity levels, farmers can intervene promptly in case of health issues or emergencies, thereby reducing mortality rates and improving overall herd management. The likelihood of different kettles and livestock to get affected with many dangerous and lethal diseases can be mitigated by means of such advanced remote monitoring technologies . Moreover, smart agriculture extends beyond the farm gate to encompass supply chain optimization, traceability, and market intelligence. IoT-enabled devices and blockchain technology enable real-time monitoring and tracking of agricultural products from farm to fork, ensuring transparency, quality control, and food safety compliance. By digitizing supply chain processes and leveraging data analytics, stakeholders can streamline logistics, minimize food waste, and respond swiftly to market demands, thereby enhancing the efficiency and resilience of the agricultural value chain. The enhancement of such efficacy related to such domain has been proved to be much responsive. Dhanarajuet al., (2022) stated the IOT application in the smart farming with all sort of security measures to be applied fopr the sustainability of the farming in long run.

Connected Vehicles and Intelligent Transportation Systems (ITS)

IoT-enabled electronic devices in vehicles, road infrastructure, and traffic management systems collect data on vehicle speed, location, traffic congestion, road conditions, and driver behaviour. This data is utilized for real-time traffic management, route optimization, predictive maintenance of vehicles, and enhancing road safety through collision avoidance systems and vehicle-to-vehicle communication. Connected vehicles and Intelligent Transportation Systems (ITS) stand at the forefront of transforming the landscape of transportation, offering a comprehensive approach to optimizing efficiency, safety, and sustainability. Connected vehicles utilize cutting-edge communication technologies such as V2V and V2I to establish seamless data exchange between vehicles and infrastructure, enabling a wide array of applications aimed at enhancing road safety, traffic flow, and navigation. These vehicles can communicate critical information in real-time, including traffic conditions, potential hazards, and weather alerts, allowing for proactive decision-making and mitigating risks on the road. The rush driving of the motor cars or motor cycles can lead to lot more causalities which can easily be controlled by investigating the behavioural analysis of the riders or the pilots and tracking their expressions. Sometimes the riders of such vehicles may appear to intoxicated and thereby increase the risk of accidents on the road. The system can be capable of monitoring such situation and easily grab the concept in faithful manner to achieve the real scenarios. Intelligent Transportation Systems, encompassing a myriad of advanced technologies and strategies, complement the capabilities of connected vehicles by providing comprehensive traffic management solutions. ITS integrates sensors, cameras, and sophisticated analytics to monitor and analyze traffic patterns, optimize signal timings, and implement dynamic routing strategies. This data-driven approach enables transportation agencies to effectively manage congestion, reduce travel times, and enhance the overall efficiency of transportation networks. The overall knowledge of the traffic management can easily be utilised to get some inputs of the j traffic congestions or to get some probabilistic scenarios of the incidence. Furthermore, the integration of connected vehicles and ITS lays the foundation for the deployment of autonomous vehicles and advanced driver assistance systems (ADAS), promising even greater improvements in safety and mobility. Autonomous vehicles leverage the connectivity and data exchange capabilities of connected systems to navigate roads autonomously, while ADAS features enhance driver awareness and responsiveness, ultimately reducing the likelihood of accidents and improving road safety for all users .Additionally, collaboration between government agencies, industry stakeholders, and researchers is crucial to address challenges related to data privacy, cyber security, and interoperability. Despite these challenges, the benefits of connected vehicles and ITS are far-reaching. By optimizing transportation systems, these technologies can reduce traffic congestion, lower emissions, and improve the overall quality of life in urban areas. Moreover, they have the potential to revolutionize mobility by providing more accessible, efficient, and sustainable transportation options for individuals and communities worldwide. As advancements continue and adoption grows, connected vehicles and ITS will play an increasingly pivotal role in shaping the future of transportation, paving the way towards safer, smarter, and more resilient cities. In this regard, the optimisation of the traffic management can easily be handled using the rudimentary behavioural aspects and its faithful applications in some optimum way. Ushakov, et al., (2022) discussed about the global transportation system emphasising on the influences of other factors like the weather impacts etc on the smooth transportation system and its optimisation.

Environmental Monitoring

Electronic sensors deployed in environmental monitoring applications collect data on air quality, water quality, noise levels, pollution levels, and weather conditions. This data is used for environmental assessment, pollution control, disaster management, and climate change mitigation. IoT-enabled environmental monitoring systems provide policymakers, researchers, and the public with valuable insights into environmental trends and challenges.Environmental monitoring through the utilization of Internet of Things (IoT) technologies represents a pivotal approach in understanding and mitigating environmental challenges. IoT-enabled sensors and devices are deployed across various ecosystems, including air, water, and soil, to collect real-time data on environmental parameters such as temperature, humidity, pollution levels, and biodiversity. These sensors can be deployed in remote or hard-to-reach locations, providing insights into environmental conditions that were previously inaccessible. By continuously monitoring environmental variables, IoT technologies offer unparalleled opportunities for early detection of environmental threats, such as pollution events, natural disasters, and climate change impacts. Furthermore, IoT-based environmental monitoring systems facilitate data-driven decision-making and adaptive management strategies. Through the integration of cloud computing and data analytics, large volumes of sensor data can be processed, analyzed, and visualized in near real-time, enabling stakeholders to identify trends, patterns, and anomalies. This information can inform policymakers, resource managers, and the public, empowering them to take proactive measures to protect and preserve ecosystem, safeguard public health, and mitigate the impacts of environmental degradation. The overall protection of the data and its further utilisation can be of immense help in making the good predictive model.

Moreover, IoT technologies enhance the efficiency and cost-effectiveness of environmental monitoring efforts. Traditional monitoring methods often involve manual data collection, which can be labor-intensive, time-consuming, and expensive. IoT-enabled sensors automate data collection processes, reducing the need for human intervention and enabling continuous monitoring over extended periods. This scalability and accessibility make IoT-based environmental monitoring systems suitable for a wide range of applications, from monitoring air quality in urban areas to tracking water quality in rivers and lakes. The current day scenarios of pollution in both in the air and water has led to application of many new kinds of technology involving AI, machine learning and IOT . The real boosting of the result has come to the rescue of the environment from many different hazards. Cunin et al., (2018) stated about the environmental parametric values like temperature, humidity, pollution, air pressure etc. The inevitable role of the IOT devices has been discussed in the paper.

However, the widespread adoption of IoT technologies for environmental monitoring also poses challenges, including data privacy, security, and interoperability concerns. Ensuring the integrity and confidentiality of environmental data is essential to maintaining public trust and maximizing the impact of monitoring efforts. Additionally, standardization of sensor technologies and data formats is necessary to facilitate data sharing and interoperability among different monitoring platforms and stakeholders. The real capability of such data mixing from various different and heterogenous sources and definition of the real trends has become the key milestone for cracking such hurdle. The ongoing research work obviously shed some light on the cropping of many such problems and its related solutions.

Retail Analytics

IoT-enabled electronic devices in retail environments collect data on customer foot traffic, purchasing behaviour, product preferences, and inventory levels. This data is utilized for personalized marketing campaigns, optimizing store layouts, predicting demand, and inventory management. Retailers leverage IoT to enhance customer experiences, increase sales, and improve operational efficiency.These examples illustrate how data-oriented electronics applications through IoT leverage data collection, analysis, and utilization to enhance various aspects of our lives, including home automation, healthcare, agriculture, transportation, environmental monitoring, retail, and industrial operations. As IoT technology continues to advance, we can expect to see further innovation and adoption of data-driven electronics applications across diverse industries and sectors.

Retail analytics empowered by Internet of Things (IoT) technology is revolutionizing the way retailers understand consumer behaviour, optimize operations, and enhance customer experiences. IoT devices such as beacons, RFID tags, smart shelves, and cameras are deployed throughout retail spaces to collect vast amounts of data on customer movements, product interactions, inventory levels, and environmental conditions. This data is then analyzed using advanced analytics techniques to derive actionable insights that drive strategic decision-making and operational improvements. The real challenge is lying hidden in the exposition of good knowledge and its wide applications to the solutions of myriad number of applications. The smart concepts of society and its application has become the great mean of future research towards the biggest goal achieved so far.One of the primary applications of IoT in retail analytics is in-store analytics, which enables retailers to track customer traffic patterns, dwell times, and navigation paths within physical stores. By analyzing this data, retailers can optimize store layouts, product placements, and promotional strategies to maximize sales and improve customer satisfaction. For example, heatmaps generated from IoT sensors can identify high-traffic areas within a store, allowing retailers to strategically position products and signage to capture shoppers' attention and drive impulse purchases. The understanding of the data pattern and its proper utilisation can lead to delving out so many hidden information apt for the future prediction in a big way.

The good prediction of the instock, overstock and out of the stock scenarios can better be handled by the IOT technology. Moreover, IoT-powered retail analytics facilitates personalized marketing and customer engagement initiatives. By leveraging data from IoT devices, retailers can deliver targeted and context-aware marketing messages, promotions, and recommendations to customers based on their preferences, purchase history, and current location within the store. For example, beacons can send personalized offers to shoppers' smartphones as they browse specific aisles or approach certain products, enhancing the relevance and effectiveness of marketing campaigns.The personalization issues also help in understanding the customer behaviour and its implications for retaining the customer base and to attract the good size of customer pool.The AI based approach for the customer tracking has resulted into the good and consolidated response towards the customisation of the technology and the product as well. Furthermore, IoT technology enhances the overall shopping experience by enabling innovative solutions such as cashier-less checkout systems and smart fitting rooms. Automated checkout systems equipped with IoT sensors and cameras enable frictionless shopping experiences, allowing customers to scan items and pay without the need for traditional checkout queues. Smart fitting rooms equipped with RFID-enabled mirrors and interactive displays provide personalized product recommendations, sizes, and styling suggestions based on customers' preferences and selections, enhancing convenience and satisfaction. The enactment of market security depending on observation of the customer base has produced the solid base of glimpse in such subject.

However, the adoption of IoT technology in retail analytics also presents challenges related to data privacy, security, and integration with existing IT infrastructure. Retailers must ensure the protection of customer data and compliance with regulatory requirements such as GDPR and CCPA to maintain trust and confidence among shoppers. Additionally, integrating IoT devices and analytics platforms with legacy systems and processes requires careful planning and investment in IT infrastructure and talent. The IOT based device application can also provide the strong sense of authentication and security towards the data enhancement for getting wide success in the data in the centralised and the distributed environment as well. Kaur et al.,(2022) stated the service oriented approach and other shopper;s activities in te computerised systems.

IOT security

The rapid growth of the IOT and machine learning technology results into myriad number if security threats. The concerned devices sometimes fall prey to the vulnerabilities of cvber criminals. Some of the important lackings on the security aspects if the IOT devices are the inadequate security due to the weak password, lack of encryption. Lack of standardisation and leads to the vulnerabilities on much larger scales. Insufficient update mechanisms, poor communication and the very weak authentication.

Some of the emerging solutions to such security vulnerabilities has been stated as the AI enablement in the security framework with the e enhanced response towards the p roper classification and identification of the threat type. The advent of the block chain in the device to device communication offers the sufficient data integrity. The IOT security gateway can also provide the good enough control mechanisms to monitor the traffics and additional layer of protection. The continuous monitoring of the e data traffics between the IOT devices and external networks can be monitored also to ensure the good and strong security to the systems.

Case study

The following case study is concerned to the IOT sensor based temperature and humidity data reading. The data driven IOT application involves the specified steps like gathering data from sensor, processing of the collected data and finally interaction with the other system. The sensor read data and its variation over some certain clock time duration has been shown in the graph. The temperature classification has been done as the high and normal depending on the threshold temperature value of 25 degree Celsius. The data collected has been continually being sent to the IOT device and the variation of the temperature and humidity value has been measured gradually. The series of graphs have been drawn with the different values of the temperature. The graphs have been plotted with the increasing values of the temperature scale plotted along the X axis of the graph. The time of recording of different temperature values have been started from 7:20 am and the time span is being increased in the intervals of some step seconds. Some of the instantaneous graphs have been provided in the Figure 1.

Figure 1. Variation of the IOT device collected temperature data with respect to time

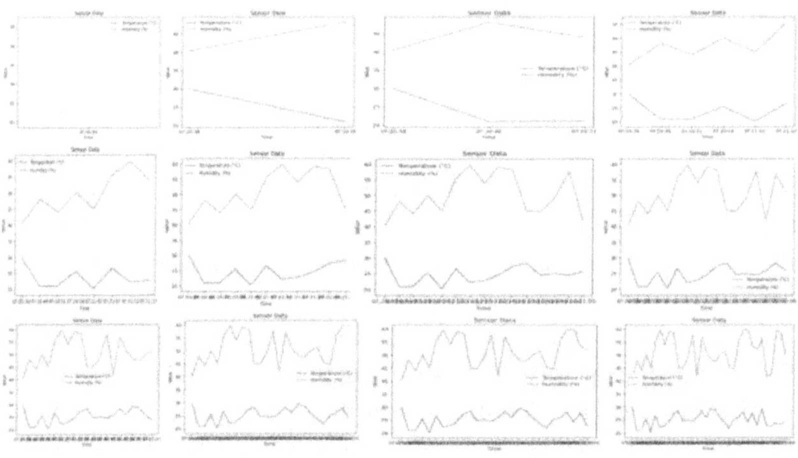

The variation of the temperature as being measured by the IOT device can get affected due tothe presence of the dust particle. The humidity level of the place concerned has been taken as constant over time. The high specific heat content of the dust particle retains the heat and thus affects the temperature variation. The sensor based sensed temperature variation in this scenario has been shown in graph in Figure 2.

Figure 2. The variation of temperature with increasing dust level

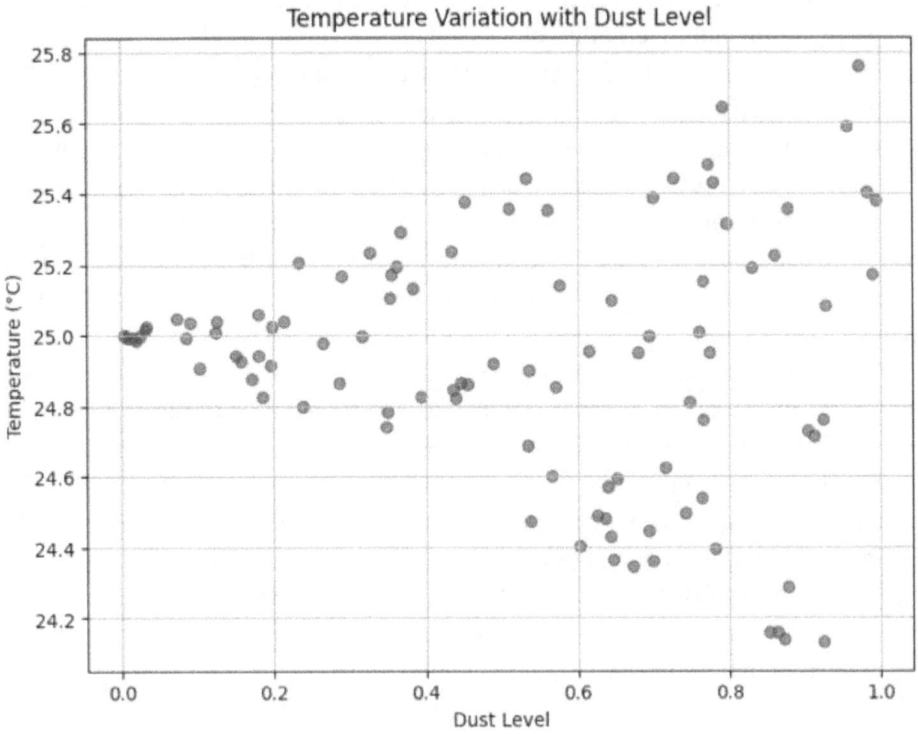

The scatter plot shown in the fig depicts the wide variation of temperature with the presence of increasing level of dust particles. The graphical representation confirms the gradual variation of the temperature range depending upon the increasing levels of the dust particles present in any place having considerably less amount of pollution level.

FUTURE RESEARCH DIRECTION

There are many future scopes for the Real-Time Monitoring and Control IoT-enabledelectronicsbased research domain. The edge computing is one such significant area where the decision making principles have been used to obtain the reduced latency. Moreover the fair chance of predictive analysis along with the anomaly detection can be fruitfully carried out as the outcome of implementation of more scaled up intelligent and control systems. Maintenance of security and privacy policy of the data can be another future research scopes. The robust communication systems with the high end intrusion detection approach can easily lay the path for the IOT enabled monitoring and the control systems. Another challenge lies in seamless integration of different IOT devices obtained from different manufacturers. The addressing of the interoperability of such challenge can be another future research scope to be addressed. Efficient energy utilisation and its optimisationcan help reducing the overall operational cost. The scopes of adaptive power management along with the integration of renewable source of energy sources to the power IT devices. The attached devices can easily be made scalable with the application of the fault tolerant protocols and decentralized control mechanisms to ensure the robust IOT monitoring approach .Contextaware monitoring is also another aspect of operation of the IOT devices and adaption of typical behaviour of data.The context aware technique can lead to the leveraging the sensor data related to the environment and its overall optimisation. Another area of research to be connected to such technology trend is the investigation of voice, commands and recognition of the gestures with the application of m augmented reality and the intuitive control mechanism based on the IOT devices. Integration of the Cyberphisical system towards the real benefits of smart cities, healthcare systems, and exploring the hidden synergies between the IOT backed electronics and physical systems in order to weave an integrated and intelligent ecosystems. The ethical and legal maintenance of the pervasive real time monitoring can be enhanced with the IOT enabled electronics for mitigation of the potential risks in the society.

CONCLUSION

The innovations in the field of AI, ML and IOT are of great significance as it has come to solve many different issues to establish the strong foundation of automation. The AI enabled approach along with the machine learning concept has helped in refurbishing the automation to the great alter of standard.The remote monitoringmanagement specially in the fieldof health sector has laid the path for the advancement in detection and diagnostics of different diseases. The contribution in the business field is also not very nominal and is gradually improving towards the achievement of automation in the real time data. The different electronic devices like switch etc and the integration of the IOT concept to such devices can bring about the large paradigm shift. The predictive model development for the perfect precision has become a challenging task as the accuracy value of the model performance depends upon the training data size and its wideness for inclusions of the varieties of data of heterogeneous nature. The extensive discussion of the IOT application in different segments of domain has been elaborated. The future research of such work will encompass the in-depth AI applications in the field of IOT for providing the real automation in the device level for overcoming the challenges concerned to the solution of real time problems in possible areas and the interoperability of the data integration among the mutually interacting devices.

REFERENCES

Alqurashi, D. R., Alkhaffaf, M., Daoud, M. K., Al-Gasawneh, J. A., & Alghizzawi, M. (2023). Exploring the Impact of Artificial Intelligence in Personalized Content Marketing: A Contemporary Digital Marketing. *Migration Letters : An International Journal of Migration Studies*, *20*(S8), 548–560.

Chataut, R., &Phoummalayvane, A. (2023). Unleashing the Power of IoT: A Comprehensive Review of IoT Applications, Advancements, and Future Prospects in Healthcare, Agriculture, Smart Homes, Smart Cities, and Industry 4.0.

Condry, M. W., & Nelson, C. B. (2016). Using smart edge IoT devices for safer, rapid response with industry IoT control operations. *Proceedings of the IEEE*, *104*(5), 938–946. DOI: 10.1109/JPROC.2015.2513672

Cunin, T., McGrath, S., & MacNamee, C. (2018, December). Environmental monitoring based on Internet of Things Technology. In *2018 12th International Conference on Sensing Technology (ICST)* (pp. 31-34). IEEE. DOI: 10.1109/ICSensT.2018.8603658

Dave, D. M. K., & Mittapally, B. K. DATA INTEGRATION AND INTEROPERABILITY IN IOT: CHALLENGES, STRATEGIES AND FUTURE DIRECTION

Daniele, L., Solanki, M., den Hartog, F., & Roes, J. (2016). Interoperability for smart appliances in the IoT world. In The Semantic Web–ISWC 2016: 15th International Semantic Web Conference, Kobe, Japan, October 17–21, 2016 [Springer International Publishing.]. *Proceedings*, *15*(Part II), 21–29.

Deshpande, S. N., & Jogdand, R. M. (2020). A survey on internet of things (iot), industrial iot (iiot) and industry 4.0. *International Journal of Computer Applications*, *175*(27), 20–27. DOI: 10.5120/ijca2020920790

Dhanaraju, M., Chenniappan, P., Ramalingam, K., Pazhanivelan, S., & Kaliaperumal, R. (2022). Smart farming: Internet of Things (IoT)-based sustainable agriculture. *Agriculture*, *12*(10), 1745. DOI: 10.3390/agriculture12101745

Dui, H., Zhang, S., Liu, M., Dong, X., & Bai, G. (2024). IoT-Enabled Real-Time Traffic Monitoring and Control Management for Intelligent Transportation Systems. *IEEE Internet of Things Journal*, *11*(9), 15842–15854. DOI: 10.1109/JIOT.2024.3351908

Hossain, M. S., Rahman, M., Sarker, M. T., Haque, M. E., & Jahid, A. (2019). A smart IoT based system for monitoring and controlling the sub-station equipment. *Internet of Things : Engineering Cyber Physical Human Systems*, *7*, 100085. DOI: 10.1016/j.iot.2019.100085

Hussain, S. Z. R., Osman, A., Moin, M. A., & Memon, J. A. (2021, June). IoT enabled real-time energy monitoring and control system. In *2021 9th International Conference on Smart Grid (icSmartGrid)* (pp. 97-102). IEEE. DOI: 10.1109/icSmartGrid52357.2021.9551208

Iranpak, S., Shahbahrami, A., & Shakeri, H. (2021). Remote patient monitoring and classifying using the internet of things platform combined with cloud computing. *Journal of Big Data*, *8*(1), 120. DOI: 10.1186/s40537-021-00507-w

Islam, M. R., Kabir, M. M., Mridha, M. F., Alfarhood, S., Safran, M., & Che, D. (2023). Deep learning-based IoT system for remote monitoring and early detection of health issues in real-time. *Sensors (Basel)*, *23*(11), 5204. DOI: 10.3390/s23115204 PMID: 37299933

Kaur, J., Santhoshkumar, N., Nomani, M. Z. M., Sharma, D. K., Maroor, J. P., & Dhiman, V. (2022). Impact of Internets of Things (IOT) in retail sector. *Materials Today: Proceedings*, *51*, 26–30. DOI: 10.1016/j.matpr.2021.04.246

Kelly, J. T., Campbell, K. L., Gong, E., & Scuffham, P. (2020). The Internet of Things: Impact and implications for health care delivery. *Journal of Medical Internet Research*, *22*(11), e20135. DOI: 10.2196/20135 PMID: 33170132

Kumar, V., & Garg, M. L. (2018). Predictive analytics: A review of trends and techniques. *International Journal of Computer Applications*, *182*(1), 31–37. DOI: 10.5120/ijca2018917434

Li, X., Liu, H., Wang, W., Zheng, Y., Lv, H., & Lv, Z. (2022). Big data analysis of the internet of things in the digital twins of smart city based on deep learning. *Future Generation Computer Systems*, *128*, 167–177. DOI: 10.1016/j.future.2021.10.006

Maghsudi, S., Lan, A., Xu, J., & van Der Schaar, M. (2021). Personalized education in the artificial intelligence era: What to expect next. *IEEE Signal Processing Magazine*, *38*(3), 37–50. DOI: 10.1109/MSP.2021.3055032

Maurya, P., Hazra, A., & Awasthi, L. K. (2023). Exploring IoT Communication Technologies and Data-Driven Solutions. In *Learning Techniques for the Internet of Things* (pp. 79–103). Springer Nature Switzerland.

Mishra, R., Desai, V. V., Krishnamoorthy, R., Begum, M. A., Ranga, J., & Taqui, S. N. (2023, November). Energy-Efficient IoT with Deep Learning: Optimizing Resource Allocation in Smart Grids. In *2023 9th International Conference on Smart Structures and Systems (ICSSS)* (pp. 1-6). IEEE

Murtaza, M., Ahmed, Y., Shamsi, J. A., Sherwani, F., & Usman, M. (2022). AI-based personalized e-learning systems: Issues, challenges, and solutions. *IEEE Access : Practical Innovations, Open Solutions, 10*, 81323–81342. DOI: 10.1109/ACCESS.2022.3193938

Nasution, T. H., Muchtar, M. A., & Simon, A. (2019, October). Designing an IoT-based air quality monitoring system. [). IOP Publishing.]. *IOP Conference Series. Materials Science and Engineering, 648*(1), 012037. DOI: 10.1088/1757-899X/648/1/012037

Nižetić, S., Šolić, P., Gonzalez-De, D. L. D. I., & Patrono, L. (2020). Internet of Things (IoT): Opportunities, issues and challenges towards a smart and sustainable future. *Journal of Cleaner Production, 274*, 122877. DOI: 10.1016/j.jclepro.2020.122877 PMID: 32834567

Pazos, N., Müller, M., Aeberli, M., & Ouerhani, N. (2015, December). ConnectOpen-automatic integration of IoT devices. In *2015 IEEE 2nd World Forum on Internet of Things (WF-IoT)* (pp. 640-644). IEEE. DOI: 10.1109/WF-IoT.2015.7389129

Petrasch, R. J., & Petrasch, R. R. (2022). Data integration and interoperability: Towards a model-driven and pattern-oriented approach. *Modelling, 3*(1), 105–126. DOI: 10.3390/modelling3010008

Prasad, V. K., Dansana, D., Bhavsar, M. D., Acharya, B., Gerogiannis, V. C., & Kanavos, A. (2023). Efficient Resource Utilization in IoT and Cloud Computing. *Information (Basel), 14*(11), 619. DOI: 10.3390/info14110619

Qian, L. P., Wu, Y., Ji, B., Huang, L., & Tsang, D. H. (2019). HybridIoT: Integration of hierarchical multiple access and computation offloading for IoT-based smart cities. *IEEE Network, 33*(2), 6–13. DOI: 10.1109/MNET.2019.1800149

Rahmani, H., Shetty, D., Wagih, M., Ghasempour, Y., Palazzi, V., Carvalho, N. B., Correia, R., Costanzo, A., Vital, D., Alimenti, F., Kettle, J., Masotti, D., Mezzanotte, P., Roselli, L., & Grosinger, J. (2023). Next-generation IoT devices: Sustainable eco-friendly manufacturing, energy harvesting, and wireless connectivity. *IEEE Journal of Microwaves, 3*(1), 237–255. DOI: 10.1109/JMW.2022.3228683

Rubí, J. N. S., & de Lira Gondim, P. R. (2021). IoT-based platform for environment data sharing in smart cities. *International Journal of Communication Systems, 34*(2), e4515. DOI: 10.1002/dac.4515

Ruiz-Guirola, D. E., Raghuwanshi, P., Mikhaylov, K., Lovén, L., & Iyer, S. (2023). Energy-sustainable iot connectivity: Vision, technological enablers, challenges, and future directions. *IEEE Open Journal of the Communications Society.*

Shafique, K., Khawaja, B. A., Sabir, F., Qazi, S., & Mustaqim, M. (2020). Internet of things (IoT) for next-generation smart systems: A review of current challenges, future trends and prospects for emerging 5G-IoT scenarios. *IEEE Access : Practical Innovations, Open Solutions*, *8*, 23022–23040. DOI: 10.1109/ACCESS.2020.2970118

Sheng, Z., Wang, H., Yin, C., Hu, X., Yang, S., & Leung, V. C. (2015). Lightweight management of resource-constrained sensor devices in internet of things. *IEEE Internet of Things Journal*, *2*(5), 402–411. DOI: 10.1109/JIOT.2015.2419740

Shorfuzzaman, M., & Hossain, M. S. (2021). Predictive Analytics of Energy Usage by IoT-Based Smart Home Appliances for Green Urban Development. *ACM Transactions on Internet Technology*, *22*(2), 1–26. DOI: 10.1145/3426970

Singh, U., & Ansari, M. A. (2019, October). Smart home automation system using Internet of Things. In *2019 2nd International Conference on Power Energy, Environment and Intelligent Control (PEEIC)* (pp. 144-149). IEEE. DOI: 10.1109/PEEIC47157.2019.8976842

Subramaniam, E. V. D., Srinivasan, K., Qaisar, S. M., & Pławiak, P. (2023). Interoperable IoMT approach for remote diagnosis with privacy-preservation perspective in edge systems. *Sensors (Basel)*, *23*(17), 7474. DOI: 10.3390/s23177474 PMID: 37687933

Taneja, T., Jatain, A., & Bajaj, S. B. (2017, May). Predictive analytics on IoT. In *2017 International Conference on Computing, Communication and Automation (ICCCA)* (pp. 1312-1317). IEEE DOI: 10.1109/CCAA.2017.8230000

Totade, R. S., & Kulkarni, H. H. (2022). Remote Monitoring and Controlling of Household Devices Using IoT. In *Intelligent Communication Technologies and Virtual Mobile Networks* [Singapore: Springer Nature Singapore.]. *Proceedings of ICICV*, *2022*, 769–777.

Tripathi, S., & De, S. (2019). Data-driven optimizations in IoT: A new frontier of challenges and opportunities. *CSI Transactions on ICT*, *7*(1), 35–43. DOI: 10.1007/s40012-019-00217-y

Ushakov, D., Dudukalov, E., Kozlova, E., & Shatila, K. (2022). The Internet of Things impact on smart public transportation. *Transportation Research Procedia*, *63*, 2392–2400. DOI: 10.1016/j.trpro.2022.06.275

Wu, Z., Li, Y., Tsung, F., & Pan, E. (2023). Real-time monitoring and diagnosis scheme for IoT-enabled devices using multivariate SPC techniques. *IISE Transactions*, *55*(4), 348–362. DOI: 10.1080/24725854.2021.2000681

Chapter 9
Advancements in the Internet of Medical Things:
Technology, Integration, and Challenges

Archana Kedar Chaudhari
https://orcid.org/0000-0002-3304-1461
Vishwakarma Institute of Technology, Pune, India

ABSTRACT

Internet of Medical Things is application of Internet of Things to medical and health care fields. Internet of Medical Things has created a revolution in the health care industry by integrating technology in the medical devices, enabling remote monitoring, data collection and real time data analytics. IOMT make use of variety of sensors for collection of patient data along with different communication devices. Upcoming technologies such as artificial intelligence and edge computing are being exploited to enhance IOMT capabilities. The research work focusses on sensors, protocols, and communication technologies in IOMT. Real-time case studies highlighting the practical applications of IoMT and showcasing its ability to monitor vital signs, manage chronic conditions, and improve patient outcomes are also explored. While IoMT has an enormous potential to transform healthcare, addressing integration challenges is crucial for understanding its benefits. The work discusses the challenges related to cybersecurity and interoperability, to ensure integration of IOMT into clinical practice.

DOI: 10.4018/979-8-3693-5448-3.ch009

1. INTRODUCTION

A. Background

The Coronavirus pandemic has shown the world the vulnerability of modern healthcare and medical systems. Cardiovascular disease and cancer disease still pose high risk to human life and medical systems are under enormous pressure to protect the health of individuals in society. There is therefore a necessity to build new technologies to detect, prevent and diagnose the health of individuals to expand-pliability of medical systems to such health calamities as cited in study by Huang et al. (2023), Garrido et al. (2008) and Sorenson et al. (2013).

B. Need of Internet of Medical Things (IOMT)

In the past few years, the Internet of Medical Things (IOMT) has developed as an innovative technology that enables monitoring the health of individuals, diagnosis of diseases from symptoms and possible ways of treatment. Thus, the Internet of Medical Things has emerged as a revolution in the way healthcare is delivered. In Internet of Medical Things (IOMT), Internet of Things is applied to the field of medicine. Medical systems in IOMT can be connected to various wearable sensors, and smart devices. These smart devices consist of sensors for observing the health parameters of the body like heart pulse, temperature of the body and many more. Health sensors gather information and transfer the data to the server. The nodes thus help in establishing an information platform for health diagnosis.

In IOMT, health care providers in different roles such as doctors, nurses are connected with various devices and objects. Therefore, connection between people to people, for example patient and doctor, people to things, for example monitoring health statistics of the patient using different sensors and things to things example patient information platform of different patients sent to server can be established. Thus, sensors or devices in the medical field act as *Things* and are connected to provide an information platform. The things in the medical scenario can include traditional medical instruments and tools, medical instruments such as CT and MRI machines, and few hospital facilities like medicine bottles and medical records as stated by Huang et al. (2023). In current years development in the medical field has led to the invention of wearable's and near-field communication devices. These are wearable devices on the body of the patient or an individual for monitoring of health parameters. The wearable sensors also act as *Things* for information gathering.

In the IOMT, not only the communication devices play an important role but the information technology platforms are also very crucial for health monitoring, diagnosis and treatment. All the data gathered using the connected devices is aggregated. This data is further transmitted from the information platform to the server. The server is an essential component of any IOMT system.

In a typical IOMT system, the sensors, wearable devices or near-field devices communicate the gathered data to the server via a gateway. The diagnosis of the information is carried out at the server and further sent to the physicians for making medical decisions. Fig. 1 presents a typical scenario of the IOMT system.

Aman et al. (2021) discussed literature review of IOMT. The authors discussed the theory of IOMT along with various architectures, technologies, domains of deployment along with applications. Aman et al. discussedstate of art review of the IOMT ecosystem along with its applications and usefulness in context to COVID-19. Singh et al. (2020) explored the orthopedic applications of IOMT in the COVID-19 pandemic. In their work, Islam et al. (2015) highlight the IOMT architecture, technology along with applications and security challenges. Awad et al. (2023) presented a review of the use of mobile edge computing technology for internet of medical things applications. Verma et al. (2022) explored the various IOT architectures and communication protocols for healthcare monitoring systems. In Verma et al. (2022) the authors illustrates a review of IOMT in healthcare management.

Figure 1. Typical Internet of Medical Things (IOMT) system

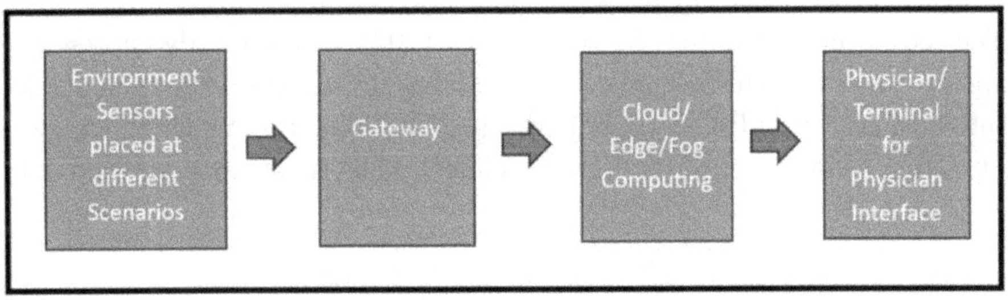

The work presents a review of the architecture, sensors, protocols, and communication technologies. It also discusses the advancements in IOMT along with real time case studies implemented using IOMT.

Section 1 presents the introduction to IOMT and its typical scenario. Section 2 reviews the different architectures used for IOMT. Section 3 discusses the basic and advanced technologies used for IOMT. Section 4 presents the deployment environments of IOMT with real time scenarios. Section 5 presents the challenges faced during integration and security issues in IOMT. Section 6 presents the conclusions and future scope.

2. VARIOUS IOMT ARCHITECTURES

Internet of Medical Things refers to applications of Internet of Things in the medical domain. Therefore, IOMT architecture is very much alike the Internet of Things architecture. The IOMT architecture is derived using the IOT architecture. IOMT architecture consists of the following different layers; Perception layer known as physical layer or environment aware layer, network layer or gateway layer also includes middleware layer in some architecture, service layer consisting the cloud services and lastly the function or the application layer.

Many researchers have proposed IoMT reference models. Zhang et al. (2018) propose four-layer architecture with cloud, edge base station and sensing layer. The authors, Ma et al. (2017) explored four layered architectures involving perception, transport, cloud and cloud to end fusion. Din et al. suggested sensing, network, cloud and application layers for IOMT architecture. Din et al. (2018) proposed a four-layer architect with sensors, network, analysis and application layers with a focus on remote monitoring.

Liu et al. (2018) proposed an architecture consisting of five layers with sensing, communication, network transmission, data integration, and lastly the application service layer. While Noor et al. (2019) discussed three-layer architecture by combining medical specific elements to the application layer.

The four-layer architecture by Aman et al. (2021) is the mostly chosen, as it comprises of classification of application, virtual platform, network communication, and physical aspects of IoT. Fig. 2 presents the IOMT Architecture with four layers. The function and features of the IOMT architecture layers are elaborated in detail in the further subsections.

Figure 2. Layers of the Internet of Medical Things Architecture

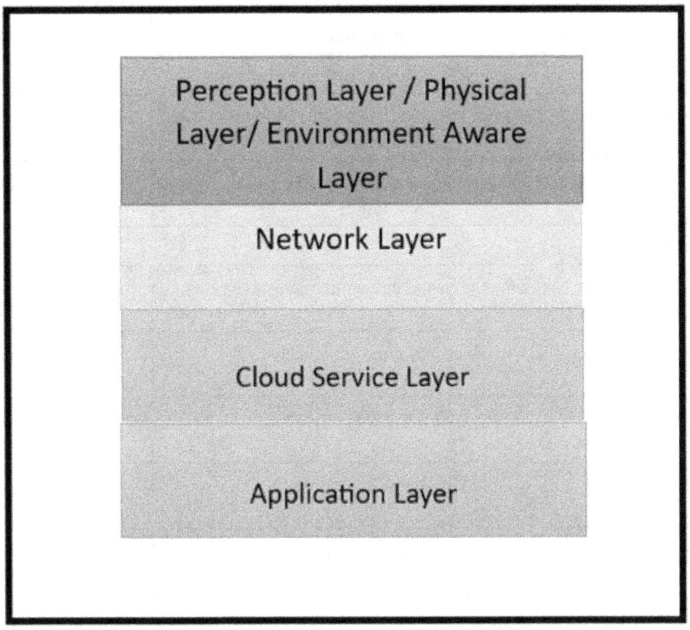

2.1 Physical or Environment Aware Layer

This physical layer consists of the medical examination instruments, bio-medical sensors and the wearable body detectors. They act as nodes for data gathering and transmission of information to the further layers. These nodes are placed in different environments for gathering the health parameters of the patients or individual. Table 1 exhibits the four different types of sensors along with their applications, and the environments where they are typically placed within the context of Internet of Medical Things as stated in Huang et al. (2023), Aman et al. (2021), Singh et al. (2020), Awad et al. (2023), Ejiyi et al. (2023), Rizvanov (2018) and Nanayakkara et al. (2019).

Table 1. Different Sensors along with applications and the placement environment in IOMT

Type of Sensor	Application	Environment for Placement
Body Sensors		
Wearable sensors	Gather physiological indicators like blood pressure, heart rate	On the body (e.g., wristbands)
Non-wearable sensors	Measure physiological indicators like blood oxygen saturation	Fixed positions near the body
Object Sensors		
Gravity sensors	Collect physical indicators like gravity	Various objects or environments
Pressure sensors	Measure pressure levels	Various objects or environments
Velocity sensors	Measure velocity	Various objects or environments
Acceleration sensors	Measure acceleration	Various objects or environments
Temperature sensors	Measure temperature	Various objects or environments
Humidity sensors	Measure humidity	Various objects or environments
Light intensity sensors	Measure light intensity	Various objects or environments
Noise level sensors	Measure noise levels	Various objects or environments
Medical Examination Instruments		
Blood cell counting systems	Test and analyze body fluids	Medical laboratories, hospitals
CT scanning systems	Imaging examinations	Hospitals, diagnostic centers
Body Detectors		
Visible light cameras	Monitor and analyze expressions, emotions, postures, movements	Indoor/outdoor, hospitals
Infrared detectors	Detect body temperature, sense human activity	Hospitals, transport hubs
Indoor radar	Detect human activity, indoor positioning, indoor modeling, fall detection	Hospitals, nursing homes, rehab centers
Hospital Assets		
Electronic health records	Manage patient health status, predict health risks	Hospitals, health centers
RFID/NFC-enabled items	Collect information and network (infusion bottles, medicine trolleys, etc.)	Hospitals
Upgraded ambulances	Deliver improved patient care and emergency treatment	Ambulances

2.2 Network layer or gateway layer

In some MIOT or IOT architectures, the network and the middleware layer is combined to form the network or gateway. The data which is gathered from environment-aware layer is sent to the gateway layer. The data from all the nodes is aggregated at this layer. Using the middleware the data is pre-processed and converted to standard required format. The middle ware also provides a platform to integrate data if connected devices are not of the same type as stated by Georgi et al. (2019). Hence middleware acts as a service layer and then data is transmitted using the gateway. The data can be transmitted using wired or wireless mode over the network using WI-FI, Bluetooth, Zigbee, 4G, 5G technology or satellite.

Zigbee, Wi-Fi and Bluetooth are short range communication modes. The long-range communication includes 4G, 5G technology or satellite. These are useful to senddataamong the gateway and the cloud server. A brief description of the communication technologies tabulated in Table 2 as stated in Huang et al. (2023), Aman et al. (2021), Awad et al. (2023), Ahad et al. (2019), Sengupta et al. (2019), Burrman et al. (2020) and Gu et al. (2020).

Table 2. Description of Communication Technologies in IOMT

Technology	Range	Applications	Use of Bluetooth
Bluetooth	Short-range (up to 100 meters)	Wireless data and voice communication, connecting fixed and mobile devices	Establishes near-field wireless connections, frequency-hopping for security and anti-interference.
Zigbee	Short-range (10-100 meters)	Low-speed, low-power wireless network protocol for IoT applications	Not directly used but can complement Bluetooth in IoT ecosystems.
RFID	Short to medium range (up to 1 meter or more)	Automatic identification and data exchange using electronic tags	Not directly related, but both RFID and Bluetooth can be used for tracking and data transfer.
NFC	Very short-range (up to 10 cm)	Contactless payments, data exchange, and access control	Similar to Bluetooth but with much shorter range and specific to close proximity interactions.
WiFi	Medium-range (up to 100 meters indoors, more outdoors)	Wireless LAN, internet access for various devices	Can work in conjunction with Bluetooth for enhanced connectivity.
4G	Wide-range (city-wide coverage)	High-speed mobile internet, remote multimedia applications	Typically, not directly related, but devices may use Bluetooth for local connections.
5G	Wide-range (city-wide coverage)	Ultra-high-speed mobile internet, IoT, low latency applications	Devices with 5G may also use Bluetooth for local device interconnectivity.
Satellite	Global coverage	Communication in remote areas, global broadcasting	Not directly related, but Bluetooth can be used for local communication in satellite-enabled devices.

2.3 Cloud Service Layer

The cloud service layer offers an interface to store the data gathered from all the sensor nodes and aggregated at the network layer. This data is also converted to standard format and can be accessed from the cloud service layer. These layers therefore consist of servers, web portal, programming and database interface.

Cloud computing, fog computing, edge computing and grid computing are the recent trends in cloud service layers in IOT as well as IOMT. Two approaches can be implemented for storage on the cloud: centralized and distributed edge cloud storage. Centralized storage is manageable and offers flexibility and scalability but it has certain concerns with privacy and security. In comparison the edge cloud storage can be said to be more reliable and secure and reduces the data load.

A few examples of cloud platforms are Amazon Web Services, Alibaba Cloud, IBM Cloud, and Microsoft Azure. They provide services like storage, data processing, messaging and analytics for IoMT applications as stated by Pace et al. (2019). Table 3 presents the applications, strengths and weaknesses of edge and fog computing as stated by Awad et al. (2023).

Table 3. Comparison of Cloud, Grid, Fog and Edge Computing for IOMT Applications

Aspect	Grid Computing	Cloud Computing	Fog Computing	Edge Computing
Latency	High	High	Low	Ultra-low
Scalability	Moderate	High	Moderate	Moderate
Bandwidth	High	High	Low	Low
Security	Moderate	Moderate	High	High
Complexity	High	Moderate	High	High
Resource Use	Distributed	Centralized	Distributed (local + cloud)	Localized (edge devices)
Applications	Scientific computing	Web services, data storage	IoT, smart cities	Real-time IoT, IoMT

In Health care or medical systems, computing at the cloud can be a challenge for critical care situations. A practical approach is processing of critical situation data at the network edge rather than the cloud. In Awad et al. (2023) mobile Edge computing (MEC) is the recent advancement that brings cloud servers to the edge of mobile networks to allow real-time subscriber and IoT data processing.

In cloud computing clients initiate requests and remote servers process these requests. This follows standard request response protocol. On the other hand, mobile edge computing follows a three-tier model with the edge server added along with the client and the server. Microservices, serverless computing and DevOps are examples that facilitate this model.

2.4 Function or Application Layer

It is the top layer in the IOMT architecture and uses the information in the cloud service layer to build various applications such as remote monitoring of health care parameters, diagnosis of disease, smart hospitals and many more. This layer uses human computer interaction and artificial intelligence for learning the data, analyzing the data and making predictions.

Recent advancements in the function layer includes virtual counseling of the patients by physicians and diagnostics using telemedicine as stated by Iwendi et al. (2020), Khan et al. (2020), Fouad et al. (2020), Alsiddiky et al. (2020) and Dong et al. (2020).

3. PROTOCOLS USED FOR IOMT

The sensors and hardware connected to the IOMT can communicate with servers using the protocols. The protocols are a way to communicate the data gathered by millions of sensors to the server and thus form the backbone of the IOMT. Table 4 presents an overview of the different types of protocols used for IOT and specifies the applications for each of the protocols as stated by Huang et al. (2023), Colakovic et al. (2018) and Bayilmis et al. (2022). In recent years TCP/IP protocol suite is used for network communication and many IoT devices can be connected to the devices through the internet.

Table 4. overview of the different types of protocols used for IOT

Protocol	Characteristics	Strengths	Weaknesses	Applications
TCP/IP	Standard protocol suite for network communication.	Widely used, robust, enables billions of IoT devices to connect through the internet.	Heavy and power-consuming for IoT use cases.	General internet communication for IoT devices
CoAP (Constrained Application Protocol)	Translates HTTP model for restrictive devices and network environments, uses UDP, leverages DTLS for security.	Efficient for constrained devices, lightweight, secure communication.	Limited to devices and environments suited for UDP and DTLS.	Smart cities, home automation, industrial IoT
MQTT (Message Queuing Telemetry Transport)	Lightweight publish/ subscribe messaging protocol, designed for battery-powered devices, works over TCP.	Simple, lightweight, low power consumption, ideal for battery-powered devices.	Increases overhead with text for topic names, not suitable for all IoT applications.	Environmental monitoring, healthcare, wearables
6LoWPAN (IPv6 Low Power Wireless Personal Area Network)	Functions on IEEE 802.15.4, network stack with IP connectivity (IPv6).	Low power consumption, longer battery life, lower bandwidth consumption, connects many devices.	Requires effective physical layer standards, suited for specific environments.	Home automation, industrial monitoring
BLE (Bluetooth Low Energy)	Standard for low power wireless communication (Bluetooth 4.0).	Low energy consumption, suitable for short-range communication, widely supported.	Limited range and data rate.	Fitness trackers, proximity sensing
ZigBee	Based on IEEE 802.15.4, designed for low-power, low-bandwidth wireless communication.	Low power consumption, supports mesh networking, suitable for sensor networks.	Limited data rate and range.	Home automation, smart lighting, sensor networks
Z-Wave	Proprietary protocol designed for home automation, low-power, low-bandwidth.	Reliable, low power consumption, good for home automation.	Proprietary, limited to Z-Wave alliance devices.	Home automation, security systems
NFC (Near Field Communication)	Short-range wireless communication technology.	Very low power, suitable for secure transactions, easy to use.	Very limited range (a few centimeters).	Contactless payments, access control

All the protocols are based on the IEEE 802.15.4, which specifies the standard for low-rate wireless personal area networks. The advantage offered by this standard is low power consumption, suitable for sensor networks but its limited bandwidth, may not be suitable for high data rate applications.

4. APPLICATIONS, CASE STUDIES AND USE CASES OF IOMT

The section discusses some case studies, use cases and applications of IOMT in real time scenarios. The Table 5 exhibits an overview of the IOMT applications based on the following: Objectives, Functionality, scenario where it is being deployed, target users and domain.

Table 5. Overview of the IOMT applications

Category	Functionality	Deployment Scenarios	Target Users	Objectives	Domain
Remote Monitoring	Remote collection of individuals' physiological data	Individual level	Patients and healthy populations	Health monitoring	On-Body, In-Home, In-Hospital
Intelligent Disease Diagnosis	Supporting healthcare professionals in clinical analysis and providing early disease warnings	Clinical settings (hospitals)	Individuals seeking medical consultation	Clinical diagnosis and early warning	In-Hospital, In-Home
Smart Hospital	Enhancing operational effectiveness and management in healthcare institutions	Healthcare system	Healthcare institutions	Improving operational efficiency and management	In-Hospital, Out-Door
Patient-Centric IoMT	Measuring efficacy of medical interventions, collecting patient feedback, help to reduce patient pain and anxiety	Patient level	Patients	Improve patient experience during the healthcare process	In-Home, On-Body, In-Hospital

The remote monitoring applications include physiological parameter monitoring. These are sensors which can be placed on the body of the patient in hospital or at home. These are wearable devices. Some of the devices such as glucose level sensing or electrocardiogram monitoring can be placed at home or hospitals near the patients for remote monitoring. As per Ting et al. (2020) the blood pressure and heart-rate can send to the cloud for analysis, thus avoiding exposure of health care workers to infections in case of pandemics and infectious disease transmission. In some cases, as stated by Song et al. (2020) IOMT has previously been successful in recognizing and tracking the source of the outbreak of infectious disease and confirmingconfinement of diseased persons. IOMT is playing a key role in remote monitoring of health in the elderly. Duran-Vega et al. (2019) have proposed systems that remotely monitor blood pressure, heart rate, temperature and oxygen levels in the elderly. Few other use cases developed by Garbhapu et al. (2017), Sacco et al. (2018), Verma et al. (2018), Hamim et al. (2019), Haur et al. (2019), Colakovic (2020) Khan and Algarni (2020), Wang et al. (2020), Zhang et al. (2020), Fallah et al. (2017), Wei et al. (2020), Aufar et al. (2021), Tan et al. (2020) and Hariri (2021) are proposed for remote monitoring of health parameters using IOMT. This demonstrates a shift of IOMT towards patient centric care enhancing the quality of life.

IOMT finds wide applications in the field of SMART Hospitals. The applications of IOMT vary from patient monitoring and patient health record management to drug supply chain management. A well-known use case in case of SMART hospitals is the Electrocardiogram (ECG) monitoring. Including IOMT in ECG monitoring helps to detect abnormality in the cardiac function at an early stage for better treatment as stated by Yang et al. (2014), Castillejo et al. (2013), Sonawane et al. (2018), Fajingbesi et al. (2017) and Sundarasekar et al. (2018).

In the area of disease diagnosis, using machine learning and Deep Learning methods, diseases like stroke, leukemia, brain tumor, heart failure, pneumonia are being diagnosed and classified. In Rajan and Nadar (2019), Deep Neural Networks are used to excerpt the features from EEG, ECG, Temperature, and Pulse Rate to find the abnormality. Deep learning and machine learning approaches are extensively used to predict abnormal or disease at an early-stages. Datasets of diabetes, temperature, Tumor, Cancer, kidney are used for the same. Features are extracted for early disease prediction and treatment by Rajan and Nadar (2019), Kumar et al. (2018), Tuli et al. (2020), Tuli et al. (2020), Felisberto et al. (2015), Tan and Halim (2019), Yuzer et al. (2020), Vineetha et al. (2021), Bibi et al. (2020), Han et a.l (2020), Souza et al. (2020), Ahmed et al. (2021), Tai et al. (2021), Khan et al. (2020), Chindambaranathan et al. (2021), Xu et al. (2021), Pan et al. (2020), Ning et al. (2021), Raj et al. (2020), Khamparia et al. (2021), Wang et al. (2020), Xuan et al. (2020) and Sadad et al. (2021).

The field of IOMT also covers areas from Smart ambulances to sanitization robots and medicine delivery drones. IOMT can also be useful to provide different health care services. Table 6 summarizes the different services along with use cases.

Table 6. IOMT Services Applications and Use Cases

Service	Description
Ambient Assisted Living	IoMT platform empowered by AI to aid elderly and incapacitated individuals, extending their autonomy and safety at home. Solutions provide confidence, autonomy, and human-servant-like assistance in case of challenges as stated in Sadad et al. (2021), Murabet et al. (2020),Maskelinuas et al. (2019), Blasco et al. (2014), Ganesan et al (2019), Vimarlund and Wass (2014), Yilmaz (2019), Thankur and Han (2021), Alosaimi et al. (2021), Shahamabadi et al. (2013), Dohr et al. (2010)Sankar et al. (2018), Wan et al. (2017), Vijayalakshmi and Jose (2020), Al-Sit et al. (2023).
Adverse Drug Reaction	IoT-based service to manage adverse reactions to drugs through intelligent pharmaceutical systems, identifying drug compatibility with allergy profiles and electronic health records as stated by Istepanian et al. (2008), Jara et al. (2010) and Catherwood et al. (2018).
Community Healthcare	IoMT-based network covering healthcare services for local communities, integrating multiple WBANs to create a "virtual hospital." Includes a resident health information service platform and data sharing techniques among medical facilities as stated by Zahoor and Mir (2021), Gope et al. (2021), Yang et al. (2021), Rohokale et al. (2011), You et al. (2011), Wang et al. (2012) and Yang et al. (2014).
Children Health Information	IoMT service designed to provide health information and support for children, often through interactive totems in pediatric wards. Aims to educate, amuse, and empower hospitalized children, and encourage good eating habits with the aid of teachers and parents as stated by Vicini et al. (2012), Seapuddin et al. (2018), Ukachi et al. (2019), Gale (2017) and Ismail et al. (2017).
Indirect Emergency Healthcare	Dedicated IoMT service for managing emergencies such as transport accidents, adverse weather, fires, and collapses. Provides solutions for information availability, post-accident actions, alert notifications, and record-keeping to assist emergency workers as stated by Kaur (2021), Pawar et al. (2017), Selvakanmani and Sumathi (2021), Mechael (2005),and Sheth et al. (2007).

A global perspective on the adoption and implementation of IOMT, highlighting differences in various regions and healthcare systems can be discussed as following real time cases:

1. In Apollo Hospital's in India, for real time rapid response patient monitoring an automated AI based system has been launched. The nurses, pre-medical team and also clinicians can access the platform using their mobile devices from the command center. The AI enabled early warning alert system aids in timely intervention by medical team providing timely patients care (Apollo Hospitals).
2. Hello Doctor Initiative in South Africa is a mobile IOMT platform that provides healthcare services using the mobile phone (Hello Doctor). This helps patients in remote areas to access health monitoring services and medical consultations.
3. In America Veterans Health Administration provides integrated health care system using IOMT for remote patient monitoring to manage diabetes and hypertension. The IOMT devices collect real-time data thus reducing hospital re-admission and improving patient outcomes (Veterans).

4. NHS England IOMT Test Beds: In the Technology Integrated Health Management (TIHM) case study, sensors, wearable's and monitoring devices are provided for health monitoring (NHS). The data from the sensors is used for well-being of elderly and enable care givers and social care workers to deliver responsive services. The objective of the project is to lessen the need for unplanned hospital re-admissions.

5. CHALLENGES IN IOMT RELATED TO INTEGRATION AND SECURITY

Following are the challenges faced in IOMT along with few probable solutions. Most of the solutions to these challenges will depend on the application scenario of the IOMT.

1. Data Security and Privacy: It is essential to keep the information of individuals or patients secure and private. Every IOMT technology needs to ensure data authenticity, integrity and confidentiality. This necessitates a robust encryption and secure access control to prevent data breaches and unauthorized access. There needs to be a security risk and mitigation plan along with security policies in place for sensitive data integrated to the cloud.
2. Device and Protocol Integration: The healthcare sector uses a diverse array of devices and communication protocols. Integrating these various systems into a cohesive IOMT framework is challenging due to compatibility issues and the need for standardized protocols to ensure seamless communication and interoperability.
3. Network Longevity: IOMT devices need to operate over long periods without frequent maintenance. Ensuring a long network lifespan requires efficient power management and reliable hardware, which can be difficult given the high demand for continuous operation and real-time information processing.
4. Data Volume and Precision: The huge volume of information created by IoT devices can lead to data overload, making it tough to manage and evaluate effectively. Additionally, ensuring the accuracy of this data is critical for making informed healthcare decisions, which requires sophisticated algorithms and data validation techniques.
5. Cost Considerations: Implementing IoT solutions in healthcare can be expensive. The costs associated with purchasing, installing, and maintaining IoT devices, along with the necessary infrastructure and security measures, can be a significant financial burden for healthcare providers.

6. Real-world challenges faced by healthcare providers when implementing IOMT solutions include the data security and patient privacy, inter-operability among diverse medical devices high cost for technology adoption. Also, integration of IMOT with existing healthcare setups requires technical expertise. The management of huge and complex data generated by IOMT devices is also a significant challenge.

6. CONCLUSIONS AND FUTURE SCOPE

In summary, the Internet of Medical Things (IoMT) has great potential to transform healthcare by enabling advanced remote monitoring, data collection, and real-time analytics. However, it is important to address integration challenges to fully realize its benefits. This research emphasizes the significant role of IoMT in enhancing patient care through the use of various sensors, communication protocols, and emerging technologies such as artificial intelligence and edge computing. Overcoming issues related to cybersecurity, interoperability, and collaboration among healthcare providers and technology developers is essential for seamlessly integrating IoMT into clinical practice. By addressing these challenges, the healthcare industry can fully harness the transformative power of IoMT, leading to improved patient outcomes and more efficient healthcare delivery.

Looking ahead, IoMT is set to expand further with advancements in 5G technology, providing faster and more reliable connectivity for IoMT devices. Additionally, the integration of more sophisticated AI algorithms and machine learning models will enable even more precise and personalized healthcare solutions. Continued research and development in these areas will drive the evolution of IoMT, making healthcare more proactive, predictive, and accessible to a broader population.

REFERENCES

Ahad, A., Tahir, M., & Yau, K. A. (2019). 5G-based smart healthcare network: Architecture, taxonomy, challenges and future research directions. *IEEE Access : Practical Innovations, Open Solutions*, *7*, 100747–100762. DOI: 10.1109/ACCESS.2019.2930628

Ahmed, I., Ahmad, A., & Jeon, G. (2021). An IoT-based deep learning framework for early assessment of COVID-19. *IEEE Internet of Things Journal*, *8*(21), 15855–15862. DOI: 10.1109/JIOT.2020.3034074 PMID: 35782174

Al-Sit, W. T., Al-Dmour, N. A., Ghazal, T. M., & Issa, G. F. (2023). IoMT-based healthcare framework for ambient assisted living using a convolutional neural network. *Computers, Materials & Continua*, *74*(1), 57–72. DOI: 10.32604/cmc.2023.034952

Alosaimi, W., Ansari, M. T. J., Alharbi, A., Alyami, H., Seh, A. H., Pandey, A. K., Agrawal, A., & Khan, R. A. (2021). Evaluating the impact of different symmetrical models of ambient assisted living systems. *Symmetry*, *13*(3), 450. DOI: 10.3390/sym13030450

Alsiddiky, A., Awwad, W., Bakarman, K., Fouad, H., & Mahmoud, N. M. (2020). Magnetic resonance imaging evaluation of vertebral tumor prediction using hierarchical hidden Markov random field model on Internet of Medical Things (IoMT) platform. *Measurement*, *159*, 107772. DOI: 10.1016/j.measurement.2020.107772

Aufar, F.. (2021). Design of non-contact thermometer using thermal camera for detecting people with fever. In *2021 International Conference on Computer Science and Engineering (IC2SE)* (pp. 1–5). DOI: 10.1109/IC2SE52832.2021.9792101

Awad, A. I., Fouda, M. M., Khashaba, M. M., Mohamed, E. R., & Hosny, K. M. (2023). Utilization of mobile edge computing on the Internet of Medical Things: A survey. *ICT Express*, *9*(3), 473–485. DOI: 10.1016/j.icte.2022.05.006

Bayılmış, C., Ebleme, M. A., Çavuşoğlu, Ü., Küçük, K., & Sevin, A. (2022). A survey on communication protocols and performance evaluations for Internet of Things. *Digital Communications and Networks*, *8*(6), 1094–1104. DOI: 10.1016/j.dcan.2022.03.013

Bibi, N., Sikandar, M., Ud Din, I., Almogren, A., Ali, S., & Nazir, S. (2020). IoMT-based automated detection and classification of leukemia using deep learning. *Journal of Healthcare Engineering*, *2020*, 1–12. DOI: 10.1155/2020/6648574 PMID: 33343851

Blasco, R., Marco, Á., Casas, R., Cirujano, D., & Picking, R. (2014). A smart kitchen for ambient assisted living. *Sensors (Basel), 14*(1), 1629–1653. DOI: 10.3390/s140101629 PMID: 24445412

Buurman, B., Kamruzzaman, J., Karmakar, G., & Islam, S. (2020). Low-power wide-area networks: Design goals, architecture, suitability to use cases and research challenges. *IEEE Access : Practical Innovations, Open Solutions, 8*, 17179–17220. DOI: 10.1109/ACCESS.2020.2968057

Cantarelli, C. C., Flyvbjerg, B., Molin, E. J. E., & van Wee, B. (2018). Cost overruns in large-scale transport infrastructure projects. *Automation in Construction, 94*, 1–15. DOI: 10.1016/j.autcon.2018.05.028

Castillejo, P., Martinez, J. F., Rodriguez-Molina, J., & Cuerva, A. (2013). Integration of wearable devices in a wireless sensor network for an E-health application. *IEEE Wireless Communications, 20*(4), 38–49. Advance online publication. DOI: 10.1109/MWC.2013.6590049

Catherwood, P. A., Steele, D., Little, M., McComb, S., & McLaughlin, J. (2018). A community-based IoT personalized wireless healthcare solution trial. *IEEE Journal of Translational Engineering in Health and Medicine, 6*, 2800111. DOI: 10.1109/JTEHM.2018.2822302 PMID: 29888145

Chidambaranathan, S.. (2021). Optimal SVM based brain tumor MRI image classification in cloud Internet of Medical Things. In Hassanien, A. E. (Ed.), *Cognitive Internet of Medical Things for Smart Healthcare: Services and Applications* (pp. 87–103). Springer International Publishing. DOI: 10.1007/978-3-030-55833-8_5

Čolaković, A., & Hadžialić, M. (2018). Internet of Things (IoT): A review of enabling technologies, challenges, and open research issues. *Computer Networks, 144*, 17–39. DOI: 10.1016/j.comnet.2018.07.017

Çolaković, D. (2020). Development of edge-IoMT computing architecture for smart healthcare monitoring platform. In *2020 4th International Symposium on Multidisciplinary Studies and Innovative Technologies (ISMSIT)* (pp. 1-4).

Din, I. U., Guizani, M., Hassan, S., Kim, B. S., Khan, M. K., & Atiquzzaman, M. (2018). The Internet of Things: A review of enabled technologies and future challenges. *IEEE Access : Practical Innovations, Open Solutions, 7*, 7606–7640. DOI: 10.1109/ACCESS.2018.2886601

Dohr, A., Modre-Osprian, R., Drobics, M., Hayn, D., & Schreier, G. (2010). The internet of things for ambient assisted living. *Proceedings of the 2010 Seventh International Conference on Information Technology: New Generations*, 804-809. DOI: 10.1109/ITNG.2010.104

Dong, P., Ning, Z., Obaidat, M. S., Jiang, X., Guo, Y., Hu, X., Hu, B., & Sadoun, B. (2020). Edge computing-based healthcare systems: Enabling decentralized health monitoring in Internet of Medical Things. *IEEE Network*, *34*(5), 254–261. DOI: 10.1109/MNET.011.1900636

Durán-Vega, L. A., Santana-Mancilla, P. C., Buenrostro-Mariscal, R., Contreras-Castillo, J., Anido-Rifón, L. E., García-Ruiz, M. A., Montesinos-López, O. A., & Estrada-González, F. (2019). An IoT system for remote health monitoring in elderly adults through a wearable device and mobile application. *Geriatrics (Basel, Switzerland)*, *4*(2), 34. DOI: 10.3390/geriatrics4020034 PMID: 31067819

Ejiyi, C., Qin, Z., Ejiyi, M. B., Nneji, G. U., Monday, H. N., Agu, F. A., Ejiyi, T. U., Diokpo, C., & Orakwue, C. O. (2023). The Internet of Medical Things in healthcare management: A review. *Journal of Digital Health*, *2*(1), 30–62. DOI: 10.55976/jdh.22023116330-62 PMID: 38264696

El Murabet, A., Abtoy, A., Touhafi, A., & Tahiri, A. (2020). Ambient assisted living system's models and architectures: A survey of the state of the art. *Journal of King Saud University. Computer and Information Sciences*, *32*(1), 1–10. Advance online publication. DOI: 10.1016/j.jksuci.2018.04.009

Fajingbesi, F. E., Olanrewaju, R. F., Rasool Pampori, B., Khan, S., & Yacoob, M. (2017). Real time telemedical health care systems with wearable sensors. *Asian Journal of Pharmaceutical Research and Health Care*. https://doi.org/.DOI: 10.18311/ajprhc/2017/149

Fallah-Haghmohammadi, H.. (2017). Fever detection for dynamic human environment using sensor fusion. In *2017 International Conference on Optimization of Electrical and Electronic Equipment (OPTIM) & 2017 International Aegean Conference on Electrical Machines and Power Electronics (ACEMP)* (pp. 881–886). DOI: 10.1109/OPTIM.2017.7975081

Felisberto, F., Laza, R., Fdez-Riverola, F., & Pereira, A. (2015). A distributed multiagent system architecture for body area networks applied to healthcare monitoring. *BioMed Research International*, *2015*, 1–17. Advance online publication. DOI: 10.1155/2015/192454 PMID: 25874202

Fouad, H., Hassanein, A. S., Soliman, A. M., & Al-Feel, H. (2020). Internet of Medical Things (IoMT) assisted vertebral tumor prediction using heuristic Hock transformation based Gautschi model: A numerical approach. *IEEE Access : Practical Innovations, Open Solutions, 8*, 17299–17309. DOI: 10.1109/ACCESS.2020.2966272

Gale, R. (2017). National child health information programme. *International Journal of Integrated Care, 17*(5), 468. DOI: 10.5334/ijic.3161

Ganesan, B., Gowda, T., Al-Jumaily, A., Fong, K. N. K., Meena, S. K., & Tong, R. K. Y. (2019). Ambient assisted living technologies for older adults with cognitive and physical impairments: A review. *European Review for Medical and Pharmacological Sciences*. Advance online publication. DOI: 10.26355/eurrev_201912_19686 PMID: 31841201

Garbhapu, V. V., & Gopalan, S. (2017). IoT based low-cost single sensor node remote health monitoring system. *Procedia Computer Science, 113*, 408–415. DOI: 10.1016/j.procs.2017.08.357

Garrido, M. V.. (2008). *Health technology assessment and health policy-making in Europe: Current status, challenges and potential*. WHO Regional Office Europe.

Georgi, N., Corvol, A., & Jeannas, R. L. B. (2018). Middleware architecture for health sensors interoperability. *IEEE Access : Practical Innovations, Open Solutions, 6*, 26283–26291. DOI: 10.1109/ACCESS.2018.2835644

Gope, P., Gheraibia, Y., Kabir, S., & Sikdar, B. (2021). A secure IoT-based modern healthcare system with fault-tolerant decision making process. *IEEE Journal of Biomedical and Health Informatics, 25*(4), 862–873. DOI: 10.1109/JBHI.2020.3007488 PMID: 32749985

Gu, F., Niu, J., Jiang, L., Liu, X., & Atiquzzaman, M. (2020). Survey of the low power wide area network technologies. *Journal of Network and Computer Applications, 149*, 102459. DOI: 10.1016/j.jnca.2019.102459

Hamim, M.. (2019). IoT based remote health monitoring system for patients and elderly people. In *2019 International Conference on Robotics, Electrical and Signal Processing Techniques (ICREST)* (pp. 533–538). DOI: 10.1109/ICREST.2019.8644514

Han, T., Nunes, V. X., De Freitas Souza, L. F., Marques, A. G., Silva, I. C. L., Junior, M. A. A. F., Sun, J., & Filho, P. P. R. (2020). Internet of medical things—Based on deep learning techniques for segmentation of lung and stroke regions in CT scans. *IEEE Access : Practical Innovations, Open Solutions, 8*, 71117–71135. DOI: 10.1109/ACCESS.2020.2987932

Hariri, W. (2021). Efficient masked face recognition method during the COVID-19 pandemic. *Signal, Image and Video Processing*, *16*(3), 605–612. DOI: 10.1007/s11760-021-02050-w PMID: 34804243

Hello Doctor Hello Doctor | The Health Foundation

Huang, C., Wang, J., Wang, S., & Zhang, Y. (2023). Internet of medical things: A systematic review. *Neurocomputing*, *557*, 126719. DOI: 10.1016/j.neucom.2023.126719

Islam, S. M. R., Kwak, D., Kabir, M. H., Hossain, M., & Kwak, K. (2015). The internet of things for health care: A comprehensive survey. *IEEE Access : Practical Innovations, Open Solutions*, *3*, 678–708. DOI: 10.1109/ACCESS.2015.2437951

Ismail, S., Alshmari, M., Latif, K., & Ahmad, H. F. (2017). A granular ontology model for maternal and child health information system. *Journal of Healthcare Engineering*, *2017*, 1–9. Advance online publication. DOI: 10.1155/2017/9519321 PMID: 29065669

Istepanian, R., Jovanov, E., & Zhang, Y. T. (2008). Guest Editorial Introduction to the Special Section on M-Health: Beyond Seamless Mobility and Global Wireless Health-Care Connectivity. *IEEE Transactions on Information Technology in Biomedicine*, *12*(1), 1–5. DOI: 10.1109/TITB.2007.908759 PMID: 18270031

Iwendi, C., Khan, S., Anajemba, J. H., Bashir, A. K., & Noor, F. (2020). Realizing an efficient IoMT-assisted patient diet recommendation system through machine learning model. *IEEE Access : Practical Innovations, Open Solutions*, *8*, 28462–28474. DOI: 10.1109/ACCESS.2020.2968537

Jara, A. J., Belchi, F. J., Alcolea, A. F., Santa, J., Zamora-Izquierdo, M. A., & Gómez-Skarmeta, A. F. (2010). A pharmaceutical intelligent information system to detect allergies and adverse drugs reactions based on internet of things. *Proceedings of the 2010 IEEE International Conference on Pervasive Computing and Communications Workshops*, 809-814. DOI: 10.1109/PERCOMW.2010.5470547

Kaur, B. (2021). IoT framework for heart diseases prediction using machine learning. *International Journal of Advanced Trends in Computer Science and Engineering*, *10*(3), 2420–2426. DOI: 10.30534/ijatcse/2021/781032021

Kaur, P., Kumar, R., & Kumar, M. (2019). A healthcare monitoring system using random forest and Internet of Things (IoT). *Multimedia Tools and Applications*, *78*(14), 19905–19916. DOI: 10.1007/s11042-019-7327-8

Khamparia, A., Singh, P. K., Rani, P., Samanta, D., Khanna, A., & Bhushan, B. (2021). An Internet of Health Things-driven deep learning framework for detection and classification of skin cancer using transfer learning. *Transactions on Emerging Telecommunications Technologies*, *32*(7), e3963. DOI: 10.1002/ett.3963

Khan, M. A., & Algarni, F. (2020). A healthcare monitoring system for the diagnosis of heart disease in the IoMT cloud environment using MSSO-ANFIS. *IEEE Access : Practical Innovations, Open Solutions*, *8*, 122259–122269. DOI: 10.1109/ACCESS.2020.3006424

Khan, S. R., Sikandar, M., Almogren, A., Din, I. U., Fortino, G., & Guerrieri, A. (2020). IoMT-based computational approach for detecting brain tumor. *Future Generation Computer Systems*, *109*, 360–367. DOI: 10.1016/j.future.2020.03.054

Khan, S. R., Sikandar, M., Almogren, A., Ud Din, I., Guerrieri, A., & Fortino, G. (2020). IoMT-based computational approach for detecting brain tumor. *Future Generation Computer Systems*, *109*, 360–367. DOI: 10.1016/j.future.2020.03.054

Kumar, P. M., Lokesh, S., Varatharajan, R., Chandra Babu, G., & Parthasarathy, P. (2018). Cloud and IoT based disease prediction and diagnosis system for healthcare using fuzzy neural classifier. *Future Generation Computer Systems*, *86*, 527–534. DOI: 10.1016/j.future.2018.04.036

Liu, C., Chen, F., Zhao, C., Wang, T., Zhang, C., & Zhang, Z. (2018). IPv6-based architecture of community medical Internet of Things. *IEEE Access : Practical Innovations, Open Solutions*, *6*, 7897–7910. DOI: 10.1109/ACCESS.2018.2801563

Ma, Y., Wang, Y., Yang, J., Miao, Y., & Li, W. (2017). Big health application system based on health internet of things and big data. *IEEE Access : Practical Innovations, Open Solutions*, *5*, 7885–7897. DOI: 10.1109/ACCESS.2016.2638449

Maskeliunas, R., Damaševicius, R., & Segal, S. (2019). A review of internet of things technologies for ambient assisted living environments. *Future Internet*, *11*(12), 259. Advance online publication. DOI: 10.3390/fi11120259

Mechael, P. (2005). Case study from Egypt: Mobile phones for mother and child care. *i4d - The First Monthly Magazine on ICT4D*, 3(1), 22-24.

Mohd Aman, A. H., Hassan, W. H., Sameen, S., Attarbashi, Z. S., Alizadeh, M., & Abdul Latiff, L. (2021). IoMT amid COVID-19 pandemic: Application, architecture, technology, and security. *Journal of Network and Computer Applications*, *174*, 102886. DOI: 10.1016/j.jnca.2020.102886 PMID: 34173428

Nanayakkara, N., Halgamuge, M., & Syed, A. (2019). Security and privacy of Internet of Medical Things (IoMT) based healthcare applications. *RE:view*.

Ning, W., Li, S., Wei, D., Guo, L. Z., & Chen, H. (2021). Automatic detection of congestive heart failure based on a hybrid deep learning algorithm in the Internet of Medical Things. *IEEE Internet of Things Journal, 8*(16), 12550–12558. DOI: 10.1109/JIOT.2020.3023105

Noor, M. M., & Hassan, W. H. (2019). Current research on Internet of Things (IoT) security: A survey. *Computer Networks, 148*, 283–294. DOI: 10.1016/j.comnet.2018.11.025

Pace, P., Aloi, G., Gravina, R., Caliciuri, G., Fortino, G., & Liotta, A. (2019). An edge-based architecture to support efficient applications for healthcare industry 4.0. *IEEE Transactions on Industrial Informatics, 15*(1), 481–489. DOI: 10.1109/TII.2018.2843169

Pan, Y., Fu, M., Cheng, B., Tao, X., & Guo, J. (2020). Enhanced deep learning assisted convolutional neural network for heart disease prediction on the Internet of Medical Things platform. *IEEE Access : Practical Innovations, Open Solutions, 8*, 189503–189512. DOI: 10.1109/ACCESS.2020.3026214

Pawar, A. B., & Ghumbre, S. (2017). A survey on IoT applications security challenges and counter measures. *Proceedings of the 2016 International Conference on Advances in Computing, Communications and Informatics (ICACCI)*, 315-320. DOI: 10.1109/CAST.2016.7914983

Raj, R. J. S., Shobana, S. J., Pustokhina, I. V., Pustokhin, D. A., Gupta, D., & Shankar, K. (2020). Optimal feature selection-based medical image classification using deep learning model in Internet of Medical Things. *IEEE Access : Practical Innovations, Open Solutions, 8*, 58006–58017. DOI: 10.1109/ACCESS.2020.2981337

Rajan Jeyaraj, P., & Nadar, E. R. S. (2019). Smart-monitor: Patient monitoring system for IoT-based healthcare system using deep learning. *Journal of the Institution of Electronics and Telecommunication Engineers*. Advance online publication. DOI: 10.1080/03772063.2019.1649215

Rizvanov, A. A. (2018). *Advances in biomedical engineering and technology*. Lecture Notes in Bioengineering.

Rohokale, V. M., Prasad, N. R., & Prasad, R. (2011). A cooperative Internet of Things (IoT) for rural healthcare monitoring and control. *Proceedings of the 2nd International Conference on Wireless Communications, Vehicular Technology, Information Theory and Aerospace &Electronic Systems Technology (Wireless VITAE)*, 1-6. DOI: 10.1109/WIRELESSVITAE.2011.5940920

Sacco, G.. (2018). A radar system for indoor human localization and breath monitoring. In *2018 IEEE International Symposium on Medical Measurements and Applications (MeMeA)* (pp. 1–6). DOI: 10.1109/MeMeA.2018.8438759

Sadad, T., Khan, A. R., Hussain, A., Tariq, U., Fati, S. M., Bahaj, S. A., & Munir, A. (2021). Internet of Medical Things embedding deep learning with data augmentation for mammogram density classification. *Microscopy Research and Technique*, *84*(9), 2186–2194. DOI: 10.1002/jemt.23773 PMID: 33908111

Saepuddin, E., Rizal, E., & Rusmana, A. (2018). Posyandu roles as mothers and child health information center. *Record and Library Journal*, *4*(2), 201–208. DOI: 10.20473/rlj.V3-I2.2017.201-208

Sankar, S., Srinivasan, P., & Saravanakumar, R. (2018). Internet of things based ambient assisted living for elderly people health monitoring. *Research Journal of Pharmacy and Technology*, *11*(6), 2473–2478. DOI: 10.5958/0974-360X.2018.00715.1

Selvakanmani, S., & Sumathi, M. (2021). Fuzzy assisted fog and cloud computing with MIoT system for performance analysis of health surveillance system. *Journal of Ambient Intelligence and Humanized Computing*, *12*(1), 357–366. DOI: 10.1007/s12652-020-02156-y

Sengupta, J., Ruj, S., & Bit, S. D. (2019). A comprehensive survey on attacks, security issues and blockchain solutions for IoT and IIoT. *Journal of Network and Computer Applications*, *149*, 102481. DOI: 10.1016/j.jnca.2019.102481

Shahamabadi, M. S., Ali, B. B. M., Varahram, P., & Jara, A. J. (2013). A network mobility solution based on 6LoWPAN hospital wireless sensor network (NEMO-HWSN). *Proceedings of the 2013 International Conference on Information Management and Engineering*, 157-162. DOI: 10.1109/IMIS.2013.157

Sheth, A. P.. (2007). Transforming big data into smart data: Deriving value via harnessing volume, variety & velocity using semantics and semantic web. *Proceedings of the 2007 International Conference on Semantic Computing (ICSC)*, 1-8. https://doi.org/DOI: 10.1109/ICSC.2007.4357934

Singh, R. P., Javaid, M., Haleem, A., Vaishya, R., & Ali, S. (2020). Internet of Medical Things (IoMT) for orthopaedic in COVID-19 pandemic: Roles, challenges, and applications. *Journal of Clinical Orthopaedics and Trauma*, *11*(4), 713–717. DOI: 10.1016/j.jcot.2020.05.011 PMID: 32425428

Sonawane, P. D., & Sutar, R. G. (2018). A schematic review on body area networks for E-health systems. *Proceedings of the 2017 IEEE 2nd International Conference on Computing and Communications Technologies (I2C2).* https://doi.org/DOI: 10.1109/I2C2.2017.8321822

Song, Y., Jiang, J., Wang, X., Yang, D., & Bai, C. (2020). Prospect and application of Internet of Things technology for prevention of SARIs. *Clinical eHealth, 3,* 1–4. DOI: 10.1016/j.ceh.2020.02.001

Sorenson, C., Drummond, , & Bhuiyan-Khan, . (2013). Medical technology as a key driver of rising health expenditure: Disentangling the relationship. *ClinicoEconomics and Outcomes Research, 5*, 223. DOI: 10.2147/CEOR.S39634 PMID: 23807855

Souza, L. F., Silva, I. C. L., Marques, A. G., Silva, F. H. S., Nunes, V. X., Hassan, M. M., Albuquerque, V. H. C., & Filho, P. P. R. (2020). Internet of medical things: An effective and fully automatic IoT approach using deep learning and fine-tuning to lung CT segmentation. *Sensors (Basel), 20*(23), 6711. DOI: 10.3390/s20236711 PMID: 33255308

Sundarasekar, R., Thanjaivadivel, M., Manogaran, G., Kumar, P. M., Varatharajan, R., Chilamkurti, N., & Hsu, C.-H. (2018). Internet of things with maximal overlap discrete wavelet transform for remote health monitoring of abnormal ECG signals. *Journal of Medical Systems, 42*(11), 228. DOI: 10.1007/s10916-018-1093-4 PMID: 30311011

Tai, Y., Gao, B., Li, Q., Yu, Z., Zhu, C., & Chang, V. (2021). Trustworthy and intelligent COVID-19 diagnostic IoMT through XR and deep-learning-based clinic data access. *IEEE Internet of Things Journal, 8*(21), 15965–15976. DOI: 10.1109/JIOT.2021.3055804 PMID: 35782175

Tan, E. T., & Halim, Z. A. (2019). Health care monitoring system and analytics based on Internet of Things framework. *Journal of the Institution of Electronics and Telecommunication Engineers, 65*(5), 653–660. DOI: 10.1080/03772063.2018.1447402

Tan, W., ... (2020). Application of face recognition in tracing COVID-19 fever patients and close contacts. In *2020 19th IEEE International Conference on Machine Learning and Applications (ICMLA)* (pp. 1112–1116). DOI: 10.1109/ICMLA51294.2020.00179

Thakur, N., & Han, C. Y. (2021). Multimodal approaches for indoor localization for ambient assisted living in smart homes. *Information (Basel), 12*(3), 114. DOI: 10.3390/info12030114

Ting, D. S. W., Carin, L., Dzau, V., & Wong, T. Y. (2020). Digital technology and COVID-19. *Nature Medicine*, *26*(7), 459–461. DOI: 10.1038/s41591-020-0824-5 PMID: 32284618

Tuli, S., Basumatary, N., Gill, S. S., Kahani, M., Arya, R. C., Wander, G. S., & Buyya, R. (2020). HealthFog: An ensemble deep learning based smart healthcare system for automatic diagnosis of heart diseases in integrated IoT and fog computing environments. *Future Generation Computer Systems*, *104*, 187–200. DOI: 10.1016/j.future.2019.10.043

Tuli, S., Basumatary, N., Gill, S. S., Kahani, M., Arya, R. C., Wander, G. S., & Buyya, R. (2020). HealthFog: An ensemble deep learning based smart healthcare system for automatic diagnosis of heart diseases in integrated IoT and fog computing environments. *Future Generation Computer Systems*, *104*, 187–200. DOI: 10.1016/j.future.2019.10.043

Ukachi, N. B., & Anasi, S. N. I. (2019). Information and communication technologies and access to maternal and children health information: Implications for sustainable development. *Information Development*, *35*(4), 610–624. DOI: 10.1177/0266666918767482

Verma, N., Singh, S., & Prasad, D. (2022). A review on existing IoT architecture and communication protocols used in healthcare monitoring system. *Journal of the Institution of Engineers (India): Series B, 103,* 245–257. https://doi.org/DOI: 10.1007/s40031-021-00632-3

Verma, P., & Sood, S. K. (2018). Fog assisted-IoT enabled patient health monitoring in smart homes. *IEEE Internet of Things Journal*, *5*(3), 1789–1796. DOI: 10.1109/JIOT.2018.2803201

Veterans Veterans Health Administration (va.gov)

Vicini, S., Bellini, S., Rosi, A., & Sanna, A. (2012). An internet of things enabled interactive totem for children in a living lab setting. *Proceedings of the 18th International Conference on Engineering, Technology and Innovation (ICE)*, 1-8. DOI: 10.1109/ICE.2012.6297713

Vijayalakshmi, A., & Jose, D. V. (2020). Internet of Things for ambient-assisted living—An overview. In *Internet of Things Use Cases for the Healthcare Industry* (pp. 209–228). Springer., DOI: 10.1007/978-3-030-37526-3_10

Vimarlund, V., & Wass, S. (2014). Big data, smart homes and ambient assisted living. *Yearbook of Medical Informatics*, *23*(1), 143–149. DOI: 10.15265/IY-2014-0011 PMID: 25123734

Vineetha, V., Misra, Y., & Kishore, K. K. (2021). A real time IoT based patient health monitoring system using machine learning algorithms. *European Journal of Molecular and Clinical Medicine*, *7*, 2912–2925.

Wan, J., Gu, X., Chen, L., & Wang, J. (2017). Internet of Things for ambient assisted living: Challenges and future opportunities. *Proceedings of the 2017 International Conference on Cyber-Enabled Distributed Computing and Knowledge Discovery*, 83-90. DOI: 10.1109/CyberC.2017.83

Wang, B. O., Guo, L., Zhang, H., & Guo, Y.-X. (2020). A millimetre-wave radar-based fall detection method using line kernel convolutional neural network. *IEEE Sensors Journal*, *20*(22), 13364–13370. DOI: 10.1109/JSEN.2020.3006918

Wang, E. K., Chen, C.-M., Hassan, M. M., & Almogren, A. (2020). A deep learning-based medical image segmentation technique in the Internet-of-Medical-Things domain. *Future Generation Computer Systems*, *108*, 135–144. DOI: 10.1016/j.future.2020.02.054

Wang, W., Li, J., Wang, L., & Zhao, W. (2012). The internet of things for resident health information service platform research. *Proceedings of the IET International Conference on Communication Technology and Application (ICCTA 2011)*, 411-414. DOI: 10.1049/cp.2011.0745

Wei, P., ... (2020). Low-cost multi-person continuous skin temperature sensing system for fever detection: Poster abstract. In *Proceedings of the 18th Conference on Embedded Networked Sensor Systems* (pp. 705–706). Association for Computing Machinery: Virtual Event, Japan. DOI: 10.1145/3384419.3430398

Xu, G., Lan, Y., Zhou, W., Huang, C., & Li, W.. (2019). An IoT-based framework of WebVR visualization for medical big data in connected health. *IEEE Access : Practical Innovations, Open Solutions*, *7*, 173866–173874. DOI: 10.1109/ACCESS.2019.2957149

Xu, Y., Holanda, G., De Freitas Souza, L. F., Silva, H., Gomes, A., Silva, I., Ferreira, M., Jia, C., Han, T., De Albuquerque, V. H. C., & Filho, P. P. R. (2021). Deep learning-enhanced Internet of Medical Things to analyze brain CT scans of hemorrhagic stroke patients: A new approach. *IEEE Sensors Journal*, *21*(22), 24941–24951. DOI: 10.1109/JSEN.2020.3032897

Xuan, W., & You, G. (2020). Detection and diagnosis of pancreatic tumor using deep learning-based hierarchical convolutional neural network on the Internet of Medical Things platform. *Future Generation Computer Systems*, *111*, 132–142. DOI: 10.1016/j.future.2020.04.037

Yang, G., Xie, L., Mäntysalo, M., Zhou, X., Pang, Z., Xu, L. D., Kao-Walter, S., Chen, Q., & Zheng, L.-R. (2014). A Health-IoT platform based on the integration of intelligent packaging, unobtrusive bio-sensor, and intelligent medicine box. *IEEE Transactions on Industrial Informatics, 10*(4), 2180–2191. Advance online publication. DOI: 10.1109/TII.2014.2307795

Yang, G., Xie, L., Mäntysalo, M., Zhou, X., Pang, Z., Xu, L. D., Kao-Walter, S., Chen, Q., & Zheng, L.-R. (2014). A Health-IoT platform based on the integration of intelligent packaging, unobtrusive bio-sensor, and intelligent medicine box. *IEEE Transactions on Industrial Informatics, 10*(4), 2180–2191. DOI: 10.1109/TII.2014.2307795

Yang, X., Nazir, S., Khan, H. U., Shafiq, M., & Mukhtar, N. (2021). Parallel computing for efficient and intelligent industrial internet of health things: An overview. *Complexity*. Advance online publication. DOI: 10.1155/2021/5554685

Yilmaz, Ö. (2019). An ambient assisted living system for dementia patients. *Turkish Journal of Electrical Engineering and Computer Sciences, 27*(3), 2361–2378. Advance online publication. DOI: 10.3906/elk-1806-124

You, L., Liu, C., & Tong, S. (2011). Community medical network (CMN): Architecture and implementation. *Proceedings of the 2011 Global Mobile Congress (GMC)*, 17-21. DOI: 10.1109/GMC.2011.6103930

Yüzer, A. H., Sümbül, H., Nour, M., & Polat, K. (2020). A different sleep apnea classification system with neural network based on the acceleration signals. *Applied Acoustics, 163*, 107225. DOI: 10.1016/j.apacoust.2020.107225

Zahoor, S., & Mir, R. N. (2021). Resource management in pervasive Internet of Things: A survey. *Journal of King Saud University. Computer and Information Sciences, 33*(3), 321–333. DOI: 10.1016/j.jksuci.2018.08.014

Zhang, H., Li, J., Wen, B., Xun, Y., & Liu, J. (2018). Connecting intelligent things in smart hospitals using NB-IoT. *IEEE Internet of Things Journal, 5*(3), 1550–1560. DOI: 10.1109/JIOT.2018.2792423

Zhang, Y., Chen, G., Du, H., Yuan, X., Kadoch, M., & Cheriet, M. (2020). Real-time remote health monitoring system driven by 5G MEC-IoT. *Electronics (Basel), 9*(11), 1753. DOI: 10.3390/electronics9111753

Chapter 10
Communication Technologies and Protocols in IoT Systems:
Principles, Standards, and Security

Noura Ibrahim Ganaou
Abdou Moumouni University, Niger

Allassane Issa Salaou
https://orcid.org/0009-0005-6068-6037
Oran 1 University, Algeria

ABSTRACT

Internet of Things (IoT) technology has revolutionized the way devices interact and communicate, enabling seamless connectivity and automation across various domains including computer systems, networking and embedded systems. Communication is one of the most important fundamental aspects of any digital system. In IoT system, apart from modern paradigms, many classical computer systems communication technologies are used to achieve the main purpose of information acquisition and processing. The actual proposed chapter aims to provide and in-depth and comprehensive exploration of communication technologies and protocols used in IoT environments according to different context and use cases. This exploration will also take into account the data acquisition, storage, processing and transmission technics, internal and external electronic architecture and communication technics.

DOI: 10.4018/979-8-3693-5448-3.ch010

COMMUNICATION TECHNOLOGIES AND PROTOCOLS IN IOT SYSTEMS

The integration of communication technologies and protocols is fundamental in establishing the foundation for effective IoT systems. IoT devices rely on robust communication channels to exchange data, receive commands, and facilitate interactions in real-time. Also, the synergy between IoT and embedded systems lies at the heart of today's interconnected world, where smart devices communicate, collect data, and perform actions autonomously. Embedded systems serve as the backbone of IoT infrastructure, enabling devices to interact and process information.

In this context, many research works highlights challenges, development and evolution of IoT systems notably in terms of communications technologies, communication protocols and embedded systems design. These studies explored a wide array of wireless and wired communication solutions, highlighting their respective strengths and limitations in supporting the diverse requirements of IoT deployments. Moreover hardware-software co-design strategies, security concerns and trends technologies that optimize the performance and reliability of IoT systems like edge computing, cloud computing, data analytics and machine learning was a focus for many researches.

Connectivity between IoT devices and IoT platforms at the infrastructural level relies on classic communication technologies used in Internet such as Wi-Fi, Ethernet, Cellular Networks and Satellites. Also, protocols like HTTP (Hyper-Text Transfer Protocol), TCP (Transport Control Protocol), TLS (Transport Layer Socket), IP (Internet Protocol) assure structured and secure communication establishment at high level. However, many communication technologies and protocols have evolved to satisfy IoT specific requirements in terms of optimization. This chapter explore several aspects of communication technologies and protocols in IoT systems including the following:

- IoT systems description and their interconnected nature with embedded systems.
- IoT systems communication technologies exploration, including wireless, wired, and hybrid solutions.
- Investigation of the significance of protocols in establishing efficient and secure communication channels within IoT networks.
- Analysis of the impact of communication technologies and protocols on the performance, scalability, and security of IoT deployments.
- Challenges and security considerations in designing and integrating IoT solutions.

By the end of the chapter, readers would acquire a comprehensive understanding of the diverse communication technologies and protocols available for IoT systems and a deeper understanding of its symbiotic relationship with embedded systems. Furthermore, insights into best practices, emerging trends and future directions in IoT communication technologies are covered in this chapter.

OVERVIEW OF IOT AND EMBEDDED SYSTEMS

IoT: Concepts, Characteristics and Architecture

Internet of Things (IoT) is a generic term describing a system in which a set of networked devices communicating over the Internet or any Internet-based computer network. These devices are particular in the way that they are not general-purpose computers, but real-world objects equipped with internal specific-purpose computers with software systems called embedded systems. Each IoT device has one or more embedded systems (Perry, 2018).

Thus, smart vehicles, smart watches, digital assistants, smart traffic lights, IP cameras, smart medical equipment, smart sensors, industrial machines and so on, are some examples of objects or IoT devices that are able to communicate with other devices (*D2D: Device-To-Device communication*), machines like web-servers (*D2M: Device-To-Machine communication*) or other systems like a cloud system (*D2C: Device-To-Cloud communication*) over the Internet or Intranet network.

Thus, each IoT device is physically identified with a unique ID (MAC: Medium Access Control Address, Serial Number, UUID: Universal Unique IDentifier, IMEI: International Mobile Equipment Identity) and logically with and IP (Internet Protocol, IPv6) address. Also, cloud applications serve as storage, processing and analytic platforms enabling artificial intelligence on IoT systems (Mohammad et al., 2023).

However, IoT aims to enable a large-scale digital system in which almost everything is connected and able to be controlled remotely, share information and consume services from other things and/or systems. The nature of communication depends to the underlying devices and context of usage, but wireless communications are mostly adopted in major use cases.

Figure 1. IoT System Overview

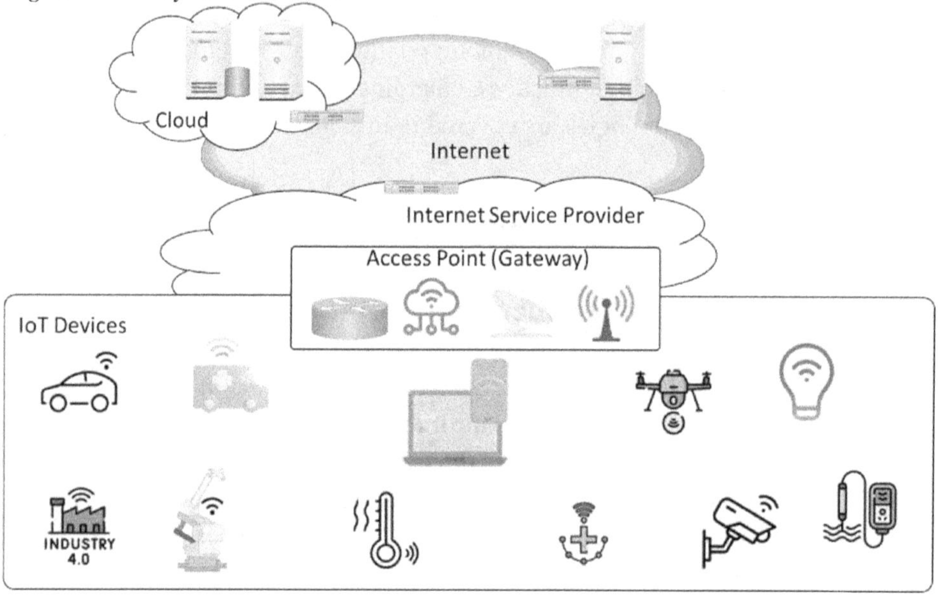

Hence, the functional architecture of IoT system consists of four (4) main logical layers:

- **Perception Layer**: Responsible for data acquisition and physical action performing. It allows interaction with the physical environment and contains IoT hardware devices for perception, known as sensors and actuators.
- **Network Layer**: Corresponds to the level of connectivity with the Local and Internet networks acting as a gateway for data transmission from IoT devices to other systems and vice versa in some use cases.
- **Computing Layer**: Collected data from IoT devices needs to be stored and processed. This is done using high-capacity servers at edge and/or cloud computation levels. The layer may add some business-oriented models for data analytics using data science and artificial intelligence algorithms like machine learning.
- **Application Layer**: Corresponds to applications using the subsequent systems to deliver service to the final users either related to the underlying IoT devices or to other systems.

Figure 2. IoT Functional Architecture

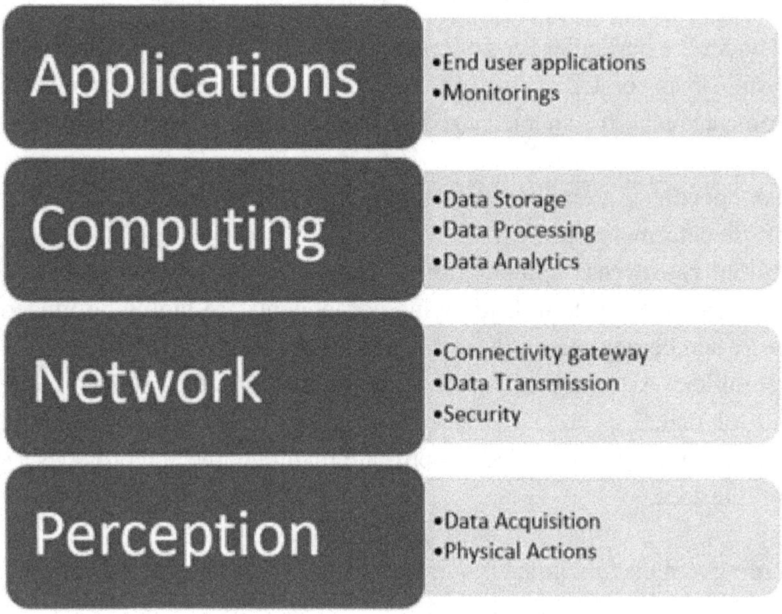

IoT have several use cases in various domains of socio-economic life such as:

- **Agriculture**: IoT contributes to smart farming with applications like precision agriculture, smart irrigation, yield monitoring, harvesting robots, etc.
- **Transportation & Logistics**: IoT enables intelligent transportation and logistics with applications like fleet management, connected vehicles, connected airplanes, drone delivery, smart shops, etc.
- **Healthcare**: Health monitoring, telemedicine, smart hospitals are examples of IoT applications in for Health domain.
- **Industry**: IoT systems that involves devices for industrial machinery is known as IIoT (Industrial Internet of Things) and uses smart sensors and actuators as leveraging technology for the fourth generation of industry, Industry 4.0 (Shrey, 2024). The goal is to simplify industrial processes, increase quality and extends innovation possibilities with a reduced cost. IIoT enables intelligent industrial system and services like smart manufacturing, digital twins and connected factories.
- **General-Purpose Applications**: Some applications of IoT like smart cities, smart buildings, smart education, security, defense and strategy involve many domains and propose a general solution for a better well-being.

IoT Devices

IoT devices are particular type of embedded systems. An embedded system is a digital system incorporated in a lager real-world system which is in general electronic or electromechanical. It's characterized by the main following aspects:

- **Task specific**: An embedded system is designed to perform limited specific tasks in determined context.
- **Limited resources**: Computational and power capacity are general limited in an embedded system. Underlying components are highly optimized to be energy and computation efficient.
- **Reliability**: An embedded system is highly efficient, stable, secure and fault-tolerant.
- **Maintainability**: Embedded systems are maintainable and provides a simple user interface.

According to their functional requirements, there is four (4) types of embedded systems:

- **Standalone**: Works independently with no interconnection with other systems.
- **Mobile**: Embedded systems that are portable.
- **Networked**: Connected to other systems through a communication network.
- **Realtime**: Response time of application is considered in performing tasks.

The cited types could be small-scale with fewer computation resources like simple microcontroller, medium-scale with general-purpose CPU (Central Processing Unit) or large-scale with multiple CPU.

IoT devices are considered as networked embedded systems at any scale that could be real-time or not, mobile or fixed.

Hardware and Functional Architecture

Now what are the architecture of embedded systems and functionalities? A classic embedded system has three principal component categories:

- **Electromechanical Hardware**: Sensors and actuators. Sensors are used to measure and acquire data whereas actuators serve for performing physical actions. Components of this category interact with the physical environment.

- **Electronical**: Electronic components included a micro-computer or embedded computer. Components of this category form the cyber environment.
- **Logical**: Software systems and/or applications ties to the embedded computer for performing tasks and controlling the system.

Figure 3. Embedded System Architecture

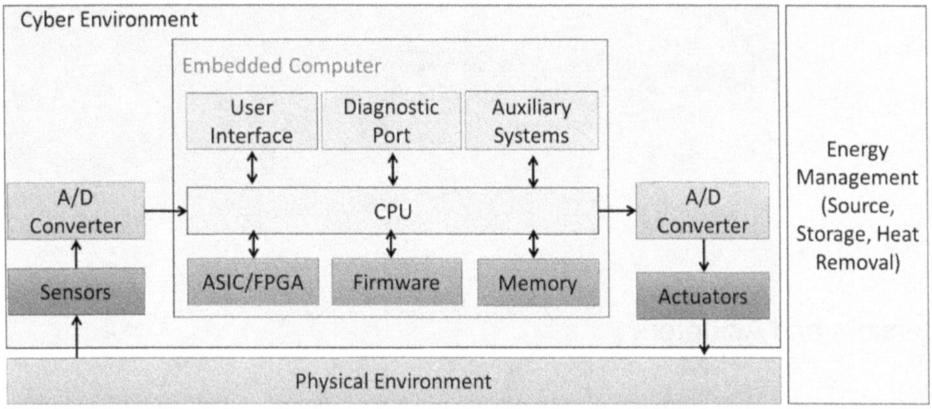

Information processing in embedded systems involves five (5) types of operations:

- **Singal/Data acquisition**: Information is collected using sensors or transmission channels.
- **Signal transduction**: Signal conversion from a form or energy to another form of energy. For example, transforming a mechanical signal to electric signal is a kind of transduction.
- **Signal conditioning**: Consists of modifying or filtering a signal to optimize it for processing.
- **Signal/Data processing**: Processing signal using ASIC (Application Specific Integrated-Circuits) or data using the embedded computer.
- **Signal/Data transmission**: Process that allows to send signal/data to another system via a communication channel.

Figure 4. Information Flow in Embedded Systems

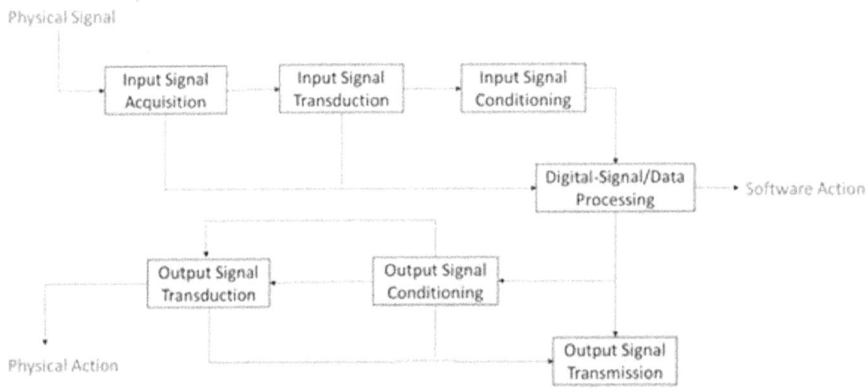

Sensors and Actuators

Both sensors and actuators are transducers. A transducer is a physical hardware which function is to convert signal/energy from one form to another. The type of energy might be electric, electromagnetic, mechanical, chemical, thermal or nuclear.

The transformed signal/energy carry information about a physical capacity like pressure, temperature, heat, current power/intensity, permittivity, radiation, refraction, polarization, chemical composition, etc.

A transducer could be analog (with continuous values) or digital (with discrete values) and takes as input a signal S_{in} to produce an output signal S_{out} such that $S_{out} = f(S_{in})$.

- For analog transducers:

$$S_{out} = C + m.S_{in}$$

where C is the neutral stimuli value

- For digital transducers:

$$S_{out} = \begin{cases} 0\, for\, S_{in} \leq 0 \\ 1\, for\, S_{in} > 0 \end{cases}$$

The value measurement inside transducers takes into account four (4) main parameters:

- **Resolution**: Defines the smallest value that is perceptible for a measurement.
- **Range**: Interval between minimal and maximal value that could be measured.
- **Accuracy**: The degree of precision in measurement. It is determined by calculated systematic and/or statistical errors between real values and measured values.
- **Sensitivity**: Indicates the degree of output variation according to the input.

A sensor is a transducer which function is to perform data acquisition by reading input signals. The figure 5 gives an overview of some sensors. The most used sensors produce electric signals as output.

Figure 5. Example of sensors

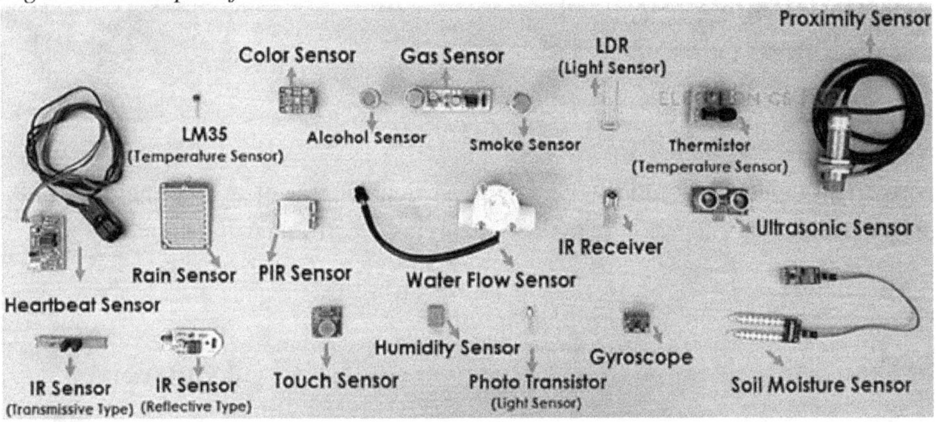

An actuator is a transducer used to perform physical actions. Actuators could be electric, pneumatic or hydrolic. The firt level actuators in embedded systems take in the most use cases an electric signal as input.

Figure 6. Example of actuators

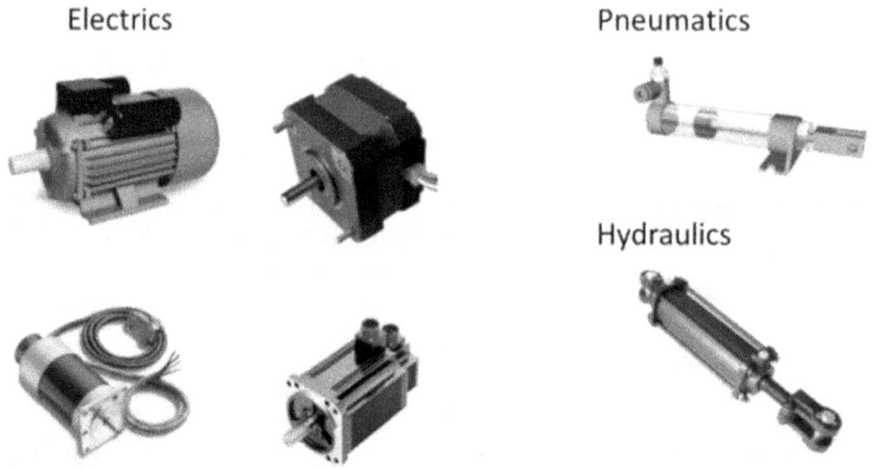

A transceiver (transmitter-receiver) is particular transducer that includes both sensor and actuator for trasmitting and receiving electromagnetic signals.

Figure 7. Transceiver (sensor and actuator)

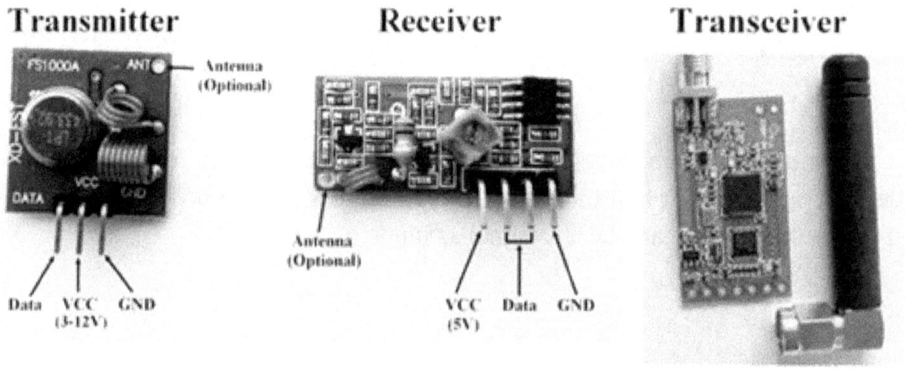

Improvements in manufacturing lead to data processing inside sensors and actuators. Thus, they are smart sensors and actuators and acts as independent IoT devices with an embedded system.

Firmware

The software integrated in embedded systems (or IoT devices) is known as Firmware. The firmware is a set of applications stored in a memory (Read-Only Memory (ROM) or Flash Memory).

Depending on the scale of the system, the firmware could be loop-oriented tasks or based on a light-weight operating system.

The first approach consists of coding sequential tasks that will be performed continuously and repetitively in a loop. This kind of firmware is used in microcontrollers (AVR/PIC) and are programmed using platforms like Arduino, MPLAB with languages like C, C++, Python, Rust and JavaScript.

The second approach allows developers to program applications based on a CGPOS (Customized General-Purpose Operating System) and package them as Firmware.

CGPOS are in the most cases RTOS (Real-Time operating Systems) allowing to handle and perform tasks in quasi-instant responding time. The most used are ARM-Mbed, QNX, VxWorks, ThreadX, MicroC/OS-II, Embedded-Linux and ROS (Robot Operating System).

Device Lifecycle Management

From their design, production, deployment to their end of service, IoT devices require a set of operations such as monitoring, maintenance, firmware updates and security patches. These tasks are shared between product vendors, system administrators (IT managers) and end users. This process is known as IoT devices lifecycle management and is composed of several stages:

- **Planning**: This phase corresponds to the conception and design of IoT devices in which organizations defines their IoT devices lifecycle strategies in terms of requirements, maintenance, security protocols, monitoring and scalability management to ensure efficient deployment and longevity of IoT devices.
- **Manufacturing**: Involves the production process management including automation, functionality test, validation and configuration. In this step, a unique identifier has been assigned to each IoT device and eventually the vendor's signature.
- **Provisioning**: When the product is ready, the distribution process adds some additional parameters allowing to monitor geographically the product distribution. Safely packaging IoT devices before sending it to organizations is also crucial and is part of provisioning process.

- **Deployment**: After their provisioning, IoT devices are deployed in real applications and configured for secure access to the network. This marks the beginning of service journey of an IoT device. During the first configuration, the system of an IoT device system needs to be updated and /or upgraded to ensure a secure deployment.
- **Maintenance**: While in service, IoT device needs to be regularly maintained in terms of its hardware and firmware and the underlying software. In this step, efficient monitoring and anticipative maintenance plans should be implemented to reduce downtime and make sure that new updates, upgrades and security patches are applied as early as possible.
- **Decommissioning**: This is the last IoT device lifecycle process phase. It ensures that when an IoT device is obsolete, it's safely placed out of service without security and environmental risks. However, it implies saving ad wiping all data associated to the devices before sending it to decommissioning service where it will be either destructed or recycled taking into account environmental aspects.

Figure 8. IoT Device Lifecycle Management

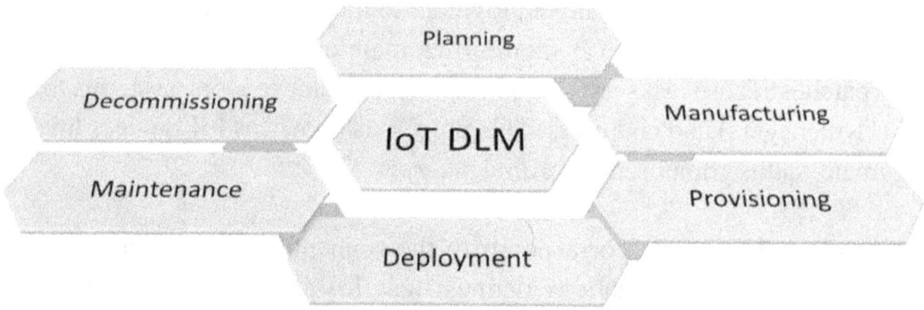

Consistent IoT devices lifecycle management process has a lot of benefits for vendors, organizations and end users such as cost savings, enhanced productivity and security, better customer satisfaction and less environmental impacts. Nonetheless, implementing this process requires for organization to evaluate their IoT device needs and goals, choose the right IoT devices and establish a robust and efficient system for monitoring these devices.

IoT Networks

IoT systems uses an aggregation of networks in a synchronized manner for communication and data transmission. There are three (3) main networks in an IoT system:

- Device Network: A set of local networks interconnecting IoT devices. The junction point of all device networks is known as gateway. Device networks are in the most cases wireless networks.
- Core Network: A private or public wide area network aggregating all device networks. It often contains a network of servers for edge and fog levels computing.
- Internet: The lager network used by IoT as a backbone for interconnecting IoT core network and end-user applications to cloud networks and services.

Figure 9. IoT Networks

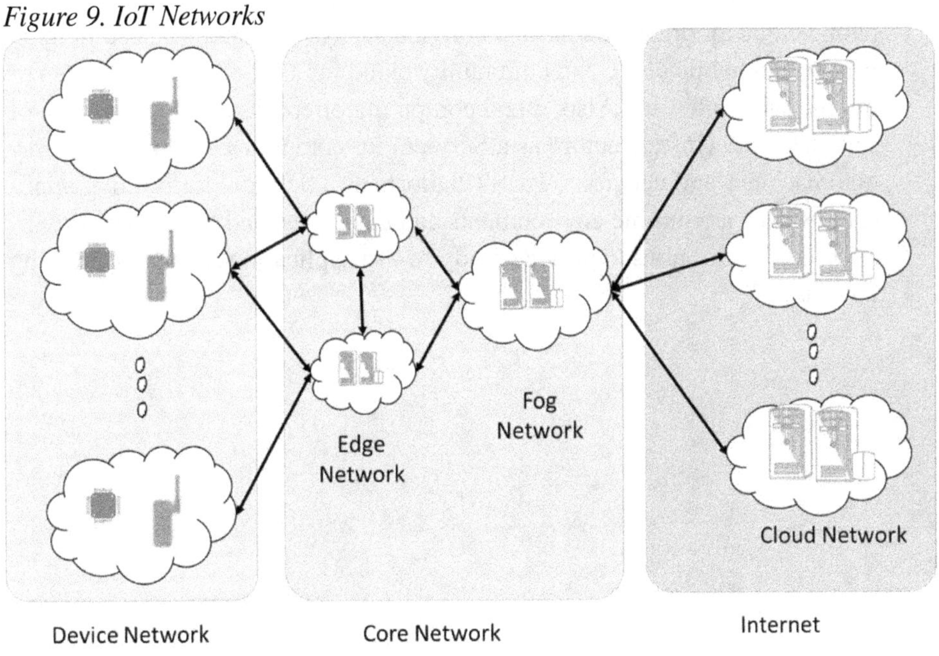

IoT Platforms

IoT platforms represents the hardware and software system that manages IoT applications used for computing and data analysis. This computing is often distributed in three levels:

- **Edge computing**: Distributed computing services located near device network used for specific local data storage and processing useful for standalone real-time applications such as monitoring in an airplane.
- **Fog computing**: An extension of edge computing for more complex and interdependent data that are not frequently used at local level but necessary for certain types of processing. This type of computing is based on distributed applications in order to decentralized processing.
- **Cloud computing**: Virtualization solution used to delocalize complex computing tasks towards datacenters accessible through the Internet. A could be deployed in four (4) deployment models: private cloud for an organization, public cloud for organizations and individuals, hybrid cloud when combining private and public cloud and community cloud for a set of organization sharing the same interests. Also, cloud computing offers three (3) main types of services: IaaS (Infrastructure as a Service) for computing resources like virtual machines and networks, PaaS (Platform as a Service) for running applications such as runtime environments and databases, and SaaS (Software as a Service) which aims to provide ready-to-use applications and programming interfaces.

Figure 10. IoT Computing Types

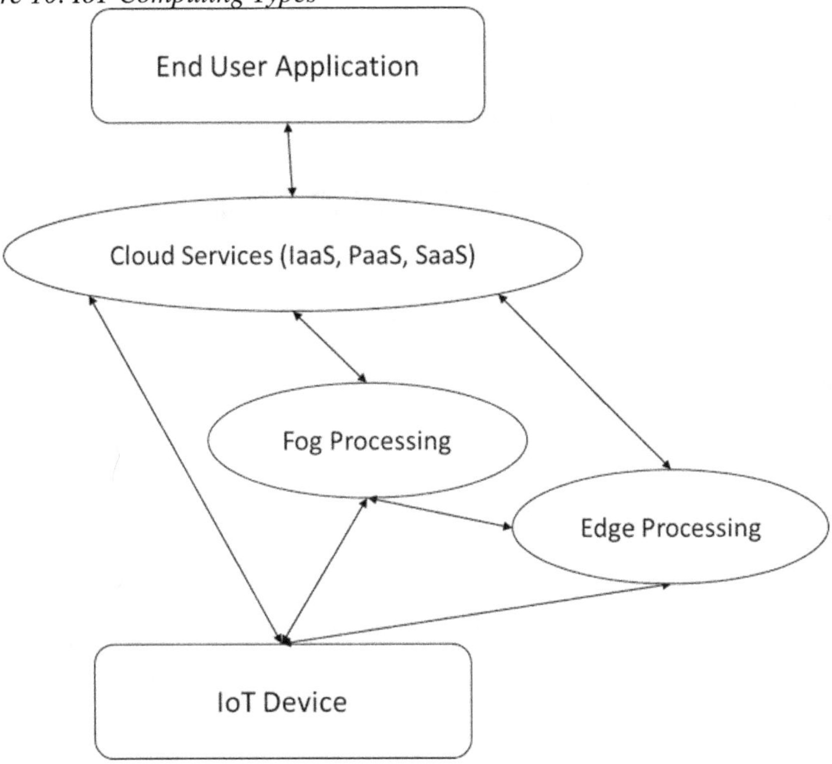

IOT SYSTEMS COMMUNICATION TECHNOLOGIES

Overview of IoT Communication Technologies

Communication models

IoT involves three (3) main communication models (Bhawani et al., 2019): Device-To-Device (D2D), Device-To-Machine (D2M) and Device-To-Cloud (D2C).

D2D refers to a communication in which an IoT device is connected with a direct link to another device without any central infrastructure. It's a kind of ad hoc network for IoT devices.

D2M consists of communication between a IoT device and a processing machine like a web server through Internet or a local network. This communication may imply a gateway.

D2C is a communication model defines an interaction between an IoT device and a cloud system.

Figure 11. IoT Systems Communication Models

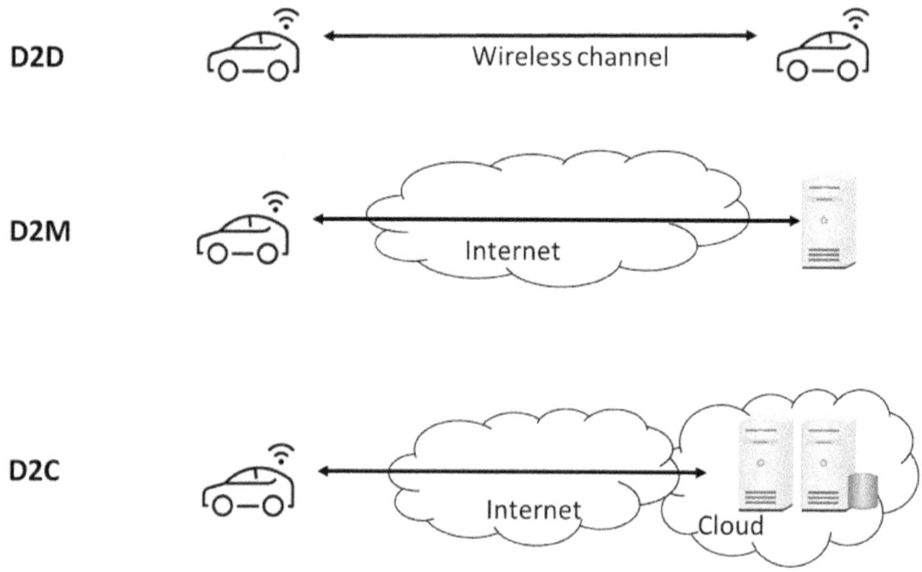

Types of Communication Technologies

Communication technologies vary according to the underlying model.

For a D2D communication short-range LoWPAN (Low-Power Wireless Personal Area Networks) wireless technologies are used like Bluetooth, NFC (Near Field Communication), Wi-Fi Direct and UWB (Ultra-WideBand), RFID (Radio Frequency IDentification).

For D2M and D2C communication models, long-range wireless LPWAN (Low-Power Wide Area Networks) technologies like Wi-Fi, ZigBee, Z-Wave, Classic Cellular Networks (GSM: Global System for Mobile, 2G, 3G, 4G, 5G) and LPWAN (Low-Power Wide Area Networks like NB-IoT (Narrow-Band IoT), LoRaWAN (Long-Range Wide Area Network), SigFox and Satellite (GPS: Global Positioning System, Galileo, StarLink, IntelSat, …) are used.

A combination of many technologies is might be applied in most of scenarios and sometimes wired network such as Ethernet, USB (Universal Serial Bus), I2C (Inter-Integrated Circuit), SPI (Serial Peripheral Interface) is used as gateways for D2M/D2C or for a direct communication in fixed industrial machinery.

Table 1. IoT Systems Communication Technologies

	D2D	D2M/D2C
Wireless	- *Bluetooth* - *Wi-Fi Direct* - *UWB* - *NFC* - *RFID* - *InfraRed*	- *Cellular: GSM, 3G, 4G, 5G, NB-IoT* - *Wi-Fi* - *ZigBee* - *SigFox* - *LoRaWAN* - *Satellite: StarLink, IntelSat, ...*
Wired	- *USB* - *Ethernet* - *I2C* - *SPI* - *UART*	- *USB* - *Ethernet*

WiMax is another wireless communication like Wi-Fi not widely used because its energy consumption and deployment cost. Li-Fi (Light-Fidelity) is also a novel wireless optical technology allowing to transmit signal like Wi-Fi using visible light.

Signals and Transmission Techniques

Wired communication is done by using cables whereas as far as wireless communication, the principle consists of using radio waves (RF: Radio-Frequency) to as signals careers by creating electromagnetic propagation. In the first case electric signals are transmitted, for the second one, electromagnetic signals are transmitted.

A signal is a physical quantity that vary in function of time having three (3) main properties: frequency (number of oscillations per second: *Hz*), amplitude (its value at a given time) and power (the amount of its energy in *Watt* or *Volt*).

According to their frequency, there are two (2) main categories of signals: narrow-banded signals and wide-banded signals.

Let say F_{min} (*resp.* F_{max}) is the minimal (*resp.* maximal) frequency of a signal $s(t)$, the spectral bandwidth is denoted by $\text{Ä}F = F_{max} - F_{min}$ and the average frequency or career frequency by $f_c = \text{Ä}F/2$, and the ratio $\alpha = \text{Ä}F/f_c$. However, if $\alpha \ll 1$, the signal is narrow-banded, else its wide-banded.

Narrow-banded signals are classified according to their career frequency f_c as indicated in *table 2*.

Table 2 . Narrow-Banded Signal Classification

Signal	f_c
Low-Frequency (LF)	< 250 KHz
High-Frequency (HF)	[250 KHz, 30 MHz [
Very-High-Frequency (VHF)	[30 MHz, 300 MHz [
Ultra-High-Frequency (UHF)	[300 MHz, 3 GHz [
Super-High-Frequency (SHF)	> 3 GHz

As far as wide-banded signals, the career wavelength (in *nm*: nanometers) denoted by $\lambda_c = 1/f_c$ is used to facilitate the classification as in *table 3*:

Table 3. Wide-Banded Signal Classification

Signal	λ_c
InfraRed (IR)	[700 nm, 1000 nm [
Visible Light	[400 nm, 700 nm [
Ultra-Violet (UV)	[10 nm, 400 nm [
X-ray	[0.01 nm, 10 nm [
Gamma-Ray (ã-ray)	> 0.01 nm

All transmission channels have the following principal properties:

- Attenuation: Expresses the weakness of the signal strength as it propagates through the channel. It's determined by the attenuation constant *A* using the signal input/output power P_{in} and P_{out}.

$$A = 10\log_{10}\left(\frac{P_{in}}{P_{out}}\right)$$

- Noise: Unwanted signals that can modify or interfere with the transferred signal. It's measured with the *SNR* (Signal/Noise Ratio) by:

$$SNR = 10\log_{10}\left(\frac{S}{N}\right)$$

- Bandwidth: Corresponds to the frequency band *B* for which the *A* is less than a fixed threshold (3dB or 6dB in general), $B \approx$ ÄF.

- Capacity: Corresponds to the maximal amount of information per unit of time D_{max} that could be send through a given transmission channel. It depends on the *SNR*.

$$D_{max} = B\log_2\left(1 + \frac{S}{N}\right)$$

The usage of these frequencies depends on the contexts and underlying technologies. The table 4 indicates some use cases of different frequencies:

Table 4. Frequency Band Usage in Telecoms

Name	Frequency	Wavelength	Applications
ELF (Extremely Low-Frequency)	30 to 300 Hz	1000 to 10000 Km	Power Line
VF (Voice Frequency)	300 to 3000 Hz	100 to 1000 Km	Telephony
VLF (Very Low-Frequency)	30 to 300 KHz	10 to 100 Km	Marine Communications
LF (Low-Frequency)	30 to 300 KHz	1 to 10 Km	Marine Communications
MF (Medium-Frequency)	300 to 3000 KHz	100 to 1000 m	AM Broadcasting
HF (High-Frequency)	30 to 300 MHz	10 to 100 m	Long-range Aircraft/Ship communications
VHF (Very High-Frequency)	30 to 300 MHz	1 to 10 m	FM Broadcasting
UHF (Ultra High-Frequency)	300 to 3000 MHz	10 to 100 cm	Cellular Telephony
SHF (Super High-Frequency)	3 to 30 GHz	1 to 10 cm	Satellites, Microwave
EHF (Extremely High-Frequency)	30 to 300 GHz	1 to 10 mm	Wireless Local Loop
InfraRed	300 GHz to 400 THz	770 nm to 1 mm	Consumer Electronics
Visible Light	400 to 900 THz	330 to 770 nm	Optical (Fiber)

Signal transmission modes vary according to the network type and the transmission channel:

- **Synchronous/Asynchronous**: Synchronous transmission involves strict timing coordination from both sender and receiver sides using clock signal. Asynchronous transmission doesn't require a clock signal.
- **Baseband/Wideband**: Baseband transmission consists of encoding/decoding physical signal to binary and vice versa without further processing. Wideband

transmission on the other hand, requires formatting signal to adapt it to transmission channel. This process is known as modulation/demodulation.

- **Simplex/Half-Duplex/Full-Duplex**: A Simplex transmission is a unidirectional transmission. It's Half-Duplex when it's bidirectional in an alternate manner, i.e. not in both direction at the same time. A Full-Duplex transmission is a parallel bidirectional transmission.
- **Anycast/Multicast/Broadcast**: Anycast transmission is a point-to-point or one-to-one transmission. Multicast transmission means one-to-many with group identification. A Broadcast transmission is a one-to-many without group identification.

Wireless Networks

Bluetooth

Bluetooth (*bluetooth.com*) is short-range wireless network for Personal Area Networks (PANs) introduced by Bluetooth Special Interest Group (Bluetooth SIG) since 1998. It's standardized as IEEE 802.15.1 (Institute of Electrical and Electronics Engineers, 2005; IEEE Std 802.15.1-2005), and operates in 2.4 GHz ISM (Industrial, Scientific, and Medical) band.

The classic version of Bluetooth (lower than v4.0) has an average data rate up to 1 Mbps with a range up to 100 m and uses FHSS (Frequency-Hopping Spread Spectrum) modulation and a protocol called BR/EDR (Bluetooth Basic Rate/Enhanced Data Rate). FHSS is based on rapid carrier frequency changing using a code shared by transmitter and receiver. BR/EDR chunks data into packed and divides frequency band in 79 channels. Each channel has a bandwidth of 1 MHz and each packet is transmitted on one channel.

The newer version of Bluetooth (from v4.0) is called Bluetooth LE or BLE (Bluetooth Low-Energy) and operates in the same frequency band as BR/EDR but with DSSS (Direct-Sequence Spread Spectrum) modulation that consist to make the transmitted signal wider in bandwidth than the information bandwidth. BLE divides frequency band in 40 channels of 2 MHz bandwidth and transmits packets in each channel using Gaussian frequency shift modulation.

BLE has a maximum transmit power of 10 mW (100 mW in Bluetooth 5) and is right suitable for limited resource IoT devices, this is why it's used massively in IoT. It has a data rate up to 1-2 Mbps in a range of 100 m depending on versions.

Bluetooth is based on Master/Slave architecture with two (2) types of networks: Piconet and Scatternet:

- **Piconet**: contains one primary node called Master and seven (7) active secondary nodes called slaves. A piconet may also have 255 not active secondary nodes called parked nodes. Communication is one-to-one between Master and Slave and could be one-to-many from Master to Slave. Slave to Slave communication is not possible.
- **Scatternet**: A scatternet is formed by two (2) or more piconets. Secondary nodes in the intersection of two (2) piconets can act as primary node in one of the two (2) piconets.

Figure 12. Bluetooth Network Topology

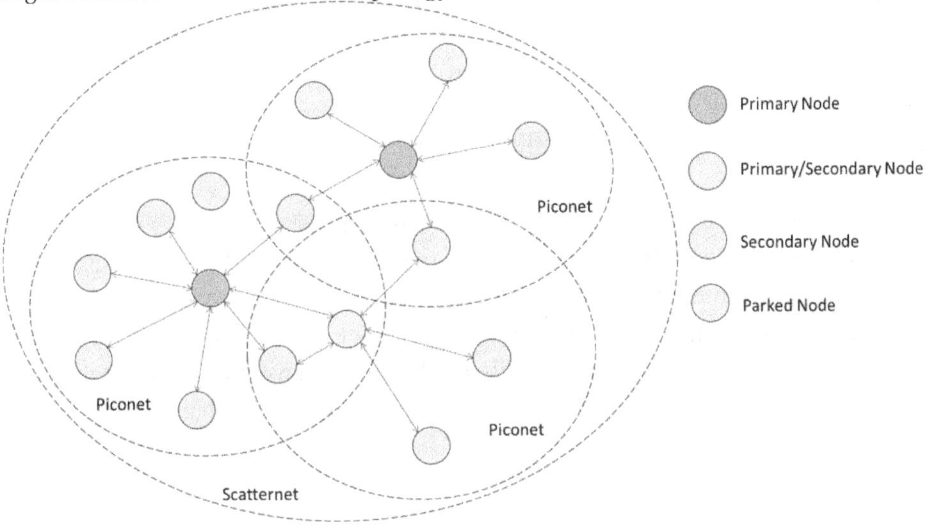

Bluetooth is massively used in mobile phones, stereo audio streaming, smart homes, wearables, automotive, healthcare, industrial and security appliances. Bluetooth applications are based on several profiles (e.g. DUN: Dial-up Networking Profile, GATT: Generic Attribute Profile, HDP: Health Device Profile, HFP: Hands-Free Profile, HSP: Headset Profile) that define the specific configuration to use in terms of protocol, power consumption, scalability, and range. The maximum permitted power consumption is limited using four (4) types of classes as indicated in the table 5

Table 5: Bluetooth Classes for Permitted Transmit Power

Class	mW	dBm
1	10 to 100	+10 to +20
1.5	2.5 to 10	+4 to +10
2	1 to 2.5	0 to +4
3	0.01 to 1	-20 to 0

Bluetooth uses 128-bit AES (Advanced Encryption Standard), a symmetric cryptographic algorithm, to encrypt data transmitted over its channels. For that matter, a Link Key is first securely shared between sender and receiver. After that, each packet is encrypted using the Link Key before being sent. The receiver uses the same Link Key to decrypt its data. In some use cases, application security layer is often added for a highly efficient data protection.

RFID (Radio-Frequency Identification)

RFID (*rfidinc.com*) is a communication technology used to identify object using electromagnetic fields. The process consists of storing and tracking tags wirelessly electromagnetic waves. There are two (2) types of RFID (Active and Passive) and they operate differently. Active RFID have their own power source and support a long-range up to 100 m, whereas passive RFID doesn't have power source and can support only distance less than 25 m and uses several frequency bands as indicated in the table 6 (Perry, 2018).

Table 6. RFID Frequencies and Properties

Band	Range	Data Speed	Tags
LF: 125-134.2 KHz	10 m	low	passive
HF: 13.56 MHz	10 cm – 1 m	low to moderate	passive
UHF: 433 MHz	1-100 m	moderate	passive or active
UHF: 856 - 960 MHz	1-12 m	low to moderate	passive or active
Microwave: 2.45 - 5.8 GHz	1-2 m	high	active
Microwave: 3.1 - 10 GHz	< 200 m	high	active

A typical RFID system consists of three (3) main components:

- **RFID Tag**: A small electronic device attached to an object that stores and transmits data by emitting a modulated signal containing stored data information. An RFID typically contains 96-128 bits of data, and may be greater than 128 bits in some use cases.

- **RFID Reader**: A device that emits radio waves and receives responses from RFID tags. If the RFID tag is passive, is uses the received energy from the reader/scanner to send back data in a modulated signal.
- **Processing System**: A system that processes and stores the data received from RFID readers.

Figure 13. RFID System Architecture

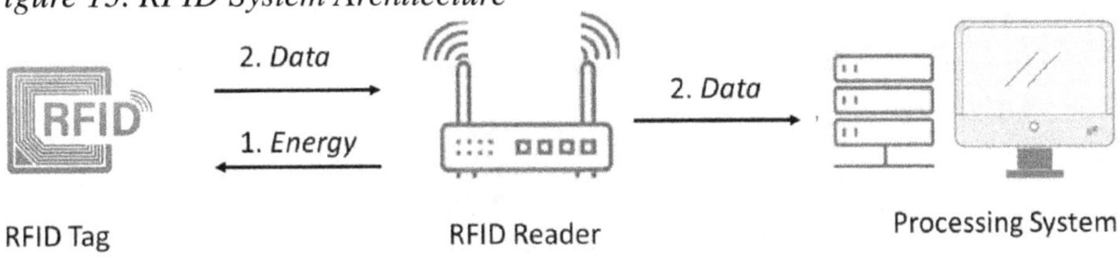

RFID is used in many domains for patient, asset, good and animal identification, production line monitoring, etc.

NFC (Near-Field Communication)

NFC (*nfc-forum.org*) is another short-range wireless communication technology like RFID allowing D2D data exchange. It operates only in HF RFID frequency band (13.56 MHz) in a range of 10 cm and acts as active and passive RFID at once.

NFC combines both tag and reader and could read and store tags using peer-to-peer communication. It's more secure than RFID and may replace some RFID short-range applications. NFC is massively used for electronic payments and data sharing applications.

Wi-Fi

Wi-Fi is a classic computer network communication technology maintained by Wi-Fi Alliance (*wi-fi.org*) based on CSMA/CA (Carrier Sense Multiple-Access with Collision Avoidance) and half-duplex over-the-air modulation techniques including DSSS (Direct-Sequence Spread Spectrum) and OFDM (Orthogonal Frequency

Division Multiplexing). It's normalized by IEEE (Institute of Electrical and Electronics Engineers) as 802.11x standard family and massively used in IoT ecosystem.

Earlier Wi-Fi standards are Wi-Fil (IEEE, 1999; IEEE Std 802.11b-1999), Wi-Fi2 (IEEE, 1999; IEEE Std 802.11a-1999) and Wi-Fi3 (IEEE, 2003; IEEE Std 802.11b-2003) operating respectively in 5 GHz, 2.4 GHz, 2.4 GHz with data rates up to 54 Mbps, 11 Mbps and 54 Mbps. However, practical rates are about 25-40 Mbps, 5-10 Mbps and 20-30 Mbps.

Recent improvements introduce Wi-Fi4 (IEEE, 2009; IEEE Std 802.11n), Wi-Fi5 (IEEE, 2013; IEEE Std 802.11ac), Wi-Fi6 (IEEE, 2021; IEEE Std 802.11ax) operating in both 2.4 GHz and 5 GHz with respective typical data rates up to 600 Mbps, 1.3 Gbps and 9.6 Gbps.

The typical range for Wi-Fi is about 45 to 90 m for 2.4 GHz band and 30 to 60 m for 5 GHz band.

Wi-Fi architecture is formed of an AP (Access Point) and BS (Base Station). The AP is the central node through which all communications pass. It's identified by a its identifier or name called SSID (Service Set IDentifier). This network service is known as BSS (Base Station Service).

Wi-Fi allows also an ad hoc configuration in which all Base Stations communicate directly without any AP. The service called IBSS (Independent Base Station Service) improve Wi-Fi mobility and flexibility.

Also, for scalability reasons, many BSS could be connected together to form one network called ESS (Extended Service Set) identified by its ESSID (Extended Service Set IDentifier). Each BSS may contain about 25 to 50 devices for earlier versions and 100 to more than 200 devices for newer versions. IBSS on the other hand is very limited to 4 to 8 devices in terms of scalability due to complexity and performance limitation.

Figure 14. Wi-Fi Network Topology

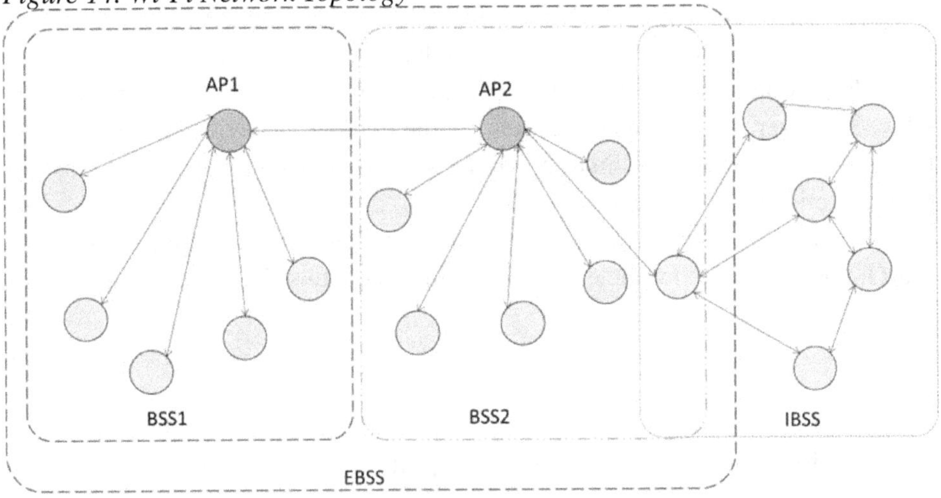

Wi-Fi transmits data in a structured format called frame. The simplified Wi-Fi MAC (Medium Access Control) frame format is described by the figure 15:

Figure 15. Wi-Fi IEEE 802.11 Frame Basic Structure

802.11 Mac Frame Structure

<table>
<tr><td>Frame Control
(2 bytes)</td><td>Duration/ID
(2 bytes)</td><td>Address 1
(6 bytes)</td><td>Address 2
(6 bytes)</td></tr>
<tr><td colspan="2">Address 3
(6 bytes)</td><td>Sequence Control
(2 bytes, optional)</td><td>Address 2
(6 bytes)</td></tr>
<tr><td>QoS Control
(optional, 2 bytes)</td><td>HT Control
(optional, 2 bytes)</td><td>Data
(Variable Length)</td><td>Frame Check Sequence
(4 bytes)</td></tr>
</table>

- Duration/ID : Estimated medium occupation time (or ID if it's a control Frame)
- Address 1 : Destination MAC Address
- Address 2 : Source MAC Address
- Address 3: Destination Access Point MAC in Distribution System
- HT Control: High Throughput Control
- Address 4: Source Access Point MAC or Address BSSID in Distribution System

Wi-Fi employs several mechanisms to ensure data protection including authentication and encryption using WEP (Wired Equivalent Protocol) and WPA (Wi-Fi Protected Access). However, WEP is obsolete and uses a manually fixed 64-bit or 128-bit encryption key whereas WPA is based on TKIP (Temporal Key Integrity Protocol) that employs a 128-bit per-packet key generated dynamically for each packet transmission. WPA uses also another protocol called CCMP (Counter Mode Cipher Block Chaining Message Authentication Code Protocol) to improve confidentiality and authentication using the symmetric cryptographic algorithm AES (Advanced Encryption Standard). Even though it presents some security weakness, Wi-Fi security is actively improved with modern efficient security protocols.

Wi-Fi power consumption is relatively moderate, about 2 to 20 watts-hour (Wh) AP with an average of 6 Wh depending on the number of devices and the traffic density. A client device consumes about 0.5 to 5 Wh depending on the device type and how actively it uses the Wi-Fi.

Wi-Fi is widely used in IoT ecosystem notably in home automation appliances, smart homes, connected vehicles, smart lighting and industrial process automation.

ZigBee

ZigBee is a low-power and low-rate mesh ad hoc network introduced by CSA (Connectivity Standards Alliance, *csa-iot.org*). It's designed for WPANs (Wireless Personal Area Networks) such as sensor networks, home automation, medical device and industrial control system networks and uses AODV (Ad hoc On-Demand Distance Vector) protocol for network packets routing. Like Bluetooth, ZigBee is also based on IEEE 802.15.4 standards and operates in 2.4 GHz by default, 868 MHz (Europe), 915 MHz (United States and Australia), and 784 MHz (China). ZigBee offers a data rate from 20 kbit/s for 868 MHz to 250 Kbit/s for 2.4 GHz in a range of about 10 to 100 m.

Zigbee is based on mesh topology and support both tree and star topology for scalability. It contains three (3) types of devices:

- **Zigbee Coordinator (ZC)**: The central node of a ZigBee network that coordinates operations in a network and acts as a network bridge and a security authority for storing encryption keys. There is one ZC in each ZigBee networks.
- **ZigBee Router (ZR)**: Assures application functions and message delivery from/to end devices.
- **ZigBee End Device (ZED)**: The end device that send or receive data either from/to ZC or ZR.

Figure 16. ZigBee Network Topology

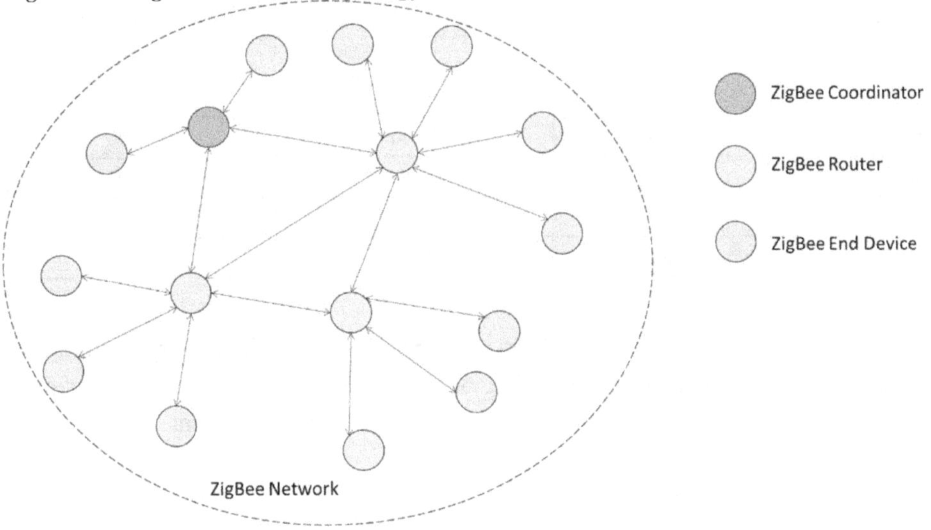

ZigBee protocol stack contains two (2) main components: ZDO (ZigBee Device Object) and APS (Application Support Sublayer). The ZDO manages the overall device and is responsible for security key, policies and defining device role as either coordinator or end device. It also intervenes in new devices and services discovery on the network. As far as the APS, it acts as a bridge between application and network layer. It offers interfaces and services for managing addressing, routing, transmission, reception and security.

Objects in ZigBee network are supervised by ZDO and communicate using APS interfaces. A device can contain up to 240 application objects, identified with numbers from 1 to 240. 0 is reserved for the ZDO data interface, 255 for broadcast and 241 to 254 range are not used. While the theoretical limit for a ZigBee network is 65,535 devices, practical limitations often restrict the number of devices to a much lower number of about 200 to 250 devices.

ZigBee employ is optimized for less power consumption. Typical power consumption of ZigBee devices according to operating modes is described in the table 7.

Table 7. ZigBee Typical Power Consumption

Device Type	Operating Mode	Typical Current (mA)	Typical Power (mW)
ZC	Active (Tx/Rx)	20-50	200-500
ZC	Idle	5-15	50-150
ZR	Active (Tx/Rx)	10-30	100-300
ZR	Sleep	1-5	10-50
ZED	Active (Tx/Rx)	5-15	50-150
ZED	Sleep	0.1-1	1-10

ZigBee uses AES (Advanced Encryption Standard) for data encryption and is based on Network Key, Link Key and devices addresses for authentication using a Trust Center which is responsible for network security management.

Z-Wave

Z-Wave is another wireless technology used primarily for smart building appliance such as home automation, smart lightening and security systems. Z-Wave is introduced by Silicon Labs and its core (MAC/PHY) is standardized by the International Telecommunication Union as ITU 9959 radio. It operates in unlicensed ISM (Industrial, Scientific and Medical) band (868.42 MHz for United States and Canada, 908.42 MHz for Europe) with about 100 Kbps in a range up to 100 m. Data transmission is based on FSK (Frequency-Shift Keying) with Manchester encoding.

A Z-Wave network is based on a mesh topology and has one central node which called controller and supports up to 232 devices. Intermediary devices could be used as repeaters to relay signal to other devices.

Z-Wave LR (Long-Range) is designed as Z-Wave improvements to cover longer distance (up to) in a star topology with up to 4000 devices in one network. It represents an extra 100 Kbps DSSS (Direct-Sequence Spread Spectrum) OQPSK (Offset Quadrature Phase-Shift Keying) modulation addition to the Z-Wave protocol with a maximum output power of 30 dBm optimizing power consumption.

Figure 17. Z-Wave and Z-Wave LR Network Topology

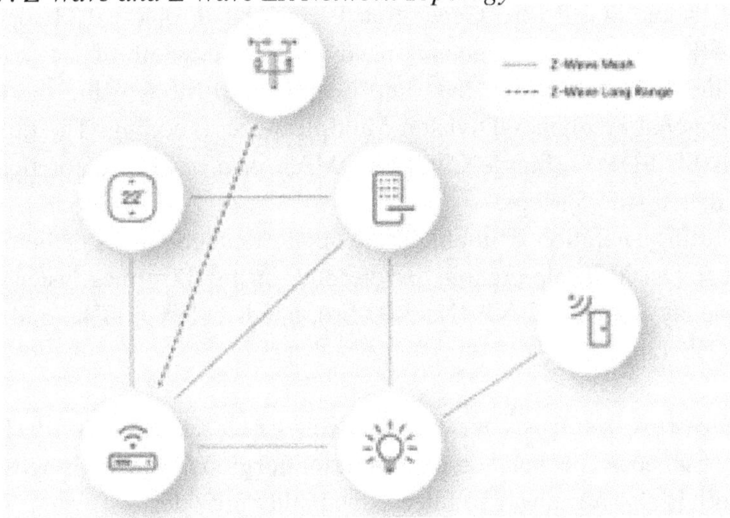

Cellular Networks

Classical Cellular networks are also used by IoT systems. These technologies are GSM (Global System for Mobile)/GPRS (General Packet Radio Service)/EDGE (Enhanced Data-rates for GSM Evolution, 2G), UMTS/HSPA (Universal Mobile Telecommunication System/High-Speed DownLink Packet Access, 3G), LTE (Long-Term Evolution, 4G), 5G:

- **GPRS** (General Packet Radio Service, 2G): An extension of GSM: Allows for packet-based data transfer. It has a theoretical maximum data rate up to 114 Kbps and up to 40 Kbps of practical data rate.
- **EDGE** (Enhanced Data rates for Global Evolution): Is another GSM extension that increases GPRS data rates fourfold and improves the mobility management. The theoretical maximum data rate is up to 384 Kbps (practical data rate is 144 Kbps).
- **UMTS** (Universal Mobile Telecommunication System, 3G): Third generation of mobile networks: It supports voice and data transmission with a data rate up to 144 Kbps (full coverage for mobile use), 384 Kbps (medium coverage for pedestrian use), 2 Mbps (fixed coverage). It operates in 1.88 to 2.025 GHz and 2.11 to 2.2 GHz frequency bands and uses the W-CDMA coding. UMTS is enhanced by HSDPA (High-Speed DownLink Packet Access) which increases data rates to 8-10 Mbps.

- **LTE** (Long Term Evolution): Is the 4th generation of mobile networks defined by the 3GPP (3rd Generation Partnership Project). LTE operates in 450 MHz to 3.8 GHz frequency band with a bandwidth of 1.4 to 20 MHz. The theoretical downlink data rate is up to 300 Mbps using the OFDMA (Orthogonal Frequency-Division Multiple Access) coding. The uplink coding is SC-FDMA (Single Carrier FDMA). And reach a theoretical uplink data date up to 75 Mbps.
- **5G**: Fifth generation of mobile networks: Standards are still being tested. Data Rates: Multiple gigabits per second, up to 100 times faster than 4G. Frequency Bands: 5G NR (New Radio), bands below 6 GHz, and the 24 – 100 GHz range.

A mobile network consists of two (2) networks: radio access network (RAN) for connecting user equipment and core network for the global provider's network. The core network is also a gateway to Internet and other providers.

The general architecture of a cellular network is described in the figure 17.

Figure 18. General Architecture of a Cellular Network

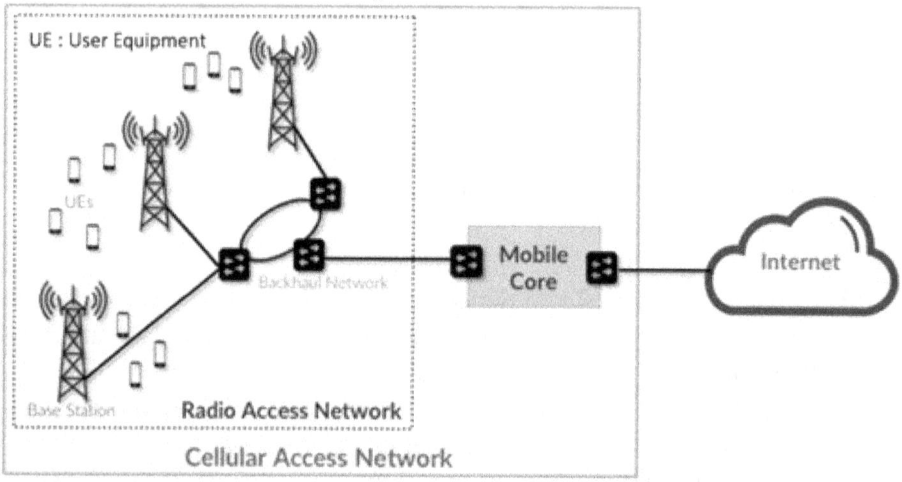

Source: (Larry Peterson & Oguz Sunay, github.com/SystemApproach/5G)

Base Station is known as BTS (Base Transceiver Station)-BSC (Base Station Controller) for GSM, NodeB (NB) for 3G, eNodeB (eNB) for 4G and gNodeB (gNB) for 5G.

Using cellular networks for IoT is part of MTC (Machine Type Communication) and is known as CIoT (Cellular IoT) or MIoT (Mobile IoT) and refers to three main technologies:

- EC-GSM (Enhanced Coverage GSM) for IoT over GSM operating in 900 and 1800 MHz frequency bands with a data rate up to 20 Kbps.
- LTE-M (LTE for Machines) for IoT over LTE using typically 800, 900, 1800 and 2100 MHz frequency bands and 16-QAM (16 Quadrature Amplitude Modulation) or QPSK (Quadrature Phase Shift Keying) modulation with a data rate up to 1 Mbps.
- NB-IoT for IoT over LTE and 5G as independent technology. In 5G ecosystem, MIoT refers also to 5G Massive IoT. New NB-IoT based technology for 5G NR (New Radio) is introduced in 3GPP release 17.

NB-IoT (Narrowband IoT)

NB-IoT (Narrowband IoT, *narrowband.com*) is a 4G/5G technology for IoT introduced by 3GPP as a new radio interface. 3GPP has also introduced different data rates suitable for NB-IoT that range from few tens (10s) of Kbps in 180 KHz bandwidth (LTE Cat-NB1) up to few hundreds (100s) of Kbps (LTE Cat-NB2) and ranges typically in 100 Kbps to 1 Mbps using QPSK (Quadrature Phase Shift Keying) modulation.

NB-IoT is designed to handle a large number of devices (up to 52547) and could work with the 4G LTE as well (Fattah, 2019). It's also optimized for low-power consumption, cost-effective and long-range coverage with efficient data rates. The frequency band for NB-IoT vary depending on the provider and the geographic location. Most used band are Band 3 (1800 MHz), Band 8 (900 MHz), Band 20 (800 MHz) and Band 28 (700 MHz).

Table 8. NB-IoT frequency Bands

Band Number	Uplink frequencies (MHz)	Downlink frequencies (MHz)
1	1920 - 1980	2110 - 2170
2	1850 - 1910	1930 - 1990
3	1710 - 1785	1805 - 1880
5	824 - 849	869 - 894
8	880 - 915	925 - 960
12	699 - 716	729 - 746
13	777 - 787	746 - 756
17	704 - 716	734 - 746
18	815 - 830	860 - 875
19	830 - 845	875 - 890

continued on following page

Table 8. Continued

Band Number	Uplink frequencies (MHz)	Downlink frequencies (MHz)
20	832 - 862	791 - 821
26	814 - 849	859 - 894
28	703 - 748	758 - 803
66	1710 - 1780	2110 - 2200

NB-IoT offers built-in security and authentication functionalities and could be used in many application scenarios like smart city, smart transportation and smart farming.

However, NB-IoT doesn't support D2D communication and QoS (Quality of Service) because data transmitted over NB-IoT are delay sensitive.

Figure 19. NB-IoT Architecture

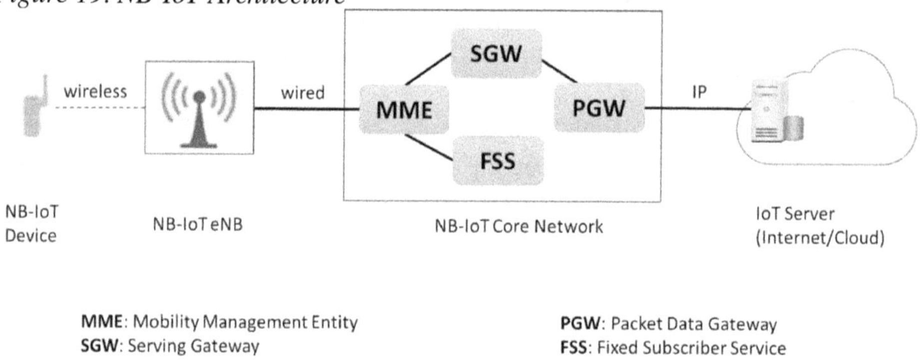

NB-IoT occupies 180 KHz of spectrum band and used resource blocks have 200 KHz bands. 3GPP specifications for NB-IoT introduces three (3) deployment or operation modes:

- **In-Band**: using LTE carrier resource blocks.
- **Guard-Band**: using unused LTE carrier resources.
- **Standalone**: using dedicated spectrum like GSM carriers.

Figure 20. NB-IoT Operation Modes

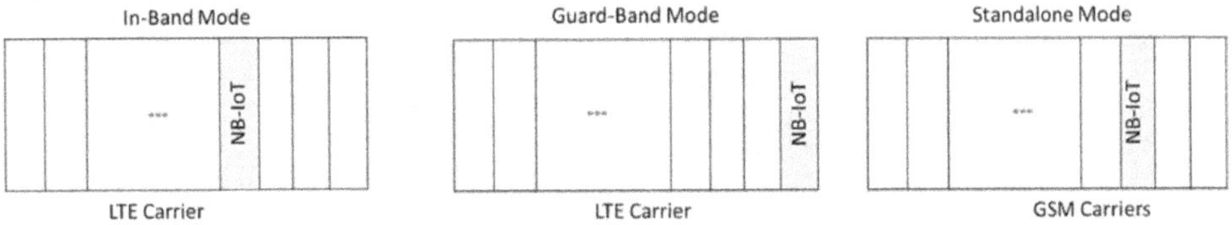

LoRaWAN (Long-Range Wide Area Network)

LoRaWAN (Long-Range Wide Area Network, *lora-alliance.com*) or simply LoRa is a long-range and low-power wireless communication technology designed to support a large number of IoT devices. The term LoRa references the extremely Long-Range data links it offers (up to 5 Km in urban areas and 15 Km in rural areas). Furthermore, it offers high-capacity bandwidth with low-power consumption and requires minimal infrastructure for deployment.

The main components of a typical LoRa network architecture are:

- **End Devices**: An IoT devices (sensor or actuator) with lora-enabled equipment.
- **Gateway**: The Radio Frequency (RF) access point enabling end devices.
- **Network Server**: Server that controls the entire network and allows end-to-end secure connection establishment between all LoRa entities (from end devices to cloud application and vice versa) using IP (Internet Protocol) and 128-bits AES (Advanced Encryption Standard).
- **Join Server**: Server that manages the over-the-air activation process for enabling new end devices access to the network. It determines which application server could be connected to a specific server and maintains encryption keys and other information like: DevEUI (end-device serial unique identifier), AppEUI (application serial unique identifier), AppKey (application encryption key), AppSKey (application session key), NwkKey (network session key), Application Server identifier and End-Device Service Profile. The Network Session Key is communicated to the Network Server and Application Session Key to the Application Server).
- **Application Server**: Server responsible for securely handling and processing end devices application data.

Figure 21. LoRaWan Network Architecture

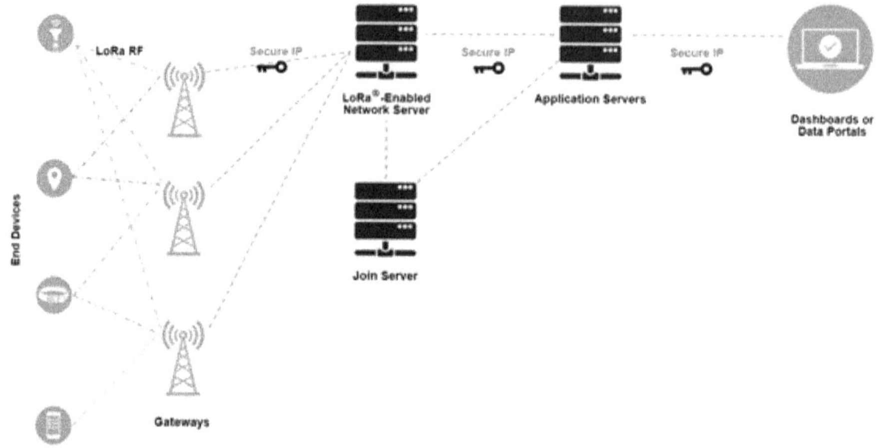

(Source: semtech.com/LoRa)

LoRa uses the unlicensed frequency bands that vary across region and countries. In Asia and Australia for example the frequency band is 923 MHz and is encoded by AS923. LoRa is based on a spread-spectrum modulation proprietary technique that is an optimization of the CSS (Chirp Spread Spectrum) and uses a technique called spreading factor (SF) for data transfer and determines the data rate that varies from SF7 for up to 50 Kbps to SF12 for up to 15.625 Mbps in a range up to 15 Km using 125/500 MHz bands for uplinks channels and 500 MHz for downlinks channels. There are several levels of data rates according to spreading factors (semtech.com/LoRa) as indicated in table 9:

Table 9. LoRa Data Rate Levels

Data Rate (DR)	Spreading Factor (SF)	Channel Frequency (KHz)	Channel (UpLink or DownLink)	Bitrate (Bits/ Sec)	Max User Payload Size (Bytes)
0	SF10	125	UL	980	11
1	SF9	125	UL	1 760	53
2	SF8	125	UL	3 125	125
3	SF7	125	UL	5 470	242
4	SF8	500	UL	12 500	242
5 – 7	-	-	-	-	-
8	SF12	500	DL	980	53

continued on following page

Table 9. Continued

Data Rate (DR)	Spreading Factor (SF)	Channel Frequency (KHz)	Channel (UpLink or DownLink)	Bitrate (Bits/ Sec)	Max User Payload Size (Bytes)
9	SF11	500	DL	1 760	129
10	SF10	500	DL	3 125	242
11	SF9	500	DL	5 470	242
12	SF8	500	DL	12 500	242
13	SF8	500	DL	21 900	242

The range and the latency are also associated to the SF. Table 10 shows range values and TOA (Time-On-Air) variation for an Uplink 11-bytes payload at 125 KHz according to SF, though the range values depend on the terrain:

Table 10. LoRa Ranges Variation

Spreading Factor (SF)	Range (km)	TOA (ms)
SF10	8	371
SF9	6	185
SF8	4	103
SF7	2	61

SigFox

SigFox (*sigfox.com*) is a proprietary LPWAN that uses UNB (Ultra-Narrow-Band) frequency and optimized end devices for less power and resource consumption while achieving long-range distance as LoRa. For that matter, it's based on moderate data rates and delegates complex processing to servers running into cloud infrastructure. SigFox operates in ISM sub-GHz bands (below 1 GHz) and uses BPSK (Binary Phase Shift Keying) modulation for bidirectional half-duplex (Tx then Rx) transmission to achieve a data rate of about 100 bps in a range up to 10 Km. The daily maximum traffic capacity is about 144 messages per device with a 12 bytes payload and 100 KHz bandwidth for uplink channels. The downlink channel capacity is limited at 4 messages per day.

There are four (4) main components in SigFox architecture:

- **End devices**: IoT devices with SigFox-enabled modules.
- **Base Stations**: The gateway to network offering access point for end devices.

- **SigFox Cloud**: The core system of a SigFox network that manages and controls network operations like data processing and message routing from application to end devices and vice versa.
- **IoT platform**: Servers running SigFox IoT applications.

Security mechanisms in SigFox include authentication and encryption using AES (Advanced Encryption Standard) with an option to encrypt the payload and without over-the-air key transmission. It also offers three (3) levels of security:

- Medium Level: End device stores security credentials.
- High Level: A software-based storage is used to store security credentials managed by KMS (Key Management System).
- Very High Level: A secure element can store security credentials.

Figure 22. SigFox Network Architecture

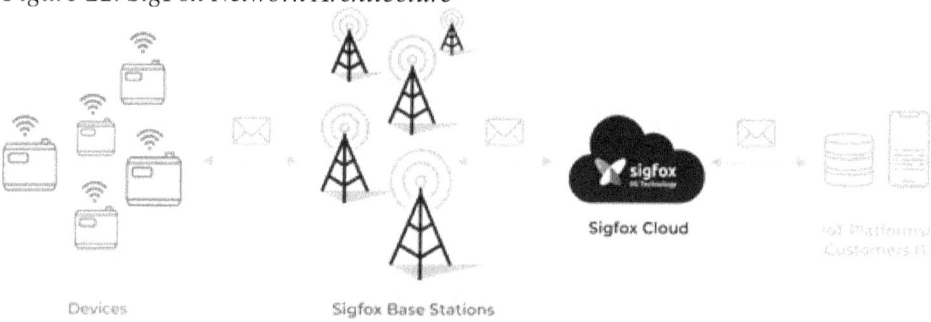

(Source: sigfox.com)

The table 11 (Perry, 2018) illustrated some key characteristics comparing common IoT communication technologies.

Table 11. Comparison of some wireless IoT communication technologies

Technology	Standard	Frequency	Range	Data Rate
Bluetooth	Bluetooth 4.2	2.4 GHz	50-150 m	1 Mbps
HF-RFID / NFC	ISO/IEC 18000-3	13.56 MHz	10 cm	100-420 Kbps
ZigBee	IEEE 802.15.4 (ZigBee 3.0)	2.4 GHz	10-100 m	250 Kbps

continued on following page

Table 11. Continued

Technology	Standard	Frequency	Range	Data Rate
Z-Wave	Z-Wave Alliance ZAD12837/ ITU-T G.9959	868.42 MHz and 908.42 MHz	<100 m	<100 Kbps
Wi-Fi	IEEE 802.11a/b/g/n/ac	2.4 GHz and 5 GHz	~50 m	<1 Gbps
Li-Fi	Relative to IEEE 802.11*	400-800 THz	<10 m	<224 Gbps
GSM/GPRS/EDGE (2G)	3GPP standards	900, 1800, 1900 and 2100 MHz	<200 Km	<500 Kbps
UMTS/HPSA (3G)	3GPP standards	1.88-2.025 GHz and 2.11-2.2 GHz	<200 Km	<2 Mbps
LTE (4G)	3GPP standards	450-3.8 GHz	<200 Km	<10 Mbps
5G	3GPP standards	5.25, 26,4 and 58.68 GHz	<200 Km	<100 Mbps
NB-IoT	3GPP standards	Several licensed and ISM Bands	1-3 Km	100 Kbps to 1 Mbps
LoRa	LoRa-Alliance	ISM Bands	<15 Km	50 Kbps to 15.625 Mbps
SigFox	SigFox-0G Technology	<1 GHz ISM	10 Km	100 bps

Wired Communication

In some use case scenarios, IoT devices needs to be connected using a wired network for security or optimization raisons, notably for D2D communications. This is applicable only for fixed systems like industrial machinery and for internal communication between different modules of one or multiple assembled IoT device systems. GPS (Global Positioning System), Network modules such as Bluetooth or Wi-Fi, Consoles, Screens, Sensors and Actuators are examples of modules that uses internal wired communication.

The most used wired technologies are Ethernet, SPI (Serial Peripheral Interface), UART (Universal Asynchronous Receiver-Transceiver), USB (Universal Serial Bus) and I2C (Inter-Integrated Circuit). USB is particularly a well-known ubiquitous communication standard present in almost all recent electronic devices including IoT devices.

Ethernet

Ethernet is most used wired communication technology in TCP/IP computer networks. It's based on physical cables (coaxial, twisted pairs, optical fiber) and could reach data rates up to 10 Mbps, 1000 Mbps, 10 Gbps or 10 Gbps according to the underlying transmission channel.

Ethernet is standardized as IEEE 802.3 and uses CSMA/CD (Carier Sense Multiple Access with Collision Detection) technique for transmission channel access. This means devices emits on the same physical support and are able to detect interferences allowing signal retransmission and error correction.

I2C (Inter-Integrated Circuit)

I2C is a serial synchronous circuit used for information transmission between modules or systems in IoT devices. It's introduced in 1980 by Philips Semiconductors actual NXP Semiconductors and operates in half-duplex mode for data rates up to 100 Kbps (Standard mode), 400 Kbps (Fast mode), 3.4 Mbps (High-Speed mode), and 5 Mbps (Ultra-Fast mode).

The I2C circuit architecture follows controller-peripheral pattern in which the controller is usually a microcontroller where peripherals are sensors, actuators or others modules/devices. and is composed of two (2) lines: SCL (Serial Clock Line), unidirectional from controller to peripheral, used by controller with a clock for transmission synchronization and SDA (Serial DAta), bidirectional, for data transmission using Big Endian technique.

Figure 23. I2C General Architecture

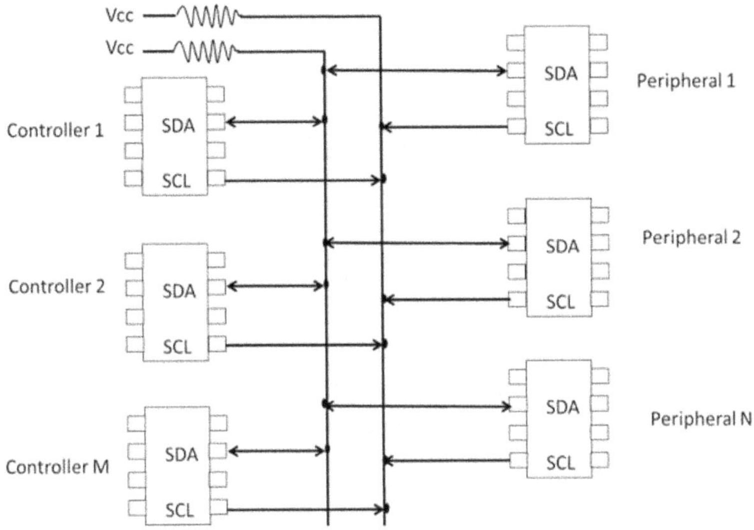

The supported maximal number of controllers is unlimited whereas the one for peripherals is fixed to 1008. However, a controller could manage multiple peripherals and the peripheral could be managed by multiple controllers.

In I2C, information is transferred in structured messages. A message contains several blocks:

- **Start** (1 bit): Indicates the start condition, i.e. the SDA switches from 1 to 0 level voltage then SCL switches from 1 to 0.
- **Address Frame** (7 or 10 bits): Binary sequence identifying each peripheral.
- **Read/Write** (1 bit): Indicates the data transmission (0) from controller to peripheral or controller request (1) for data transmission from peripheral.
- **Data Frame** (8 bits): Represents a data fragment to transfer in a message. A message contains one or more data frames.
- **ACK/NACK** (1 bit): Indicates if a frame (Address/Data Frame) is received successfully with ACK (Acknowledgement, 0) or considered as NACK (No-Acknowledgement, 1) if the peripheral can't respond.
- **Stop** (1 bit): Marks the end condition of message transmission, i.e. the SCL switches from 0 to 1 level voltage then SDA switches from 1 to 0.

Figure 24. I2C Message Structure

More Data Frames

Start (1 bit)	Address Frame (7 or 10 bits)	Read/Write (1 bit)	Data Frame 1 (8 bits)	ACK/NACK (1 bit)	...	Stop (1 bit)

I2C Message

The transmission protocol respects the following rules:

- The controller transmits a start condition to each peripheral;
- The controller transmits the address of the peripheral with which it wants to communicate on SDA (i.e. to each peripheral);
- Each peripheral compares the received address with its address. If it matches, the peripheral sends back an ACK (1);
- The controller begins sending/receiving data from/to the peripheral;
- After each data frame reception by the peripheral, it sends to the controller an ACK if possible;
- When the controller can't receive the ACK, for a data frame, it's considered as not received and retransmitted.
- The controller ends the transmission by sending a stop condition.

SPI (Serial Peripheral Interface)

SPI is another serial communication interface used in major IoT devices. It operates in synchronous half-duplex mode with data rates reaching up to 10 Mbps.

Figure 25. SPI Architecture

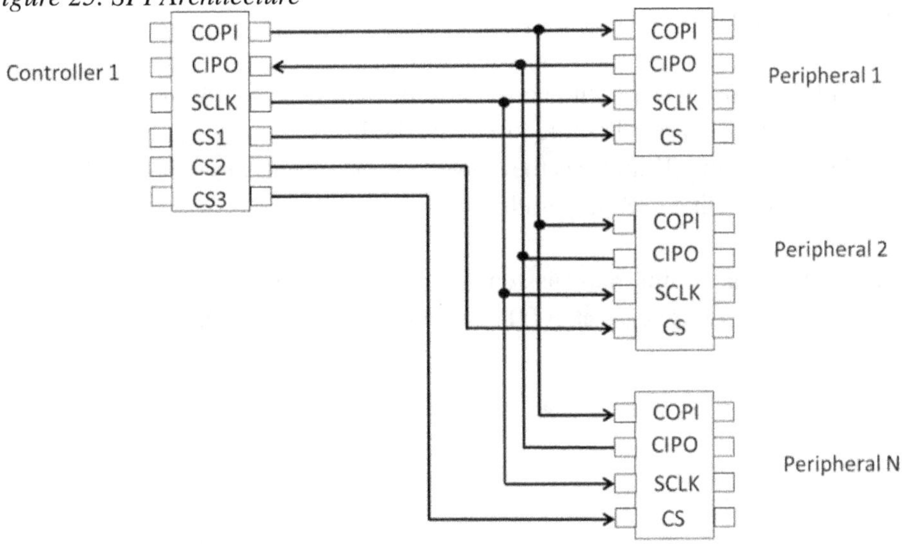

Like I2C, SPI is based on controller-peripheral architecture and supports one controller managing several peripherals. The theorical peripheral maximal number is unbounded, even though fixed in practice. There are four (4) lines in SPI circuit:

- **COPI** (Controller Output / Peripheral Input): Used by controller to send data from controller to peripherals.
- **CIPO** (Controller Input / Peripheral Output): Used by peripherals to send data to the controller.
- **SCLK** (Clock): Used by the controller to synchronize data sampling from peripheral to controller.
- **CS** (Chip Select): Used by the controller to select the peripheral with which it wants to communicates. There are usually as many CS as there are peripherals. However, with one CS, peripherals could be interconnected using daisy-chain links.

SPI transmits continuously data bit by bit as stream without any concept of packet.

UART (Universal Asynchronous Receiver-Transceiver)

UART is a circuit for serial asynchronous simplex/half-duplex/full-duplex communication between two (2) UART modules using two (2) lines: Rx (Reception) line for receiving data and Tx (Transmission) line for sending data. Each UART is connected to microcontroller by an 8-bit parallel bus and performs multiplexing (or demultiplexing) when sending (or receiving) data:

- Multiplexing (Mux) is the conversion of many entries to one entry, in this context, an 8-bit parallel line to 1-bit serial line.
- Demultiplexing (Demux) is the conversion of one entry to many entries, here, a 1-bit serial line to 8-bit parallel line.

The UART maximum transmission speed is 115200 baud (bps: bit per second) and is fixed in most use case at 9600 baud (bps: bit per second).

Figure 26. UART Communication Overview

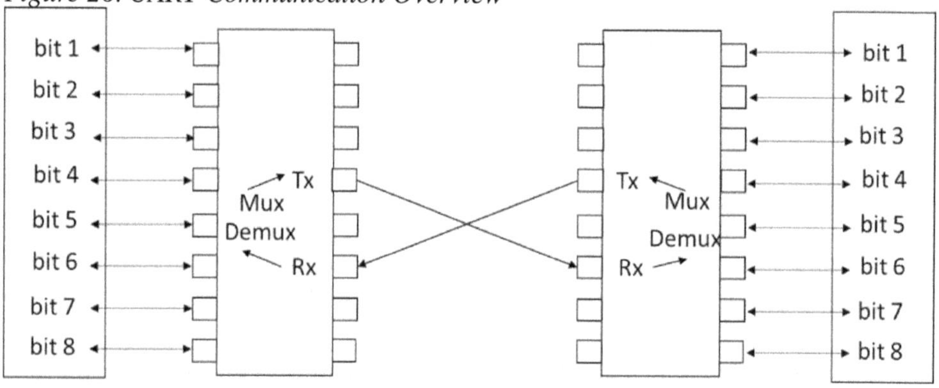

Data transmission is organized using packets. An UART packet is composed of several segments:

- **Start** (1 bit): Indicates transmission beginning, i.e. switching data line from 1 to 0;
- **Data Frame** (5-9 bits): Contains the data to transmit.
- **Parity bits** (0-1 bit): Represents the parity bit allowing to detect transmission error. It's optional, and when used, the maximum data frame size is 8 bits.
- **Stop** (1-2 bits): Indicates the end of data transmission, i.e. switching data line from 0 to 1.

Figure 27. UART Packet Structure

Start (1 bit)	Data Frame (5-9 bits)	Parity Bit (0-1 bit)	Stop (1-2 bits)

UART Packet

COMMUNICATION PROTOCOLS IN IOT SYSTEMS

Overview of Communication Protocols

A part from network communication technologies, protocols are the second most important part of communication. A communication protocol defines a set of rules that should be respected in a communication between two (2) systems or applications. Every layer of a network has its own protocols stacks used to synchronize and secure data transfer.

IoT systems uses Internet protocols for communication and some new specific protocols designed IoT devices communication.

The network models are the TCP/IP (Transmission Control Protocol/Internet Protocol) for Internet and the ISO/OSI (International Standardization Organization / Open System Interconnection) reference model for a general representation of network.

Figure 28. Network Models and IoT Protocols

	ISO/OSI		TCP/IP	Protocols
7	Application	4	Application	HTTP, MQTT, XMPP, CoAP, WebSocket, ...
6	Presentation			
5	Session			
4	Transport	3	Transport	TCP, UDP, SSL, TLS, 6LoWPAN ...
3	Network	2	Network	IPv4, IPv6, RPL, ...
2	Data Link	1	Physical (Network Access)	MAC, Bluetooth, ZigBee, LoRaWAN, LTE-M, NB-IoT...
1	Physical			

Physical layer protocols are related to the network technology as developed in communication technology section (Bluetooth, ZigBee, Z-Wave, Wi-Fi, LoRaWAN, SigFox). Most short-range networks like Bluetooth, ZigBee are based on IEEE (Institute of Electrical and Electronics Engineers) 802.15.4 standard.

Network and transport layers protocols remains likely the same as classic Internet protocols (IPv4, IPv6, TCP, UDP: User Datagram Protocol, SSL: Secure Socket Layer, TLS: Transport Layer Security) and routing protocols. However, newer protocols like RPL (Routing Protocol for Low-Power and Lossy Networks) and 6LoWPAN (IPv6 over Low-Power Wireless Personal Area Network) and ad hoc routing protocols like DSR (Dynamic Source Routing) and AODV (Ad hoc On-Demand Distance Vector) are used for D2D communications.

Application protocols for IoT includes several protocols like HTTP (Hyper-Text Transfer Protocol), CoAP (Constrained Application Protocol), MQTT (Message Queuing Telemetry Transport), XMPP (Extensible Messaging and Presence Protocol).

TCP/IP model related standards and recommendations are defined by IETF (Internet Engineering Task Force) in documents called RFC (Request For Comments) each referenced using an identification number.

IEEE 802.15.4 Standard

IEEE 802.15.4 is a standard for low-rates WPANs (Wireless Personal Area Networks) used in IoT that defines TCP/IP data access (physical) layer specifications using two (2) logical sublayers: MAC (Medium Access Control) and PHY (Physical layers). PHY defines frequency band, modulation scheme, and transmission power whereas MAC defines flow control politics.

Figure 29. IEEE 802.15.4 Standard Overview

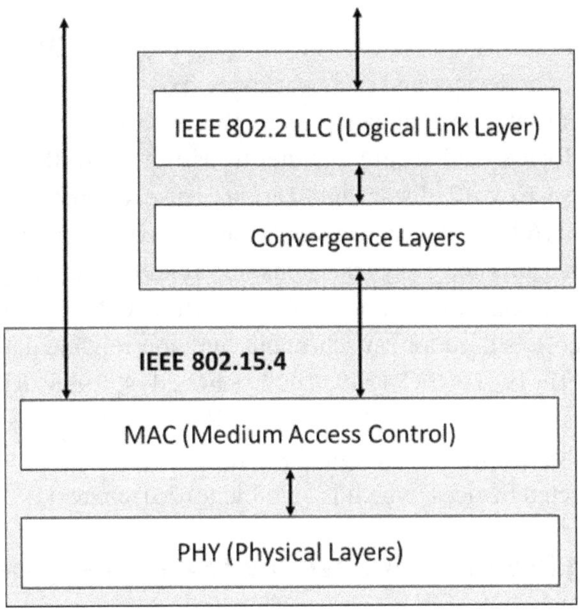

A IEEE 802.15.4 standard based network operates in ISM band with low duty cycle (<1%) and minimum power level of -3 dBm or 0.5 mW in a range of 10-75 m. It uses DSSS (Direct Sequence Spread Spectrum) modulation or Binary Phase Shift Keying (BPSK) for low data-rates versions and O-QPSK (Offset-Quadrature Phase-Shift Keying) for high data-rates versions. The support access is based on CSMA-CA (Carrier Sense Multiple Access with Collision Avoidance) techniques. Figure 29 shows the IEEE 802.15.4 frame structure at PHY and MAC levels. At MAC level, 48-bits physical addresses allow devices identification.

Figure 30. IEEE 802.15.4 Frame structure

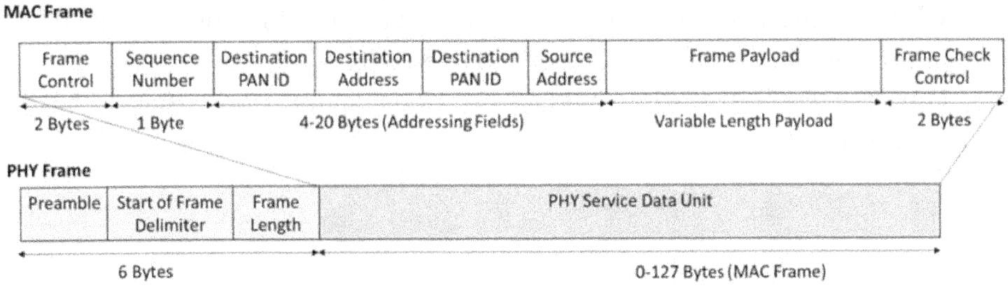

TCP (Transmission Control Protocol) and IP (Internet Protocol)

TCP and IP protocols are the core of the Internet. While IP is used to identify logically computer or devices and route packages, TCP is used to assure end-to-end data transmission from one application to another across the network.

TCP creates logical and secure channel to assure controlled end-to-end data transmission between two (2) distant applications. It uses a combination of IP address and port number (16-bits integer) to identify an application process from a distant machine. TCP transmission is synchronous and reliable including data integrity, flow and congestion controls. In some applications the TCP counterpart, UDP (User Datagram Protocol), is used for asynchronous and non-reliable data transmission.

IPv4 (RFC 791, 1981; RFC 1349, 1992; RFC 2474, 1998; RFC 6864, 2013) for 32-bits addressing and IPv6 (RFC 2460, 1998) 128-bits addressing are two (2) versions of IP. IPv6 is specially designed as an extension of IPv4 to handle a large number of connected devices, which is suitable for IoT context.

RPL (Routing Protocol for Low-Power and Lossy Networks)

RPL (RFC 6550, 2012) is a proactive distance-vector routing protocol designed for wireless low-power networks. It operates particularly on IEEE 802.15.4 networks with a multi-hop and many-to-one or one-to-one communication.

RPL represents network topology as a DAG (Directed Acyclic Graph) called DODAG (Destination Oriented DAG). The sink node or the last destination is the DODAG root and extends progressively the graph till the last farther node.

The RPL principles is to build a global DODAG associating to each node a rank. The father the node is, the highest is its rank. It defines three (3) control messages (RFC 4443, 2006):

- **DIS** (DODAG Information Solicitation): Used to solicit information from other nodes mainly when a node vows to join an existing DODAG.
- **DIO** (DODAG Information Object): Response to DIS containing DODAG information like DODAG ID and NODE RANK.
- **DAO** (DODAG Advertisement Object): Update information from child node to parent node.

Figure 31. RPL Routing Principle

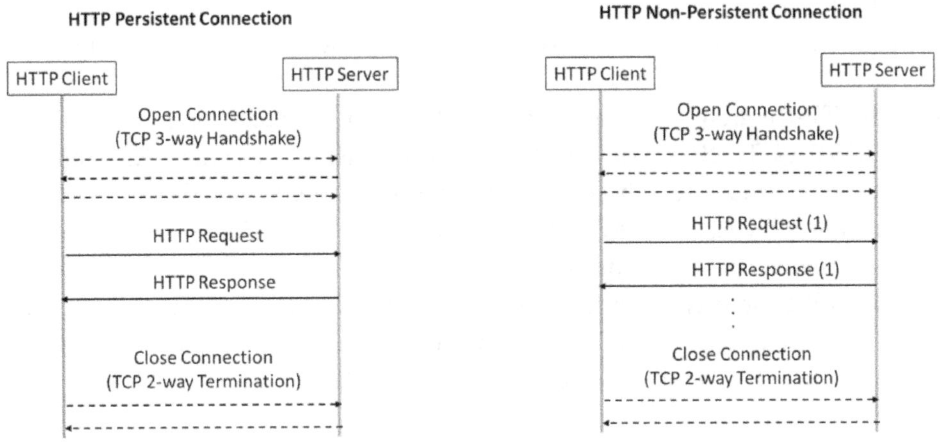

6LoWPAN (IPv6 over Low-Power Wireless Personal Area Network)

6LoWPAN (RFC 4944, 2007; RFC 6282, 2011) is a network level standard that allows IPv6 encapsulation over IEEE 802.15.4 networks. However, IPv6 headers are long (128-bits) and the underlying devices have limited processing capabilities. To achieving this purpose, 6LoWPAN reduces the maximum IPv6 payload from 1024 bytes to 127 bytes. It uses also header compression reducing transmission overhead and fragmentation with a multi-hop delivery support. 6LoWPAN defines four (4) types of frames headers:

- **No 6loWPAN header** (00) to discard frames that doesn't follow 6LoWPAN specifications;
- **Dispatch header** (01) for multicast and IPv6 compression;
- **Mesh header** (10) for broadcasting;
- **Fragmentation header** (11) for fragmenting IPv6 headers into 128-bytes maximum length fragments.

HTTP (Hyper-Text Transfer Protocol)

HTTP (RFC 9110, 2022; RFC 9113, 2022; RFC 9114, 2022) is the most used protocol is the web communication context at application level. It allows data transfer between two (2) remote machines over TCP connection according to a client-server model. For now, there are four (3) main stable versions of HTTP protocol: HTTP/1.0, HTTP/1.1, HTTP/2, HTTP/3.

The communication model of HTTP is based on two (2) types of messages:

- **HTTP-Request**: Message sent by a client that contains a command (GET, POST, PUT, DELETE, OPTIONS, ...) addressing a specific URI (Uniform Resource Identifier), headers (Content-Type, Date, Status, Connection, ...) and in some case a body carrying request data. These data may also be sent via URI in a GET command.
- **HTTP-Response**: Message sent by the server in response to a request. It contains a status code (number between 100 to 599) that indicates the response type, headers and response body carrying the data.

The server sends HTTP-Cookie (key-value data) regularly to the client in order to maintain its information using a HTTP-Session object.

HTTP supports basic authentication and propose new version, HTTPS (HTTP Secure), for security matters. HTTPS rely on transport layer security (SSL/TLS) protocol to secure its communication channel.

The communication principle consists of the following steps:

- Server listens on port 80 for HTTP and 443 for HTTPS and waiting for new connection;
- Connection establishment between server and client on client demand;
- Client sends request;
- Server sends response;
- Server closes connection.

HTTP supports two (2) connection modes: non-persistent and persistent connection. Non-persistent connection is the default mode and allows only one request-response messages. The persistent-mode uses one connection for multiple request-response messages.

Figure 32 . HTTP communication mode

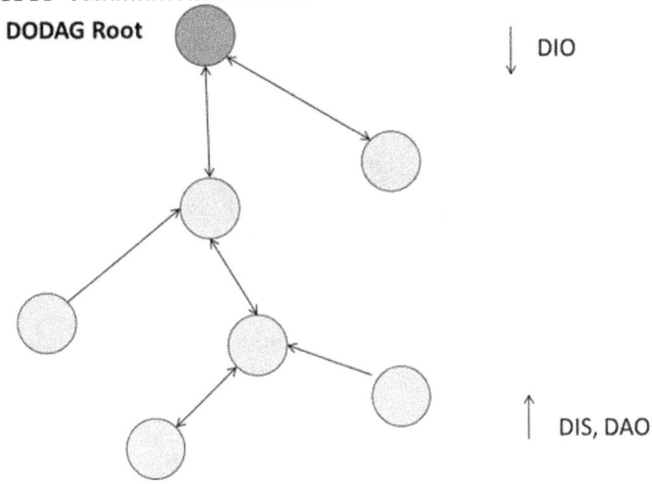

In IoT context, the main applications of HTTP is REST (REpresentational State Transfer) APIs. REST API allows to access web services using HTTP request-response messages. The response is in general a JSON (JavaScript Object Notation) or XML (eXtensible Markup Language).

WebSocket

WebSocket (RFC 6455, 2011) introduced by IETF (Internet Engineering Task Force) is a protocol that works over HTTP and create a full-duplex continuous communication between client and server.

HTTP Upgrade Header is used to switch from HTTP mode to WebSocket mode. It allows to transfer real-time data using HTTP request-response messages using *ws://* or *wss://* as URI scheme.

Figure 33. WebSocket communication mode

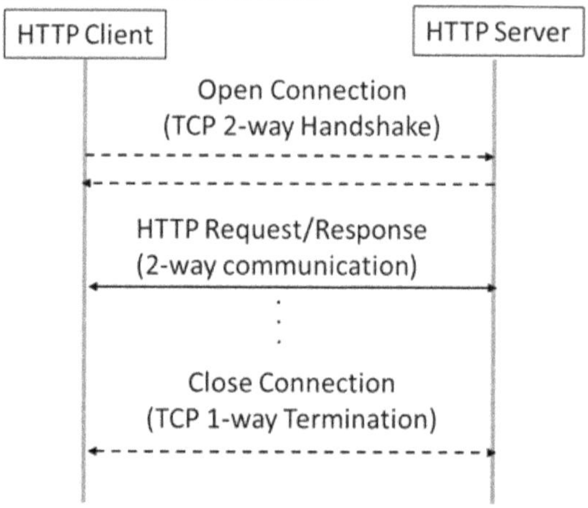

CoAP (Constrained Application Protocol)

CoAP (RFC 7252, 2014) is another UDP-based and simplified HTTP-like protocol designed for RESTful environments in networks with resource limitations (low-power, low-bandwidth, …) like IoT devices networks for both D2D and D2M communications. It's designed to overcome 6LoWPANs networks that have high packet error rates.

It offers a lightweight RESTful standard interface between servers and clients, and is easily interoperable with HTTP. At transport level, CoAP uses UDP and DTLS (Datagram Transport Layer Security) to enable secure communication.

CoAP uses the same request/response messaging model as HTPP and has four (4) message types identified by a 2-bit code:

- **Confirmable Message** (CON, 00): Reliable request/response with acknowledgement;
- **Non-confirmable Message** (NON, 01): Non-reliable request with no ACK;
- **Acknowledgement Message** (ACK, 10): Acknowlegment (ACK) response to a confirmable request/response;
- **Reset Message** (RST, 11): Indicates that a request is received with no response.

There are four (4) CoAP message delivery scenarios based on these message types: Non reliable message, Reliable message, Piggybacked responses (responses with ACK) and Separate Response as described in figure 33.

Figure 34. CoAP Message Delivery Types

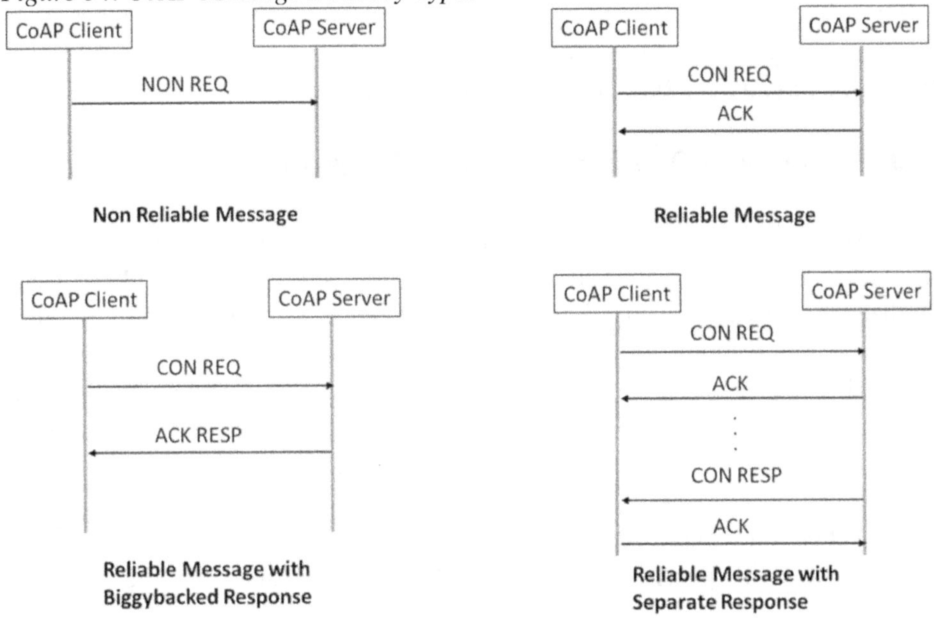

CoAP supports multicast request and offers reliability supports despite it uses a non-reliable transport protocol, UDP. It supports four (4) security levels:

- **NoSec**: No security support, i.e. DTLS is disabled.
- **PreSharedKey**: DTLS with pre-shared keys including accessible device list.
- **RawPublicKey**: DTLS with asymmetric key pair without certificate.
- **Certificate**: DTLS with PKI (Public Key Infrastructure) X.509 (asymmetric key pair and certificate).

XMPP (Extensible Messaging and Presence Protocol)

XMPP (RFC 6120, 2011) is an open standard XML-based communication protocol designed primarily for instant messaging and presence information using TCP or via HTTP in some use cases.

It allows secure real-time data transfer and sharing and support a distributed environment like IoT systems using and TLS for encryption and SASL (Simple Authentication and Security Layer) for authentication.

In XMPP, each user/device has an email-like (*username@domain.com*) unique ID called JID and multiple locations could be created by adding a repertory to the JID (like *username@domain.com/resource*).

XMPP is based on request-response and publish-subscribe messaging model with an extensible XML-format message but doesn't guaranty QoS (Quality of Service). There are many XMPP message extension standards called XEP (XMPP Extension Protocol) defined by XSF (XMPP Standard Foundation).

MQTT (Message Queuing Telemetry Transport)

MQTT (*mqtt.org*) is a lightweight open protocol optimized for networks constrained of resource limitations like IoT networks introduced by IBM in 1999, standardized by OASIS (Organization for the Advancement of Structured Information Standards) and recommended by ISO consortium.

MQTT is based on publish-subscribe messaging model based on architecture and a reliable 2-way transport protocol typically TCP. Its architecture has two (2) types of nodes:

- **Broker**: The central server that receive store and dispatch messages.
- **Client**: Either publisher (producer), subscriber (consumer) or both.

Figure 35. MQTT Architecture

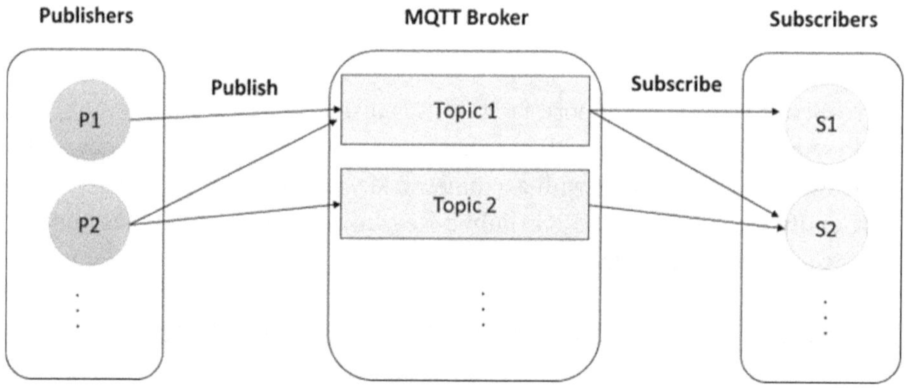

MQTT workflow consists of a number of steps:

- A device (client) publishes messages to broker for a specific topic;
- Another device (client) subscribes to the topic
- The broker publishes messages to all subscribers as far as messages are published in topics.

There are three (3) levels of QoS in MQTT connections: Fire and Forget (At most once: message is sent once) as QoS 0, Acknowledged Delivery (at least once: message is retransmitted until ACK reception) as QoS 1 and Assured Delivery (exactly once: 2-way handshake allows to send-receive only one copy of message) as QoS 2. MQTT has TLS (Transport Layer Security) encryption with certificate and connection authentication (username, password) security supports.

Figure 36 . MQTT QoS 2 Workflow Example

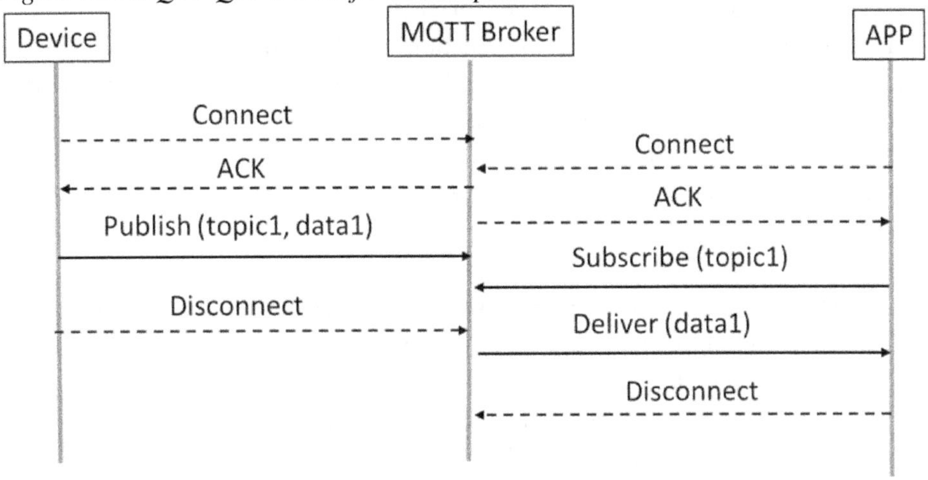

AMQP (*amqp.org*) is another OASIS open standard message-oriented protocol more complex than MQTT. It's designed for Enterprise/Industrial-Level applications and is used for IoT complex applications as well.

Table 12 . Comparison of some IoT Protocols

	WebSocket	**CoAP**	**XMPP**	**MQTT**
Messaging	Req-Resp/ Pub-Sub	Req-Resp	Pub-Sub	Pub-Sub
Transport	TCP	UDP	TCP	TCP
Security	SSL/TLS	DTLS	SSL	SSL
QoS	Low	Moderate	High	High
Scalability	Moderate-High	High	Moderate	Very High
Power	Moderate	Very-Low	Moderate	Very Low
Use Cases	Web applications, real-time communication	IoT devices, sensor networks	Chat, messaging, presence	IoT devices, industrial automation

SECURITY IOT ENVIRONMENTS

Cybersecurity Principles and Standards

Cybersecurity is an interdisciplinary domain which aims to apply technologies, processes, practices and measures to protect electronic data/information (from corruption, unauthorized access or exfiltration), secure computer systems and networks (from attacks, compromission, unauthorized access) by guarantying quality of service (QoS) and ensuring end users privacy.

This domain implies securing information in all its forms and places, i.e. in acquisition, storage, processing and transmission. It hence involves many other domains like business management, computer science security, behavioral and ethics studies and low enforcement.

Securing an information system in electronic environment requires the following operations:

- Identify and secure information sources and service endpoints in the system;
- Identify and authenticate all entities trying to access to an information or service in the system;
- Allow only entities authorized to access to a given information or service using access politics and strategies;
- Supervise and control entities operations when accessing information or using services to prevent unauthorized access and/or operations;
- Securing information acquisition, storage, processing on computers and its transmission through the computer networks using encryption.
- Ensure confidentiality and personal data protection to preserve user privacy.
- Identify vulnerabilities and threats to anticipate, manage and mitigate eventual risks.

In simple terms, it's about protecting information, securing systems and applications on computers/end-devices and securing systems and services on network as well as the underlying physical infrastructures (William, 2023).

Cybersecurity Fundamentals

Security fundamental attributes in digital environments are confidentiality, integrity and availability. However, it has others principles as described below:

- **Integrity**: Assures the quality and accuracy of data. Data shouldn't be corrupted or altered during storage and/or transmission. Many techniques such as cryptographic techniques like checksums, hashing and digital signature are used to verify data integrity.

- **Authentication**: Involves identity verification of users, applications and/or devices accessing a system. MFA (Multi-Factor Authentication), ZTA (Zero-Trust Architecture), PKI (Public Key Infrastructure) are some solutions allowing to authenticate entities in a system.
- **Authorization**: Consist to ensure that only authorized users, applications and/or devices have access to a specific resource (data or service) in a system. AC (Access Control) principles and policies are employed in order to guaranty only authorized entities have access to the right resource at the right moment and in the right context.
- **Confidentiality**: Implies protecting sensitive information is accessible only to authorized users, applications, systems or devices namely when in transit through the network. Authentication and authorization principles combined information encryption solutions are employed to assure confidentiality.
- **Privacy**: Involves users' sensitive data protection against unauthorized access, manipulation and misuse. Privacy solutions involve authentication, authorization, and confidentiality principles and law enforcement.
- **Non-Repudiation**: Consists of identifying the origin of any action in a system to prevent denial of action. This fact is assured using accountability principles that includes logging and activities monitoring.
- **Availability**: Corresponds to the fact of ensuring data and services are accessible at the right time when needed. Techniques such as redundancy, scalability, interoperability, failover mechanisms, disaster recovery policies, fault-tolerance applications and systems, and infrastructure security allow availability of system functions.
- **Resilience**: Represents the ability of a system to identify, manage, defend and mitigate risks, threats, vulnerabilities, attacks and failures for a better recovery. Resilience assurance involve using many security mechanisms such as dynamic security policies, risk management, firewalls, IDS (Intrusion Detection Systems), defense strategy, and incident response policies.

Cybersecurity Techniques

Many techniques are employed to secure computer systems, networks and information:

- **Encryption**: Process that consists to transform information structure in order to protect it from undesirable access. The inverse operation is called decryption, and allows to find the original information. An encryption/decryption system (or cryptosystem) is composed of the following elements:

- An information X or the original message called *PlainText*;
- A cryptogram Y or the encrypted message called *CypherText*;
- A generated encryption/decryption key K. Some systems have two (2) keys K for encryption and K' for decryption and some have no key;
- An encryption function or algorithm $D(x,y)$ such that $Y = E(X, K)$.
- A decryption function or algorithm $E(x, y)$ such that $X = D(Y, K')$.

If $K = K$', then the algorithm is said symmetric and the key is shared between the sender and the receiver, else, it's said asymmetric, K key is public and the K' is private. If there is no key, the original message can't be found back anymore, the algorithm is then unidirectional and called hashing algorithm and has the same output hash size.

The most common symmetric encryption algorithms are DES (Digital/Data Encryption Standard) with 56-bit key, 3DES (Triple DES) with 112 and 168-bit keys, and AES (Advanced Encryption Standard) with 128, 192 and 256-bit keys.

The most common asymmetric encryption algorithms are RSA (Rivest-Shamir-Adleman), DSS (Digital Signature Standard).

The most common hashing algorithms are MD5 (Message Digest 5) and SHA (Secure Hash Algorithm).

Encryption is applied in several levels to secure an information system:

- ***Shared Key Exchange***: Using asymmetric algorithm to shared symmetric key algorithm.
- ***Digital Signature***: Using hashing or asymmetric algorithm to authenticate a message.
- ***Public Key Infrastructure***: System that authenticate users and manages public key distribution with a digital signature called digital certificate.
    - **Authentication**: Identifying entities in system is based on three (3) elements: either something the entity properties such as biometrics, or something it possesses physically like key, electronic cards or something knows like passwords, response to specific questions. Combining those elements together enforce authentication efficiency and is known as MFA (Multi-Factor Authentication). ZTA (Zero Trust Architecture) defines mechanisms to always authenticate users in several level of system access.
    - **Access Control Principles**: Ensure efficient authorization in a system, i.e. actions that an entity can perform and data that is can access:

- *Least privileges*: Granting the minimum privileges necessary to perform a specific task;
- *Separation of Duties*: Dividing critical tasks among multiple individuals to prevent fraud and errors.
- *Defense in Depth*: Concept: Implementing multiple layers of security controls to protect resources.
- *Audit*: Tracing and logging activities for investigation on security incidents and compliance monitoring.

Access control principles are implemented through various mechanisms, including:

- *Access Control Lists (ACLs)*: Define which users or groups have access to specific resources.
- *Role-Based Access Control (RBAC)*: Assigns permissions based on a user's role within an organization.
- *Attribute-Based Access Control (ABAC)*: Uses attributes of users, resources, and environments to determine access.
    - **Firewalls**: Systems that analyzing network traffic and filter it according to a predefined policies and rules for outgoing and incoming traffic.
    - **Intrusion Detection System (IDS)**: Systems that analyzing network traffic to continually identifying eventual threats against the network. It's often based on Machine Leaning applications.

Cybersecurity Frameworks

Several Cybersecurity Frameworks (CSF) or Control Frameworks (CF) are introduced to provide necessary controls to protect information from cyber threats. These frameworks are coupled to Risk Management Frameworks as enforcement in order to reduce and mitigate risks (Anne et al., 2016):

- **NIST** (National Institute of Standardizations and Technology) CSF/RMF: Provides IT (Information Technology) security standards with risk management for information systems.
- **ISO** (International Standardization Organization) CSF/RMF: Provides an international standard for ISMS (Information Security Management System) ISO/IEC 27001 and for organization governance ISO 31000.
- **COBIT** (Control Objectives for Information and Related Technologies) CSF: Provides process and practices for designing, managing and improving IT governance with business goals alignment.

- **CIS-CSC** (Center of Internet Security / Critical Security Controls) CSF: Provides guidelines and best practices for information and IT systems security.
- **MITRE CSF**: Provides two (2) CSF with necessary tools for understanding threats and securing information systems MITRE ATT&KS (Adversarial Tactics, Techniques and Common Knowledge) and MITRE ATLAS (Adversarial Tactics, Language for AI Systems) specifically designed for AI security.
- **COSO** (Committee of Sponsoring Organizations) RMF: Provides a Framework for financial operations and COSO ERM (Enterprise Risk Management).
- **FAIR** (Factor Analysis of Information Risk) CSF: Quantitative risk analysis for understanding and evaluating information risk.

A risk corresponds to the potential damage that could be the consequences of an exploitation. It has four (4) factors: vulnerability, threat, likelihood and impact:

- **Vulnerability**: Weakness of a system that could be used as a vector of exploitation. Misconfiguration of security parameters or using weak passwords, outdated software or unsecure protocol are examples of vulnerabilities.
- **Threat**: Potential danger that could exploit a vulnerability to harm a system. Cyber-attacks, Insider misuse, Human errors and natural disaster are examples of threats.
- **Likelihood**: Probability of an exploitation to take place.
- **Impact**: The damage or consequences caused by an exploitation, it's the realization of a risk. A data breach, operational disruption and financial lost are examples of impacts.

There are two (2) categories of cyberattacks in terms of their nature: passive or active. Passive attacks involve a hacker eavesdropping or spying on network in order to gain discretely sensitive information without altering or damaging it. On the other hand, active attacks imply attacker trying to disrupt or intrude in a system in order to gain unauthorized access, manipulate data, cause damage or preventing service availability (William, 2023). The common cyber-attacks methods are:

- **Malwares**: Malicious programs (viruses, worms, trojan horses, ransomwares, spywares) hidden somewhere in a system to perform a cyberattack.
- **Phishing**: Tactics employing technical and social engineering skills in order rip off, manipulate or gain sensitive information from individuals.

- **MITM** (Man-In-The-Middle): Intercepting communication between two (2) systems to eavesdrop or manipulate information.
- **DoS** (Denial of Service): Disrupting a system with inutile traffic to disrupt its services or make it unavailable to legitimate users. It's known as DDoS (Distributed Denial of Service) when many compromised systems may be employed for that matter.
- **SQL Injection**: Injecting SQL (Structured Query Language) query from web application endpoints to gain unauthorized access to database.
- **XSS** (Cross-Site Scripting): Injecting malicious code in web scripts allowing to stole data from HTTP sessions or cookies.

IoT Security: Challenges and Threats

IoT, as its name suggests, use the Internet as its backbone. It's an aggregation of several networks IoT devices connected to multiple servers running applications and services in the cloud, fog or edge environments. In addition, these IoT services are in the most use cases AI-enabled systems and usually consumed by others computers (Laptop, Tablets, Smartphones, etc.) also connected to the Internet.

However, apart from protecting IoT devices, securing IoT systems involves Internet security, network security, cloud security and application security with more complexity. All threats appliable to IT systems could also be applied to IoT systems with a larger attack surface and risks due to its direct proximity with vital world environments like healthcare, transportation and industrial systems.

ENISA (European Union Agency for Network and Information Security) proposes a baseline security recommendation for IoT with a set of issues in the development of secure IoT ecosystems among them:

- **Very large attack surfaces**: IoT systems presents a security high risk with so many points of vulnerabilities that could be exploited to compromise data or services.
- **Limited device resources**: IoT devices are typically constrained devices, with limited power supply, memory, processing capabilities making it difficult to apply advanced security controls.
- **Complex ecosystem**: IoT involves a large number of devices, interconnections, communications and dependencies of various systems increasing the complexity of risk management tasks.
- **Fragmentation of standards and regulations**: Lack of efficient standards and regulations for security guidance in IoT environments.

- **Widespread deployment, Low cost**: IoT deployment is rapidly growing in commercial and industrial systems due to its low cost despite the lack of expertise like comprehensive risk assessment and security planning.
- **Security integration complexity**: IoT devices use wide variety of communications technologies and protocols complexing security integration solutions and security updates.
- **Safety aspects**: Security threats in IoT can become safety threats because IoT devices interact with physical environment, raising the need for high efficiency in security solutions.

IoT Security Solutions

Security solutions in IoT systems focus in securing IoT devices, networks, gateway and applications. NISTIR 8200 (Interagency Report on Status of International Cybersecurity Standardization for the Internet of Things) lists the following security objectives for IoT:

- Logical access restriction to the IoT network using for example unidirectional gateways, firewalls and isolated authentication mechanisms and credentials for users;
- Physical access restriction to IoT network and components using combination of physical access controls such as locks, card readers and/or guards;
- Protecting individual IoT components from exploitation by deploying security patches, disabling all unused ports and services, applying security controls using antivirus software and file integrity checking, restricting IoT user privileges with PLP (Principle of Least Privilege), etc.
- Preventing unauthorized modification of data in transit, in processing and in storage;
- Detecting security events and incidents early as they occur to break the attack chain before eventual exploitation. Mechanisms of tamper resistance and detection could be applied for that matter. A tamper corresponds to each unauthorized action that could lead to system function or data alteration or compromission;
- Maintaining availability and functionality during critical conditions by increasing fault tolerance and robustness using backups, redundancy and replication;
- Restoring the system after an incident by establishing an incident response plan for better recovery.

IoT Device Security

There are two (2) types mains types of IoT devices in terms of resources performance and capacity: constrained and unconstrained devices. Constrained devices have less performance, less supplied with power and less intelligent whereas unconstrained devices are more intelligent with sufficient resource capabilities.

The constrained character of some IoT devices make it complicated to implement complex security protocols and operations. However, at the gateway level protocols like SSL, TLS/DTLS and IPsec are used for security issues. In this scenario, security operations may be delegated to the gateway at the edge or fog level on behalf of constrained IoT devices by performing protocol conversion operations.

IoT devices security involves implementing a robust firmware with identification and authentication high capabilities and regular updates to prevent local threats.

Figure 37. IoT Devices Security Overview

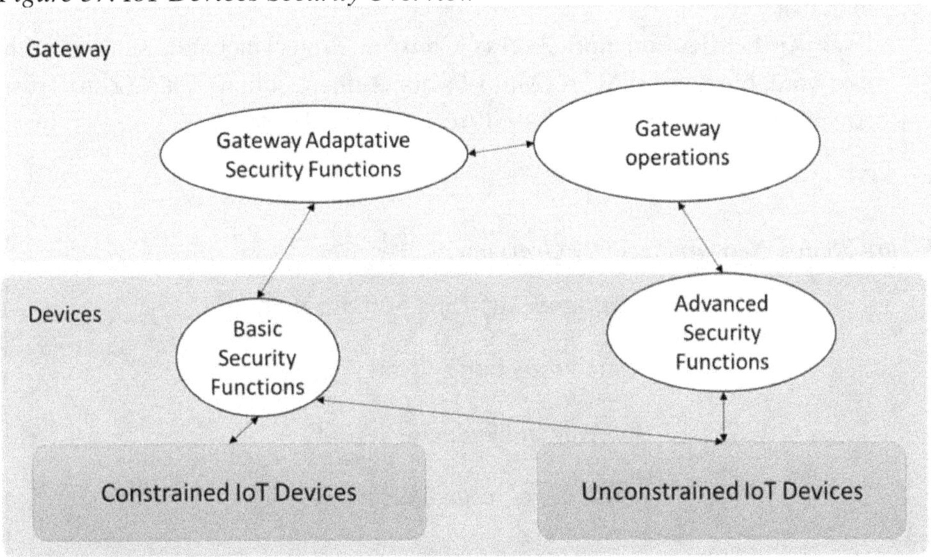

IoT Network Security

IoT network is composed of aggregation of device network, edge network, fog network, core network and cloud network. IoT network security involves the following operations:

- **Network Access Control**: Securing physical infrastructure and transmission channel namely for wireless networks with mechanisms like CSMA/CA and secure point-to-point protocols.
- **Network Protocol Security**: Using secure protocols for packet routing and transport like IPsec, SSL, TLS/DTLS.
- **Gateway and Endpoint Security**: IoT network security involves also securing its neuralgy point. IoT gateway is the central hub that acts as a bridge between IoT devices and the rest of the IoT ecosystem namely cloud services. Network services are accessed from endpoints The common security solutions are Firewalls, DMS (Demilitarized Zone), IDS (Intrusion Detection Systems) and proxies.
- **Data Protection and Confidentiality**: Using secure protocols and advanced cryptographic algorithms and security mechanisms like AES (Advanced Encryption Standard), RSA (Rivest-Shamir-Adleman), SHA (Secure Hash Algorithm), PKI (Public Key Infrastructure) and SSH (Secure Shell) for data in transit.
- **User Authentication and Access Control**: Protect network service with password, biometrics, MFA (Multi-Factor Authentication), ZTA (Zero Trust Architecture), RBAC (Role-Based Access Control), etc.

Figure 38. IoT Network Security Overview

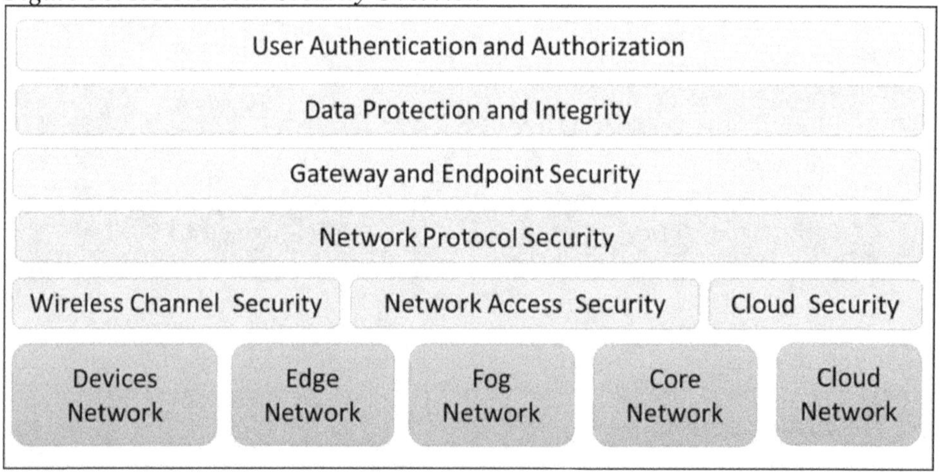

IoT Platform Security

IoT platform corresponds to software services running on top level of IoT layers stack. These services include processing services, end user applications and business analytic applications. The main security mechanisms for securing IoT platform are the following:

- User Authentication and Access Control as in network security;
- Data Protection and Confidentiality as in network security;
- Secure development practices using software engineering methods, AI and Big Data engineering principles and DevOps best practices to reduce vulnerabilities and secure application endpoints;
- Security auditing to assure security specifications are implemented.

IoT platform operations arise some privacy issues notably when it comes to AI and analytic applications. However, collected sensitive data could be used for analysis purpose and need to be protected against misuses and manipulation. In this case, sensitive data masking and cybersecurity law enforcement should be implemented to guaranty privacy and confidentiality in use data usage.

PRIVACY AND ETHICAL CONCERNS IN IOT ECOSYSTEM

With its promise of interconnected devices and automated processes, IoT presents a complex landscape of privacy and ethical concerns. The pervasive collection of personal data by IoT devices, including location, habits, and preferences, raises anxieties about data security and potential misuse. Security vulnerabilities inherent in many IoT devices increase the risk of data breaches, exposing sensitive information to malicious actors. Furthermore, a lack of transparency regarding data collection practices and limited user control over data sharing erode trust and raise concerns about surveillance and behavioral monitoring. The potential for algorithmic bias and discriminatory outcomes, coupled with unclear accountability for data breaches and malicious device use, necessitates a proactive approach to mitigating these risks. Developing robust security standards, promoting user-centric design, and establishing clear ethical guidelines are crucial for fostering a responsible and beneficial IoT ecosystem.

Addressing privacy and ethical concerns in the IoT requires a multi-pronged approach. This includes prioritizing security and privacy by design through robust standards and encryption, empowering users with control over their data and transparency into its use, promoting ethical AI development with bias mitigation and explainability, establishing clear regulatory frameworks for accountability and liability, and improving public awareness and education. By combining technological advancements, regulatory measures, and a user-centric design philosophy, we can pave the way for a responsible and beneficial IoT future that respects individual privacy and promotes societal well-being.

CONCLUSION

In this work we discussed the IoT global concepts and its underlying associated technologies and paradigms like embedded systems and their architecture. An in-depth overview communication principles, technologies and protocols according to ISO/OSI and TCP/IP models were provided with a focus on available standards for an efficient IoT systems deployment. Security threats challenges and IoT system implementations were also covered in this chapter with sufficient details allowing a good understanding of security controls and risk management in the overall IoT ecosystem.

REFERENCES

Anne, K., Dan, S., & Ken, S. (2016). *The Complete Guide to Cybersecurity Risks and Controls*. CRC Press.

Bhawani, S. C., Faisal, K. S., & Naeem, A. M. (Eds.). (2019). *IoT Architectures, Models, and Platforms for Smart City Applications*. IGI Global ISBN 9781799812548, ISSN 2327-039X, eISSN 2327-0403.

Fattah, H. (2019). *5G LTE Narrowband Internet of Things*. NB-IoT.

Hu, F. (Ed.). (2016). *Security and Privacy in Internet of Things (IoTs): Models, Algorithms, and Implementations*. CRC Press. DOI: 10.1201/b19516

Mohammad, S. A., Shamim, K. T., & Nilanjan, D. M. (Eds.). (2023). *Proceedings of the 2nd International Conference on Big Data, IoT and Machine Learning*. 867. Springer DOI: 10.1007/978-981-99-8937-9

Perry, X. (2018). *Designing Embedded Systems and the Internet of Things (IoT) with the ARM® Mbed*. Wiley. DOI: 10.1002/9781119364009

Shrey, S. (2024). *Mastering IoT for Industrial Environments: Unlock the IoT Landscape for Industrial Environments with Industry 4.0, Covering Architecture, Protocols like MQTT, and Advancements with ESP-IDF*. Orange Education Pvt Ltd, AVA™ ISBN 978-81-97081-97-2.

William, S. (2023). *Cryptography and network Security: Principles and Practice, eighth edition, global edition,* 22-42, 534-742. Pearson Education Limited, ISBN 10: 1-292-43748-0.

ADDITIONAL READING

Alex, K., Vugar, A., Vladimir, H., & Vrushank, S. (Eds.). (2024). *Advanced IoT Technologies and Applications in the Industry 4.0 Digital Economy*. CRC Press., DOI: 10.1201/9781003434269

Amit, K. T. (2022). *Internet of Things Theory and Practice: Build Smarter Projects to Explore the IoT Architecture and Applications*. BDP Online.

Anand, T. (2022). *Build Your Own IoT Platform: Develop a Flexible and Scalable Internet of Things Platform Second Edition*. APress DOI: 10.1007/978-1-4842-8073-7

Gnana Swathika, O. V., Karthikeyan, K., & Sanjeevikumar, P. (Eds.). (2024). *IoT and Analytics in Renewable Energy Systems (Volume 1): Sustainable Smart Grids & Renewable Energy Systems*. CRC Press DOI: DOI: 10.1201/9781003331117

Kamlesh, L., Hemant, K. G., Joseph, K. W., & Kamal, K. H. (2020). *Internet of Things (IoT): Principles, Paradigms and Applications of IoT*. BPB Publications.

Nidhi, G. (Ed.). (2024). *Applications of Optimization and Machine Learning in Image Processing and IoT*. CRC Press.

Srikanta, P., Ishwar, K. S., & Xiaolong, L. (Eds.). (2016). *Internet of Things (IoT) in 5G Mobile Technologies: Modeling and Optimization in Science and Technologies* (Vol. 8). Springer., DOI: 10.1007/978-3-319-30913-2

Tripathy, B. K., & Anuradha, J. (Eds.). (2018). *Internet of Things (IoT): Technologies, Applications, Challenges, and Solutions*. CRC Press.

KEY TERMS AND DEFINITIONS

Computer Networks: Interconnected electronic devices (computer especially) that communicate together to share resources and provide services to end-users. Interconnexions between these computers are physical (wired via cables or wireless via radio waves).

Communication Protocols: Set of communication rules between computers allowing to synchronize and secure communication between devices, systems and/or applications.

Distributed Computer Systems: Collection of interconnected hardware and software components that cooperate together to acquire, store and process data in order to achieve a specific goal. Internet is an example of computer network and system.

Cloud Computing: Distributed computer system paradigm that provides computation services (network, data processing and applications) from datacenters to several end-user devices through a computer network (like Internet) using virtualization techniques. In a cloud, computation resources are virtualized as virtual machines, virtual networks.

Edge Computing: Decentralized distributed computer system paradigm that aims to bring computation resources closer source of data (i.e devices that acquire/produce information) instead of transferring the processing to the cloud.

Embedded Systems: Specialized computer systems that are integrated to a lager system (electronic and/or electromechanical) to perform a specific function using sensors for data acquisition and actuators for physical actions.

Internet of Things (IoT): Network of interconnected embedded system devices called objects (like vehicles, home appliances, etc.) that are able to communicate via Internet and other computer networks to share information and resources.

Data Analytics: Process of examining and applying mathematical and statistical models over raw data to extract meaningful insights.

Machine Learning: Artificial intelligence programming paradigm that enables programs to automatically learn performing tasks from existing data, based on a set of model-driven logics and patterns without being explicitly programmed.

Chapter 11
Field to Cloud:
Unveiling IoT Triumphs in Agricultural Evolution

Mrutyunjay Padhiary
https://orcid.org/0000-0002-2236-568X
Assam University, India

ABSTRACT

The integration of Internet of Things (IoT) into the agricultural sector exhibits the potential for modernising conventional farming and tackling various obstacles faced by the community. This chapter focuses on the transformative capacity of the IoT in the agricultural sector, stressing significant insights and discoveries obtained from several case studies. The IoT enables farmers to empower for efficient farming operations and risk reduction by monitoring and analysing data in real-time. Real world applications show IoT's precise agriculture solutions, which contribute to sustainable farm mechanisation, enhanced livestock management, food safety, blockchain technology, and supply chain visibility. Collaboration among stakeholders, giving priorities for research and innovation, and encouraging technology adoption are essential for progress. By adopting the IoT and incorporating research inputs, the agricultural industry can initiate an approach towards efficient and sustainable automation, thereby guaranteeing a prosperous future for farmers and communities on a global scale.

DOI: 10.4018/979-8-3693-5448-3.ch011

INTRODUCTION

Background

The Internet of Things (IoT) refers to a network of interconnected devices that are capable of engaging with the Internet, facilitating the exchange of data, and enabling uninterrupted connectivity across various platforms and devices. The system consists of a range of devices, including sensors, actuators, controllers, and intelligent gadgets, which are endowed with computing capabilities and connect to the internet. These gadgets employ sensors to gather data from their surroundings and transmit it across the internet to centralized platforms for the purpose of analysis and subsequent action. Cisco defines the IoT as a network of various devices such as appliances, cars, buildings, and other objects that facilitate the collection, exchange, and connectivity of data (Gupta & Quamara, 2020).

Historical Development of IoT

IoT technology emerged in 2000 with the introduction of RFID (Radio Frequency Identification) and sensor networks (Duroc, 2022). The widespread use of wireless communication technologies like Wi-Fi, Bluetooth, and cellular networks, along with miniaturization and energy efficiency, significantly accelerated IoT's growth across various industries. It is significantly transforming the agricultural sector through its ability to facilitate precision farming, intelligent decision-making, and real-time monitoring of critical parameters such as soil moisture, temperature, crop health, and animal behavior (Lambrinos, 2019). This data improves agronomic techniques such as tillage, irrigation, fertilization, and pest control, hence increasing productivity, optimization of resources, and promotion of sustainability. In recent years, the integration of IoT with automatic all-terrain vehicles has significantly advanced precision farming practices, providing new avenues for enhancing agricultural efficiency and productivity (Padhiary, Kumar, et al., 2024). The integration of multiple agricultural systems, facilitation of smooth data interchange, assurance of traceability and transparency, and adherence to legal requirements are all dependent on the utilization of IoT technology.

Aim and Scope of Study

This chapter examines the profound influence of the IoT on the agricultural sector, with a specific emphasis on the effective integration of this technology in various industries. This elucidates how IoT technologies augment productivity, sustainability, and profitability. The utilization of IoT technology in precision agriculture, data-driven decision-making, and automation of important agricultural activities is exemplified through various case studies. The chapter additionally examines the potential for further progress in agriculture enabled by the IoT, with a specific emphasis on emerging trends, challenges, and opportunities. This study offers valuable perspectives on the research, innovation, and implementation of IoT technology in the agricultural sector, empowering stakeholders to effectively employ sustainable and financially viable farming methods.

IOT IN AGRICULTURE

This section presents a definition of IoT and offers a concise explanation of its various components, including sensors, actuators, and connectivity. These components collaborate to establish a linked agricultural ecosystem. Gaining a thorough comprehension of these basic principles is essential as we enter into the integration of IoT into different agricultural activities.

Components of IoT Systems

Sensors

The IoT systems incorporate a range of essential components to obtain, transmit, analyze, and utilize data about the physical world. The efficient operation of IoT solutions necessitates the utilization of sensors, actuators, communication technologies, and control systems. Motion, pressure, temperature, humidity, light, and chemical composition are quantified and characterized using sensors. IoT devices gather data from their immediate environment. Sensors can be employed to monitor and gather real-time data for various products, machinery, and systems. IoT applications utilize accelerometers, temperature sensors, humidity sensors, GPS (Global Positioning System) modules, and cameras (Padhiary, Sethi, et al., 2024) for efficient farming in hilly terrains. Actuators facilitate the conversion of electrical impulses into physical responses. Data-driven control and manipulation are employed to establish a connection between IoT devices and the physical world. Actuators carry out instructions that are driven by sensors. Relays, motors, switches, valves, and servos are examples of actuators. Actuators are employed in the field of agriculture to regulate factors such as cattle ventilation, machinery, and irrigation.

Connectivity

Connectivity technologies enable the easy sharing of data among IoT devices, sensors, actuators, and centralized platforms. These technologies facilitate uninterrupted connectivity over both wired and wireless networks (Shafique et al., 2020), enabling IoT equipment to transmit data over varying distances, whether they are short or long. Common networking technologies in the IoT include Wi-Fi, Bluetooth, Zigbee, RFID, cellular networks (3G/4G/5G), LoRaWAN, and satellite communication. The selection of connectivity technology is contingent upon factors such as distance, data transmission capacity, energy consumption, and operational coverage. The sensor input is processed, analyzed, and subsequently utilized by control systems to make conclusions or execute algorithms. These systems confer intelligence and autonomy to objects and applications, serving as the central processing units of IoT solutions (Fortino et al., 2021). The utilization of embedded software, microcontrollers, edge computing devices, and cloud platforms is prevalent in control systems designed for real-time data processing. They enhance operational efficiency and enable real-time responsiveness of IoT systems.

Middleware and Actuators

Middleware and software platforms facilitate the administration, orchestration, and integration of IoT devices, data, and applications through the provision of infrastructure and tools (Aafreen et al., 2019). These systems enable the management, preservation, processing, and display of information, as well as hardware control, security, and inspection. Middleware solutions facilitate the connection between IoT devices and backend systems, allowing for the seamless flow and exchange of data. The middleware and platforms used for the IoT encompass AWS, Azure, and Google Cloud, while the edge computing platforms include MQTT, CoAP, and OPC UA (De Nardis et al., 2022). **Fig. 1** shows various systems and their components working in tandem, with sensors collecting data, middleware providing the platform for control, actuators executing tasks, and interconnections facilitating feedback, data management, storage, human input, and weather data.

Figure 1. IoT System components

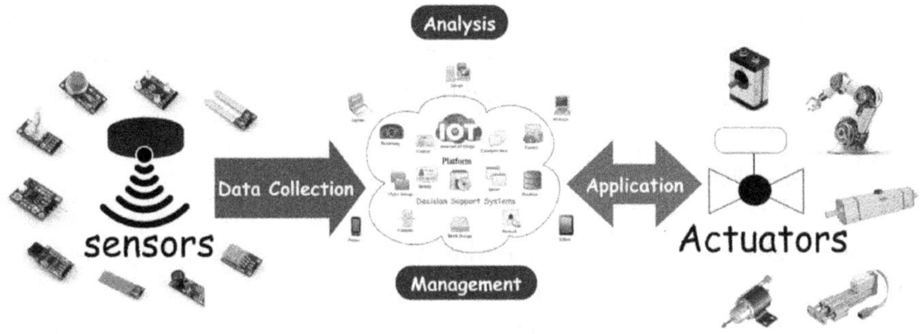

IoT technologies provide interconnected and intelligent methodologies for the surveillance, regulation, and enhancement of mobility, production, healthcare, automated agriculture, and intelligent urban areas. IoT solutions enable enterprises and individuals to make data-driven decisions, enhance efficiency, and identify novel development and innovation prospects through the use of sensors, actuators, connection technologies, control systems, and middleware platforms.

Data Collection and Management

The capture, storage, processing, and exploitation of data from linked devices and sensors are essential components of IoT systems, involving data gathering and management (Ouafiq et al., 2022). In the context of IoT applications, the process of data collecting entails the acquisition of information from diverse sources inside the physical world (Tripathi et al., 2019). On the contrary, data management covers the systematic arrangement, storage, analysis, and extraction of practical insights from this data. Below is an elaborate summary of data collecting and management in the IoT:

Data Collection

IoT systems employ sensors and gadgets to collect up-to-the-minute data from diverse industries, including agriculture, healthcare, industry, transportation, and smart cities (Morchid et al., 2024). Communication technologies such as Wi-Fi, Bluetooth, Zigbee, cellular networks, LoRaWAN, and satellite communication are utilized to deliver data to centralized platforms or cloud servers. This facilitates the ability to monitor remotely and share data instantaneously. Network-edge computing is a technology that aims to minimize latency and bandwidth, hence improving the ability to make real-time decisions in scenarios where connectivity is intermittent or has low latency. In IoT contexts, the fusion and integration of data from many sources and formats are imperative to guarantee coherence, precision, and comprehensiveness. This methodology employs sensors and cutting-edge technologies to acquire a thorough comprehension of the immediate surroundings.

Data Management

IoT data is stored in databases, data lakes, and distributed storage for security and analysis. These systems use cloud computing, NoSQL databases, time-series analysis, and object storage for scalability and durability (Azzabi et al., 2024). Data management systems evaluate IoT data and make educated decisions using real-time streaming, batch processing, CEP, machine learning, and predictive analytics. Encryption, authentication, access control, and anonymization are prioritized for security and privacy. Lawful and ethical data processing requires GDPR, HIPAA, and CCPA. IoT data lifecycle management collects, converts, stores, and protects data, focusing on data quality, durability, and conservation for data governance (A. Zhang et al., 2020). Regulations on collecting, disposing, and meeting corporate and regulatory requirements are crucial. Networked devices and sensors in the IoT improve operational efficiency, innovation, and insight. Data collection, management, and security can alter several areas.

Data-Driven Systems in IoT

IoT devices and sensors generate a substantial volume of data, and analytics as well as decision support systems are crucial in extracting practical insights from this data. These technologies facilitate the ability of organizations to make well-informed decisions, enhance operational efficiency, and drive innovation in diverse sectors such as agriculture, healthcare, manufacturing, transportation, and smart cities. Below is an elaborate summary of analytics and decision support systems in the IoT:

Data Analytics

In the field of descriptive analytics, historical data is employed to examine and interpret trends, patterns, and occurrences. Historical data provides significant insights into past occurrences and serves as the basis for advanced analytical methodologies. Descriptive analytics involves many tools such as data visualization, dashboards, reports, and summary statistics. The utilization of these strategies enhances stakeholders' understanding of the problem.

The field of predictive analytics utilizes statistical algorithms, machine learning models, and data mining techniques to analyze historical data and make projections about future events and patterns. This technology enables organizations to engage in forecasting, risk identification, and proactive decision-making. Predictive analytics is utilized in IoT applications such as predictive maintenance, demand forecasting, risk management, and anomaly detection. These software applications have the capacity to improve reliability, efficacy, and productivity.

Prescriptive analytics surpasses predictive analytics by offering suggestions for the most effective ways to achieve goals. The system utilizes advanced algorithms, optimization techniques, and simulation models to evaluate decision scenarios and determine the most effective alternative. Prescriptive analytics in the IoT has various uses, including optimizing resources, optimizing processes, managing supply chains, and scheduling dynamically. These applications enable organizations to improve their operations, efficiency, and strategic decision-making.

Decision Support Systems (DSS)

Within the framework of the IoT, real-time decision support systems play a crucial role in enabling stakeholders to make timely and spatially informed decisions (Dewi & Chen, 2020). This is achieved through the utilization of real-time data streams, analytics models, and decision rules. These systems include functionalities such as dynamic pricing, fraud detection, proactive maintenance alerts, and emergency response devices. They enable organizations to adapt to evolving conditions and capitalize on opportunities for business growth. Context-aware decision support systems leverage user profiles, location data, and sensor data to tailor recommendations and decision-making procedures based on individual requirements. These applications demonstrate practicality in various domains, including personalized healthcare (Prasad et al., 2025), location-based marketing, intelligent transportation, and intelligent home automation, thereby enhancing customer satisfaction.

The setup of effective communication channels, exchange of data, and employment of analytical tools are essential for promoting collaborative decision-making, with collaboration among stakeholders playing a significant role in this process. The aforementioned technologies enable instantaneous contact, information exchange, and coordination of decision-making processes, regardless of geographical or organizational constraints. Incorporating the Internet of Things (IoT) into decision support systems, such as virtual collaboration platforms and multi-stage environmental monitoring programs, enables stakeholders to enhance transparency, foster collaboration, and reach consensus.

SUCCESSFUL APPLICATIONS

This section examines the impact of IoT on several farming operations, such as soil analysis and crop monitoring. IoT is modernizing these processes by providing real-time data and automation options (Hoque & Padhiary, 2024). The use of case studies and examples illustrates the concrete advantages of IoT in improving agricultural processes.

Tillage

A large-scale wheat farm in the Great Plains, USA, uses IoT sensors on tillage machinery to track soil moisture in real-time (Balivada et al., 2022). These sensors are installed strategically throughout fields and wirelessly transferred to a central control system. The data is then examined and utilized to dynamically adjust the depth and intensity of tillage (**Fig. 2a),** optimizing soil preparation and reducing erosion and compaction. This approach leads to higher crop yields and higher-quality wheat. 3D printing technology is revolutionizing the design of all-terrain vehicles used in precision agriculture, allowing for customized solutions that improve the efficiency of tillage operations (Padhiary & Roy, 2024).

In India, a smallholder farm uses GPS-guided precision tillage to optimize resource use and crop yields in a region prone to erratic rainfall and soil erosion. GPS mapping tools accurately delineate field boundaries and identify variations in topography and soil types (Bhattacharyya et al., 2022). This data is integrated into GPS-guided tillage equipment, which automatically adjusts tillage depth and intensity based on real-time location data.

A winery in France uses sensor-equipped cultivators to maximize the amount of soil prepared between vine rows (Sozzi et al., 2020). The sensors track soil moisture content and compaction in real-time, allowing cultivators to adjust their operations to minimize soil compaction and ensure optimal moisture levels for vine growth. This precision farming method enhances vine growth and improves grape quality while implementing sustainable soil management practices using IoT technology. A sustainable Australian organic farm uses IoT-enabled soil health monitoring to guide tillage decisions. The sensors measure pH, nutrient concentrations, and organic matter content of the soil, guiding tillage practices to maximize nutrient cycling and soil structure (Toselli et al., 2023).

Figure 2. Application of IoT in farming operations: (a) Soil moisture sensor (Balivada et al., 2022), (b) Vegetable plug transplanter (Toselli et al., 2023)

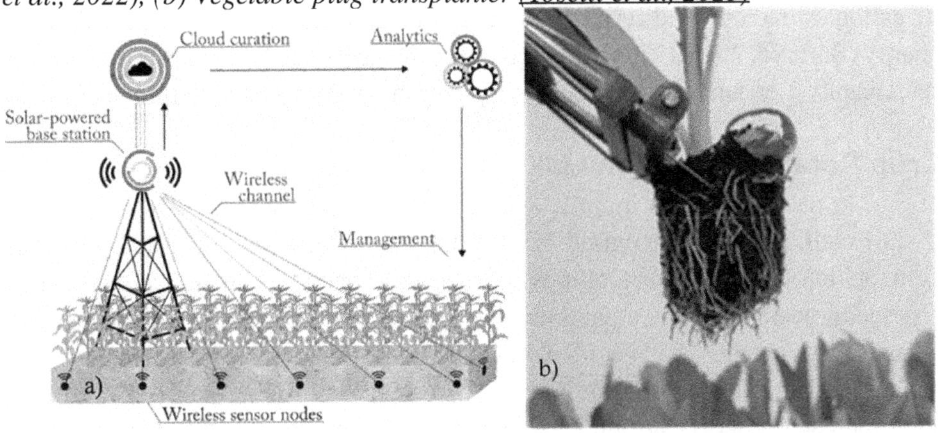

Irrigation

In the context of water scarcity and fluctuating weather conditions, several innovative irrigation systems have been implemented. In California, an almond orchard used IoT-enabled soil moisture sensors to monitor moisture levels, adjusting irrigation schedules and water delivery based on these readings (Gillispie, 2021). This approach improved water use efficiency and minimized water waste, leading to improved almond yields and quality.

In Vietnam, a rice farm used IoT-enabled weather-based irrigation scheduling to optimize water use efficiency (Rupavatharam et al., 2023), reduce the risk of overwatering or underwatering, and enhance rice yields and quality. This approach conserves water resources and mitigates the impact of climate variability on crop production.

In Spain, a vegetable farm implemented an IoT-enabled drip irrigation system with remote monitoring to address water scarcity and improve water use efficiency (Seyar et al., 2023). The system delivers water directly to the root zone of crops, minimizing water loss and monitoring soil moisture levels and system performance. This smart irrigation approach improved crop yields, quality, and resource use efficiency, enhancing profitability and sustainability.

In India, a cotton farm used an IoT-enabled mobile app-controlled irrigation system to improve water use efficiency and cotton yields (Mandi & Patnaik, 2019). The app allowed farmers to remotely monitor soil moisture levels and manage irrigation schedules, providing real-time data on soil conditions. This technology-driven approach enhanced crop performance and resilience in water-stressed environments, reducing labor and water costs. In South Africa, a winery used precision irrigation technology and aerial drone monitoring to optimize water use and grape quality (Barnard, 2018). This approach improved grape quality, yield, and vineyard sustainability, enhancing the winery's reputation for premium wine production.

Crop Rotation Decision Making

In Brazil, a diverse farm used IoT weather data (Garbelini et al., 2022) to plan crop rotation, ensuring crop rotations matched weather patterns to increase water use efficiency, pest management, and farm resilience. In the UK, an organic vegetable farm used IoT-enabled soil health monitoring to inform crop rotation decisions, enhancing soil fertility, nutrient cycling, and pest suppression (Postolache et al., 2023). This approach supports sustainable organic farming practices and reduces reliance on synthetic inputs. In Canada, a grain farm implemented IoT sensors to monitor pest populations, disease incidence, and beneficial insect activity in a region with diverse pest and disease pressures (Ali et al., 2023). Real-time data was analyzed using predictive modeling and machine learning algorithms to forecast outbreaks, promoting sustainable pest management practices.

In Kenya, a smallholder farm cooperative developed an IoT-enabled knowledge-sharing platform for crop rotation planning, allowing farmers to exchange experiences, best practices, and local insights on crop rotation strategies (Mwangi et al., 2021). This participatory approach strengthened community ties, promoted peer learning, and enhanced farm productivity and sustainability. Overall, these examples demonstrate the potential of IoT-enabled technology in enhancing crop rotation decisions, soil health, and farm productivity.

Seeding, Sowing, and Transplanting

A large-scale corn farm in the Midwest, USA, implemented IoT-enabled precision planting equipment to measure soil moisture, temperature, and texture in real-time (Maestrini & Basso, 2018). This data was integrated into a centralized control system, adjusting seed placement and depth based on soil conditions. This method improved crop emergence, uniform stand establishment, and yields while reducing input costs and enhancing farm profitability.

A diverse crop farm in India uses smart seeding drones equipped with IoT technology for sowing operations. Farmers remotely monitor drone flights and seeding operations via a mobile application connected to IoT-enabled controllers (Kose et al., 2022). This approach optimizes crop establishment, minimizes seed wastage, and improves crop yields and resilience, especially in remote or inaccessible areas.

An Italian organic vegetable farm has adopted automated transplanting robots equipped with IoT sensors for sustainable agriculture. These robots use cameras and sensors to identify suitable planting locations and assess soil conditions in real-time. IoT-enabled controllers guide the robots to transplant seedlings at optimal depths and spacing (**Fig. 2b**), reducing labor costs and minimizing transplant shock (Jorg et al., 2021).

A Chinese rice paddy farm implemented GPS-guided rice transplanters with IoT technology to monitor soil moisture levels and seedling quality in rice paddies (Masud Cheema et al., 2023). The data is transmitted wirelessly to a centralized control system, providing feedback on optimal transplanting locations and densities. In Spain, sensor-equipped tree-planting machines with IoT technology were used for transplanting (Cañadas-Aránega et al., 2024). These devices monitored soil, root health, and planting depth in real-time, allowing automatic controls to ensure tree placement and establishment.

Weeding and Weed Management

The use of IoT technology in agriculture has impacted in a positive way, including vegetable farming, irrigation, and weed management. In California, a vegetable farm used robotic weeders equipped with cameras and AI algorithms to manage weeds in real-time (Vijayakumar et al., 2023). These weeders monitored weed density and distribution, autonomously targeting and removing weeds while avoiding crop plants. This approach reduced the use of herbicides and manual labor, resulting in effective weed control.

In Kenya, a maize farm implemented an IoT-enabled drip irrigation system with herbicide injection for integrated weed management (Waweru et al., 2024). This system delivered water directly to the root zones of maize plants, minimizing water wastage and weed growth. Integrated IoT controllers targeted weeds while sparing crop plants, optimizing resource use efficiency, and reducing labor requirements. In France, a vineyard used smart weed mapping drones equipped with IoT technology for precise weed management (Abdulsalam & Aouf, 2020). These drones captured aerial imagery of the vineyard, identifying weed-infested areas using AI algorithms. Integrated IoT controllers directed targeted herbicide applications to weed hotspots while avoiding grapevines.

In Vietnam, a rice paddy farm used GPS-guided mechanical weeders with IoT technology for mechanical weed control. These weeders monitored weed density and crop health in real-time, providing feedback to operators on optimal weeding paths and speeds. By precisely targeting weeds and minimizing crop damage, the farm achieved effective weed control while reducing labor costs and herbicide usage. In Australia, a wheat farm used IoT-enabled autonomous drones with thermal imaging technology for weed detection and management (Vijayalakshmi et al., 2023). These drones captured thermal images that highlighted weed-infested areas based on temperature differences, directing targeted herbicide application or mechanical weeding in identified weed hotspots.

Fertilizing

In Iowa, a corn farm used variable rate fertilization using IoT technology and soil nutrient mapping (Karthika, 2024). IoT sensors collected data on soil nutrient levels, pH, and organic matter content, which were integrated into a precision agriculture platform. This data was then used to generate soil nutrient maps for different field zones, allowing variable-rate fertilizer applicators to adjust application rates in real-time. This improved nutrient efficiency reduced environmental impact, and enhanced corn yields.

In Thailand, a rice farm implemented a smart fertigation system with IoT technology for precise nutrient management (Suanpang & Jamjuntr, 2019). The system combined irrigation and fertilization, delivering nutrients directly to the root zones of rice plants. IoT sensors monitor soil moisture, nutrient levels, and environmental conditions in real-time. Integrated IoT controllers adjusted fertigation rates based on sensor data, ensuring optimal nutrient supply while minimizing leaching and runoff. In Spain, a citrus orchard farm used sensor-based nutrient monitoring systems with automated fertilizer injection for precise fertilization (Acosta et al., 2023). The farm improved citrus tree health, productivity, and fruit quality by adjusting nutrient supply for crop needs and soil conditions.

Argentina's agricultural sector used satellite imaging and IoT technologies to enhance the efficiency of variable-rate fertilization in wheat crops (de Lara et al., 2023). The customized application of fertilizer increased nutrient utilization efficiency and increased wheat yields. In Brazil, a coffee plantation implemented a soil moisture-based drip fertigation system for precise nutrient delivery (Swett, 2020). IoT sensors monitored soil moisture levels in real-time, and integrated IoT controllers regulated the flow of nutrient solutions through the drip irrigation system.

Spraying and Herbicide Application

In California, a vineyard used precision sprayers equipped with IoT technology for herbicide application (Papadopoulos et al., 2024). The sprayers were integrated with sensors that monitored vine health indicators in real-time, allowing them to adjust herbicide application rates and spray patterns to target weeds while minimizing herbicide drift and vine exposure. This resulted in effective weed control while preserving vine health and wine quality. The development of IoT-based solenoid-controlled pressure regulation systems has provided precision sprayers with enhanced control, reducing waste and improving the accuracy of herbicide applications (Saha et al., 2023).

In Brazil, a soybean farm used unmanned aerial vehicles (UAVs) equipped with IoT technology for aerial spraying and weed detection (Cunha & Silva, 2021). The UAVs captured high-resolution images of the field, processed them using IoT analytics software to identify weed-infested areas, and autonomously sprayed herbicides with precision. This led to improved soybean yields and profitability. In Iowa, a corn farm used sensor-guided spot sprayers equipped with IoT technology for targeted herbicide application (Dean et al., 2021). The sensors continuously scanned the field for weed presence and density, and the data was processed in real-time to create weed distribution maps. The spot sprayers autonomously targeted and sprayed herbicides only on weed-infested areas, minimizing herbicide usage and off-target effects. Semi-autonomous vehicle sprayers, powered by IoT technology, represent a significant advancement in the precise and efficient application of agricultural inputs, particularly herbicides (Padhiary, Tikute, et al., 2024).

Harvesting

A Washington State apple orchard has implemented automated harvesting robots equipped with IoT technology for efficient fruit picking (Silwal et al., 2016). The robots use cameras and sensors to identify ripe apples based on color, size, and firmness. They assess fruit quality in real-time, sorting apples into different grades for packaging and processing. This automation optimizes labor efficiency and minimizes post-harvest losses, leading to improved profitability. Machine learning and AI vision applications are integral to the automation of all-terrain vehicles in agriculture, significantly enhancing the precision and efficiency of harvesting operations (Padhiary, Saha, et al., 2024).

In Saskatchewan, Canada, a wheat farm uses GPS-guided combine harvesters with IoT technology for precision harvesting (Cheema et al., 2023). These combine harvesters are equipped with GPS receivers and sensors that monitor crop yield and moisture content in real-time. Integrated IoT controllers generate yield maps, indicating variations in crop yield across the field. The combine harvesters autonomously adjust harvesting parameters, optimizing grain collection and minimizing losses.

In Italy, a grape vineyard uses automated harvesters equipped with IoT technology for grape picking (Mezzasalma et al., 2018). The harvesters use cameras and sensors to assess grape ripeness and quality parameters, such as sugar content and acidity levels. Integrated IoT controllers sort grapes in real-time based on quality criteria, directing them to designated bins for winemaking or table grapes.

In Vietnam, a rice farm uses sensor-guided rice combine harvesters with IoT technology for efficient rice harvesting. The combined harvesters adjust harvesting parameters based on sensor feedback, optimizing grain retention and quality (Salam, 2020). By synchronizing harvesting operations with rice moisture levels, the farm minimizes losses and post-harvest drying requirements, improving overall rice quality and market value. In Chile, a blueberry farm uses automated picking robots equipped with IoT technology for efficient berry harvesting (Galeas et al., 2019). The robots use vision systems and sensors to identify ripe berries based on color, size, and sweetness.

Crop Monitoring

In Iowa, USA, farmers used satellite imagery to monitor crops, assessing crop health, growth phases, and field variability (Kim & Lee, 2016). Integrated IoT analytics software created crop health maps and identified stress and pest infestations from satellite data, enhancing irrigation schedule, fertilizer application, and insect control. This led to increased corn yields and profitability. In Australia, a wheat farm used unmanned aerial vehicles (UAVs) with high-resolution cameras for crop monitoring and scouting. The images provided detailed information on crop health, pest incidence, and weed infestation (Nduku et al., 2023). Integrated IoT analytics software processed the images, enabling farmers to identify and assess crop issues in real time. Regular UAV-based crop scouting missions detected potential problems early, optimizing management practices and maximizing wheat yields.

In India, a rice farm deployed IoT sensors for crop monitoring, with data accessible through a mobile app. Farmers could remotely access real-time sensor data and receive alerts or recommendations (Sahoo, 2022). This led to optimized rice yields and resource use efficiency. In France, a vineyard winery used a sensor network to monitor climate and disease. The sensors measured temperature, humidity, and leaf wetness, and the integrated IoT analytics program predicted disease outbreaks and grape quality using sensor data (Marcu et al., 2022). This allowed the winery to apply fungicides or regulate canopies, maintaining grape quality and vineyard sustainability. The use of Python-based Raspberry Pi image processing for real-time plant target identification has emerged as a cost-effective solution for precision agriculture, enhancing crop monitoring and management practices (Padhiary et al., 2023).

Soil Analysis

The soybean farm in Iowa, USA, implemented a soil sensor network equipped with IoT technology for comprehensive soil analysis. The sensors measured soil parameters like moisture content, pH levels, nutrient levels, and percentage compaction (Rankoth et al., 2021), allowing farmers to access real-time data and historical trends via a mobile application. This led to optimized soybean yields and resource use efficiency. In Australia, a farm used portable soil testing devices connected to IoT technology for on-the-go soil analysis (Afridi et al., 2023). Farmers collected soil samples from different locations and used these devices to measure key soil parameters like pH, nutrient levels, and organic matter content. The data was uploaded to a cloud-based platform for storage and analysis.

In Brazil, a corn farm used soil spectroscopy technology with IoT integration for rapid soil analysis. Handheld soil spectrometers collected spectral data from soil samples across the farm (Cassim et al., 2022). Integrated IoT analytics software processed the data and applied machine learning algorithms to predict soil properties, leading to improved yields and sustainability. In the Netherlands, a potato farm used soil moisture and nutrient probes equipped with IoT technology for comprehensive soil analysis (Plauborg et al., 2022). The probes monitored soil moisture levels and nutrient concentrations, and integrated IoT controllers adjusted irrigation and fertigation rates based on real-time sensor data. This optimized potato yields and tuber quality while conserving water and nutrients. Recent studies have demonstrated the potential of advanced electrode materials in improving the accuracy of in-situ soil fertilizer concentration measurements, which is crucial for optimizing fertilizer application and ensuring soil health (Padhiary, Kyndiah, et al., 2024).

Livestock Management

This dairy farm in Wisconsin implemented RFID ear tags equipped with IoT technology for individual animal monitoring. Each cow was fitted with an RFID ear tag that transmitted unique identification and health data to a centralized system. Integrated IoT software analyzed the data in real-time, tracking each cow's activity, milk production, health status, and reproductive cycle (Vu et al., 2023). By monitoring individual cows' health and behavior, the farm optimized feeding, breeding, and healthcare management practices, leading to improved milk yield, herd health, and farm profitability.

A Queensland beef cattle ranch managed livestock utilizing GPS-enabled smart collars and IoT technology (Manning et al., 2017). Cattle had smart collars that tracked their position, activity, and grazing in real-time. Temperature, humidity, and fodder quality were monitored via IoT sensors. The ranch adjusted grazing rotations, pasture management, and water supplies by assessing animal and environmental data, improving cow health, weight increase, and land sustainability.

Nestled in the poultry farms of the Netherlands, this farm deployed automated feeding systems equipped with IoT technology for poultry management (Chinaeke-Ogbuka et al., 2021). The feeding systems were programmed to dispense feed automatically based on predefined schedules or individual bird requirements. Integrated IoT sensors monitored feed consumption and bird weight in real-time. By adjusting feed rationing and monitoring bird growth, the farm optimized feed efficiency, flock uniformity, and meat quality leading to improved poultry production outcomes and profitability.

A sheep farm in New Zealand's hilly landscape used IoT-enabled wearable health monitoring for proactive livestock management. Wearable equipment monitored sheep's heart rate, body temperature, and activity levels in real-time. Integrated IoT analytics software uses predictive algorithms to detect early disease or distress in health data (Herlin et al., 2021). The farm reduced disease spread, mortality, and flock productivity and welfare by proactively addressing health issues.

Situated in the swine farms of Iowa, this farm deployed environmental sensors equipped with IoT technology for swine management. The sensors monitored air quality parameters such as temperature, humidity, and ammonia levels in swine barns in real-time (Pandey et al., 2021). Integrated IoT controllers regulated ventilation systems based on sensor data, optimizing indoor air quality and thermal comfort for pigs. By maintaining optimal environmental conditions, the farm reduced stress, improved growth rates, and enhanced swine health and productivity.

Post-Harvest Operations, Storage, and Marketing

In Saskatchewan's grain farms, smart grain storage bins equipped with IoT technology optimize post-harvest operations. These storage bins are fitted with IoT sensors that continuously monitor grain temperature and moisture levels in real-time (Richardson, 2023). Integrated IoT controllers adjust aeration and cooling systems based on sensor data, preventing grain spoilage and insect infestation during storage. By maintaining optimal storage conditions, the grain farm preserves grain quality, minimizes losses, and ensures the marketability of stored grain.

IoT-enabled cold chain monitoring sensors keep California fruit plantations fresh throughout post-harvest operations and shipping. IoT sensors in refrigerated storage and transport vehicles monitor temperature and humidity from farm to market. The integrated IoT software alerts for temperature or equipment changes to prevent fruit from rotting (Lamberty & Kreyenschmidt, 2022). Fruit orchard storage and transportation efficiency improves market competitiveness and customer satisfaction by preserving fruit quality and shelf life.

Automated packing lines with IoT technology streamline post-harvest operations and marketing in Netherland vegetable farms. These packing lines use IoT sensors for quality grading and traceability, ensuring product quality and food safety. Integrated IoT software assigns batch numbers and records harvest date, variety, and grower information for each product from field to packaging line (Blom et al., 2022).

The vegetable farm increases product value, meets market demands, and creates consumer trust by ensuring traceability and quality.

New Zealand dairy farms use IoT milk quality monitoring devices to improve post-harvest operations and marketing. Milking parlors use IoT sensors to monitor temperature, fat content, and somatic cell count in real-time (McCoard et al., 2023). Integrated IoT analytics software analyses milk quality data and generates reports to assure dairy sector quality and compliance. Through high-quality milk production and monitoring, the dairy farm maintains product integrity, enhances marketability, and earns premium rates.

IoT-enabled barrel management systems boost post-harvest operations and wine marketing at a French winery. During wine maturation, IoT sensors in barrels monitor temperature, humidity, and oxygen levels. Winemakers may assess wine quality and development using IoT software to track barrel movements, age times, and tasting notes (Sánchez-Gil et al., 2024). The winery improves wine quality, market differentiation, and brand reputation by optimizing wine aging and tracking product origin.

Transport, Value Addition, and Economics

South Africa's fruit export industry faces challenges due to temperature breaks, affecting fruit quality, market credibility, and financial losses. The Cape Town Container Terminal (CTCT) handles 70% of South African fruit exports. 81% of temperature breaks in fruit reefer containers originate within CTCT, with an average time of 1 hour and 52 minutes. Operational issues include increased genset use, improved scheduling, and training port personnel (Goedhals-Gerber et al., 2017). Collaboration between producers, exporters, logistics providers, CTCT, and shipping lines is also needed.

Blockchain and IoT technology boost Ethiopian coffee bean value through shipping and marketing. IoT-enabled coffee bags may track their whereabouts from fields to processing factories and export destinations and monitor environmental conditions (Abebe & Semegn, 2021). Data and transactions are recorded unambiguously and irreversibly using blockchain technology. Consumers may check coffee origin, quality, and sustainability. Ethically grown coffee beans have market access, product differentiation, and premium prices due to supply chain transparency and traceability.

Norwegian aquaculture farms use IoT-based real-time water quality monitoring devices to improve live fish transportation. Transport tanks use IoT sensors to monitor temperature, dissolved oxygen, and pH in real time (H. Zhang & Gui, 2023). Integrated IoT software alerts transportation operators to water quality abnormalities, allowing them to take immediate action to keep seafood fresh and healthy.

Maintaining high water quality helps the aquaculture farm increase the quality and value of live fish for home and international markets.

In Kansas' grain mills, automated grain handling systems equipped with IoT technology add value to grains during processing and transport. These systems incorporate IoT sensors and automation technology to monitor and control grain-handling processes such as cleaning, drying, and packaging. Integrated IoT software tracks grain quality parameters and processing metrics in real-time, ensuring consistent product quality and safety standards. By optimizing grain processing and handling operations, the grain mill enhances product value, meets customer specifications, and expands market opportunities for value-added grain products (Sherin et al., 2024).

A California vegetable processing factory uses IoT-enabled automated sorting and packing lines to improve fresh produce quality during processing and shipment (Williamson et al., 2018). IoT sensors in manufacturing lines monitor product quality, including size, color, and defects, in real-time. The integrated IoT software analyses sensor data and automates sorting and packing to meet quality standards, ensuring product quality and appearance. Simplifying processing and packaging at the vegetable processing facility improves product attractiveness, shelf life, and market acceptance.

ECONOMIC BENEFITS

The economic implications of IoT technology in agriculture are significant, affecting productivity, cost-effectiveness, and overall profitability. Various studies have demonstrated that the implementation of IoT technology in precision agriculture can result in decreased expenses for inputs like seeds, water, and fertilizers, while also boosting crop productivity. IoT use in agriculture provides a strong return on investment and significant economic advantages (Papadopoulos et al., 2024). Farmers may increase output, cut expenses, and improve overall farm management by utilizing IoT technologies. The financial benefits of IoT in agriculture are illustrated, backed up by statistics and case studies, and it also covers the knowledge and expertise needed for successful adoption.

Enhanced Efficiency and Productivity

IoT technologies enable precision agriculture techniques that make the best use of resources like water, fertilizer, and pesticides. Farmers are empowered to make more educated decisions that lead to increased crop yields and improved quality thanks to real-time monitoring and data analysis. For example, a case study on a California vineyard showed that the use of IoT-based irrigation systems resulted in a 15% reduction in water usage and a 20% increase in grape output (Papadopoulos et al., 2024).

Lowering of Expenses

IoT devices improve resource allocation and minimize waste, which lowers operating expenses. Pest control, fertilizer, and irrigation systems that are automated guarantee that inputs are only applied when and where they are needed. After implementing IoT-driven precision agriculture techniques, a large-scale corn farm in Iowa had a 25% reduction in fertilizer expenditures and a 30% decrease in labor expenses, according to research (Karthika, 2024).

Improved Crop Quality and Marketability

IoT technology improves crop quality by lowering insect and disease incidence and preserving ideal growth conditions. Improved quality makes agricultural products more profitable and marketable. According to Guo et al. (2021), an IoT-enabled smart farming initiative in the Netherlands saw a 15% increase in the market price of vegetables as a result of improved quality control (Guo et al., 2021).

Improved Risk Administration

IoT devices enable farmers to take preventative action by offering early warning of unfavorable weather, pest infestations, and problems with the health of the soil. This lowers crop losses and lowers the chance of financial loss. IoT-based weather monitoring systems helped reduce crop loss by 20% during extreme weather events, according to a case study of an Australian wheat farm (Ukhurebor et al., 2022).

Return on Investment,

The scale of operations, the kind of crops grown, and the particular IoT solutions used all have an impact on the return on investment (ROI) of IoT investments in agriculture. Based on research findings, farmers can anticipate a return on investment (ROI) of 20% to 40% within the initial three years of IoT implementation. After installing automated feeding systems and IoT-based livestock monitoring, a dairy farm in New Zealand saw a 35% return on investment in just two years (Vázquez-López et al., 2021).

CONCERNS AND CHALLENGES

The security of IoT systems in agriculture is of utmost importance, considering the sensitive nature of the data gathered from fields, machinery, and livestock. With the increasing prevalence of IoT devices, the likelihood of cyber threats rises, making it imperative to implement strong security measures. Incorporating multi-factor authentication (MFA) and biometric security methods can effectively prevent illegal access to IoT systems, ensuring the protection of data integrity and system operation (Ashitha et al., 2024). The increasing sophistication of IoT networks also brings forth new weaknesses, such as denial-of-service (DoS) assaults and the possibility of data breaches.

Data Privacy and Security

IoT systems capture, transmit, and store sensitive data across networked devices and networks, making data privacy and security crucial (Jaime et al., 2023). Establish and implement strong protocols to secure data integrity, confidentiality, and availability, decreasing the risks of unauthorized access, manipulation, and disclosure. Data privacy and security in IoT are examined in detail below:

Authentication and Encryption

Symmetric and asymmetric encryption algorithms protect data sent and stored in IoT devices. Data encryption protects sensitive data from unauthorized access or decryption during transport or storage (Diro et al., 2020). Data transferred via communication channels is often encrypted using Transport Layer Security (TLS) and Secure Sockets Layer (SSL) protocols. However, data is encrypted using AES and RSA. Authentication procedures are essential for verifying the identities of people, devices, and services accessing IoT networks. This prevents unlawful entry and ensures accountability. Computers use passwords, biometric authentication, digital certificates, and multifactor authentication. Users and devices must authenticate to access IoT resources and policies and permissions limit access. To prevent data access, manipulation, and harm, robust authentication procedures are used.

Regulators Compliance

IoT systems must comply with data protection regulations like the General Data Protection Regulation (GDPR) for obtaining, manipulating, and storing personal data of EU and EEA residents. Personal data management organizations must develop procedures for confidentiality, openness, and user authorization (Lee et al., 2019).

The GDPR requires data encryption, anonymization, breach notification, and, if necessary, Data Protection Officers (DPOs). IoT technologies in healthcare must comply with HIPAA regulations to secure ePHI and guarantee patient anonymity. To prevent unauthorized access, disclosure, and modification of electronic protected health information (ePHI), covered companies and their business associates must take administrative, physical, and technical precautions. Risk assessments, access restrictions, audit trails, and data encryption are HIPAA compliance requirements.

Mitigation Strategies

Threat modeling helps IoT systems identify security vulnerabilities, weaknesses, and attack tactics. Threat modeling helps businesses prioritize security mitigation by assessing the probability and effect of security concerns. Unauthorized access, data breaches, DoS attacks, and malware outbreaks are common IoT hazards (Mahadevappa et al., 2024). The development lifecycle of IoT systems requires a comprehensive approach to security, incorporating security by design principles. This involves implementing secure coding techniques, vulnerability assessments, and penetration testing to detect and resolve security vulnerabilities early. Regular firmware updates and secure configuration management further reduce the risk of exploitation from unaddressed vulnerabilities (Alsabbagh et al., 2024). IoT systems need reliable incident response and recovery procedures to quickly identify, contain, and resolve security incidents. Companies should form crisis response teams, establish escalation policies, and practice drills to evaluate their response skills. Rapid incident response helps mitigate security breaches and restore systems and data faster.

Issues of Interoperability

The concept of interoperability holds significant importance within the context of IoT ecosystems, as it enables seamless communication and effective exchange of data among diverse devices, platforms, and systems. It is necessary to undertake standardization projects, make use of middleware technologies for seamless integration, and cultivate collaborative environments and collaborations to effectively address interoperability difficulties. Standardization organizations, such as the Institute of Electrical and Electronics Engineers (IEEE) and the Internet Engineering Task Force (IETF), are responsible for improving the process of developing interoperability standards and protocols for the IoT. The establishment of shared data formats, communication protocols, and interfaces is made possible by standards such as MQTT (Message Queuing Telemetry Transport), CoAP (Constrained Application Protocol), and OPC UA (Open Platform Communications Unified Architecture). These standards make interoperability easier to achieve (Abdelghaffar & Abousteit,

2021). When industry standards are adhered to, it ensures that Internet of Things devices and platforms are interoperable with one another and can function together without any problems.

Middleware solutions provide seamless connection and communication between diverse IoT devices. This ensures the seamless execution of the process, devoid of any complications. Middleware systems provide a wide range of diverse services. Data normalization, protocol translation, and mediation are among the most widely used techniques. The utilization of services facilitating communication between devices employing diverse data formats and transmission methods enhances their interoperability. IoT gateways, IoT message brokers, and IoT platforms are software applications that facilitate the collection, modification, and transmission of data over IoT networks. Collaborative ecosystems and partnerships can boost interoperability. Facilitating stakeholder convergence from various industries, disciplines, and technical stacks achieves this. Collaborations like industrial consortia, open-source communities, and interoperability testing labs promote cross-platform compatibility and standardization. Collaborations between device, software, and service companies promote the development and testing of compatible solutions, which drives innovation and market acceptance of interoperable Internet of Things technologies.

Scalability

The capacity to scale is a crucial factor for IoT systems to effectively handle increasing amounts of data, devices, and users, all while ensuring optimal performance and dependability. To effectively tackle scaling difficulties, it is imperative to employ cost-effective sensor networks, edge computing, fog computing, and hybrid cloud architectures (Thakur et al., 2024). Cost-effective sensor networks enable global IoT system implementation by sensing and collecting data. The causes are downsizing, energy efficiency improvements, and sensor technology. Cost-effective and energy-efficient sensors are created for IoT applications. Edge and cloud computing solutions reduce latency and bandwidth by processing tasks near the data source. Edge devices like gateways and servers can process, analyze, and decide in real time, boosting system reactivity and scalability. Fog computing extends edge computing, providing scalable and dispersed IoT computing.

The integration of public, private, and on-premises infrastructure in hybrid cloud architectures enables enhanced scalability, flexibility, and cost-effectiveness (**Fig. 3**). These solutions implement cloud resources to allocate compute, storage, and networking resources as required, while IoT applications adapt resource allocation in real-time to accommodate demand and workload requirements.

Figure 3. Concerns and future trends of IoT in agriculture

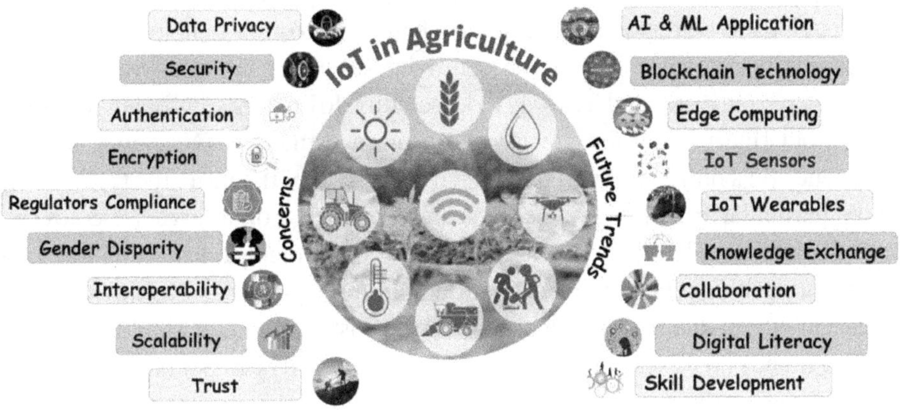

Integration Challenges

To ensure successful adoption, a number of challenges related to the integration of IoT systems into existing agricultural operations must be addressed. Due to the presence of multiple protocols and standards, attaining smooth integration might be challenging. Middleware solutions and continuous standardization initiatives are essential for overcoming these obstacles, allowing various IoT components to interact and operate efficiently as part of a unified system (Carvajal-Flores et al., 2024). It is crucial to acknowledge the significant difficulty of incorporating IoT into existing legacy systems. A significant number of farms depend on outdated machinery and infrastructure that were not originally designed to include IoT technology. Retrofit solutions provide a means to connect old equipment with new IoT systems, thereby enhancing their functionality and minimizing the need for expensive upgrades. To fully utilize the advantages of IoT in agriculture, it is crucial to tackle these integration obstacles. This will guarantee seamless collaboration between new and existing technology.

High Initial Costs

A substantial upfront investment is needed for sensors, networking infrastructure, data management systems, and training before IoT technologies can be implemented in agriculture (Fuentes-Peñailillo et al., 2024). It can be difficult for small-scale farmers to cover these expenses, which can prevent adoption from spreading widely. For instance, the high cost of IoT equipment and their ongoing maintenance requirements posed challenges for an Indian precision agriculture project. Many smallholder farmers were deterred from engaging by the initial financial strain, even though there were potential long-term benefits. In order to help farmers adopt IoT technologies, government subsidies, low-interest loans, and cooperative finance methods can be investigated as a way to lessen this. IoT offers significant long-term benefits, but figuring out the precise return on investment (ROI) can be difficult and depends on a number of variables, such as the size of the farm, the type of crop, and the particular IoT applications. Research conducted on a large-scale soybean farm in Brazil revealed that, despite high initial investment in IoT technology, the farm had a positive return on investment within three years after achieving a 30% increase in output and a 20% reduction in input costs (Prando et al., 2024). Before spending money on IoT solutions, farmers should carefully consider the cost-benefit ratio, as these results are not guaranteed.

Gender Disparity in Rural Regions

The digital divide in rural areas is an urgent matter that necessitates the establishment of infrastructure, enhancement of capabilities, and implementation of IoT systems directed by the community (Rich & Pather, 2021). The objective of these programs is to diminish inequality and promote digital inclusivity. Rural areas get enhanced broadband connectivity, network coverage, and digital infrastructure as a result of infrastructure development initiatives. In order to facilitate socioeconomic development, capacity building, and training programs provide communities with the necessary knowledge, skills, and resources to effectively adopt and utilize IoT technology. These initiatives impart knowledge on the utilization, administration, and upkeep of IoT systems, promote rural entrepreneurship, and augment awareness of IoT. Community-driven IoT deployments entail the active participation of local communities in the process of developing, constructing, and overseeing IoT systems that align with their specific needs and challenges.

Skills and Training Required for Effective IoT Implementation

Need for Technical Skills

A variety of technical skills are necessary for the effective application of IoT in agriculture, including system integration and data analysis expertise, as well as the capacity to install, run, and maintain IoT devices (Khan et al., 2021). It is necessary to provide farmers and agricultural workers with training on how to use IoT platforms, interpret sensor data, and make decisions based on current knowledge.

To guarantee continued operation and prevent potential downtime due to technical issues, it is also essential to understand how to debug and manage IoT systems.

Educational Programs and Workshops

Universities and agricultural institutions have created a variety of educational programs, workshops, and resources to close the skills gap. Universities and agricultural institutions offer specialized courses in data analytics, IoT technologies, and precision agriculture. Programs such as the "Smart Farming Training Initiative (SFTI)" in Europe, for example, offer thorough training modules on the use of IoT in agriculture (Stępień et al., 2023). These programs frequently include in-person seminars that allow participants to engage with IoT devices, discover their features, and practice using them in mock farming scenarios.

Resources for Skill Development

A number of online learning environments, including Coursera and edX, provide classes on the principles of the IoT and how it applies to agriculture (Liapis et al., 2023). Furthermore, to assist farmers and agribusiness professionals in staying current with the newest IoT technology and best practices, industry groups and agricultural extension services host field demonstrations and training sessions. For the agricultural workforce to be prepared to use IoT for increased productivity and sustainability, access to these tools is crucial.

FUTURE DIRECTIONS AND OPPORTUNITIES

As the IoT advances, trends, discoveries, and technical breakthroughs could benefit agriculture. Recognizing these opportunities, stakeholders and decision-makers can use the IoT to boost agricultural resilience and growth. This analysis examines future courses, patterns, and recommendations for important parties.

Emerging Trends in IoT Agriculture

The incorporation of Artificial Intelligence (AI) and machine learning algorithms into IoT systems enable the successful deployment of advanced analytics, predictive modelling, and decision automation in the agriculture industry (Khan et al., 2021). AI-powered systems have the capability to analyze large amounts of sensor data, identify patterns, and improve agricultural methods, leading to increased crop productivity, prioritized investment, and quality. The implementation of blockchain technology in the agriculture sector offers decentralized and transparent data management solutions, hence improving supply chain traceability, food safety, and market access. By offering total visibility and authenticity throughout the entire process, it guarantees secure transaction documentation, enhances transparency, reduces fraudulent activity, and encourages fair trade practices (**Figure 3).**

The utilization of edge computing and AI at the edge facilitates instantaneous data processing, inference, and decision-making on IoT devices, resulting in decreased latency, bandwidth usage, and dependence on centralized cloud resources (Walia et al., 2024). By enabling local analysis and response, this technology enhances the scalability, stability, and autonomy of IoT agriculture applications, particularly in remote or resource-constrained areas. Consequently, it diminishes the reliance on centralized cloud resources. The current status of farm automation in India highlights significant advancements and trends, showcasing the scope for further innovations and the critical role of IoT in transforming agricultural practices (Padhiary, 2024).

Innovation and Technological Progress

Advanced IoT-enabled technologies in precision agriculture offer farmers vital information and decision support for resource management, crop monitoring, and pest control. Precision agriculture technology including unmanned aerial vehicles, satellite imaging, and robotic systems improve data collection, analysis, and mechanization, enabling targeted management and customized agronomic interventions (Hoque et al., 2025).

IoT sensors, wearables, and smart devices provide real-time monitoring of animal health, behaviours, and environmental conditions in smart farming and livestock management conditions in smart farming and livestock management solutions (Barenkamp, 2020). The implementation of this technology has resulted in enhanced animal welfare increased productivity, and higher profitability for farmers by means of predictive modelling, early disease detection, and precise feeding strategies.

Stakeholder and Decision-Maker Recommendations

Invest in Research and Development: Stakeholders and decision-makers should fund IoT agriculture research and development. This method promotes creativity and technology transfer. IoT-enabled solutions for farmers, agribusinesses, and rural communities are developed through research collaborations, funding efforts, and incubation centers.

Facilitate Collaboration and Knowledge Exchange: Fostering collaboration and knowledge exchange among varied stakeholders, such as farmers, researchers, industry professionals, and policymakers, is essential for the widespread adoption and dissemination of IoT agriculture technology (Lambrinos, 2019). Workshops, networks, and knowledge forums are crucial for collaboration, technique exchange, and research dissemination. These methods accelerate technological adoption

Promotion of digital literacy and skills development: Digital literacy and skills development are essential for farmers, agronomists, and rural communities to employ IoT agriculture technologies. Capability-building initiatives, instructional seminars, and extension services educate stakeholders with the knowledge and skills to adopt, manage, and maintain IoT capabilities. Decision-makers may empower individuals and communities to use IoT for sustainable agriculture and rural development by investing in education and training.

CONCLUSION

Throughout the course of our investigation, noteworthy findings have emerged, indicating the potential for improvement in the agricultural industry with IoT technology. Farmers can improve the efficiency of field operations by automation and risk reduction with real-time data analysis and making intelligent decisions. IoT technology can improve precision agriculture by enabling targeted interventions and sustainable crop production as well as livestock productivity and well-being, which depend on better management. Blockchain technology in the supply chain improves food safety and transparency, boosting consumer confidence. However, stakeholders and decision-makers must collaborate to optimize IoT in agriculture. Progress includes funding innovation and research, facilitating technological adoption, and encouraging stakeholder collaboration. IoT technology is predicted to improve resilience, sustainability, and efficiency, providing a bright future for farmers and communities worldwide.

STATEMENTS AND DECLARATIONS

The authors would like to state that there is no conflict of interest and that no funding has been invested in this research.

REFERENCES

Aafreen, R., Neyaz, S. Y., Shamim, R., & Beg, M. S. (2019). An IoT based system for telemetry and control of Greenhouse environment. *2019 International Conference on Electrical, Electronics and Computer Engineering (UPCON)*, 1–6. DOI: 10.1109/UPCON47278.2019.8980258

Abdelghaffar, H., & Abousteit, M. (2021). Internet of Things (IoT) Interoperability Success Criteria. *International Journal of Enterprise Information Systems*, *17*(1), 85–105. DOI: 10.4018/IJEIS.2021010105

Abdulsalam, M., & Aouf, N. (2020). Deep Weed Detector/Classifier Network for Precision Agriculture. *2020 28th Mediterranean Conference on Control and Automation (MED)*, 1087–1092. DOI: 10.1109/MED48518.2020.9183325

Abebe, T. M., & Semegn, A. M. (2021). Blockchain Based Green Coffee Supply Chain Management to Improve Traceability and Transparency (Case Study on Sidama Coffee). In *Advances of Science and Technology. ICAST 2020.* (Vol. 384).

Acosta, M., Quiñones, A., Munera, S., De Paz, J. M., & Blasco, J. (2023). Rapid Prediction of Nutrient Concentration in Citrus Leaves Using Vis-NIR Spectroscopy. *Sensors (Basel)*, *23*(14), 6530. DOI: 10.3390/s23146530 PMID: 37514824

Afridi, W. A. K., Akhter, F., Vitoria, I., & Mukhopadhyay, S. C. (2023). A Technology Review and Field Testing of a Soil Water Quality Monitoring System. In N. K. Suryadevara, B. George, K. P. Jayasundera, & S. C. Mukhopadhyay (Eds.), *Sensing Technology* (Vol. 1035, pp. 460–475). Springer Nature Switzerland. DOI: 10.1007/978-3-031-29871-4_47

Ali, A., Hussain, T., Tantashutikun, N., Hussain, N., & Cocetta, G. (2023). Application of Smart Techniques, Internet of Things and Data Mining for Resource Use Efficient and Sustainable Crop Production. *Agriculture*, *13*(2), 397. DOI: 10.3390/agriculture13020397

Alsabbagh, W., Kim, C., & Langendörfer, P. (2024). Investigating the Security of OpenPLC: Vulnerabilities, Attacks, and Mitigation Solutions. *IEEE Access : Practical Innovations, Open Solutions*, *12*, 11561–11583. DOI: 10.1109/ACCESS.2024.3356051

Ashitha, U., Amulya, A. H. L. S., & Sushma, G., Irfana, Syed., & Saibaba, Ch. M. H. (2024). Strengthening Cloud Security Through Multi-Factor Authentication: A Comprehensive Integration of Diverse Approaches. *2024 International Conference on Electronics, Computing, Communication and Control Technology (ICECCC)*, 1–8. DOI: 10.1109/ICECCC61767.2024.10593923

Azzabi, S., Alfughi, Z., & Ouda, A. (2024). Data Lakes: A Survey of Concepts and Architectures. *Computers*, *13*(7), 183. DOI: 10.3390/computers13070183

Balivada, S., Grant, G., Zhang, X., Ghosh, M., Guha, S., & Matamala, R. (2022). A Wireless Underground Sensor Network Field Pilot for Agriculture and Ecology: Soil Moisture Mapping Using Signal Attenuation. *Sensors (Basel)*, *22*(10), 3913. DOI: 10.3390/s22103913 PMID: 35632322

Barenkamp, M. (2020). A New IoT Gateway for Artificial Intelligence in Agriculture. *2020 International Conference on Electrical, Communication, and Computer Engineering (ICECCE)*, 1–5. DOI: 10.1109/ICECCE49384.2020.9179418

Barnard, Y. (2018). *The use of technology to improve current precision viticulture practices: Predicting vineyard performance* [Stellenbosch : Stellenbosch University]. http://hdl.handle.net/10019.1/104893

Bhattacharyya, P., Chakraborty, K., Molla, K. A., Poonam, A., Bhaduri, D., Sah, R. P., Paul, S., & Hanjagi, P. S. (2022). *Climate Resilient Technologies for Rice based Production Systems in Eastern India*. ICAR-National Rice Research Institute.

Blom, T., Jenkins, A., Pulselli, R. M., & Van Den Dobbelsteen, A. A. J. F. (2022). The embodied carbon emissions of lettuce production in vertical farming, greenhouse horticulture, and open-field farming in the Netherlands. *Journal of Cleaner Production*, *377*, 134443. DOI: 10.1016/j.jclepro.2022.134443

Cañadas-Aránega, F., Blanco-Claraco, J. L., Moreno, J. C., & Rodriguez, F. (2024). The GREENBOT dataset: Multimodal mobile robotic dataset for a typical Mediterranean greenhouse. *Sensors (Basel)*, *24*(6), 1874. DOI: 10.3390/s24061874 PMID: 38544137

Carvajal-Flores, D. F., Abril-Jiménez, P., Buhid, E., Fico, G., & Cabrera Umpiérrez, M. F. (2024). Enhancing Industrial Digitalisation through an Adaptable Component for Bridging Semantic Interoperability Gaps. *Applied Sciences (Basel, Switzerland)*, *14*(6), 2309. DOI: 10.3390/app14062309

Cassim, B. M. A. R., Besen, M. R., Kachinski, W. D., Macon, C. R., De Almeida, J. H. V.Junior, Sakurada, R., Inoue, T. T., & Batista, M. A. (2022). Nitrogen Fertilizers Technologies for Corn in Two Yield Environments in South Brazil. *Plants*, *11*(14), 1890. DOI: 10.3390/plants11141890 PMID: 35890524

Cheema, M. J. M., Iqbal, T., Daccache, A., Hussain, S., & Awais, M. (2023). Precision agriculture technologies: Present adoption and future strategies. In *Precision Agriculture* (pp. 231–250). Elsevier., https://www.sciencedirect.com/science/article/pii/B9780443189531000118 DOI: 10.1016/B978-0-443-18953-1.00011-8

Chinaeke-Ogbuka, I., Anoliefo, E., Ajibo, A., & Ogbuka, C. (2021). *Design and Implementation of an Automated Feeding System for Poultry Farms.*

da Cunha, J. P. A. R., & da Silva, M. R. A. (2021). Deposition of spray applied to a soybean crop using an unmanned aerial vehicle. *International Journal of Precision Agricultural Aviation, 4*(2), 2. http://www.ijpaa.org/index.php/ijpaa/article/view/167

de Lara, A., Mieno, T., Luck, J. D., & Puntel, L. A. (2023). Predicting site-specific economic optimal nitrogen rate using machine learning methods and on-farm precision experimentation. *Precision Agriculture, 24*(5), 1792–1812. DOI: 10.1007/s11119-023-10018-8

De Nardis, L., Mohammadpour, A., Caso, G., Ali, U., & Di Benedetto, M.-G. (2022). Internet of Things Platforms for Academic Research and Development: A Critical Review. *Applied Sciences (Basel, Switzerland), 12*(4), 2172. DOI: 10.3390/app12042172

Dean, A. N., Hodgson, E. W., Rieck-Hinz, A., & Anderson, M. (2021). Needs Assessment for Corn Insect Pest Management in Iowa. *Journal of Integrated Pest Management, 12*(1), 27. DOI: 10.1093/jipm/pmab021

Dewi, C., & Chen, R.-C. (2020). Decision Making Based on IoT Data Collection for Precision Agriculture. In Huk, M., Maleszka, M., & Szczerbicki, E. (Eds.), *Intelligent Information and Database Systems: Recent Developments* (Vol. 830, pp. 31–42). Springer International Publishing., DOI: 10.1007/978-3-030-14132-5_3

Diro, A., Reda, H., Chilamkurti, N., Mahmood, A., Zaman, N., & Nam, Y. (2020). Lightweight Authenticated-Encryption Scheme for Internet of Things Based on Publish-Subscribe Communication. *IEEE Access : Practical Innovations, Open Solutions, 8*, 60539–60551. DOI: 10.1109/ACCESS.2020.2983117

Duroc, Y. (2022). From Identification to Sensing: RFID Is One of the Key Technologies in the IoT Field. *Sensors (Basel), 22*(19), 7523. DOI: 10.3390/s22197523 PMID: 36236622

Fortino, G., Savaglio, C., Spezzano, G., & Zhou, M. (2021). Internet of Things as System of Systems: A Review of Methodologies, Frameworks, Platforms, and Tools. *IEEE Transactions on Systems, Man, and Cybernetics. Systems, 51*(1), 223–236. DOI: 10.1109/TSMC.2020.3042898

Fuentes-Peñailillo, F., Gutter, K., Vega, R., & Silva, G. C. (2024). Transformative Technologies in Digital Agriculture: Leveraging Internet of Things, Remote Sensing, and Artificial Intelligence for Smart Crop Management. *Journal of Sensor and Actuator Networks, 13*(4), 39. DOI: 10.3390/jsan13040039

Galeas, P., Muñoz, C., Huircan, J., Fernandez, M., Segura-Ponce, L. A., & Duran-Faundez, C. (2019). Smartbins: Using Intelligent Harvest Baskets to Estimate the Stages of Berry Harvesting. *Sensors (Basel), 19*(6), 6. Advance online publication. DOI: 10.3390/s19061361 PMID: 30893779

Garbelini, L. G., Debiasi, H., Junior, A. A. B., Franchini, J. C., Coelho, A. E., & Telles, T. S. (2022). Diversified crop rotations increase the yield and economic efficiency of grain production systems. *European Journal of Agronomy, 137*, 126528. DOI: 10.1016/j.eja.2022.126528

Gillispie, G. (2021). *Internet of Things (IoT) Controlled Irrigation for Almond Orchardists in California: A Multiple Case Study*. Capella University.

Goedhals-Gerber, L. L., Stander, C., & Van Dyk, F. E. (2017). Maintaining cold chain integrity: Temperature breaks within fruit reefer containers in the Cape Town Container Terminal. *Southern African Business Review, 21*(1), 362–384.

Guo, X., Snels, J. C. M. A., & Tromp, S. (2021). *Quality-controlled logistics with internet of things: A conceptual framework* (2131; p.). Wageningen Food & Biobased Research. https://library.wur.nl/WebQuery/wurpubs/579167

Gupta, B. B., & Quamara, M. (2020). An overview of Internet of Things (IoT): Architectural aspects, challenges, and protocols. *Concurrency and Computation, 32*(21), e4946. DOI: 10.1002/cpe.4946

Herlin, A., Brunberg, E., Hultgren, J., Högberg, N., Rydberg, A., & Skarin, A. (2021). Animal Welfare Implications of Digital Tools for Monitoring and Management of Cattle and Sheep on Pasture. *Animals (Basel), 11*(3), 3. Advance online publication. DOI: 10.3390/ani11030829 PMID: 33804235

Hoque, A., & Padhiary, M. (2024). Automation and AI in Precision Agriculture: Innovations for Enhanced Crop Management and Sustainability. *Asian Journal of Research in Computer Science, 17*(10), 95–109. https://doi.org/10.9734/ajrcos/2024/v17i10512

Hoque, A., Padhiary, M., Prasad, G., & Kumar, K. (2025). Real-Time Data Processing in Agricultural Robotics. In Bora, D. J., & Bania, R. K. (Eds.), *Computer Vision Techniques for Agricultural Advancements* (pp. 431–468). IGI Global., DOI: 10.4018/979-8-3693-8019-2.ch014

Jaime, F. J., Muñoz, A., Rodríguez-Gómez, F., & Jerez-Calero, A. (2023). Strengthening Privacy and Data Security in Biomedical Microelectromechanical Systems by IoT Communication Security and Protection in Smart Healthcare. *Sensors (Basel), 23*(21), 8944. DOI: 10.3390/s23218944 PMID: 37960646

Jorg, O. J., Sportelli, M., Fontanelli, M., Frasconi, C., Raffaelli, M., & Fantoni, G. (2021). Design, Development and Testing of Feeding Grippers for Vegetable Plug Transplanters. *AgriEngineering*, *3*(3), 3. Advance online publication. DOI: 10.3390/agriengineering3030043

Karthika, R. (2024). Role of precision agriculture in soil fertility and its application to farmers. In *Remote Sensing in Precision Agriculture* (pp. 137–147). Elsevier., https://www.sciencedirect.com/science/article/pii/B9780323910682000175 DOI: 10.1016/B978-0-323-91068-2.00017-5

Khan, N., Ray, R. L., Sargani, G. R., Ihtisham, M., Khayyam, M., & Ismail, S. (2021). Current Progress and Future Prospects of Agriculture Technology: Gateway to Sustainable Agriculture. *Sustainability (Basel)*, *13*(9), 4883. DOI: 10.3390/su13094883

Kim, N., & Lee, Y.-W. (2016). Machine Learning Approaches to Corn Yield Estimation Using Satellite Images and Climate Data: A Case of Iowa State: A Case of Iowa State. , *34*(4), 383–390.

Kose, U., Prasath, V. B. S., Mondal, M. R. H., Podder, P., & Bharati, S. (2022). *Artificial Intelligence and Smart Agriculture Applications.* CRC Press. DOI: 10.1201/9781003311782

Kulkarni, A. A., Dhanush, P., Chethan, B. S., Thammegowda, C. S., & Shrivastava, P. K. (2020). Recent Development of AI and IoT in the field of Agriculture Industries: A Review. In Pant, M., Kumar Sharma, T., Arya, R., Sahana, B. C., & Zolfagharinia, H. (Eds.), *Soft Computing: Theories and Applications* (Vol. 1154, pp. 793–798). Springer Singapore., DOI: 10.1007/978-981-15-4032-5_71

Lamberty, A., & Kreyenschmidt, J. (2022). Ambient Parameter Monitoring in Fresh Fruit and Vegetable Supply Chains Using Internet of Things-Enabled Sensor and Communication Technology. *Foods*, *11*(12), 1777. DOI: 10.3390/foods11121777 PMID: 35741974

Lambrinos, L. (2019). Internet of Things in Agriculture: A Decision Support System for Precision Farming. *2019 IEEE Intl Conf on Dependable, Autonomic and Secure Computing, Intl Conf on Pervasive Intelligence and Computing, Intl Conf on Cloud and Big Data Computing, Intl Conf on Cyber Science and Technology Congress (DASC/PiCom/CBDCom/CyberSciTech)*, 889–892. DOI: 10.1109/DASC/PiCom/CBDCom/CyberSciTech.2019.00163

Lee, G., Epiphaniou, G., Al-Khateeb, H., & Maple, C. (2019). Security and Privacy of Things: Regulatory Challenges and Gaps for the Secure Integration of Cyber-Physical Systems. In X.-S. Yang, S. Sherratt, N. Dey, & A. Joshi (Eds.), *Third International Congress on Information and Communication Technology* (Vol. 797, pp. 1–12). Springer Singapore. DOI: 10.1007/978-981-13-1165-9_1

Liapis, A., Maratou, V., Panagiotakopoulos, T., Katsanos, C., & Kameas, A. (2023). UX evaluation of open MOOC platforms: A comparative study between Moodle and Open edX combining user interaction metrics and wearable biosensors. *Interactive Learning Environments*, *31*(10), 6841–6855. DOI: 10.1080/10494820.2022.2048674

Maestrini, B., & Basso, B. (2018). Drivers of within-field spatial and temporal variability of crop yield across the US Midwest. *Scientific Reports*, *8*(1), 14833. DOI: 10.1038/s41598-018-32779-3 PMID: 30287846

Mahadevappa, P., Al-amri, R., Alkawsi, G., Alkahtani, A., Alghenaim, M., & Al-samman, M. (2024). Analyzing Threats and Attacks in Edge Data Analytics within IoT Environments. *IoT*, *5*(1), 123–154. DOI: 10.3390/iot5010007

Mandi, K., & Patnaik, N. M. (2019). Mobile apps in agriculture and allied sector: An extended arm for farmers. *Agriculture Update*, *14*(4), 334–342. DOI: 10.15740/HAS/AU/14.4/334-342

Manning, J. K., Cronin, G. M., González, L. A., Hall, E. J. S., Merchant, A., & Ingram, L. J. (2017). The effects of global navigation satellite system (GNSS) collars on cattle (*Bos taurus)* behaviour. *Applied Animal Behaviour Science*, *187*, 54–59. DOI: 10.1016/j.applanim.2016.11.013

Marcu, I., Drăgulinescu, A.-M., Oprea, C., Suciu, G., & Bălăceanu, C. (2022). Predictive Analysis and Wine-Grapes Disease Risk Assessment Based on Atmospheric Parameters and Precision Agriculture Platform. *Sustainability (Basel)*, *14*(18), 18. Advance online publication. DOI: 10.3390/su141811487

Masud Cheema, M. J., Iqbal, T., Daccache, A., Hussain, S., & Awais, M. (2023). Precision agriculture technologies: Present adoption and future strategies. In Zaman, Q. (Ed.), *Precision Agriculture* (pp. 231–250). Academic Press., DOI: 10.1016/B978-0-443-18953-1.00011-8

McCoard, S., Stevens, D., Selbie, D., Day, L., Young, W., Bekhit, A. E.-D., & Samuelsson, L. (2023). Supporting the growth of the dairy sheep industry in New Zealand–industry update and review of a programme linking industry and science. *New Zealand Journal of Agricultural Research*, 1–35. DOI: 10.1080/00288233.2023.2272594

Mezzasalma, V., Sandionigi, A., Guzzetti, L., Galimberti, A., Grando, M. S., Tardaguila, J., & Labra, M. (2018). Geographical and Cultivar Features Differentiate Grape Microbiota in Northern Italy and Spain Vineyards. *Frontiers in Microbiology*, *9*, 946. Advance online publication. DOI: 10.3389/fmicb.2018.00946 PMID: 29867854

Morchid, A., El Alami, R., Raezah, A. A., & Sabbar, Y. (2024). Applications of internet of things (IoT) and sensors technology to increase food security and agricultural Sustainability: Benefits and challenges. *Ain Shams Engineering Journal*, *15*(3), 102509. DOI: 10.1016/j.asej.2023.102509

Mwangi, V., Owuor, S., Kiteme, B., & Giger, M. (2021). Assessing Smallholder Farmer's Participation in the Wheat Value Chain in North-West Mt. Kenya. *Frontiers in Sustainable Food Systems*, *5*, 657744. Advance online publication. DOI: 10.3389/fsufs.2021.657744

Nduku, L., Munghemezulu, C., Mashaba-Munghemezulu, Z., Kalumba, A. M., Chirima, G. J., Masiza, W., & De Villiers, C. (2023). Global Research Trends for Unmanned Aerial Vehicle Remote Sensing Application in Wheat Crop Monitoring. *Geomatics*, *3*(1), 1. Advance online publication. DOI: 10.3390/geomatics3010006

Ouafiq, E. M., Saadane, R., & Chehri, A. (2022). Data Management and Integration of Low Power Consumption Embedded Devices IoT for Transforming Smart Agriculture into Actionable Knowledge. *Agriculture*, *12*(3), 329. DOI: 10.3390/agriculture12030329

Padhiary, M. (2024). Status of Farm Automation, Advances, Trends, and Scope in India. [IJSR]. *International Journal of Science and Research (Raipur, India)*, *13*(7), 737–745. DOI: 10.21275/SR24713184513

Padhiary, M., Kumar, R., & Sethi, L. N. (2024). Navigating the Future of Agriculture: A Comprehensive Review of Automatic All-Terrain Vehicles in Precision Farming. *Journal of The Institution of Engineers (India): Series A*, *105*(2). DOI: 10.1007/s40030-024-00816-2

Padhiary, M., Kyndiah, A. K., Kumar, R., & Saha, D. (2024). Exploration of electrode materials for in-situ soil fertilizer concentration measurement by electrochemical method. *International Journal of Advanced Biochemistry Research*, *8*(4), 539–544. DOI: 10.33545/26174693.2024.v8.i4g.1011

Padhiary, M., Rani, N., Saha, D., Barbhuiya, J. A., & Sethi, L. N. (2023). Efficient Precision Agriculture with Python-based Raspberry Pi Image Processing for Real-Time Plant Target Identification. *International Journal of Research and Analytical Reviews*, *10*(3), 539–545. http://doi.one/10.1729/Journal.35531

Padhiary, M., & Roy, P. (2024). Advancements in Precision Agriculture: Exploring the Role of 3D Printing in Designing All-Terrain Vehicles for Farming Applications. *International Journal of Scientific Research, 13*(5), 861–868.

Padhiary, M., Saha, D., Kumar, R., Sethi, L. N., & Kumar, A. (2024). Enhancing Precision Agriculture: A Comprehensive Review of Machine Learning and AI Vision Applications in All-Terrain Vehicle for Farm Automation. *Smart Agricultural Technology*, *8*, 100483. DOI: 10.1016/j.atech.2024.100483

Padhiary, M., Sethi, L. N., & Kumar, A. (2024). Enhancing Hill Farming Efficiency Using Unmanned Agricultural Vehicles: A Comprehensive Review. *Transactions of the Indian National Academy of Engineering : an International Journal of Engineering and Technology*, *9*(2), 253–268. DOI: 10.1007/s41403-024-00458-7

Padhiary, M., Tikute, S. V., Saha, D., Barbhuiya, J. A., & Sethi, L. N. (2024). Development of an IOT-Based Semi-Autonomous Vehicle Sprayer. *Agricultural Research*. Advance online publication. DOI: 10.1007/s40003-024-00760-4

Pandey, S., Kalwa, U., Kong, T., Guo, B., Gauger, P. C., Peters, D. J., & Yoon, K.-J. (2021). Behavioral Monitoring Tool for Pig Farmers: Ear Tag Sensors, Machine Intelligence, and Technology Adoption Roadmap. *Animals (Basel)*, *11*(9), 9. Advance online publication. DOI: 10.3390/ani11092665 PMID: 34573631

Papadopoulos, G., Arduini, S., Uyar, H., Psiroukis, V., Kasimati, A., & Fountas, S. (2024). Economic and environmental benefits of digital agricultural technologies in crop production: A review. *Smart Agricultural Technology*, *8*, 100441. DOI: 10.1016/j.atech.2024.100441

Plauborg, F., Motarjemi, S. K., Nagy, D., & Zhou, Z. (2022). Analysing potato response to subsurface drip irrigation and nitrogen fertigation regimes in a temperate environment using the Daisy model. *Field Crops Research*, *276*, 108367. DOI: 10.1016/j.fcr.2021.108367

Postolache, S., Sebastião, P., Viegas, V., Postolache, O., & Cercas, F. (2023). IoT-Based Systems for Soil Nutrients Assessment in Horticulture. *Sensors (Basel)*, *23*(1), 1. Advance online publication. DOI: 10.3390/s23010403 PMID: 36617000

Prando, A. M., Barbosa, J. Z., Oliveira, A. B. D., Nogueira, M. A., Possamai, E. J., & Hungria, M. (2024). Benefits of soybean co-inoculation with Bradyrhizobium spp. and Azospirillum brasilense: Large-scale validation with farmers in Brazil. *European Journal of Agronomy*, *155*, 127112. DOI: 10.1016/j.eja.2024.127112

Prasad, G., Padhiary, M., Hoque, A., & Kumar, K. (2025). AI-Driven Personalized Nutrition Apps and Platforms for Enhanced Diet and Wellness. In Alhussaini Hamad, A. M., & Soni, R. (Eds.), *Food in the Metaverse and Web 3.0 Era* (pp. 125–158). IGI Global., DOI: 10.4018/979-8-3693-9025-2.ch006

Rankoth, L. M., Udawatta, R. P., Anderson, S. H., Gantzer, C. J., & Alagele, S. (2021). Cover crop influence on soil water dynamics for a corn–soybean rotation. *Agrosystems, Geosciences & Environment*, *4*(3), e20175. DOI: 10.1002/agg2.20175

Rich, M. J., & Pather, S. (2021). A response to the persistent digital divide: Critical components of a community network ecosystem. *Information Development*, *37*(4), 558–578. DOI: 10.1177/0266666920924696

Richardson, O. (2023). *Digital Agriculture Technology's Capacity to Reduce On-Farm Environmental Impacts: Perspectives from Canadian Oilseed and Grain Farmers* [University of Guelph]. https://hdl.handle.net/10214/27694

Rupavatharam, S., Patil, M., Gogumalla, P., & Jat, M. L. (2023). Digital Technologies and Applications for Agricultural Transformation. *Souvenir*, 20.

Saha, D., Padhiary, M., Barbhuiya, J. A., Chakrabarty, T., & Sethi, L. N. (2023). Development of an IOT based Solenoid Controlled Pressure Regulation System for Precision Sprayer. *International Journal for Research in Applied Science and Engineering Technology*, *11*(7), 2210–2216. DOI: 10.22214/ijraset.2023.55103

Sahoo, R. N. (2022). Sensor-Based Monitoring of Soil and Crop Health for Enhancing Input Use Efficiency. In Ray, C., Muddu, S., & Sharma, S. (Eds.), *Food, Energy, and Water Nexus: A Consideration for the 21st Century* (pp. 129–147). Springer International Publishing., DOI: 10.1007/978-3-030-85728-8_7

Salam, A. (2020). Internet of Things in Agricultural Innovation and Security. In Salam, A. (Ed.), *Internet of Things for Sustainable Community Development: Wireless Communications, Sensing, and Systems* (pp. 71–112). Springer International Publishing., DOI: 10.1007/978-3-030-35291-2_3

Sánchez-Gil, J. J., Cañete-Carmona, E., Brox, M., Gersnoviez, A., Molina-Espinosa, M. Á., Gámez-Granados, J. C., & Moreno, J. (2024). An Electronic Barrel Bung to Wirelessly Monitor the Biological Aging Process of Fino Sherry Wine. *IEEE Access : Practical Innovations, Open Solutions*, *12*, 35337–35350. DOI: 10.1109/ACCESS.2024.3372000

Seyar, M. H., Kahandage, P. D., & Ahamed, T. (2023). An IoT-Based Precision Irrigation System to Optimize Plant Water Requirements for Indoor and Outdoor Farming Systems. In Ahamed, T. (Ed.), *IoT and AI in Agriculture* (pp. 47–69). Springer Nature Singapore., DOI: 10.1007/978-981-19-8113-5_4

Shafique, K., Khawaja, B. A., Sabir, F., Qazi, S., & Mustaqim, M. (2020). Internet of Things (IoT) for Next-Generation Smart Systems: A Review of Current Challenges, Future Trends and Prospects for Emerging 5G-IoT Scenarios. *IEEE Access : Practical Innovations, Open Solutions*, *8*, 23022–23040. DOI: 10.1109/ACCESS.2020.2970118

Sherin, J., Buvaneswaran, M., Sunil, C. K., Athmaselvi, K. A., Venkatachalapathy, N., Anandharamakrishnan, C., & Balasubramaniam, V. M. (2024). Emerging technologies in unit operations of food grain processing. In *Unit Operations in Food Grain Processing* (pp. 547–569). Elsevier., DOI: 10.1016/B978-0-443-18965-4.00019-4

Silwal, A., Davidson, J., Karkee, M., Mo, C., Zhang, Q., & Lewis, K. (2016). Effort towards robotic apple harvesting in Washington State. *2016 ASABE Annual International Meeting*, 1. https://elibrary.asabe.org/abstract.asp?aid=47224

Sozzi, M., Bernardi, E., Kayad, A., Marinello, F., Boscaro, D., Cogato, A., Gasparini, F., & Tomasi, D. (2020). On-the-go variable rate fertilizer application on vineyard using a proximal spectral sensor. *2020 IEEE International Workshop on Metrology for Agriculture and Forestry (MetroAgriFor)*, 343–347. DOI: 10.1109/MetroAgriFor50201.2020.9277569

Stępień, S., Smędzik-Ambroży, K., Polcyn, J., Kwiliński, A., & Maican, I. (2023). Are small farms sustainable and technologically smart? Evidence from Poland, Romania, and Lithuania. *Central European Economic Journal*, *10*(57), 116–132. DOI: 10.2478/ceej-2023-0007

Suanpang, P., & Jamjuntr, P. (2019). A smart farm prototype with an Internet of Things (IoT) case study: Thailand. *Technology*, *5*(12), 15.

Swett, C. L. (2020). Managing Crop Diseases Under Water Scarcity. *Annual Review of Phytopathology*, *58*(1), 387–406. DOI: 10.1146/annurev-phyto-030320-041421 PMID: 32580692

Thakur, R., Sikka, G., Bansal, U., Giri, J., & Mallik, S. (2024). Deadline-aware and energy efficient IoT task scheduling using fuzzy logic in fog computing. *Multimedia Tools and Applications*. Advance online publication. DOI: 10.1007/s11042-024-19509-w

Toselli, M., Baldi, E., Ferro, F., Rossi, S., & Cillis, D. (2023). Smart Farming Tool for Monitoring Nutrients in Soil and Plants for Precise Fertilization. *Horticulturae*, *9*(9), 1011. DOI: 10.3390/horticulturae9091011

Tripathi, A. D., Mishra, R., Maurya, K. K., Singh, R. B., & Wilson, D. W. (2019). Estimates for World Population and Global Food Availability for Global Health. In *The Role of Functional Food Security in Global Health* (pp. 3–24). Elsevier., DOI: 10.1016/B978-0-12-813148-0.00001-3

Ukhurebor, K. E., Adetunji, C. O., Olugbemi, O. T., Nwankwo, W., Olayinka, A. S., Umezuruike, C., & Hefft, D. I. (2022). Precision agriculture: Weather forecasting for future farming. In *AI, Edge and IoT-based Smart Agriculture* (pp. 101–121). Elsevier. DOI: 10.1016/B978-0-12-823694-9.00008-6

Vázquez-López, A., Barrasa-Rioja, M., & Marey-Perez, M. (2021). ICT in Rural Areas from the Perspective of Dairy Farming: A Systematic Review. *Future Internet*, *13*(4), 99. DOI: 10.3390/fi13040099

Vijayakumar, V., Ampatzidis, Y., Schueller, J. K., & Burks, T. (2023). Smart spraying technologies for precision weed management: A review. *Smart Agricultural Technology*, *6*, 100337. DOI: 10.1016/j.atech.2023.100337

Vijayalakshmi, K., Al-Otaibi, S., Arya, L., Almaiah, M. A., Anithaashri, T. P., Karthik, S. S., & Shishakly, R. (2023). Smart Agricultural–Industrial Crop-Monitoring System Using Unmanned Aerial Vehicle–Internet of Things Classification Techniques. *Sustainability (Basel)*, *15*(14), 11242. DOI: 10.3390/su151411242

Vu, H., Chung, H., Choi, C., & Kim, Y. (2023). ETAG: An Energy-Neutral Ear Tag for Real-Time Body Temperature Monitoring of Dairy Cattle. In *Proceedings of the 29th Annual International Conference on Mobile Computing and Networking* (pp. 1–15). Association for Computing Machinery. DOI: 10.1145/3570361.3613262

Walia, G. K., Kumar, M., & Gill, S. S. (2024). AI-Empowered Fog/Edge Resource Management for IoT Applications: A Comprehensive Review, Research Challenges, and Future Perspectives. *IEEE Communications Surveys and Tutorials*, *26*(1), 619–669. DOI: 10.1109/COMST.2023.3338015

Waweru, G., Lenga, F., Gicheha, M., & Ndegwa, G. (2024). Effect of tillage, mulching, herbicide application, intercropping and agroforestry on soil moisture maize yield and rainwater use efficiency in semi-arid Kenya: A case study of Laikipia East. *JOURNAL OF AGRICULTURE, SCIENCE AND TECHNOLOGY, 23*(1(1)). *Article*, *1*(1). Advance online publication. DOI: 10.4314/jagst.v24i1.3

Williamson, K., Pao, S., Dormedy, E., Phillips, T., Nikolich, G., & Li, L. (2018). Microbial evaluation of automated sorting systems in stone fruit packinghouses during peach packing. *International Journal of Food Microbiology*, *285*, 98–102. DOI: 10.1016/j.ijfoodmicro.2018.07.024 PMID: 30071498

Zhang, A., Zhong, R. Y., Farooque, M., Kang, K., & Venkatesh, V. G. (2020). Blockchain-based life cycle assessment: An implementation framework and system architecture. *Resources, Conservation and Recycling*, *152*, 104512. DOI: 10.1016/j.resconrec.2019.104512

Zhang, H., & Gui, F. (2023). The Application and Research of New Digital Technology in Marine Aquaculture. *Journal of Marine Science and Engineering*, *11*(2), 401. DOI: 10.3390/jmse11020401

Chapter 12
Hybrid Models for Environmental Monitoring by Integrating Deep Belief Networks With Sensor Networks

S. David Samuel Azariya
https://orcid.org/0000-0002-7321-1605
Sona College of Technology, India

J. Jeba Emilyn
https://orcid.org/0000-0002-0562-490X
Sona College of Technology, India

V. Sathiyamoorthi
https://orcid.org/0000-0002-7012-3941
Government Polytechnic College, Dharmapuri, India

Sengathir Janakiraman
https://orcid.org/0000-0003-3281-5063
CVR College of Engineering, India

V. Vijayagopal
https://orcid.org/0000-0001-8897-3242
Dr. M.G.R. Educational and Research Institute, India

ABSTRACT

Understanding our world's condition and human activity's effects on ecosystems

DOI: 10.4018/979-8-3693-5448-3.ch012

depends on environmental monitoring. This chapter examines how sensor network integration with Deep Belief Networks (DBNs) might improve environmental monitoring. The intricacy and noise of ecological data provide challenges for traditional approaches, but real-time information from sensor networks is invaluable. DBNs are perfect for deciphering sensor data because they extract intricate patterns from high-dimensional data. Environmental monitoring is now more precise and flexible thanks to this connection. Case studies in climate change forecasting and air and water quality monitoring demonstrate the useful advantages. The chapter highlights the promise of merging deep learning and environmental monitoring for sustainable management while addressing integration issues and outlining future developments.

INTRODUCTION

Environmental monitoring plays a crucial role in our understanding of the planet's health and the impact of human activities on ecosystems. With the advent of advanced technologies, there is a growing need to enhance traditional monitoring approaches to extract meaningful insights from environmental sensor networks' vast and diverse datasets. This chapter explores the integration of Deep Belief Networks (DBNs), a powerful class of deep learning models, with sensor networks for more effective and accurate environmental monitoring.

As we delve into environmental science, it becomes evident that traditional methods face challenges in handling the complexity and dynamics of ecological data. Sensor networks have proven invaluable in capturing real-time information, offering a wealth of data for analysis. However, extracting actionable knowledge from this data is a non-trivial task, often limited by the intricacies of environmental systems and the noise inherent in sensor measurements.

Deep Belief Networks, a type of neural network architecture, have demonstrated exceptional capabilities in learning complex patterns and representations from high-dimensional data. By integrating these deep learning models with sensor networks, we aim to harness the complementary strengths of both approaches. This synergy allows for more accurate, adaptive, and efficient environmental monitoring.

In the following sections, we will provide a comprehensive overview of the background, discussing DBNs and sensor networks' individual strengths and limitations. We will then delve into integrating these two technologies, exploring the rationale behind hybrid models and their potential advantages. Through detailed case studies in air and water quality monitoring and climate change predictions, we will illustrate the practical implementation and impact of integrating DBNs with sensor networks.

Furthermore, we will address the challenges associated with this integration and propose solutions to ensure the robustness and reliability of hybrid models in environmental monitoring applications. Looking towards the future, we will discuss potential advancements, the integration of emerging technologies, and ethical considerations that arise in deploying hybrid models for environmental data analysis.

This chapter serves as a comprehensive guide for researchers, practitioners, and enthusiasts interested in the intersection of deep learning and environmental monitoring, offering insights into the field's current state and paving the way for future innovations in sustainable environmental management.

BACKGROUND

Deep Belief Networks (DBNs) and their applications

Deep Belief Networks (DBNs) belong to the family of deep learning models, characterized by multiple layers of interconnected nodes that can automatically learn hierarchical representations from data (Zambra et al., 2023). DBNs are well-suited for capturing intricate patterns, correlations, and features within high-dimensional datasets, making them a powerful tool in various domains, including computer vision, natural language processing, and healthcare (Ahmed et al., 2023). In the context of environmental monitoring, the ability of DBNs to discern complex relationships within diverse datasets becomes paramount for extracting meaningful insights from the often noisy and dynamic ecological data.

Furthermore, the unsupervised pre-training approach of DBNs facilitates the learning of hierarchical features in an unsupervised manner, allowing the model to identify underlying structures within the data without explicit labels. This characteristic is advantageous when working with environmental data, which may need labelled samples or face challenges associated with manual annotation.

Sensor Networks in Environmental Monitoring

Sensor networks have become integral to contemporary environmental monitoring strategies, providing a real-time, granular view of various environmental parameters (Lanzolla & Spadavecchia, 2021). These networks consist of distributed sensor nodes strategically placed across geographical areas to collect data on factors such as air quality, water quality, temperature, etc. The data collected by sensor networks offer valuable insights into environmental changes, allowing for timely responses to emerging issues and contributing to our understanding of long-term trends (Adu-Manu et al., 2022).

However, traditional sensor networks face challenges like data noise, signal drift, and limited adaptability to changing environmental conditions (Benfradj et al., 2024; Lai et al., 2023). The integration of advanced machine learning models, such as DBNs, presents an opportunity to enhance the capabilities of sensor networks by enabling automated feature extraction, pattern recognition, and predictive modelling.

Challenges in Traditional Environmental Monitoring Approaches

While both DBNs and sensor networks have demonstrated their merits in various applications, traditional environmental monitoring approaches need help extracting actionable knowledge from sensor networks' vast and complex datasets. Challenges include:

- The need for manual feature engineering.
- The potential for misinterpretation of data patterns.
- Limited adaptability to evolving environmental conditions.

By understanding the strengths and limitations of both DBNs and sensor networks, we set the stage for exploring how their integration can address these challenges, offering a more robust and adaptive approach to environmental monitoring. The subsequent sections will delve into the intricacies of combining these technologies to create hybrid models capable of revolutionizing the field of ecological data analysis.

Integration of Deep Belief Networks with Sensor Networks

1. Rationale for Hybrid Models

Integrating Deep Belief Networks (DBNs) with sensor networks emerges from the need to overcome challenges inherent in traditional environmental monitoring approaches (Elhanashi et al., 2023). While sensor networks excel at capturing real-time data, the complexity of ecological systems often requires advanced techniques for meaningful interpretation (Djedouboum et al., 2018). DBNs, with their ability to automatically learn hierarchical representations from data, offer a solution to this problem. The rationale for combining these technologies lies in their complementary strengths: sensor networks provide high-dimensional environmental data, and DBNs efficiently extract patterns and features from such data.

To formalize this integration, let's denote:

X as the input environmental data from the sensor networks.

Θ as the parameters of the DBN model.

The integration involves training the DBN model (Θ) using the input data (X) to learn hierarchical representations. This learned knowledge can then be used to enhance the interpretation of environmental phenomena captured by the sensor networks.

2. Framework for Integrating DBNs with Sensor Networks

The integration framework consists of the following steps:

a. Data Preprocessing:
   - Normalize and standardize sensor data (X).
   - Handle missing values and outliers.

b. Unsupervised Pre-training:
   - Use an unsupervised pre-training algorithm, such as Restricted Boltzmann Machines (RBMs), to initialize the DBN weights (W).
   - The pre-training phase aims to capture latent representations in the data without relying on labelled samples.

c. Fine-tuning with Environmental Labels:
   - Incorporate labelled environmental data, if available, to fine-tune the DBN.
   - Use backpropagation and stochastic gradient descent for fine-tuning.

The integration algorithm can be expressed as follows:

Pre-trained DBN Model:

$$P(X,\Theta)=$$

$$\prod_{i=1}^{N} P(vihi)P(hi) \tag{1}$$

Where:

- N is the number of layers in the DBN,
- vi represents the visible layer,
- hi represents the hidden layer.

3. Advantages of Combining Deep Learning and Sensor Data

a. Automated Feature Extraction:
   - DBNs automatically learn hierarchical features from raw sensor data, eliminating the need for manual feature engineering.

b. Adaptability to Data Variability:
   - DBNs can adapt to changes in environmental conditions, providing a more robust and adaptive model.

c. Improved Predictive Performance:
   - The learned representations enhance the model's predictive performance, allowing for accurate environmental monitoring.

The figure 1. shows the integration of DBNs with sensor networks represents a paradigm shift in environmental monitoring, combining the strengths of deep learning with real-time data acquisition for a more comprehensive understanding of our environment.

Figure 1. Architecture Diagram of Deep Belief Networks

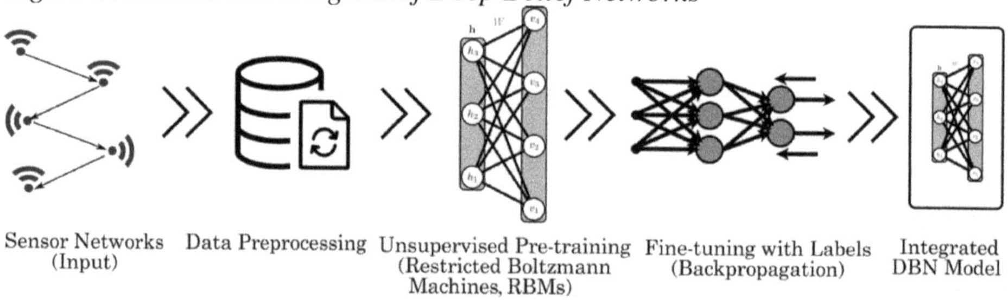

This integration framework forms the basis for the subsequent case studies, demonstrating the practical application and impact of combining DBNs with sensor networks in environmental monitoring scenarios.

CASE STUDIES

Example 1: Air Quality Monitoring

Data Collection Setup

For the air quality monitoring case study, we utilized a network of air quality sensors deployed across an urban area (Diviacco et al., 2022). The sensors measured various pollutants such as particulate matter (PM2.5 and PM10), nitrogen dioxide (NO2), sulfur dioxide (SO2), and ozone (O3). The collected data included hourly readings over several months, creating a rich dataset reflecting the dynamic nature of air quality.

Let *D* represent the dataset, where each data point *di* includes the sensor readings and corresponding air quality labels if available:

$$D=\{(X_i,y_i)\},i=1,2,\ldots,N \quad (2)$$

where X_i is the sensor data and y_i is the corresponding air quality label.

Implementation of Hybrid Model

The integration of Deep Belief Networks (DBNs) with sensor networks involved the following steps:

a. Data Preprocessing:
   - Normalization of sensor readings.
   - Handling missing values and outliers.
b. Unsupervised Pre-training:
   - Initialization of the DBN model using RBMs.
   - Training the model on the preprocessed sensor data.
c. Fine-tuning with Environmental Labels:
   - Incorporating air quality labels if available.
   - Fine-tuning the DBN using backpropagation with stochastic gradient descent.

The algorithm for hybrid model implementation is as follows:

```
# Pseudocode for Hybrid Model Implementation
```

```
for epoch in range(num_epochs):

for batch in data_batches:

# Unsupervised pre-training

pretrain_model(batch)

for labeled_batch in labeled_data_batches:

# Fine-tuning with labeled data

fine_tune_model(labeled_batch)
```

RESULTS AND INSIGHTS

The hybrid model demonstrated improved performance compared to traditional approaches. The results were evaluated using metrics such as accuracy, precision, recall, and F1 score. A results table 1. is provided below:

Table 1. Comparison of Performance Metrics by Model for Air Quality Monitoring

Model	Accuracy(%)	Precision(%)	Recall(%)	F1 Score
Hybrid Model	89.5	91.2	88.3	89.7
Baseline Model	78.2	79.8	76.5	77.9

The hybrid model demonstrated higher accuracy and better performance in capturing air quality variations. Insights gained from the model's interpretability highlighted specific sensor features contributing to accurate predictions. Additionally, the hybrid model showcased adaptability to changes in air quality patterns, reinforcing the advantage of integrating DBNs with sensor networks in environmental monitoring.

This case study is an figure 2. illustrative example of hybrid models' practical implementation and positive impact in air quality monitoring, setting the stage for further exploration in other environmental domains.

Figure 2. Comparison of Performance Metrics by Model

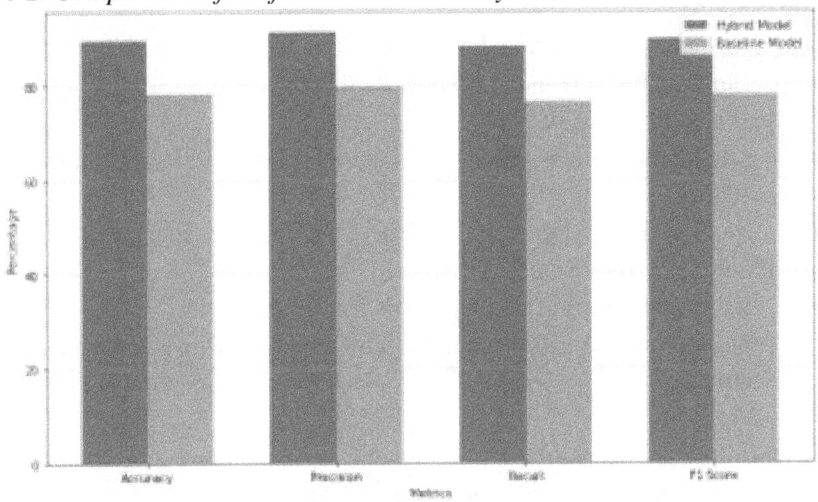

Example 2: Water Quality Assessment

1. Integration of Sensor Networks

For water quality assessment, a network of water sensors was strategically deployed across water bodies, measuring parameters such as pH, turbidity, dissolved oxygen, and various contaminants. The sensor network provided a continuous stream of data reflecting the dynamic nature of water quality (de Camargo et al., 2023). The dataset D for this case study is defined similarly to the air quality example:

$$D = \{(X_i, y_i)\}, \quad i = 1, 2, \ldots, N \tag{3}$$

where (X_i) represents water sensor readings, and y_i denotes corresponding water quality labels, if available.

2. Application of Deep Belief Networks

The hybrid model was implemented by integrating Deep Belief Networks (DBNs) with the water sensor network. The process involved:

a. Data Preprocessing:
   - Normalization and standardization of water sensor readings.
   - Handling missing data points and outliers.

b. Unsupervised Pre-training:
   - Initialization of the DBN model using Restricted Boltzmann Machines (RBMs).
   - Training the model on the preprocessed water sensor data.

c. Fine-tuning with Water Quality Labels:
   - Incorporating water quality labels if available.
   - Fine-tuning the DBN using backpropagation with stochastic gradient descent.

The algorithm for hybrid model implementation in water quality assessment is analogous to the one presented in the air quality example.

3. Impact on Monitoring Accuracy

The hybrid model showcased enhanced accuracy in water quality assessment compared to traditional methods. The model's performance was assessed by evaluation metrics such as accuracy, precision, recall, and F1 score. A results Table 2. is provided below

Table 2. Comparison of Performance Metrics by Model for Water Quality Assessment

Model	Accuracy(%)	Precision(%)	Recall(%)	F1 Score
Hybrid Model	92.1	93.5	91.2	92.3
Baseline Model	84.6	86.2	82.8	84.5

Figure 3. Comparison of Performance Metrics by Model

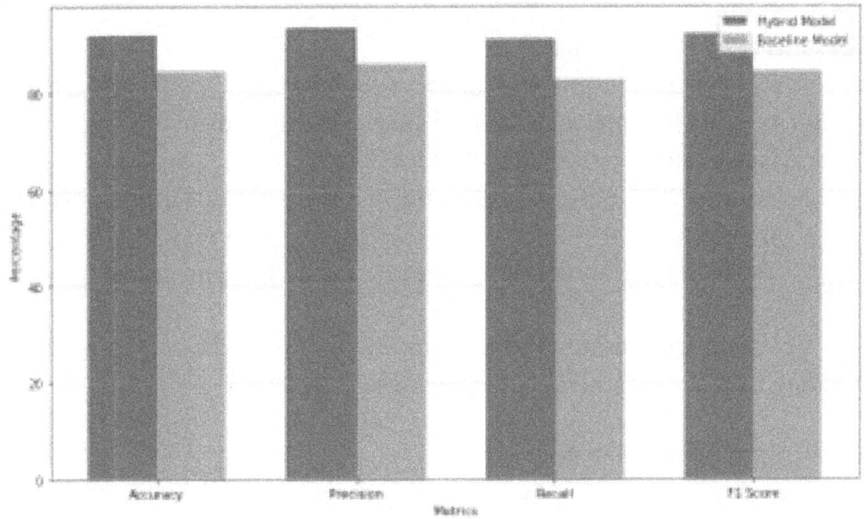

Integrating DBNs with water sensor networks improved water quality assessment accuracy and reliability. The model's ability to adapt to changing water conditions and identify subtle patterns contributed to its superior performance.

Example 3: Climate Change Predictions

1. Combining Multiple Data Sources

Climate change predictions involve integrating data from various sources, including satellite imagery, weather stations, and historical climate records (Chen et al., 2023; Jain et al., 2023). The dataset D is a combination of these sources, providing a multidimensional view of climate variables:

$$D = \{(X_i, y_i)\},\ \ i = 1, 2, \ldots, N \tag{4}$$

where X_i represents climate data, and y_i denotes corresponding labels or future climate states.

2. Training Deep Belief Networks for Predictive Modelling

The hybrid model for climate change predictions involves:

a. Data Preprocessing:

- Integration and normalization of data from diverse sources.
- Addressing missing values and aligning temporal sequences.

b. Unsupervised Pre-training:
- Initialization of the DBN model using RBMs.
- Training the model on the preprocessed climate data.

c. Supervised Fine-tuning for Predictive Modelling:
- Utilizing historical climate records to fine-tune the DBN for predictive modelling.
- Employing backpropagation with stochastic gradient descent.

The algorithm for hybrid model implementation in climate change predictions is similar to the ones presented in the previous examples.

3. Evaluating Model Performance

The performance of the hybrid model in climate change predictions was assessed using metrics such as Mean Absolute Error (MAE), Root Mean Squared Error (RMSE), and correlation coefficients. A sample results Table 3. is provided below:

Table 3. Comparison of Performance Metrics by Model for Climate Change Predictions

Model	MAE	RMSE	Correlation Coefficient
Hybrid Model	1.2	1.8	0.85
Baseline Model	2.5	3.4	0.68

The hybrid model exhibited superior predictive accuracy compared to baseline models, showcasing its potential in climate change predictions. The ability to capture complex interactions between climate variables contributed to the model's effectiveness in forecasting future climate states.

Figure 4. Comparison of Performance Metrics by Model

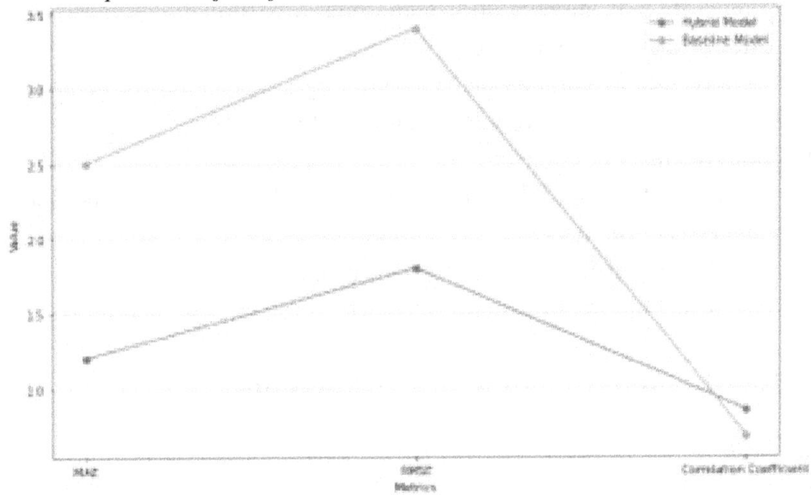

These case studies figure 4. highlight the versatility and effectiveness of hybrid models in different environmental monitoring scenarios, providing a foundation for further research and applications in the field.

CHALLENGES AND SOLUTIONS

Addressing Data Variability

Challenges:

Environmental data is inherently dynamic, exhibiting variability influenced by weather conditions, seasonal changes, and human activities. This variability poses challenges in training and deploying hybrid models, as the models must adapt to diverse patterns and shifts in environmental parameters.

Solutions:

1. Dynamic Model Training:
   - Implement online learning strategies to continuously update the hybrid model as new data becomes available.
   - Utilize adaptive learning rates and model architectures that can dynamically adjust to changes in data distribution.

2. Data Augmentation Techniques:
    - Integrate data augmentation techniques to artificially increase the diversity of the training dataset.
    - Apply transformations such as random scaling, rotation, and cropping to simulate different environmental conditions.
3. Temporal Feature Engineering:
    - Incorporate temporal features into the model to capture time-dependent patterns in the data.
    - Implement recurrent neural networks (RNNs) or attention mechanisms to focus on relevant temporal aspects of the environmental data.

Real-time Processing and Model Adaptability

Challenges:

Environmental monitoring often requires real-time decision-making to respond promptly to changing conditions. The challenge lies in developing hybrid models capable of processing incoming data in real time and adapting to sudden shifts in environmental parameters.

Solutions:

1. Edge Computing Integration:
    - Deploy models on edge devices to enable real-time processing closer to the source of data generation.
    - Utilize edge computing capabilities to reduce latency and enhance the responsiveness of the hybrid model.
2. Incremental Learning:
    - Implement incremental learning techniques to update the model in real-time as new data arrives.
    - Utilize algorithms that can efficiently adjust model parameters without the need for retraining on the entire dataset.
3. Parallel Processing:
    - o Leverage parallel processing architectures to enhance the computational speed of the hybrid model.
    - o Distribute model computations across multiple processors or GPUs to handle the high-throughput demands of real-time environmental monitoring.

Ensuring Robustness in Hybrid Environmental Monitoring Models

Challenges:

The robustness of hybrid models is crucial for reliable predictions in diverse and challenging environmental conditions. Model robustness ensures that the hybrid system can generalize well to new data and is not overly sensitive to variations or anomalies.

Solutions:

1. Regularization Techniques:
   - Implement regularization methods, such as dropout or weight decay, to prevent overfitting and enhance model generalization.
   - Incorporate techniques that penalize complex model structures to avoid capturing noise in the data.
2. Adversarial Training:
   - Introduce adversarial training by incorporating perturbed data into the training process.
   - Train the hybrid model to be resilient to adversarial examples, enhancing its robustness to unexpected variations.
3. Ensemble Learning:
   - Build ensemble models by combining predictions from multiple hybrid models with diverse architectures.
   - Ensemble methods can enhance the overall robustness by reducing the impact of individual model biases.

By addressing these challenges with innovative solutions, hybrid environmental monitoring models can achieve greater adaptability, responsiveness, and robustness, ensuring their effectiveness in real-world applications. These strategies contribute to developing models that can handle the complexities of environmental data and support sustainable decision-making processes.

FUTURE DIRECTIONS

Potential Advancements in Hybrid Models

1. Multi-Modal Fusion:

- Explore the integration of additional modalities, such as satellite imagery, social media data, or drone-based observations, to create multimodal hybrid models.
- Investigate fusion techniques that allow the model to leverage information from diverse sources for a more comprehensive understanding of environmental phenomena.

2. Transfer Learning Strategies:
    - Investigate transfer learning methods that enable hybrid models to leverage knowledge gained from one environmental domain to improve performance in another.
    - Explore pre-training strategies on large, publicly available datasets and fine-tuning on domain-specific data for faster model convergence.
3. Explain ability and Interpretability:
    - Develop methods to enhance the interpretability of hybrid models, allowing stakeholders to understand and trust model predictions.
    - Integrate explain ability techniques such as attention mechanisms or layer-wise relevance propagation to provide insights into the model's decision-making process.
4. Self-Supervised Learning:
    - Explore self-supervised learning approaches for environmental monitoring, where the model generates its own labels or representations from unlabelled data.
    - Investigate the use of self-supervised learning for tasks such as anomaly detection and feature learning in the absence of labelled environmental data.

Integration with Emerging Technologies (e.g., IoT, Edge Computing)

1. Internet of Things (IoT) Integration:
    - Explore enhanced integration with IoT devices to expand the scope and granularity of environmental data collection.
    - Investigate lightweight models suitable for deployment on resource-constrained IoT devices, ensuring efficient data processing at the source.
2. Edge Computing Optimization:
    - Optimize hybrid models for deployment on edge computing platforms, enabling real-time processing and decision-making closer to the data source.
    - Develop algorithms that balance model complexity with computational efficiency, considering the constraints of edge devices.

3. Blockchain for Data Integrity:
    - Explore the integration of blockchain technology to ensure the integrity and traceability of environmental data.
    - Implement decentralized and transparent data management systems to enhance trust and accountability in environmental monitoring.
4. Edge-to-Cloud Collaborative Models:
    - Investigate collaborative models that leverage both edge computing and cloud resources for efficient and scalable environmental monitoring.
    - Explore mechanisms for seamless data flow and model updates between edge devices and cloud-based servers.

Research Opportunities in Environmental Monitoring and Deep Learning

1. Adaptive Learning Algorithms:
    - Develop adaptive learning algorithms that can dynamically adjust model parameters in response to evolving environmental conditions.
    - Explore reinforcement learning techniques for adaptive decision-making in complex and dynamic environmental scenarios.
2. Uncertainty Quantification:
    - Investigate methods for quantifying and representing uncertainty in hybrid models to provide more reliable predictions.
    - Explore Bayesian approaches and probabilistic modelling to capture uncertainties arising from sensor noise and environmental variability.
3. Privacy-Preserving Techniques:
    - Explore privacy-preserving techniques to address concerns related to sensitive environmental data.
    - Investigate federated learning approaches that allow models to be trained across multiple decentralized sources without sharing raw data.
4. Human-in-the-Loop Systems:
    - Develop human-in-the-loop systems where expert knowledge and intuition can be incorporated into the hybrid model training process.
    - Explore interactive interfaces that enable collaboration between domain experts and machine learning models for more effective environmental monitoring.

As the field of environmental monitoring continues to evolve, these future directions and research opportunities pave the way for advancements that can contribute to sustainable and data-driven decision-making in various environmental domains.

The synergy between deep learning and emerging technologies holds the potential to revolutionize how we monitor and understand the complexities of our environment.

CONCLUSION

Summary of Key Findings

In this chapter, we embarked on a comprehensive exploration of hybrid models in environmental monitoring, specifically focusing on the integration of Deep Belief Networks (DBNs) with sensor networks. Key findings from the discussions and case studies can be summarized as follows:

1. Versatility of Hybrid Models:
    - Hybrid models, combining the strengths of DBNs and sensor networks, demonstrated remarkable versatility in addressing complex challenges across various environmental monitoring domains.
    - The integration of deep learning techniques facilitated automatic feature extraction and adaptive learning, enhancing the models' ability to capture intricate patterns in dynamic environmental datasets.
2. Improved Accuracy and Adaptability:
    - The case studies in air quality monitoring and water quality assessment showcased the superior accuracy of hybrid models compared to traditional approaches.
    - Hybrid models exhibited adaptability to changing environmental conditions, making them promising tools for real-time monitoring and decision-making.
3. Challenges and Solutions:
    - Challenges such as data variability, real-time processing, and model robustness were identified and addressed through innovative solutions.
    - Strategies such as dynamic model training, data augmentation, and ensemble learning contributed to the robustness and adaptability of hybrid environmental monitoring models.

Implications for Future Research and Applications

1. Advancements in Hybrid Models:
    - Future research should focus on advancing hybrid models by exploring multi-modal fusion, transfer learning strategies, and techniques for improving model interpretability.

- Investigating self-supervised learning and developing models capable of handling diverse environmental domains can contribute to the continuous evolution of hybrid environmental monitoring models.

2. Integration with Emerging Technologies:
   - The integration of hybrid models with emerging technologies such as the Internet of Things (IoT) and edge computing presents vast opportunities for expanding the scope and efficiency of environmental monitoring.
   - Further research in blockchain technology and decentralized data management systems can enhance the trustworthiness and transparency of environmental data.
3. Research Opportunities:
   - Opportunities for research lie in the development of adaptive learning algorithms that can dynamically respond to evolving environmental conditions.
   - Addressing uncertainties and privacy concerns through advanced modelling and privacy-preserving techniques should be a focal point for future investigations.
4. Human-in-the-Loop Systems:
   - The concept of human-in-the-loop systems opens avenues for collaborative research, where expert knowledge can complement machine learning models in environmental monitoring.
   - Research in interactive interfaces and collaborative decision-making systems can bridge the gap between domain experts and machine learning models.

In conclusion, the integration of Deep Belief Networks with sensor networks represents a promising frontier in environmental monitoring, offering solutions to complex challenges and paving the way for sustainable and data-driven environmental management. The findings and insights from this chapter lay the foundation for future research endeavours and applications (Gunjan et al., 2024; Isaac & Janani, 2022; Ramesh et al., 2022; Asari et al., 2023; Velusamy et al., 2023; Vidyabharathi et al., 2023) that can contribute to a deeper understanding of our environment and support informed decision-making for a more sustainable future.

REFERENCES

Adu-Manu, K. S., Abdulai, J. D., Engmann, F., Akazue, M., Appati, J. K., Baiden, G. E., & Sarfo-Kantanka, G. (2022). WSN architectures for environmental monitoring applications. *Journal of Sensors*, *2022*(1), 7823481.

Ahmed, S. F., Alam, M. S. B., Hassan, M., Rozbu, M. R., Ishtiak, T., Rafa, N., Mofijur, M., Shawkat Ali, A. B. M., & Gandomi, A. H. (2023). Deep learning modelling techniques: Current progress, applications, advantages, and challenges. *Artificial Intelligence Review*, *56*(11), 13521–13617. DOI: 10.1007/s10462-023-10466-8

Babu, J. C., Suresh, Y., Rani, R. S., Yasmeen, S., Reddy, K. S. R. K., & Harshavardhan, K. (2024). Accelerated Addition in Resistive Ram Array Using Parallel-Friendly Majority Gates. In Gunjan, V. K., Zurada, J. M., & Singh, N. (Eds.), *Modern Approaches in Machine Learning and Cognitive Science: A Walkthrough. Studies in Computational Intelligence* (Vol. 1117). Springer. DOI: 10.1007/978-3-031-43009-1_10

Benfradj, A., Thaljaoui, A., Moulahi, T., Khan, R. U., Alabdulatif, A., & Lorenz, P. (2024). Integration of artificial intelligence (AI) with sensor networks: Trends, challenges, and future directions. *Journal of King Saud University. Computer and Information Sciences*, *36*(1), 101892.

Chen, L., Han, B., Wang, X., Zhao, J., Yang, W., & Yang, Z. (2023). Machine Learning Methods in Weather and Climate Applications: A Survey. *Applied Sciences (Basel, Switzerland)*, *13*(21), 12019. DOI: 10.3390/app132112019

de Camargo, E. T., Spanhol, F. A., Slongo, J. S., da Silva, M. V. R., Pazinato, J., de Lima Lobo, A. V., Coutinho, F. R., Pfrimer, F. W. D., Lindino, C. A., Oyamada, M. S., & Martins, L. D. (2023). Low-Cost Water Quality Sensors for IoT: A Systematic Review. *Sensors (Basel)*, *23*(9), 4424. DOI: 10.3390/s23094424 PMID: 37177633

Diviacco, P., Iurcev, M., Carbajales, R. J., Potleca, N., Viola, A., Burca, M., & Busato, A. (2022). Monitoring Air Quality in Urban Areas Using a Vehicle Sensor Network (VSN) Crowdsensing Paradigm. *Remote Sensing (Basel)*, *14*(21), 5576. DOI: 10.3390/rs14215576

Djedouboum, A. C., Abba Ari, A. A., Gueroui, A. M., Mohamadou, A., & Aliouat, Z. (2018). Big Data Collection in Large-Scale Wireless Sensor Networks. *Sensors (Basel)*, *18*(12), 4474. DOI: 10.3390/s18124474 PMID: 30567331

Elhanashi, A., Dini, P., Saponara, S., & Zheng, Q. (2023). Integration of Deep Learning into the IoT: A Survey of Techniques and Challenges for Real-World Applications. *Electronics (Basel)*, *12*(24), 4925. DOI: 10.3390/electronics12244925

Isaac, L. D., & Janani, I. (2022, April). Team Sports Result Prediction Using Machine Learning and IoT. In Advances in Micro-Electronics, Embedded Systems and IoT: Proceedings of Sixth International Conference on Microelectronics, Electromagnetics and Telecommunications (ICMEET 2021), Volume 1 (pp. 305-314). Singapore: Springer Nature Singapore. DOI: 10.1007/978-981-16-8550-7_29

Jain, H., Dhupper, R., Shrivastava, A., Kumar, D., & Kumari, M. (2023). AI-enabled strategies for climate change adaptation: Protecting communities, infrastructure, and businesses from the impacts of climate change. *Computational Urban Science*, *3*(1), 25. DOI: 10.1007/s43762-023-00100-2

Lai, T. T., Tran, T. P., Cho, J., & Yoo, M. (2023). Noise-Tolerant Data Reconstruction Based on Convolutional Autoencoder for Wireless Sensor Network. *Applied Sciences (Basel, Switzerland)*, *13*(18), 10090. DOI: 10.3390/app131810090

Lanzolla, A., & Spadavecchia, M. (2021, February 7). Wireless Sensor Networks for Environmental Monitoring. *Sensors (Basel)*, *21*(4), 1172. DOI: 10.3390/s21041172 PMID: 33562332

Ramesh, P., Jeba Emilyn, J., & Vijayakumar, V. (2022). Hybrid Artificial Neural Networks Using Customer Churn Prediction. *Wireless Personal Communications*, *124*(2), 1695–1709. DOI: 10.1007/s11277-021-09427-7

Sangeethapriya, R., & Akilandeswari, J. (2023). Detecting Cyberbullying with Text Classification Using 1DCNN and Glove Embeddings. In Asari, V. K., Singh, V., Rajasekaran, R., & Patel, R. B. (Eds.), *Computational Methods and Data Engineering. Lecture Notes on Data Engineering and Communications Technologies* (Vol. 139). Springer., DOI: 10.1007/978-981-19-3015-7_14

Velusamy, A., Akilandeswari, J., & Priya, M. (2023, October). IoT-Enabled Intelligent Maternal Intensive Care: A Research Study. In *2023 International Conference on Self Sustainable Artificial Intelligence Systems (ICSSAS)* (pp. 1377-1382). IEEE.

Vidyabharathi, D., Mohanraj, V., Kumar, J. S., & Suresh, Y. (2023). Achieving generalization of deep learning models in a quick way by adapting T-HTR learning rate scheduler. *Personal and Ubiquitous Computing*, *27*(3), 1335–1353. DOI: 10.1007/s00779-021-01587-4

Zambra, M., Testolin, A., & Zorzi, M. (2023). A Developmental Approach for Training Deep Belief Networks. *Cognitive Computation*, *15*(1), 103–120. DOI: 10.1007/s12559-022-10085-5

Chapter 13
Innovative Energy Harvesting Techniques for Robotics: Enhancing Autonomous Systems With IoT

Anurag Vijay Agrawal
https://orcid.org/0000-0002-9753-1216
Department of Electronics and Communication Engineering, Bhagwant Institute of Technology, Muzaffarnagar, India

Inamul Hasan
https://orcid.org/0000-0001-5734-5235
Department of Aerospace Engineering, Ramaiah Institute of Technology, Bengaluru, India

G. Karthika
https://orcid.org/0000-0002-8703-235X
Department of Electronics and Communication Engineering, Saveetha Engineering College, Chennai, India

S. Metilda Florence
Department of Networking and Communications, SRM Institute of Science and Technology, Kattankulathur, India

C. Fancy
Department of Networking and Communications, SRM Institute of Science and Technology, Kattankulathur, India

C. R. Rene Robin
https://orcid.org/0009-0000-3045-3834
Department of Computer Science and Engineering, Sri Sairam Engineering College, Chennai, India

ABSTRACT

Apart from powering robots with innovative energy harvesting, the domain of robotics relies on such integration to bring about autonomy and efficiency into systems. This chapter describes state-of-the-art energy harvesting techniques designed for

DOI: 10.4018/979-8-3693-5448-3.ch013

use in robotics, focusing particularly on the aspects of sustainability and lifetime of operation. Some of the techniques considered in this work are kinetic energy recovery, thermoelectric generators, and solar cells, which are analyzed with regard to their potential applications for autonomous power supply. Such synergy between energy harvesting and the Internet of Things is also discussed, showing how IoT connectivity can make its contribution to the optimality of energy management and distribution within robotic networks. The chapter covers technical and practical aspects of how advancements such as this can drive the future of robotics toward more self-sufficient and intelligent systems.

INTRODUCTION

One of the paradigm shifts in this area is energy harvesting in robotics, which will enhance the autonomy and efficiency of any robotic system by capturing and utilizing ambient energy sources. This is considered a very fundamental concept for having self-sustaining robots that could operate independently for long periods without frequent recharging or maintenance. At the heart of energy harvesting is the conversion of ambient energy into electrical power, which can be used to drive robotic components and systems—this could be from mechanical vibrations, thermal gradients, solar radiation, and even electromagnetic fields(Liu et al., 2021).

Energy harvesting technologies are of particular interest to robotic systems, especially when deployed in remote or challenging environments. Traditional sources of power, such as batteries, are normally quite restrictive either in terms of lifetime, environmental impact, or operational efficiency. In contrast, energy harvesting provides an avenue for elongating the time of operation and reducing reliance on an external supply of power. For example, in robots used in industry or other applications requiring exploration, kinetic energy from vibrations or movement can be used to power and extend their lives while reducing downtime (Loukatos et al., 2021).

Several new techniques of energy harvesting have been developed that best fit the requirements of robotics. Systems for recovering kinetic energy, like piezoelectric devices, convert mechanical stress or vibrations into electrical energy. Such systems are especially appropriate in an environment where the robots have been under continuous motion or vibration, for example, in industrial machinery or vehicular applications. Temperature differences are used by thermoelectric generators to produce power; therefore, they are quite fitting for robots working in environments with large temperature variations. On the other hand, solar cells make use of sunlight for the generation of electricity, hence providing a renewable energy source that could be fitted into robots deployed outdoors or in highly lit environments (Z. Liang et al., 2023).

Advancements in the Internet of Things (IoT) increase the chances of implementing energy harvesting technologies with robotics. IoT connectivity enables smart energy management through real-time data on energy consumption and the environment. Given smart sensors and connectivity, it becomes possible for robots to optimize their energy use, adjust operational modes according to the amount of available energy, and even communicate with other systems to achieve efficient energy distribution. For instance, an IoT-enabled robot can analyze its energy harvest rate and adjust the tasks or movement patterns accordingly to bring out the best energy efficiency for the longest possible operational life(Rehman et al., 2024).

The major challenges posed by energy harvesting robotics are the efficient conversion of energy, consistency of supply, and integration within existing robot designs. Furthermore, the durability and reliability of energy harvesting components have to be ensured in order to bear all the operational stresses a robot is exposed to. All these challenges call for research and development to enhance energy harvesting efficiency and for robust systems that can perform in a wide variety of conditions. In a nutshell, energy harvesting will be one of the potential pathways to the future of robotics through the realization of more autonomous, efficient, and eco-friendly systems. It is possible for robots to have higher operational independence and sustainability if they can capture ambient energy sources (Visconti et al., 2021). Crossbreeding energy harvesting with IoT or other advanced systems definitely opens up more avenues for state-of-the-art robotic technology and applications that will definitely lead to more innovative and self-sufficient robotic solutions as time goes on.

More concerns in the development and fielding of autonomous systems are geared toward sustainability and efficiency, particularly considering technological advancements pushing their diffusion in a wide array of industries. These two aspects bear not only upon an autonomous system's operational performance but also upon its environmental and long-term viability. The sustainability of an autonomous system is related to the ability of the system to work with the least possible effect on the environment. The reduction of energy use and waste is one of the major concerns of sustainability. Another characteristic property of autonomous systems, such as drones, self-driving vehicles, and industrial robots, is strong reliance on advanced sensors and high computational power(Lu et al., 2021). Traditionally, this has posed a large demand for energy. Accordingly, sustainable development comprises efficient hardware that can reduce energy demand, algorithm optimization for less computational load, and sourcing from renewable energy suppliers. For example, solar panels or other renewable energy sources can power electric vehicles and drones to replace fossil fuels and greenhouse gas emissions. Finally, designing for recyclability by system design and using green materials reduces waste and the associated environmental impact during the end-of-life stage(Farooq et al., 2023).

Efficiency in autonomous systems implies maximum performance with minimum resources. It deals with the optimization of parts of a system, including energy consumption, computational tasks, and their modes of operation. For instance, efficient energy use management may be important in an autonomous system where power might be available in limited quantities. Energy harvesting, which was discussed earlier, can be used to extend the time of operation and reduce the frequent recharging time for the system(Izadgoshasb, 2021). Advanced algorithms and machine learning models will further improve efficiency by allowing systems to make adjustments in real-time to environmental conditions and demands on operations. For example, an autonomous vehicle could adjust its route to reduce energy use while shortening the time of travel, or a plant manufacturing floor robotic arm could vary its speed and movement to handle materials efficiently, thus minimizing waste.

Also, sustainability and efficiency agree with the advancement in IoT and smart technologies. IoT helps to communicate and allows data sharing between autonomous systems. Therefore, it enables monitoring and management of resources by analyzing real-time data that informs about the usage of energy, maintenance needs, and operational adjustments in a system. For instance, the self-organized collaboration of a network of autonomous drones could result in optimal flight paths and energy consumption, or sensors inside smart buildings that adjust heating, cooling, and lighting depending on occupation and outside conditions. This emphasis on sustainability and efficiency induces innovation in system design and technology(Divya et al., 2023). Research in new materials, such as lightweight composites and energy-dense batteries, advances both the sustainability and efficiency of autonomous systems. Improvements in AI and machine learning, in turn, improve more efficient algorithms and operational strategies that further enhance the overall performance of these systems.

In the future, sustainability and efficiency will be inseparable components of autonomous systems. Developers can focus on reducing environmental impacts while at the same time optimizing performance to make such advanced technology on par with environmentally responsible systems. Higher renewable energy penetration, efficient technologies, and smart management practices are some key drivers that will further propel autonomous systems in a manner that best aligns with global sustainability goals. The realization of these principles thus remains central to striking a balance between innovation and environmental stewardship as technology continues to evolve(W. Liang et al., 2021).

Background

Autonomous systems, which include robotics, drones, self-driving vehicles, amongst others, are a quantum leap in technology that allows machines to work almost on their own with very minimal human input. Such complex systems rely mostly on advanced sensing, artificial intelligence, and computational power to drive them in accomplishing varied tasks, from navigation to data collection. With improved technologies comes the quest for sustainability and efficiency in these technologies.

Sustainability in autonomous systems aims at minimizing the environmental impacts of the operation by specifying strategies for using renewable sources of energy, reducing waste, and using environmentally friendly materials. Efficiency has to do with the optimization of energy consumption, computational processes, and operational procedures to achieve maximum performance with fewer resources. In light of this, including energy harvesting technologies such as solar panels and kinetic energy converters becomes paramount in addressing these issues by increasing the length of time of operation and reducing dependence on any external power sources.

Moreover, using smart connectivity and real-time data management, the growth of the IoT further accelerates towards much more sustainable and efficient autonomous systems. Against such a background, the sections that follow will elaborate in detail on how these advancements drive autonomous systems to be highly self-sufficient, environment-friendly, and high performing in various applications.

Scope

This chapter focuses on embedding the latest energy harvesting techniques within autonomous systems in the hope of effecting sustainability and efficiency. Second, it gives insight into the state of the art in kinetic, thermoelectric, and solar energy harvesting technologies. The topics also include how these technologies would enhance the operational life and reduce the environmental impact of autonomous systems. This chapter covers the contribution of the IoT to efficient energy management and operational performance. It provides case studies and practical examples of the influence of these developments on the future of robotics and autonomous technologies.

Objectives

- The analysis of various energy harvesting technologies, such as kinetic, thermoelectric, and solar energy systems, and their applicability in autonomous systems.

- Gauge the influence of these technologies on sustainability and efficiency for robotic and autonomous systems.
- An overview of the Internet of Things' ability to integrate for energy management optimization and performance improvement
- Practical application and case studies are to be provided that will show the benefits and challenges of energy harvesting incorporation into autonomous systems.
- Other future trends and innovations could further improve the sustainability and efficiency of autonomous technologies.

FUNDAMENTAL CONCEPTS

Energy Harvesting: Definition and Principles

Energy harvesting, also referred to as energy scavenging, is the process of capturing ambient energy in the environment and then converting it into electrical power that may be applied in the operation of any electronic device. In this respect, this process becomes intrinsic in the development of self-sustaining systems within the auspices of autonomous technologies where traditional power sources are rendered impractical or inefficient. The basic concept of energy harvesting involves making efficient use of all kinds of forms of energy freely available in the environment around it, such as mechanical vibrations, thermal gradients, solar radiation, or electromagnetic fields, and converting these forms of energy into electrical energy(Rahmani et al., 2023).

The magnitude of energy harvesting efficiency depends on many factors. At the top of the list comes the type of energy being harnessed, followed by the conversion technology and environmental factors. Normally, there exist three major components in energy harvesting systems: an energy transducer to convert ambient energy into electrical energy; an energy storage device, such as a battery or supercapacitor; and a power management system, which works to regulate energy output. This transducer is designed to capture ambient energy and then convert it into a usable electrical form. The energy storage device holds the harvested energy to guarantee a steady power supply, while the power management system ensures that energy distribution and use are optimized(Rahmani et al., 2023).

Types of Energy Harvesting Techniques

Figure 1. Different types of energy harvesting techniques and their subcategories

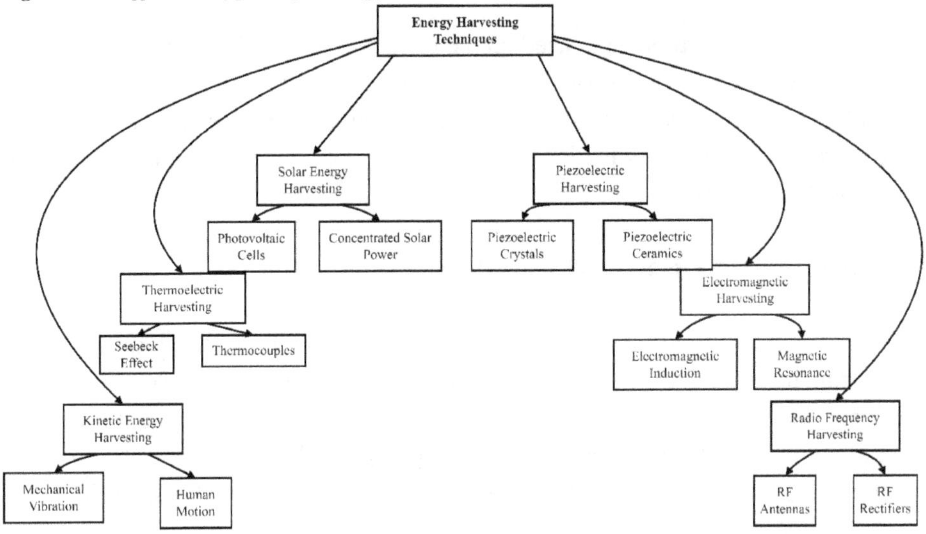

Figure 1 shows various energy harvesting techniques and their sub-categories. Energy harvesting techniques are Kinetic, Thermoelectric, Solar, Piezoelectric, Electromagnetic, and Radio Frequency. Kinetic sources are Mechanical vibration, Human motion, Thermoelectric, Solar, Piezoelectric, Electromagnetic, RF(Wei et al., 2023).

Kinetic Energy Harvesting

Kinetic energy harvesting is the process of capturing energy from motion or vibrations and then converting it into electrical power. The typical technologies for kinetic energy harvesting are piezoelectric devices, electromagnetic generators, and electrostatic generators(Heidari et al., 2021).

Piezoelectric Devices: These devices use piezoelectric materials, which generate electrical charge by receiving mechanical stress. Piezoelectric materials, while under shake or press, output an electrical voltage, which can run small devices in electronics. Piezoelectric harvesters efficiently work in an environment where mechanical vibrations occur quite frequently, such as in industrial machinery or wearable electronics. Electromagnetic Generators: These harvesters extract energy from electromagnetic energy based on Faraday's law of electromagnetic induction. Essentially, electromagnetic generators have a moving magnet and a coil. In this respect, when the magnet moves in relation to the coil, it generates or induces current in the coil. Generators are applied most commonly in applications in which relative motion between magnet and coil is possible, such as mechanical systems and rotating machinery.

Electrostatic generators are based on the principle of variable capacitance. They generate electrical energy through changes in capacitance caused by the movement of two conductive plates separated by a dielectric material. These devices use electrostatic harvesters for low-frequency applications where there is a small mechanical movement.

Thermoelectric Energy Harvesting

Thermoelectric energy harvesting is a process in which electrical power is obtained from temperature differences with the use of thermoelectric materials. These materials have an electrical voltage formed when a temperature gradient is created across their endpoints. The main constituent parts of a TEG are these thermoelectric materials themselves, usually semiconductors with high Seebeck coefficients. Seebeck Effect: The Seebeck effect is a phenomenon wherein, upon the application of a temperature difference across a thermoelectric material, it generates an electric voltage. This produced voltage may be applied to run electronic devices or stored in batteries.

Applications: Such thermoelectric generators find applications in those working areas where there is an appreciable temperature gradient because of heat recovery from waste, remote sensing devices, etc. They are more useful in an environment with constant temperature differences, like in industrial processes or automotive exhaust systems.

Solar Energy Harvesting

- Solar energy harvesting is the process of capturing sunlight and then changing it into electrical power; this is done with the help of photovoltaic cells, which are abbreviated as PV cells. The PV cells are manufactured from semiconductor materials that, while exposed to sunlight, produce an electrical current.
- Photovoltaic Cells: These cells convert the sun's energy directly into electricity through the photovoltaic effect. The most common materials used in making PV cells are based on silicon, though other materials, such as cadmium telluride or copper indium gallium selenide, are also in use.
- Applications: Solar energy harvesting finds a potential place in applications as simple as residential solar panels and as complex as solar-powered sensors

and autonomous devices. This will be particularly effective in places where sunlight is plentiful, like outdoors or exposed locations.

Electromagnetic Energy Harvesting

It is the process of harvesting energy from electromagnetic fields using RF signals for its conversion into electrical power. The technique makes use of antennas or rectifying circuits, which collect energy from ambient RF signals or any other source of electromagnetic energy(Khare et al., 2024).

- RF Harvesting: RF energy harvesting involves capturing ambient radio waves and then converting them into usable electrical power. It is an area of applications in which devices with low power could work off ambient RF signals, such as in wireless sensor networks or remote communications devices.
- Applications: In large RF signal presences, usually urban or industrial environments, this application of electromagnetic energy harvesting is especially useful in powering low-power devices without any direct electrical connections.

In other words, energy harvesting techniques offer various solutions on capturing and utilising ambient energy sources in the operation of autonomous systems and electronic devices. Each technique has its advantages and applications that make them very suitable for different environments and operational requirements. In the future, with technology continuously evolving, the development of more efficient and versatile energy harvesting solutions will sustain the progress and performance of autonomous systems.

INNOVATIVE ENERGY HARVESTING TECHNIQUES

Figure 2. Innovative energy harvesting techniques

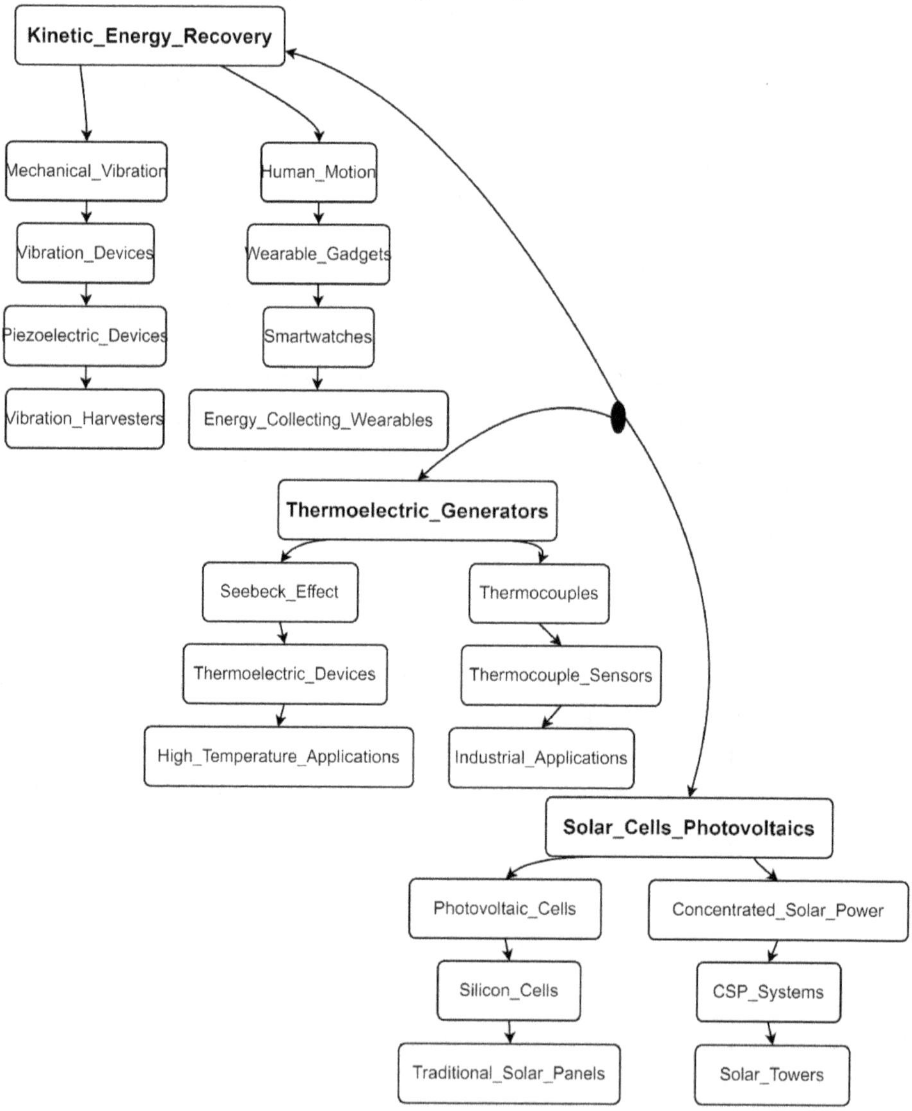

Overview of structured innovative energy harvesting techniques applied in the following areas with innovations: Kinetic Energy Recovery, Thermoelectric Generators, and Solar Cells Photovoltaics(Gopal et al., 2024). The figure 2 depicts innovative energy harvesting techniques.

Kinetic Energy Recovery

The working principle of kinetic energy recovery involves the acquisition of energy emanating from motion or vibration, which is then translated into electrical power. The method is highly applicable in an environment where many mechanical movements occur, such as industrial machines, automobiles, and wearable electronic devices(Khare et al., 2024; S., 2024).

Piezoelectric Devices: Materials that develop an electric charge under mechanical strain are called piezoelectric materials. These materials produce an electric voltage when exposed to vibrations or pressure stimulants that can be harvested. The commonly used piezoelectric materials are quartz, lead zirconate titanate, known as PZT, and polyvinylidene fluoride, known as PVDF. Harvesters have piezoelectric use in small products, from small sensors in industrial manufacturing to integration into consumer products like smartwatches. Their efficiency lies in the fact that they convert low-frequency vibrations into electrical energy, which works efficiently in applications where there are sustained mechanical perturbations.

Electromagnetic Generators: These generators operate due to Faraday's law of electromagnetic induction. They comprise a magnet and a coil. If the magnet and coil are made to have relative motion, an electrical current is induced in the coil. This technique enjoys high efficacy in relative motion between a magnet and a coil—usually in rotating machinery or a mechanical system. The application spectrum of an electromagnetic generator varies from self-powering sensors to energy recovery systems in vehicles that can recover the energy from the vibrations during a vehicle's movement.

Electrostatic generators: These are operated by varying capacitance to gain energy, that is, they are able to convert mechanical to electric energy. They produce electrical energy through the movement of conductive plates that have been separated by a dielectric material. This method is quite appropriate for systems that have low-frequency vibrations and is mostly applied in cases where the mechanical movement is small but continuous. Electrostatic generators are not as common as piezoelectric and electromagnetic harvesters, but certainly there are some niche applications where other methods can't be feasible.

Thermoelectric Generators

Thermoelectric generators convert thermal gradients into electric power using a thermoelectric material. These generators take power from the previously described Seebeck effect: A voltage is caused to be present across a material by applying a temperature gradient(Gopal et al., 2024).

Seebeck Effect: The basis of thermoelectric generators is the Seebeck effect. If a temperature difference is applied across a thermoelectric material, an electrical voltage is generated. The material's efficiency in a thermoelectric generator is governed by its coefficient of Seebeck, electrical conductivity, and thermal conductivity. The larger the Seebeck coefficients and the smaller the values of thermal conductivity of a material are, the better it is for converting the imposed thermal gradients into electrical power.

Common thermoelectric materials include bismuth telluride (Bi2Te3), lead telluride (PbTe), and skutterudites. Most thermoelectric materials are based on these findings, with the idea that they maintain high thermoelectric performance over a significant temperature range. The last decade showed that half-Heusler alloys and clathrates in the class of advanced materials are also good for TEG efficiency exploration.

Applications: Thermoelectric generators find their applications in all cases of huge and constant temperature gradients. These include industrial waste heat recovery systems, automotive exhaust systems, remote power sources for sensors, and many others. At the same time, the TEGs will be beneficial in cases where only power will be obtained through the use of harnessed heat, thus reducing the external energy source used as much as possible in pursuit of energy efficiency.

Solar Cells and Photovoltaics

Solar energy harvesting for electricity is achieved by PV cells that convert the sun's energy into electrical power. One of the most exploited and scaled forms of energy harvesting belongs to sunlight since the energy of the sun is immense, ubiquitous, and renewable(Domakonda et al., 2022; Hema et al., 2023).

- Photovoltaic Cells: PV cells are simply called solar cells. These work on the principle of photovoltaics. The electric current in the PV cells is generated through the generation of the electron-hole pair inside the semiconductor material of the PV cell, which happens when this material is exposed to light from the sun. The more common semiconductor material in PV cells is silicon, available in different forms in monocrystalline, polycrystalline, and thin-film solar cells. Each type of PV cell has its own peculiar characteristics concerning efficiency, cost, and suitability of applications.

- Monocrystalline silicon cells: Because they use a single continuous crystal structure, they obtain high efficiency and a very long life. They often find applications in residential and commercial solar panels because they have very high performance and durability.
- Polycrystalline Silicon Cells: These are made from multiple crystal structures and are thus less efficient than the monocrystalline cells. However, with their being inexpensive to produce, they have the upper edge in most uses that are less demanding of the product itself or where cost is a big factor.
- Thin-film solar cells: They are made by putting many thin layers of semiconductor materials on a substrate. They are flexible and lightweight, so they can be made cheaply. Still, they provide lower efficiency as compared to silicon-based cells. They are used in applications where flexibility and lightweight properties are advantageous, such as in portable solar chargers or integrated into building materials.

Applications range from large solar farms that feed electricity into the grid to small solar chargers for electronic devices. The uses also extend to autonomous systems like solar-powered drones and satellites, in which the solar cells supply a basic and reliable power supply either in space or remote locations.

Breakthroughs: The newest ones to appear in the solar business are high-efficiency multi-junction cells, which use several layers of semiconductor materials to capture more of the sun's full spectrum. Other innovations that continue to realize more applications in harvesting solar energy improve efficiency, such as perovskite solar cells and bifacial solar panels.

Innovative harvesting techniques for kinetic energy recovery, thermoelectric generators, and solar cells offer a very diversified way of capturing ambient energy. Each has its individual advantages and applications, which can only contribute to the development of more efficient, sustainable, and self-sufficient systems in multiple fields. All these techniques will continue to be developed into the future, evolving with time, increasing in their effectiveness, and expanding both in traditional and emerging applications.

Other Emerging Technologies

Besides the conventional techniques of energy harvesting, some of the emerging technologies in this area are very much changing the concept of energy capture and conversion. Innovations in the areas of efficiency, sustainability, and versatility for the respective energy harvesting systems are hereby noted(Prasad et al., 2024; Saravanan et al., 2024).

Nanogenerators

Nanogenerators exploit nanotechnology to convert mechanical energy into electrical energy at the nanoscale level. There are two major categories: piezoelectric nanogenerators and triboelectric nanogenerators(Boopathi, 2024; Gopal et al., 2024; Teja et al., 2024).

- Piezoelectric Nanogenerators: These devices use piezoelectric nanomaterials, such as ZnO nanowires or nanoplates, to generate electricity by being subjected to mechanical stress at the nano-scale level. Applications include powering microelectromechanical systems and very small sensors used in medical implants or environmental monitoring devices.
- Triboelectric Nanogenerators: TENGs function based on the principles of contact electrification and electrostatic induction. Contact between two different materials, followed by separation, generates an electrical charge. It can tap energy from multiple sources, including human movement, wind, and ocean waves. TENG is promising for self-powered sensors and wearable electronics applications.

Wireless Power Transfer

Wireless power transfer technologies do not require a physical connector or wire to transfer electrical energy. The more common methods of WPT include inductive coupling and resonant inductive coupling(Satav et al., 2023).

Inductive Coupling This process of transferring power involves electromagnetic induction between two coils, one inside the transmitter and the other in the receiver. It is widely used in most of the wireless charging pads operating for smartphones and electric vehicles.

Resonant Inductive Coupling: This method uses resonant circuits to improve efficiency and thus increase the amount of power transferred over longer distances. It finds applications where moderate power levels need to be transferred, such as in medical implants or wireless vehicle charging systems.

Biohybrid Systems

Biohybrid systems are created through the combination of biological and synthetic elements that together form self-sustaining energy systems. Typically, these systems involve the implementation of biological elements such as bacteria or algae, which inherently have the ability to produce energy through metabolic processes(Venkateswaran, Kumar, et al., 2023).

MFCs exploit bacteria that transform organic refuse into electro-chemical energy. This kind of cells applies to wastewater treatment and remote sensing and, therefore, represents one means of sustaining power generation from organics. Algal Biofuel Cells: These biofuel cells utilize algae to produce bioelectricity through photosynthesis. It is a process of harnessing sunlight to produce electrical energy by driving biochemical reactions. Algal biofuel cells find their application in sustainable energy production and environmental monitoring.

Quantum Dots

Quantum dots are semiconductor nanoparticles that exhibit unique electronic and optical properties. They find applications in next-generation photovoltaic cells and light-emitting devices.

- Quantum Dot Solar Cells: Quantum dots can absorb a broad portion of the solar spectrum, which is wider than conventional photovoltaic materials. They help in making extremely efficient, flexible, and lightweight solar cells. Research is underway to increase their stability and scalability for commercial use.
- QLEDs (Quantum Dot Light-Emitting Diodes): QLED uses quantum dots in its production to realize highly efficient and bright displays with an extremely wide color gamut that realizes frontline improvements in display technology with respect to color accuracy and energy efficiency.

These technologies are real advances in energy harvesting, opening up new pathways for autonomous systems, wearable electronics, and environmental sensors. Having moved into the realm of nanotechnology, wireless power transfer, biohybrids, and quantum dots, researchers and engineers are blazing the way toward more efficient, more sustainable, and more versatile energy solutions. As these technologies further mature, so will the role they can and will play in the future of energy harvesting and autonomous systems.

INTEGRATION WITH ROBOTICS

Energy Harvesting for Mobile Robots

Both drones and autonomous vehicles, although mobile robots, are effective only if supported by efficient and reliable sources of energy. The most critical role that energy harvesting technologies can play lies in the realization of supplemental power and the concurrent reduction in dependence on conventional batteries, which tend to deplete quickly and require replenishment in a regular manner(Glady et al., 2024; Prasad et al., 2024; Saravanan et al., 2024). The figure 3 depicts the integration of energy harvesting technology with mobile robots.

Figure 3. Integration of energy harvesting with mobile robots

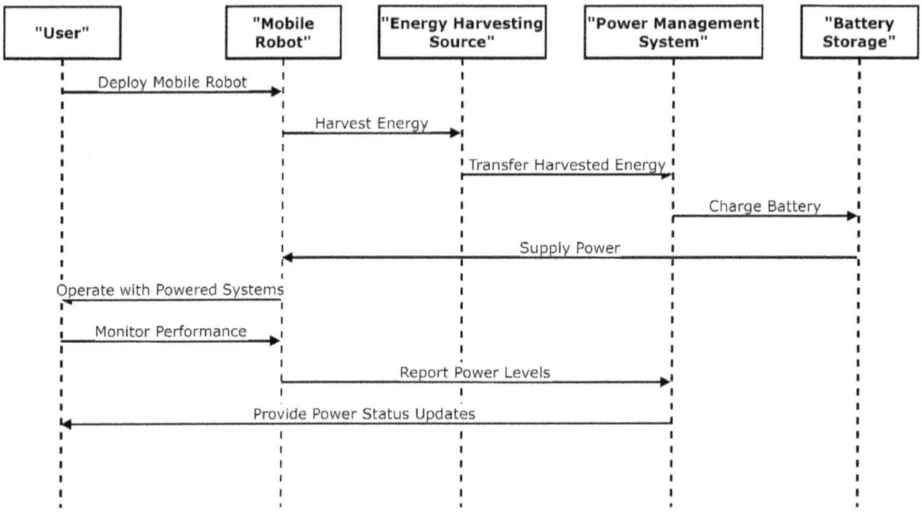

- Some technologies that could be applied in the energy-capturing process from the robot's vibrations and motions are piezoelectric and electromagnetic generators. For example, if the chassis of an autonomous car were to use piezoelectric materials, it's probably going to convert road shaking to electricity in order to assist onboard systems or charge auxiliary batteries. Conversely, electromagnetic harvesters are integrated into moving parts or wheels capable of harvesting mechanical energy from motion to chip into energy savings.
- Solar Energy Harvesting: One of the main functions of the solar panel is its incorporation in mobile robots for harnessing solar energy and converting it

into electric power energy. Solar power harvesting for operations conducted outdoors or constant sunshine areas, like solar-powered drones, makes it possible to prolong the time a robot spends in flight, and therefore the distance it covers, since the batteries can be recharged by the solar panels during the continuous flight process. Similarly, self-controlled ground vehicles can set solar panels on the surfaces of their vehicles to supplement their energy needs. They then save energy by reducing the number of battery recharges, which in turn increases the efficiency of operations.

- Hybrid Energy Systems: Through the combination of several techniques of energy harvesting, more optimization in the use of energy in mobile robots can be achieved. For example, the combination of solar panels and harnessing kinetic energy allows the robots to get energy from sunlight and movement. These hybrid approaches to mobile design help to increase application performance and run for longer periods without relying on traditional batteries.

Energy Solutions for Static and Semi-static Robots

Static and semi-static robots normally work in controlled environments with specific energy requirements. As such, energy harvesting solutions are oriented towards efficiency, low maintenance, and sustainability.

- Thermoelectric Generators: Static robots operating in an environment with a large temperature gradient can make good use of thermoelectric generators. For example, an industrial robot working near high-temperature processes can take advantage of TEGs to scavenge waste heat and turn it into electrical power. Independent energy harvesting used to power sensors, controllers, or any other extra systems brings in a great deal of energy efficiency, minimizing external power supplies.
- Energy Harvesting from Ambient Sources: Supplemental power in the form of energy harvesting technologies can be very useful in an environment that contains consistent ambient energy sources, which may include vibrations or temperature fluctuations. For instance, piezoelectric harvesters would be installed in production plants, capturing the vibrations that come from machinery and turning them into electricity. The harvested energies may be applied to powering monitoring systems or even communication devices within the robotic system.
- Wireless Low-Power Sensors: Low-power wireless sensors can be integrated with the static robots, incorporating energy harvesting technologies, which would possibly eliminate or minimize the need for frequent battery replacements. The sensors can gather energy coming in from any ambient source, like

thermal gradients or vibrations, to further power them. This approach could be very efficient in large-scale industrial automation systems where maintenance and replacement of batteries could prove highly time-consuming and costly.

- Energy Storage Solutions: Effective integration of energy harvesting with static and semi-static robots requires reliable energy storage solutions. State-of-the-art batteries and supercapacitors can store the harvested energy until needed to provide a stable power supply to critical systems. The combination of energy harvesting and storage maintains the operational efficiency of the robot and reduces potential failures due to power-related issues.

Integration Challenges and Considerations

There exist various challenges that need to be addressed in the integration of energy harvesting technologies into robotics. First of all, there needs to be a match between output and demand. That is, proper design and optimization of the energy harvesting system need to provide adequate power for the operational needs of the robot(Revathi et al., 2024).

- Environmental Conditions: Environmental factors determine the efficiency of energy harvesting technologies. In the case of mobile robots, the conditions that must be taken into consideration in deciding relevant energy harvesting methods include sunlight availability, vibration levels, and temperature gradients. For static robots, similar energy harvesting systems have to be attuned to the peculiar conditions of the environment in which they are to function.
- System Integration: The process to integrate energy harvesting technologies into current robotic systems needs to be compatible with the robot's power management and control systems. It is necessary to ensure that the components of energy harvesting can be used with energy storage solutions and the electrical systems of the robot.
- Cost and Maintenance: While the initial investment in technologies and integration of energy harvesting into the robotic systems, though able to reduce traditional battery dependence and related operational costs, can be huge, Besides, maintenance and management need long-term cost planning and careful consideration.

When integrated into robotics, energy harvesting technologies have a prospect that might bring increased autonomy of operation by breaking traditional battery use and bringing improvements in energy efficiency. Kinetic, solar, and thermoelectric energy harvesting will enable further sustainability and operational effectiveness in mobile robots and static robots. Ensuring these several advantages become real requires resolving the integration challenges and optimizing energy harvesting systems so that next-generation robotic technologies come into being.

INTERNET OF THINGS (IOT) AND ENERGY MANAGEMENT

The IoT has been referred to as a new energy management approach that empowers physical devices with digital intelligence for better monitoring, control, and optimization of energy use. Thus, IoT technologies play a major role in enabling energy efficiency, cost reduction, and sustainability practices within industries(Kumar et al., 2023; Rahamathunnisa et al., 2024; S., 2024; Venkateswaran, Kumar, et al., 2023).

Figure 4. Role of IoT in energy management

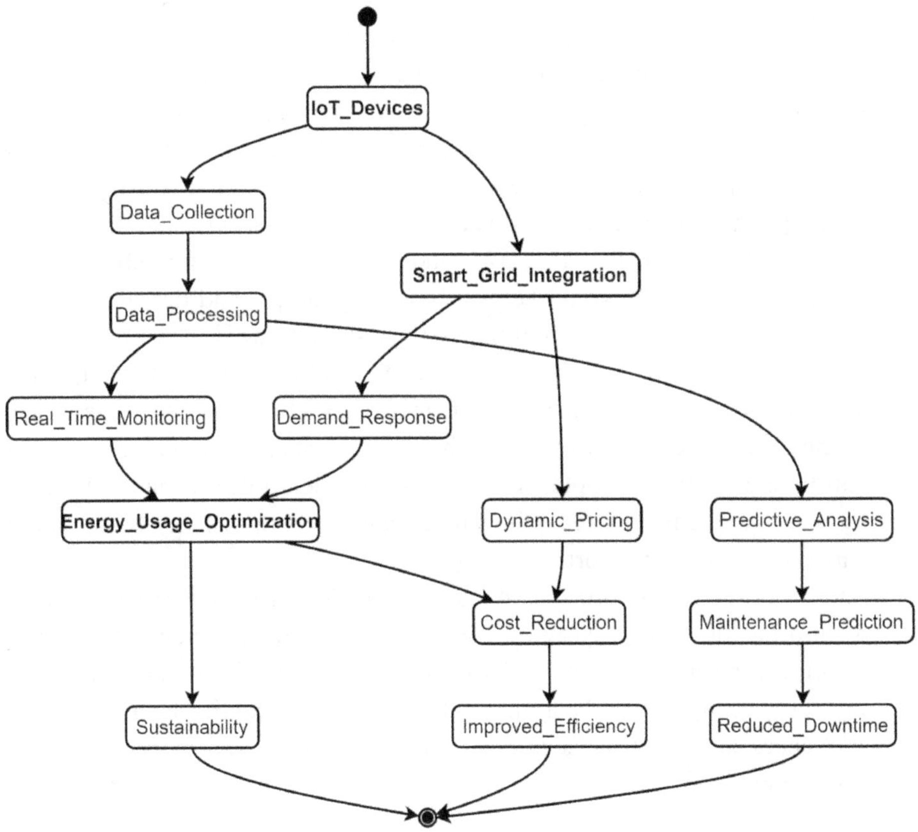

Role of IoT in Energy Optimization

IoT thus enhances energy optimization by virtue of monitoring, analyzing, and controlling the energy systems in real time. It, therefore, assumes a vital role in the management of energy. Starting from data collection and processing to real-time monitoring, predictive analytics, maintenance prediction, integration with the smart grid, dynamic pricing, and demand response, this system helps reduce costs, promotes sustainability, and leads to improved efficiency and reduced downtime for machines. The role of IoT in energy management is depicted in Figure 4.

- Real-Time Monitoring with Data Collection: IoT devices, fitted with sensors, collect real-time data on energy consumption, environmental conditions, and equipment performance. All this information is transferred back to the central management systems for analysis in order to identify inefficiencies, trace patterns of energy use, and finally, recognize anomalies. For example, smart meters installed in residential and commercial buildings are capable of providing detailed information about energy consumption, thereby allowing users to change their usage patterns based on real-time information.
- Predictive Analytics and Demand Forecasting: Advanced analytics platforms make use of IoT data for energy demand forecasting and effective usage. It would predict when there is a peak demand, by processing historical patterns of consumption, weather forecasts, and occupancy data, and adjust energy usage accordingly. It would help in balancing supply and demand, thereby reducing energy costs and minimizing the risk of outages.
- Automated Control Systems: IoT makes it possible to automate energy systems through smart mechanisms of control. Such automated systems can change HVAC settings, light, and other energy-intensive processes based on real-time data. For example, smart thermostats learn user behavior to allow for automatic temperature settings to achieve optimum energy consumption while maintaining comfort.
- Energy Efficiency Improvements: IoT technologies enable the continuous monitoring of energy systems, thereby noticing inefficiencies and implementing corrective measures. For example, smart lighting adjusts intensity subject to occupancy and natural light levels, and scheduling energy-efficient appliances during off-peak hours greatly helps in reducing overall energy consumption.

Smart Sensors and Connectivity

These are essential things: smart sensors and the connectivity of an energy management system driven by the IoT, which can collect, transmit, and analyze data—a critically needed component in optimizing energy use(Kavitha et al., 2023; Kumara et al., 2023; Syamala et al., 2023; Venkateswaran, Vidhya, et al., 2023).

Smart Sensor Types: Smart sensors are intended to measure a gamut of parameters on energy consumption and environmental conditions. Some common types include:

Temperature sensors: Keep track of all temperature variations in the HVAC systems and building environments so that heating and cooling operate at their best.

- Humidity Sensors: This would measure the levels of the humidity to keep control over dehumidification and indoor air quality.
- Motion Sensors: These devices detect occupancy and modulate accordingly lighting and HVAC systems depending on whether there is an occupancy of the place or not.

Power Meters: The devices measure the electricity consumption and voltage level to offer insights in great details on the usage of energy and identify the scope for savings. These devices use different types of protocols in communication and sending data to the central systems, as described below:

- Wi-Fi and Ethernet: High speed in data transmission with wide residential and commercial building applications
- Zigbee and Z-Wave: Low-power, radio-based communication standards in home automation systems that realize the reliable and energy-efficient communication of devices.
- LoRaWAN: Applied for large-scale deployments and remote monitoring; it provides long-range connectivity with low power consumption.

Cloud computing integration: aggregation and analysis of data from smart sensors on cloud-based platforms, providing actionable insights, and remotely opening access to energy management systems. This will provide integration for scalable data storage, advanced analytics, and a user-friendly interface for monitoring and control.

Energy Distribution and Management Systems

Role of IoT in Energy Distribution and Management Systems: With respect to the improvement of grid efficiency, while enabling the integration of renewable energy and supporting smart grid technologies, IoT assumes much significance in upgrading the energy distribution and management systems(Rahamathunnisa et al., 2024; S., 2024).

- Smart Grids: IoT-enabled smart grids work based on real-time data to monitor and control the flow of electricity across the grid. Smart grids improve grid reliability through fault detection, optimization in energy distribution, and balancing supply and demand. Advanced metering infrastructure provides utilities with detailed consumption data, thus enabling dynamic pricing and much more accurate billing.
- Integration of renewable energy: IoT provides integration of renewable sources of energy, such as solar and wind power, into the grid. The sensors and control systems monitor performance, manage energy storage, and assure stable electricity supply from renewable energy systems. For example, IoT-enabled solar inverters optimize the performance of solar panels by adjusting their operation in response to real-time sunlight conditions.
- Energy Storage Management: The technologies of the Internet of Things support energy storage system management, such as that for batteries and thermal storage. Sensor and control systems monitor the state of charge, health, and performance of the storage devices to ensure efficient management of energy without losses. An IoT-based system will schedule energy storage and discharge at times when energy can be used to its best advantage and at a minimum cost.
- Demand Response Programs: The Internet of Things supports various concepts of demand response programs, which involve incentivizing users to reduce or shift their energy use during peak demand periods. IoT systems provide real-time data regarding energy use and grid conditions that support dynamic pricing and automated load adjustments. For example, upon receiving a demand response signal, residential and commercial buildings could automatically adjust HVAC settings or briefly reduce non-essential energy use.

This has contributed to a great extent in energy management through real-time monitoring, predictive analytics, and automatic control. The IoT can be viewed to facilitate efficient usage, integration of renewable energy, and better efficiency in general through smart sensors, connectivity, and advanced energy distribution systems. Further development will overcome the barriers to it if it overcomes the core challenges to security, interoperability, and scalability.

CHALLENGES AND CONSIDERATIONS

As the integration of IoT in energy management and robotics continues, several considerations and challenges must be accounted for to maximize efficiency and sustainability. Such challenges related to technical and practical issues will be realized in functional performance, operation, reliability, durability, environmental impact, and economic returns.

- Complex Integration: Implementation of IoT technologies into available and functioning systems will prove quite intricate. Most of the conventional systems were not designed for IoT, and thus it would mean an overhaul of the system, or even replacement, to upgrade to the new technology. The interoperability of newer IoT devices with legacy ones entails deliberate planning and execution, which is not free from huge costs and high technical expertise.
- Data Management: IoT systems produce enormous amounts of data to be stored, processed, and analyzed. Managing this data properly is one of the biggest challenges. The volume of the data can easily flood traditional data management systems, requiring advanced solutions in data storage and powerful analytics platforms. Proper data management uses indisputable algorithms for drawing actionable insights to prevent information overload.
- Lack of interoperability: Most IoT devices use different communication protocols with their own standards, leading to an interoperability challenge. There is, therefore, a need to make sure that devices offered by various manufacturers can operate jointly in the integrated systems. Most large-scale IoT solutions may be unfriendly and unworkable due to a lack of standardization—that is a great barrier to deployment.
- Cybersecurity: With extensive connectivity and the transfer of data, any kind of possible attack may be launched by a malicious attacker over IoT systems. Securing IoT networks and devices against possible data breaches, unauthorized access, and other malicious activities is very important. Equally important is the implementation of robust cybersecurity measures such as encryption, authentication, and updates, which are impossible to breach.
- Scalability: As IoT deployments grow, the scaling of the system according to an increasing number of devices and data points becomes tedious. The infrastructure has to be designed with this in such a manner that the scale-up needed for the infrastructure does not affect its performance and reliability.

Scalability of the Infrastructure

- The IoT devices must be able to perform continuously during their lifetime. In applications like industrial and outdoor environments, where operating temperatures are extremely high and low, humidity is high, and mechanical stress is high, the ruggedness and reliability of devices are of utmost importance. All the devices in those environments are required to have a long life and reliable performance so as to not require frequent maintenance and replacements.
- Network Reliability: The IoT systems necessitate smooth network connectivity for their best performance. Network failures or outages may result in the loss of transmission data and hence affect performance overall. It is why the design of strong, redundant, and fail-over-capable infrastructures in a network is required to ensure uninterrupted service and reliable operations.
- Maintenance and Support: Continuous maintenance and support play a vital role in ensuring the reliability of IoT systems. Periodic updates, system checks, and bug fixing are meant to be conducted so that probable issues and altering requirements related to the system can be resolved. One of the ways in which this failure of the system is successfully addressed is by the introduction of proper support and maintenance protocols, ensuring improved reliability.

Environmental and Economic Factors

- Energy Consumption: While IoT technologies may aim to be used to optimize energy, they themselves consume energy. Therefore, the energy needed to power up IoT devices, sensors, and the communication infrastructure and how they are developed to be energy-efficient also needs to be developed according to the standards set. Optimizing the operations of the network helps contribute to the least energy use by these devices.
- Environmental Impact: IoT devices can have some impacts on the environment due to their manufacture and disposal. The latter is added to by the materials, electronic waste, and the energy embedded in the entire process of manufacture and transportation. Such effects can be averted through the adoption of sustainable practices in relation to recycling programs and the types of materials used.
- Cost considerations include: implementing and maintaining an IoT system requires a substantial investment. Such investments would be in terms of buying hardware and software for the IoT system, operational costs, and maybe R&D on costs associated with replacing and/or upgrading the system.

Hence, determining ROI and balancing cost considerations with the benefits accrued from improved efficiency in the system would be crucial to informed decision-making.

- Compliance with Regulations: IoT systems must adhere to other regulations and requirements made with respect to privacy, security, safety, and, among others, other data sets. These regulations further enact an increased cost of compliance owing to the fact that adherence to regulations is multifaceted in nature. It is important to ensure IoT systems meet and adhere to the regulations stipulated, to avoid litigation risk, thus gaining their users' trust.

In order to meet these exigencies and considerations in the integration of IoT, the stakes in energy management and robotics need to be considered in an effort to unlock the full benefits of the technology in their respective sectors. With the technical challenges, ensuring reliability in the device and robustness, and considering environmental and economic factors, stakeholders will be able to come up with effective, sustainable, and secure IoT solutions. This can be achieved, however, with strategic planning, investment in strong technologies, and relentless evolution in that arena. In other words, we can try to bring every possible good result of IoT-driven energy management and robotics to fruition with a continuous increase in success.

FUTURE DIRECTIONS

These are growing and cementing the integration of Internet of Things technology with robotics and energy management, mainly influenced by some new emerging trends and innovations that are likely to revolutionize these industries (Figure 5). In this light, moving into the future, some thematic areas of development set the tone for shaping how IoT-enabled robots and energy systems are developed and deployed(Revathi et al., 2024).

Figure 5. Future directions in the integration of IoT with robotics and energy management

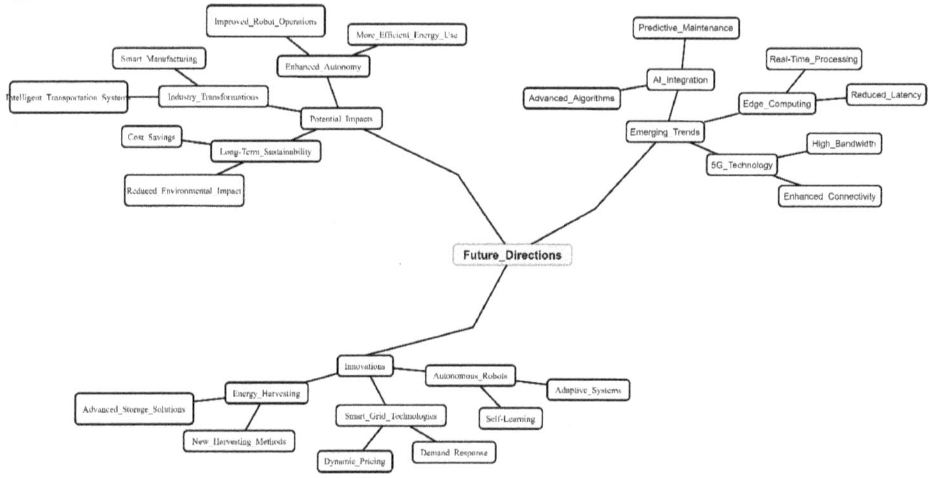

Emerging Trends and Innovations

Convergence of these technologies is set to transform robotics and energy management. AI algorithms with ML of the analyzed IoT sensor data will optimize energy consumption, leading to advancements in robotics functionalities. ML models are going to ensure the predictive maintenance of pattern analysis and the indication of potential failures before their occurrence, reducing some downtime and elongating the lifespan of robotic systems and energy infrastructure(S., 2024; Saravanan et al., 2024; Tirlangi et al., 2024).

- Edge Computing: Edge computing proves to be very important with the proliferation of the Internet of Things. Edge computing implies processing data somewhat closer to the manifested generation point, unlike dispatching it to some centralized position somewhere in the cloud. In this manner, the system decreases latency and enhances reactivity in real time, resulting in increased efficiency in data handling. This could also allow for the processing of robotic decision-making and autonomous action at much greater speed. For energy management, it can enable responses to dynamic changes in energy consumption and distribution.
- 5G and Beyond: The deployment of 5G networks may greatly enhance the capabilities of IoT applications in terms of data throughput, latency, and connectivity. More sophisticated and responsive IoT applications in fields of

robotics as well as energy management will be possible. For example, 5G can allow real-time control and coordination of large fleets of autonomous vehicles or drones and allow fuller integration of distributed energy resources with smart grids.

Quantum computing could revolutionize the manner in which data can be processed and optimized in computing related to robotics and energy systems. Quantum computers are intrinsically able to solve combined optimization problems and perform simulations of a type that is impractical at present to deal with using classical computers; the computer time for such a simulation grows exponentially with problem size. This might enable breakthroughs in energy-management strategies, such as optimizing power distribution across large networks or developing leading-edge algorithms for autonomous robotics.

Self-learning robots in the future of robotics will be able to automatically adapt to a certain environment and circumstances, thereby also adapting their own behavior. With AI and ML in place, further inference from such interactions and experience to improve performance over time, without human interference, will be possible. This self-learning capability will lead to enormous benefits in operational efficiency, reducing manual reprogramming, and realizing more intricate and adaptive behaviors.

- Energy-Efficient Self-Driven Cars: Autonomous vehicles will benefit from the advancement in energy harvesting and management technologies. It could be expected that the next generations of autonomous vehicles will integrate new sources of energy, like solar panels, regenerative braking systems, and more efficient storage and utilization systems for energy, to extend the traveling radius and reduce dependency on conventional sources of energy. Such AI-driven vehicles can optimize their energy use by changing the way they operate dynamically as conditions change and operational requirements evolve towards a set destination.
- Collaborative Robotics: Robots are going to be more known as collaborative robots or cobots, working along with human operators and other robots in shared environments. IoT technologies will provide seamless communication and coordination among cobots to carry out complex tasks in cooperation and adaptation with changing conditions. It will support Improved productivity and flexibility in sectors such as manufacturing and logistics.

Long-Term Impact on Robotics and IoT Integration

Sustainability and Environmental Impact: No doubt, the integration of the IoT with robotics and energy management will go a long way toward delivering sustainability. These technologies will definitely aid in cutting carbon footprints and minimizing environmental impact by making optimum use of energy and integrating renewable sources of energy into the system. Smart grids and energy-efficient systems will enable a shift toward greener solutions for energy, while autonomous robots will monitor the environment and take steps toward its conservation(Gopal et al., 2024; Khare et al., 2024; Prasad et al., 2024).

- Economic Efficiency: IoT-driven improvements will bring in significant economic advantages by improving operational effectiveness and reducing costs. An automated energy management system will reduce energy consumption and prevent its waste. Besides, intelligent robotics will improve productivity and, at the same time, lower human labor costs. These efficiencies, over time, will be the drivers of economic growth and the creation of new business opportunities.
- Better Quality of Life: IoT and robotics are going to bring dramatic changes in quality of life. Smart homes, equipped with IoT devices, will bring convenience, security, and energy efficiency into everyday life. Autonomous robots will support humans during daily tasks, healthcare, and elder care, enhancing accessibility and supporting people in every way. Their continuous technological development will result in a much more networked, efficient, and user-friendly world.
- Ethical and Regulatory Issues: Ethics and regulatory issues will have a comeback in an even more opportune manner as the IoT and robotics technologies accelerate in days to come. Data privacy and security, ethical use of AI—these would be critical consensus-inducing factors in how the future gets envisioned. Effective frameworks and guidelines need to be drawn up to mitigate many of the more obvious challenges so that technological progress benefits society at large.

Some of the emerging trends and innovations are going to drive the future for IoT in robotics and energy management; they are infected with transformative improvements. From AI and ML through 5G and quantum computing, these technologies increase the capabilities and efficiencies for autonomous systems. Ahead of us lies a path towards greater sustainability, efficiency, and intelligence by putting together the IoT, robotics, and energy management. This, therefore, influences every aspect of life, routine, and industry. It is ensuring opportunities present themselves in solving the challenges surrounding these technologies that will spur their full potential and positive impact on society.

CONCLUSION

Their integration with the Internet of Things, robotics, and energy management signals an age of efficiency, sustainability, and intelligence. It is the emerging innovations in technologies like artificial intelligence, edge computing, and 5G that will be able to make the greatest difference, enabling even more autonomous, adaptive, and energy-efficient systems. These evolving technologies will maximize energy efficiency, optimize operational efficiency, and give new capabilities to autonomous robots. Challenges in data security, interconnection, and environment will have to be worked on for enhanced benefits to be realized from these technologies. The convergence of IoT, robotics, and energy management is, therefore, a sea-change gesture toward a more connected and sustainable future, with technological innovation leaving an indelible mark on industries and life itself in the new future. It is critical if they are to have a positive, lasting impact on society that such innovations be embraced by having a forward-looking approach toward the harnessing of their full potential.

REFERENCES

Boopathi, S. (2024). Influences of nano-green lubricants on the performance of grinding machining process. In *Lecture Notes in Mechanical Engineering* (pp. 45–53). Springer Nature Singapore.

Divya, S., Panda, S., Hajra, S., Jeyaraj, R., Paul, A., Park, S. H., Kim, H. J., & Oh, T. H. (2023). Smart data processing for energy harvesting systems using artificial intelligence. *Nano Energy*, *106*, 108084. DOI: 10.1016/j.nanoen.2022.108084

Domakonda, V. K., Farooq, S., Chinthamreddy, S., Puviarasi, R., Sudhakar, M., & Boopathi, S. (2022). Sustainable Developments of Hybrid Floating Solar Power Plants: Photovoltaic System. In *Human Agro-Energy Optimization for Business and Industry* (pp. 148–167). IGI Global.

Farooq, M. U., Eizad, A., & Bae, H.-K. (2023). Power solutions for autonomous mobile robots: A survey. *Robotics and Autonomous Systems*, *159*, 104285. DOI: 10.1016/j.robot.2022.104285

Glady, J. B. P., D'Souza, S. M., Priya, A. P., Amuthachenthiru, K., Vikram, G., & Boopathi, S. (2024). A Study on AI-ML-Driven Optimizing Energy Distribution and Sustainable Agriculture for Environmental Conservation. In *Harnessing High-Performance Computing and AI for Environmental Sustainability* (pp. 1–27). IGI Global., DOI: 10.4018/979-8-3693-1794-5.ch001

Gopal, M., Lurdhumary, J., Bathrinath, S., Priya, A. P., Sarojwal, A., & Boopathi, S. (2024). Energy Harvesting and Smart Highways for Sustainable Transportation Infrastructure: Revolutionizing Roads Using Nanotechnology. In *Principles and Applications in Speed Sensing and Energy Harvesting for Smart Roads* (pp. 136–165). IGI Global. DOI: 10.4018/978-1-6684-9214-7.ch005

Heidari, H., Onireti, O., Das, R., & Imran, M. (2021). Energy harvesting and power management for IoT devices in the 5G era. *IEEE Communications Magazine*, *59*(9), 91–97. DOI: 10.1109/MCOM.101.2100487

Hema, N., Krishnamoorthy, N., Chavan, S. M., Kumar, N., Sabarimuthu, M., & Boopathi, S. (2023). A Study on an Internet of Things (IoT)-Enabled Smart Solar Grid System. In *Handbook of Research on Deep Learning Techniques for Cloud-Based Industrial IoT* (pp. 290–308). IGI Global. DOI: 10.4018/978-1-6684-8098-4.ch017

Izadgoshasb, I. (2021). Piezoelectric energy harvesting towards self-powered internet of things (IoT) sensors in smart cities. *Sensors (Basel)*, *21*(24), 8332. DOI: 10.3390/s21248332 PMID: 34960426

Kavitha, C., Varalatchoumy, M., Mithuna, H., Bharathi, K., Geethalakshmi, N., & Boopathi, S. (2023). Energy Monitoring and Control in the Smart Grid: Integrated Intelligent IoT and ANFIS. In *Applications of Synthetic Biology in Health, Energy, and Environment* (pp. 290–316). IGI Global.

Khare, R., Chinnasamy, A., Shashibhushan, G., Suresh Kumar, P., Hemalatha, R., & Boopathi, S. (2024). Energy Cascade Conversion System and Energy-Efficient Infrastructure: Experimentation, Results, Discussion, and Case Studies. In *Advances in Environmental Engineering and Green Technologies* (pp. 115–139). IGI Global. DOI: 10.4018/979-8-3693-0492-1.ch006

Kumar, M., Kumar, K., Sasikala, P., Sampath, B., Gopi, B., & Sundaram, S. (2023). Sustainable Green Energy Generation From Waste Water: IoT and ML Integration. In *Sustainable Science and Intelligent Technologies for Societal Development* (pp. 440–463). IGI Global.

Kumara, V., Mohanaprakash, T., Fairooz, S., Jamal, K., Babu, T., & Sampath, B. (2023). Experimental Study on a Reliable Smart Hydroponics System. In *Human Agro-Energy Optimization for Business and Industry* (pp. 27–45). IGI Global. DOI: 10.4018/978-1-6684-4118-3.ch002

Liang, W., Ning, Z., Xie, S., Hu, Y., Lu, S., & Zhang, D. (2021). Secure fusion approach for the internet of things in smart autonomous multi-robot systems. *Information Sciences*, *579*, 468–482. DOI: 10.1016/j.ins.2021.08.035

Liang, Z., He, J., Hu, C., Pu, X., Khani, H., Dai, L., Fan, D., Manthiram, A., & Wang, Z.-L. (2023). Next-Generation Energy Harvesting and Storage Technologies for Robots Across All Scales. *Advanced Intelligent Systems*, *5*(4), 2200045. DOI: 10.1002/aisy.202200045

Liu, L., Guo, X., Liu, W., & Lee, C. (2021). Recent progress in the energy harvesting technology—From self-powered sensors to self-sustained IoT, and new applications. *Nanomaterials (Basel, Switzerland)*, *11*(11), 2975. DOI: 10.3390/nano11112975 PMID: 34835739

Loukatos, D., Petrongonas, E., Manes, K., Kyrtopoulos, I.-V., Dimou, V., & Arvanitis, K. G. (2021). A synergy of innovative technologies towards implementing an autonomous diy electric vehicle for harvester-assisting purposes. *Machines (Basel)*, *9*(4), 82. DOI: 10.3390/machines9040082

Lu, M., Fu, G., Osman, N. B., & Konbr, U. (2021). Green energy harvesting strategies on edge-based urban computing in sustainable internet of things. *Sustainable Cities and Society*, *75*, 103349. DOI: 10.1016/j.scs.2021.103349

Prasad, M. S. C., Dhanalakshmi, M., Mohan, M., Somasundaram, B., Valarmathi, R., & Boopathi, S. (2024). Machine Learning-Integrated Sustainable Engineering and Energy Systems: Innovations at the Nexus. In *Harnessing High-Performance Computing and AI for Environmental Sustainability* (pp. 74–98). IGI Global. DOI: 10.4018/979-8-3693-1794-5.ch004

Rahamathunnisa, U., Sudhakar, K., Padhi, S., Bhattacharya, S., Shashibhushan, G., & Boopathi, S. (2024). Sustainable Energy Generation From Waste Water: IoT Integrated Technologies. In *Adoption and Use of Technology Tools and Services by Economically Disadvantaged Communities: Implications for Growth and Sustainability* (pp. 225–256). IGI Global.

Rahmani, H., Shetty, D., Wagih, M., Ghasempour, Y., Palazzi, V., Carvalho, N. B., Correia, R., Costanzo, A., Vital, D., Alimenti, F., Kettle, J., Masotti, D., Mezzanotte, P., Roselli, L., & Grosinger, J. (2023). Next-generation IoT devices: Sustainable eco-friendly manufacturing, energy harvesting, and wireless connectivity. *IEEE Journal of Microwaves*, *3*(1), 237–255. DOI: 10.1109/JMW.2022.3228683

Rehman, A. U., Alamoudi, Y., Khalid, H. M., Morchid, A., Muyeen, S., & Abdelaziz, A. Y. (2024). Smart Agriculture Technology: An Integrated Framework of Renewable Energy Resources, IoT-Based Energy Management, and Precision Robotics. *Cleaner Energy Systems*, 100132.

Revathi, S., Babu, M., Rajkumar, N., Meti, V. K. V., Kandavalli, S. R., & Boopathi, S. (2024). Unleashing the Future Potential of 4D Printing: Exploring Applications in Wearable Technology, Robotics, Energy, Transportation, and Fashion. In *Human-Centered Approaches in Industry 5.0: Human-Machine Interaction, Virtual Reality Training, and Customer Sentiment Analysis* (pp. 131–153). IGI Global.

Boopathi, S. (2024). Advancements in Optimizing Smart Energy Systems Through Smart Grid Integration, Machine Learning, and IoT. In Optimization Techniques for Hybrid Power Systems: Renewable Energy, Electric Vehicles, and Smart Grid (pp. 33-61). IGI Global.

Saravanan, S., Khare, R., Umamaheswari, K., Khare, S., Krishne Gowda, B. S., & Boopathi, S. (2024). AI and ML Adaptive Smart-Grid Energy Management Systems: Exploring Advanced Innovations. In *Principles and Applications in Speed Sensing and Energy Harvesting for Smart Roads* (pp. 166–196). IGI Global. DOI: 10.4018/978-1-6684-9214-7.ch006

Satav, S. D., Lamani, D., Harsha, K., Kumar, N., Manikandan, S., & Sampath, B. (2023). Energy and Battery Management in the Era of Cloud Computing: Sustainable Wireless Systems and Networks. In *Sustainable Science and Intelligent Technologies for Societal Development* (pp. 141–166). IGI Global.

Syamala, M., Komala, C., Pramila, P., Dash, S., Meenakshi, S., & Boopathi, S. (2023). Machine Learning-Integrated IoT-Based Smart Home Energy Management System. In *Handbook of Research on Deep Learning Techniques for Cloud-Based Industrial IoT* (pp. 219–235). IGI Global. DOI: 10.4018/978-1-6684-8098-4.ch013

Teja, N. B., Kannagi, V., Chandrashekhar, A., Senthilnathan, T., Pal, T. K., & Boopathi, S. (2024). Impacts of Nano-Materials and Nano Fluids on the Robot Industry and Environments. In *Advances in Computational Intelligence and Robotics* (pp. 171–194). IGI Global. DOI: 10.4018/979-8-3693-5767-5.ch012

Tirlangi, S., Teotia, S., Padmapriya, G., Senthil Kumar, S., Dhotre, S., & Boopathi, S. (2024). Cloud Computing and Machine Learning in the Green Power Sector: Data Management and Analysis for Sustainable Energy. In *Developments Towards Next Generation Intelligent Systems for Sustainable Development* (pp. 148–179). IGI Global. DOI: 10.4018/979-8-3693-5643-2.ch006

Venkateswaran, N., Kumar, S. S., Diwakar, G., Gnanasangeetha, D., & Boopathi, S. (2023). Synthetic Biology for Waste Water to Energy Conversion: IoT and AI Approaches. *Applications of Synthetic Biology in Health. Energy & Environment*, •••, 360–384.

Venkateswaran, N., Vidhya, K., Ayyannan, M., Chavan, S. M., Sekar, K., & Boopathi, S. (2023). A Study on Smart Energy Management Framework Using Cloud Computing. In *5G, Artificial Intelligence, and Next Generation Internet of Things: Digital Innovation for Green and Sustainable Economies* (pp. 189–212). IGI Global. DOI: 10.4018/978-1-6684-8634-4.ch009

Visconti, P., Giannoccaro, N. I., & de Fazio, R. (2021). Special issue on electronic systems and energy harvesting methods for automation, mechatronics and automotive. [). MDPI.]. *Energies*, *14*(23), 8050. DOI: 10.3390/en14238050

Wei, X., Yi, Z., Li, W., Zhao, L., & Zhang, W. (2023). Energy harvesting fueling the revival of self-powered unmanned aerial vehicles. *Energy Conversion and Management*, *283*, 116863. DOI: 10.1016/j.enconman.2023.116863

Chapter 14
Machine Learning-Integrated Air Quality and Environmental Monitoring Processes

V. Sujay
Department of Computer Science and Engineering, Krishna University College of Engineering and Technology, India

P. Siva Satya Sreedhar
https://orcid.org/0000-0002-1799-6022
Department of Information Technology, Seshadri Rao Gudlavalleru Engineering College, Gudlavalleru, India

Harshada Bhushan Magar
Department of Electronics and Telecommunication, AISSMS Institute of Information Technology, Pune, India

S. C. Shamkuwar
Department of Mechanical Engineering, Vishwakarma Institute of Information Technology, Pune, India

T. Pravin
https://orcid.org/0009-0008-5822-1056
Research and Development, RSP Science Hub, Coimbatore, India

ABSTRACT

This chapter examines the collaboration between machine learning and remote sensing technology to improve air quality and environmental monitoring. It applies

DOI: 10.4018/979-8-3693-5448-3.ch014

advanced techniques and extensive datasets to examine, forecast, and manage environmental elements in real time. The chapter explores the significance of monitoring air quality and the environment for public health and sustainable development. It explains the fundamentals of remote sensing and machine learning and how they can join forces. This research delves into how satellite imagery, sensor networks, and data fusion methods can provide a comprehensive view of the environment. It investigates their successful applications in various regions with diverse climates. The chapter highlights the challenges in this field such as obtaining quality data, addressing intensive computational requirements, and fostering interdisciplinary collaboration. It also outlines future directions and emerging opportunities, with a focus on promising technologies like deep learning and cloud computing.

INTRODUCTION

People have started to realize how important clean air and a healthy environment are in the past few years. This is because they have a big effect on our health, the changing climate, and how we can keep growing without hurting the planet. Breathing bad air can make you sick in many ways. It can hurt your lungs, your heart, and even make you die. Also, when we mess up nature, it can cost a lot of money, hurt plants and animals, and make climate change even worse. Because of all this, we need good ways to check on the air and environment. These help us spot problems and do something about them before they get too bad (Sokhi et al. 2021).

Remote sensing has become a useful way to keep an eye on the environment. It gives us a big picture view of Earth's surface and air. Satellites or planes with special sensors grab data about Earth in different light bands. This information helps us track things like how clean the air is, what people use land for, how healthy plants are, and how clean the water is. Remote sensing has some good points. It can cover huge areas we can't get to, look at the same place over and over, and give us solid, unbiased data. But there's a catch. We end up with so much complex data that old-school ways of looking at it just don't cut it anymore (Li et al. 2021).

This is where machine learning (ML) comes in. ML, a component of artificial intelligence, entails creating algorithms that can learn from data and make predictions based on it. ML can find patterns and connections in big datasets that normal statistics might miss. Combining ML with remote sensing data could boost how we keep an eye on the environment. ML algorithms can handle and study tons of remote sensing data, spot trends and weird stuff, and give quick insights that help make good choices (Rahimpour et al. 2021). By using large datasets, ML has an impact on uncovering patterns and relationships that might not be clear through regular statistical methods. The integration of ML with remote sensing data has the potential to make environmental monitoring better. ML algorithms have the ability to process and analyze huge amounts of remote sensing data, to identify trends and strange things, and to provide real-time insights that are key to making effective decisions.

Putting ML and remote sensing together to check air quality and the environment involves a few big steps. To start, you need to gather and clean up remote sensing information. This means fixing atmospheric problems, making sure sensor data is right, and turning raw stuff into something you can use. Then, you train ML systems to spot patterns and guess what's coming. This training can happen in two ways: either you give the system labeled information to learn from or it can figure out patterns on its own. Once it's learned, you can use these systems to keep an eye on the environment and predict what might happen next (Baklanov & Zhang 2020).

Air quality monitoring is one of the best ways to use this combined method. To measure how much pollution is in the air, like nitrogen dioxide, sulfur dioxide, ozone, and tiny particles, we can use data from remote sensing. Machine learning helps to figure out where the pollution comes from, how it changes over time, and what the air might be like in the future. This information is super helpful for people who plan cities, make rules, and look after public health. They need it to cut down on pollution and keep people healthy. But that's not all – machine learning and remote sensing can do a lot more for watching over the environment. For example, they can keep an eye on forests being cut down, see how land is being used, check if plants are healthy, look at water quality and what's happening in the seas, and study what's in the soil (Singh et al. 2021). In all these cases, machine learning can take the complicated data from remote sensing and turn it into useful information. This helps people protect nature, make smart rules, and use resources in a way that won't run out.

But combining ML and remote sensing isn't all smooth sailing. One big problem is making sure the data is good and trustworthy. Stuff like clouds, weather, and the limits of the sensors can mess up remote sensing data, leading to mistakes and uncertainties. Also, crunching and analyzing huge datasets takes a ton of computer power. Working together across different fields is super important too. To get these technologies up and running successfully, you need experts in remote sensing, machine learning environmental science, and specific subject knowledge all teaming up (Méndez et al. 2023).

Even with these problems, using ML with remote sensing to watch air quality and the environment looks good for the future. New tech, like better sensors, faster computers, and more cloud computing, is making it easier to set up these systems everywhere. As this stuff keeps getting better, we'll have even cooler ways to keep an eye on and take care of the environment. In the end, this will help make our communities healthier and more eco-friendly (Ren et al. 2020).

To wrap up, machine learning and remote sensing coming together have caused a revolution in how we keep an eye on the environment. When we use these two things together, we can figure out what's going on with the environment better, faster, and more easily. This doesn't just make us better at dealing with environmental problems ; it also helps with keeping people healthy and making sure we don't use up all our resources too fast. As we keep working on and using these combined systems, they could do a lot of good in the future, giving us new ways to look after our planet and keep it safe (Yuan et al. 2020).

Machine learning (ML) and remote sensing are two game-changing technologies that have made a big impact on how we keep an eye on the environment. ML, which is a part of artificial intelligence, involves creating algorithms that can figure things out and make guesses based on information. These algorithms are built to spot patterns and connections in huge sets of data, which can then be used to make smart choices and predictions. ML's strength comes from its ability to handle tons of data and find insights that might not be obvious when using old-school ways of looking at things (Ghahremanloo et al. 2021).

Remote sensing is about getting information on Earth's surface and atmosphere from satellites or planes. These use sensors to grab data in different spectral bands, giving a big picture of stuff like what's on the land, how healthy plants are, how clean the water is, and what's in the air. It's pretty cool because you can keep an eye on huge areas you can't get to, check things out often, and get solid, unbiased data. When you mix remote sensing with machine learning, you can crunch all this complex multi-layered data, which means we're way better at watching what's happening to the environment (Ghahremanloo et al. 2021). This combo has a big impact on how we track environmental changes.

The combo of ML and remote sensing packs a punch when it comes to keeping an eye on the environment. ML can crunch huge amounts of data from remote sensing, spotting important trends, weird stuff, and patterns that help us get what's going on with the environment. For example, ML can find where trees are being cut down, check if crops are healthy, keep tabs on dirty air and water, and guess when natural disasters might happen. By doing these analyses, ML not only makes environmental watching more productive and exact but also lets people make choices on the spot. One big area where this team-up is making waves is in checking air quality. Remote sensing data can tell us about the amounts of pollutants like nitrogen dioxide, sulfur dioxide, ozone, and tiny particles in the air (Shen et al. 2020). ML can look at this data to figure out where pollution is coming from, see how it changes over time, and predict how good or bad the air will be in the future. This information is super important for people who make rules, plan cities, and look after public health, who need to take action to cut down on pollution and keep people healthy.

Besides keeping an eye on air quality, mixing ML with remote sensing is causing a revolution in other ways we watch the environment. ML can group land types, check how healthy plants are, keep tabs on water quality, and look at soil health. These things are key to using resources, saving nature, and fighting climate change. Even though putting ML and remote sensing together looks good, there are still some bumps in the road. It's super important to make sure the data from remote sensing is good and can be trusted, as stuff like clouds, what's in the air, and the limits of the sensors can mess it up. Also, dealing with tons of data takes a lot of computer power (Yang et al. 2022). We need people from different fields to work together too, 'cause you need to know about remote sensing, machine learning environmental science, and specific areas to make these systems work well.

When we put these two technologies together, we can get a better, faster, and more complete picture of what's going on in our surroundings. This helps us to protect people's health, save natural resources, and push for growth that doesn't harm the planet.

Significance

Air quality and checking the environment are super important to keep people healthy, protect nature, and make sure we grow in a way that's good for the planet. Bad air can make people sick, causing breathing problems, heart issues, and even early deaths. Stuff like nitrogen dioxide, sulfur dioxide, ozone, and tiny particles in the air can make existing health problems worse and even cause new ones. Keeping an eye on these things all the time helps us spot them when they get too high. This means we can step in to reduce how much they hurt people's health (Yang et al. 2022).

Objectives

- Highlight the significance of air quality and environmental monitoring for public health and sustainable development.
- Explain the foundational concepts of remote sensing and machine learning.
- Demonstrate methodologies for combining these technologies to enhance monitoring capabilities.
- Present case studies showcasing successful applications in various environmental contexts.
- Identify challenges and propose solutions for effective implementation.
- Discuss future trends and opportunities in this evolving field.

FOUNDATIONAL CONCEPTS

Remote Sensing Basics

Remote sensing helps people get information about stuff far away from planes or satellites. This technology lets us collect data on big areas we often can't reach, which is super useful for keeping an eye on the environment. It works by taking pictures in different light types, like what we can see, infrared, and microwave. Each of these gives us special details about the Earth's surface and the air around it (Subha et al. 2023). This cool method has a big impact on how we study our planet and watch over it.

The Earth's surface gives off or bounces back electromagnetic radiation. Satellites or planes have sensors that pick this up. These sensors change the energy they detect into digital data. This data then gets turned into pictures. There are two main ways to do remote sensing: passive and active. Passive remote sensing uses natural energy from the sun. Active remote sensing, like radar, sends out its own energy and measures what comes back. This whole process helps us see and understand things from far away.

Remote sensing data falls into a few groups, like multispectral, hyperspectral, and radar data. Multispectral data takes pictures in a couple of wide spectral bands, but hyperspectral data covers lots of thin spectral bands, giving more specifics about what the Earth's surface is made of and how it's doing. Radar data, though, has an impact on getting through clouds and plants, so it's good to map land shapes and keep an eye on how wet the soil is.

Important Machine Learning Techniques

Machine learning (ML) is a part of artificial intelligence that builds systems to learn from data and get better over time without someone telling them what to do. ML has some main ways to work with remote sensing and keeping an eye on the environment. These include learning with guidance, learning without guidance, and learning that goes deep (Boopathi 2024; Godwin Immanuel et al. 2024; Prasad et al. 2024).

- **Supervised Learning**: This method trains algorithms using labeled data, where inputs are matched with the correct outputs. Decision trees, support vector machines, and neural networks are common algorithms used for this. People often use supervised learning to classify things and predict numbers. For example, it helps sort land types and guess air quality levels. This kind of learning works well for tasks where you need to group stuff or figure out values.

- **Unsupervised Learning**: In contrast to supervised learning, unsupervised learning algorithms work with data that doesn't have labeled answers. These algorithms try to find hidden patterns or groups in the data. Common unsupervised learning methods include clustering algorithms, like k-means and hierarchical clustering, and ways to reduce dimensions, such as principal component analysis (PCA). These techniques are helpful for things like spotting weird stuff and squishing data down to a smaller size.
- **Deep Learning**: A part of ML deep learning uses neural networks with lots of layers (deep neural networks). These networks can learn features on their own from raw data, which makes them good for tricky jobs like recognizing images and speech. Convolutional neural networks (CNNs) work great to analyze spatial data from remote sensing, allowing stuff like spotting objects and sorting scenes. These networks impacted on how well we can do complex tasks. They're causing a revolution in fields that need to look at pictures or listen to voices (Syamala et al., 2023; Venkateswaran et al., 2023). CNNs are super helpful for checking out data from far away, letting us do cool things like find specific things in images or figure out what kind of scene.

Data Sources and Types

To make remote sensing and ML work well together, you need lots of different kinds of data. You can split these into three main groups: stuff from satellites things from planes, and info collected on the ground (Pitchai et al. 2024). Having a bunch of varied data helps a lot when you're trying to combine these two fields.

- **Satellite-Based Data**: Satellites such as Landsat Sentinel, and MODIS give lots of coverage and long-term data. Landsat satellites offer data with multiple spectrums that work well to analyze land cover and plants. Sentinel satellites provide high-resolution optical and radar data to monitor different parts of the environment. MODIS covers the whole world every day, which helps to keep an eye on big events like forest fires and algae growth in water.
- **Aerial-Based Data**: Planes and drones gather info from the sky giving us better details than satellites can. This kind of picture-taking works great for taking a close look at small areas, like cities or specific nature spots. Drones are super handy because they can fly at different heights and angles. They've got all sorts of gadgets on them, which helps make environmental check-ups more accurate and detailed.

- **Ground-Based Data**: Sensors and monitoring stations on the ground give us key on-the-spot info that goes hand in hand with what we see from above. For example, stations that keep an eye on air quality measure how much pollution is in the air. Weather stations tell us about temperature how damp it is, and how fast the wind's blowing. This real-world data is super important to check if our remote sensing models are right and to make them better.

INTEGRATION OF MACHINE LEARNING AND REMOTE SENSING

Machine learning (ML) and remote sensing tech are causing a revolution in how we keep tabs on the environment. This combo helps us to analyze data in smart ways and get useful info. It makes monitoring more accurate and productive, which means we can handle environmental stuff better. Some key things about this mashup are ways to mix data, formulas for watching the environment, and the gear we use (Kamali Maskooni et al. 2020).

Figure 1. Integration of machine learning with remote sensing Process

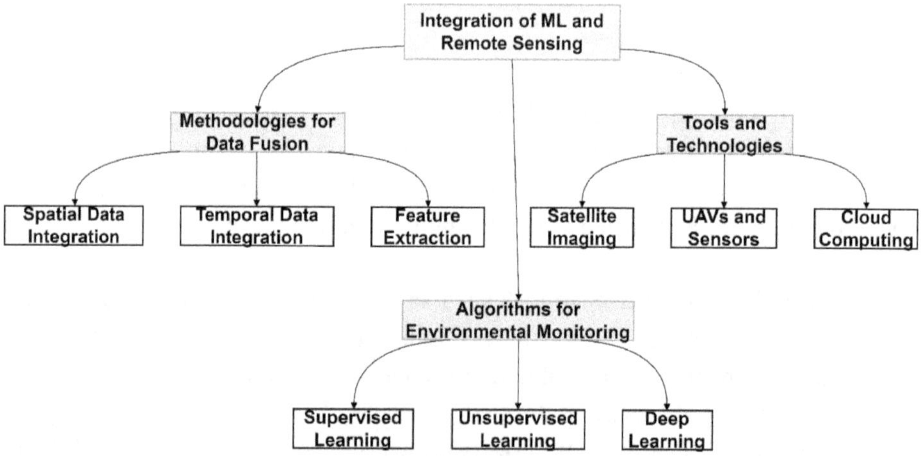

The integration of machine learning with remote sensing in environmental monitoring applications is facilitated by various methodologies, algorithms, and tools and technologies as shown in Figure 1.

Methodologies for Data Fusion

- Data fusion combines info from different places to give a better picture of what's happening in the environment. When it comes to remote sensing and machine learning, mixing data helps make it clearer and easier to understand. This leads to more accurate and dependable ways to keep an eye on the environment (Baklanov & Zhang 2020). The whole process has an impact on how well we can track environmental changes and make sense of what's going on.
- Pixel-Level Fusion: This method combines data at the pixel level mixing info from many spectral bands or sensors. For example, putting optical and radar data together can give extra info making it easier to spot things like how healthy plants are and how wet the soil is.
- Feature-Level Fusion: This way involves pulling out features from different data sources and mixing them before using ML algorithms. As an example spectral features from multispectral images and texture features from high-res images can be put together to make land cover classification more accurate.
- Decision-Level Fusion: This method combines decisions or classifications from different data sources to make a final call. It's handy when you're dealing with data from all sorts of places, like mixing satellite info with measurements from sensors on the ground. People often use ensemble learning tricks, like voting and stacking, in decision-level fusion to make predictions more reliable and accurate.
- Temporal Fusion: This type of fusion brings together data collected at different times to look at changes and trends over time. It's super important for keeping an eye on environmental stuff that's always changing, like forests disappearing, cities growing, and how air quality shifts with the seasons.

Algorithms for Environmental Monitoring

People use different ML algorithms to look at remote sensing data and keep an eye on the environment. You can put these algorithms into three big groups: supervised, unsupervised, and deep learning methods (Asha et al. 2022).

Supervised Learning Algorithms:

- o **Decision Trees and Random Forests**: People use these a lot to sort stuff like mapping land and figuring out plant types. They work well with big sets of data and give results you can make sense of.

- o **Support Vector Machines (SVMs)**: SVMs help with sorting and guessing numbers. They're good for remote sensing data with lots of parts and can handle tricky relationships between things.
- o **Neural Networks**: Old-school neural networks do all sorts of jobs, like guessing numbers and predicting what'll happen next. They can understand complex stuff in remote sensing data, which makes their guesses better.

Unsupervised Learning Algorithms:

- o **K-Means Clustering**: This method puts alike data points in groups. It's handy to sort land cover types and spot weird stuff in nature tracking.
- o **Principal Component Analysis (PCA)**: PCA cuts down the size of remote sensing info. It shows the big deal features and makes later number crunching faster.

Deep Learning Algorithms:

- **Convolutional Neural Networks (CNNs)**: CNNs rock at looking at pictures. They can find things, sort scenes, and see what's changed. They teach themselves to understand different levels of stuff in raw data, so they're great for remote sensing pics.
- **Recurrent Neural Networks (RNNs)**: People use RNNs to look at data that changes over time. This makes them great for guessing what will happen with things like air quality and weather. RNNs help to spot patterns in these areas.

Tools and Technologies

A bunch of tools and tech help bring ML and remote sensing together. These make it easier to process, analyze, and show environmental data (Asha et al. 2022; Reddy et al. 2020). This tech allows us to handle and understand info about our environment better.

1. **Software Platforms**:
    - o **Google Earth Engine (GEE)**: GEE gives you access to loads of remote sensing data and strong computing power in the cloud. You can use different ML methods on it to study the environment on a big scale.
    - o **QGIS and ArcGIS**: These GIS tools help you process, study, and show remote sensing data. You can add ML models to them and they're easy to use for keeping an eye on the environment.

- o **Python and R**: These coding languages have lots of tools for machine learning and looking at satellite images. Things like TensorFlow, scikit-learn, and GDAL (which helps work with map data) let you make and use ML models to keep an eye on the environment.

2. **Hardware**:
    - o **High-Performance Computing (HPC)**: HPC machines have the muscle to crunch through tons of satellite data using ML. You need these powerful computers to do the heavy lifting for deep learning and mixing different types of data.
    - o **Drones and UAVs**: Flying robots (UAVs) with different sensors can gather super detailed info. They're easy to use and can get specific data from certain spots.
3. **Cloud Computing**:
    - o **Amazon Web Services (AWS) and Microsoft Azure**: These big online platforms give you lots of space to store and work with huge amounts of remote sensing data. You can use them to run ML models and they have tools to help you manage and see your data better.

Machine learning and remote sensing are changing how we keep an eye on the environment. Together, they give us a better and faster way to check out what's going on with nature. By using smart ways to mix data cool math tricks, and new tech, scientists can come up with fresh ideas to tackle environmental problems and help the planet.

APPLICATIONS IN AIR QUALITY MONITORING

Keeping tabs on air quality is super important to see how pollution affects people's health and the environment. When you combine machine learning with remote sensing tech, it makes it easier to watch air quality in different places like cities, factories, and farms (El Hafyani et al., 2024).

Figure 2. Applications in Air Quality Monitoring

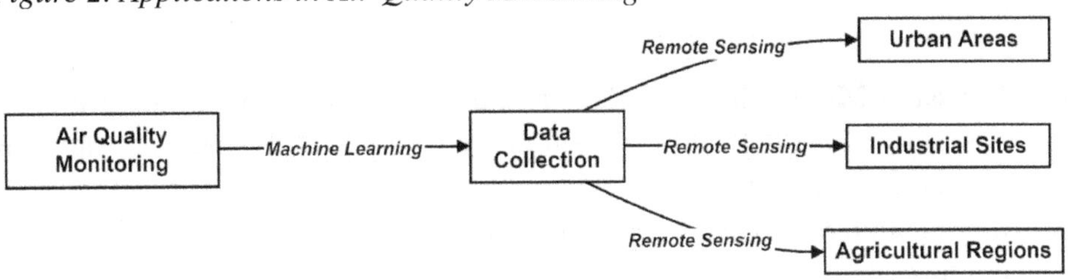

Figure 2 shows how ML can make air quality checks better by using info from space pictures, drones, and sensors on the ground.

Urban Air Quality Management

Cities have a lot of people and factories that make the air dirty. So, we need to keep an eye on the air all the time to keep people healthy and make good rules. Watching from far away gives us a big picture of bad stuff in the air like NO_2, SO_2, O_3, and tiny bits of dust. Smart computer programs look at this info to find the worst spots, figure out where the pollution comes from, and guess how bad it'll get.

- **Pollution Hotspot Identification**: Smart computer programs can spot areas with lots of pollution. They learn from pictures taken from space and info from sensors on the ground. This helps city planners and people who make rules to focus on fixing things. They might change traffic rules or control factory smoke to keep people safer from bad air.
- **Source Attribution**: Smart programs look at how pollution changes over time and in different places. They can figure out where the pollution comes from, like cars, factories, or home heating. This makes it easier to target and cut down on pollution to make the air cleaner.
- **Air Quality Forecasting**: People use ML stuff like time-series analysis and ensemble models to guess how clean the air will be. These models mash together info from weather conditions, pollution records, and past air quality data to predict future air quality. These guesses help set up early warning systems and let us take action before the air gets bad and makes people sick(Veeranjaneyulu et al., 2023).

Monitoring Industrial Emissions

Factories, power plants, and waste burners pump out a lot of air pollution. These industrial activities are big troublemakers when it comes to dirty air. To keep an eye on and control these emissions, we use cool tech like remote sensing and machine learning. These tools are super important for watching what industries are putting into the air.

- **Emission Monitoring**: Satellites and drones with sensors keep an eye on factories to spot pollutants like sulfur stuff, VOCs, and gases that warm up the planet. Smart computer programs crunch these numbers to figure out how much pollution there is, see how it changes, and make sure companies follow the rules.

- **Spotting Weird Stuff**: Smart computer models look at data from far away to find strange pollution patterns or leaks from factories. When they catch something, people can jump in to stop it from messing up the environment or causing trouble.
- **Checking if Rules are Followed**: Clever computer programs check if the pollution numbers are within what's allowed. They mix info from different places and times to give a good picture of who's following the rules. This helps the people in charge make sure everyone's doing what they should.

Managing Agricultural Air Quality

Farming stuff like raising animals, growing crops, and using bug spray can mess up the air. It lets out things like ammonia, methane, and tiny bits of dust. New tech like machine learning and looking at things from far away gives us cool ways to keep an eye on farm air and make it better (Durairaj et al. 2023; Veeranjaneyulu, Boopathi, Kumari, et al. 2023).

- **Livestock Management**: Satellites and planes keep an eye on farms to check how many animals are there and where ammonia levels are high. Computer programs look at pictures from space and info from sensors on the ground to guess how much ammonia is coming out. This helps farmers feed animals better and be kinder to the environment.
- **Crop Monitoring**: Smart computer systems look at data from space, like special pictures that show more than what we can see. These systems check how healthy plants are and see how farming affects the air. They can spot signs of trouble, like when plants don't have enough food or when bugs are attacking. This lets farmers change what they're doing to make less pollution.
- **Pesticide Use and Drift Monitoring**: Drones with sensors keep an eye on how farmers spray pesticides and where they end up. Smart computer programs look at where and when the pesticides spread out. They make maps showing where the chemicals go and figure out how they might affect the air and nature around them. This helps farmers use pesticides in a way that's better for the environment and cuts down on pollution.

When mix machine learning with remote sensing tech, we get a bunch of cool ways to check air quality in cities, factories, and farms. These tools give us quick and spot-on info about how much pollution is in the air, where it's coming from, and how it messes with the environment. This helps people make smart choices and take action to keep everyone healthy and make things better for the long run. By using fancy data crunching and new ways to keep an eye on things, we can tackle air quality problems head-on and try to make the air cleaner and safer for people all over the world.

APPLICATIONS IN ENVIRONMENTAL MONITORING

Environmental monitoring involves lots of activities to check and manage how healthy natural ecosystems and resources are. Machine learning (ML) combined with remote sensing tech has a big impact on making monitoring more accurate, bigger in scale, and more productive across different environmental areas. These include checking forest and plant health, water quality and ocean environments, and analyzing soil and land use (Pitchai et al. 2024).

Figure 3. Machine learning integrated with remote sensing technologies for environmental monitoring

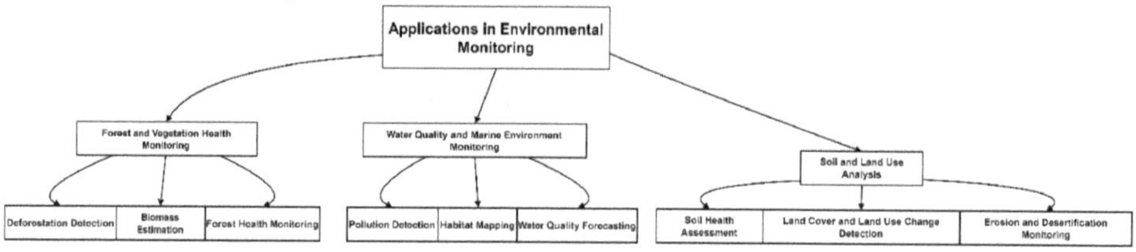

Figure 3 shows how people use machine learning with remote sensing tech to monitor different parts of the environment. This includes keeping an eye on forests and plants, checking water quality and ocean environments, and looking at soil and land use.

Forest and Vegetation Health Monitoring

Trees and plants cover huge chunks of land all over the world. They do a lot of good stuff, like helping ecosystems, soaking up carbon, and giving animals places to live. It's super important to keep an eye on how healthy they are. This helps protect all kinds of living things, fight climate change, and make sure we use forests in a way that doesn't mess them up (Boopathi 2023; Kavitha et al. 2023; Subha et al. 2023).

- **Deforestation Detection**: Satellites like Landsat and Sentinel keep an eye on forests all the time. Smart computer programs look at these satellite pictures to spot changes in how the land looks and where trees are being cut down.

These programs can tell the difference between natural changes and ones caused by people. This helps people step in and make sure rules are followed.

- **Biomass Estimation**: Clever computer methods, like random forests and CNNs (a type of artificial brain), look at different kinds of satellite data to guess how much plant stuff and carbon is in forests. When we know how much plant stuff there is, we can figure out how much carbon is stored. This helps with big projects like REDD+, which tries to cut down on carbon from cutting down trees.
- **Forest Health Monitoring**: ML models look at color patterns from satellite images to check how healthy forests are. They can spot things like stressed plants and bug attacks. When they find forest pests, diseases, or invasive species, people can step in . This helps to cut down on damage to nature and money losses.

Water Quality and Marine Environment Monitoring

Water bodies like oceans, rivers, and lakes play a key role in supporting biodiversity, farming, and people's jobs. These ecosystems are essential for life. To protect aquatic ecosystems find sources of pollution, and make sure we have safe water to drink, we need to keep an eye on water quality and marine environments (Boopathi et al. 2022). This has an impact on how we manage our water resources and keep them healthy for future generations.

- Pollution Detection: Satellites and drones with special sensors keep an eye on the water to spot bad stuff like algae blooms, oil spills, and dirt in the water. Smart computer programs look at the data to figure out what kind of pollution it is and how much there is. This helps people know how to clean it up and stop it from hurting the environment.
- Habitat Mapping: Computer programs that are good at learning use different kinds of data to make maps of where sea animals live. They can show things like coral reefs, underwater grass, and places where rivers meet the sea. These maps help people plan how to use the ocean and protect it by showing which areas need the most help to stay safe.
- Water Quality Forecasting: ML models mix remote sensing data with information about water and weather to predict water quality stuff, like how murky it is, how much oxygen is in it, and what nutrients are there. These predictions help give a heads up about bad conditions and tell people how to change things up to keep the water clean enough. This helps them stay on top of water quality rules.

Soil and Land Use Analysis

Soil and land resources play a key role in farming, keeping nature diverse, and helping ecosystems bounce back. When we keep an eye on soil traits and how land use shifts, we can make land management better, slow down damage to the environment, and push for land use that lasts (Boopathi et al., 2023; Jeevanantham et al. 2022; Jeya et al. 2023).

- **Soil Health Assessment**: Computers crunch numbers from satellite pictures to map out stuff in the soil, like how wet it is how much carbon it has, and what nutrients are there. This helps farmers be more precise, grow more food, and use less fertilizer, which is better for the environment.
- **Land Cover and Land Use Change Detection**: Satellites keep an eye on how the land looks from above. Smart computer programs figure out what's on the ground and how it's changing over time. They can spot things like cities getting bigger, forests shrinking, and farms expanding. This info helps people plan how to use land better, protect nature, and deal with climate change.
- **Erosion and Desertification Monitoring**: ML algorithms look at land shape and multi-colored data to figure out how fast soil is washing away and turning into desert. This helps spot areas that might lose their good soil and become barren. It also helps decide where to focus on fixing the land to make the soil stronger and keep the environment stable.

These tools give us useful info about how ecosystems are doing how to manage resources, and what climate change is doing. This helps people make smart choices and work towards goals that don't harm the planet. By using fancy data stuff and new ways to watch things, people can tackle environmental problems better and help nature bounce back for kids in the future.

CHALLENGES AND LIMITATIONS

Combining machine learning (ML) with remote sensing tech can shake things up for keeping an eye on and taking care of the environment. But there are some tricky parts and limits we need to figure out to make the most of these cool tools (Ingle et al. 2023; Prasad et al. 2024). Figure 4 shows the bumps in the road and possible hiccups when you try to mix ML with remote sensing stuff for watching over and managing the environment.

Figure 4. Challenges and potential limitations in integrating machine learning with remote sensing technologies for environmental monitoring and management

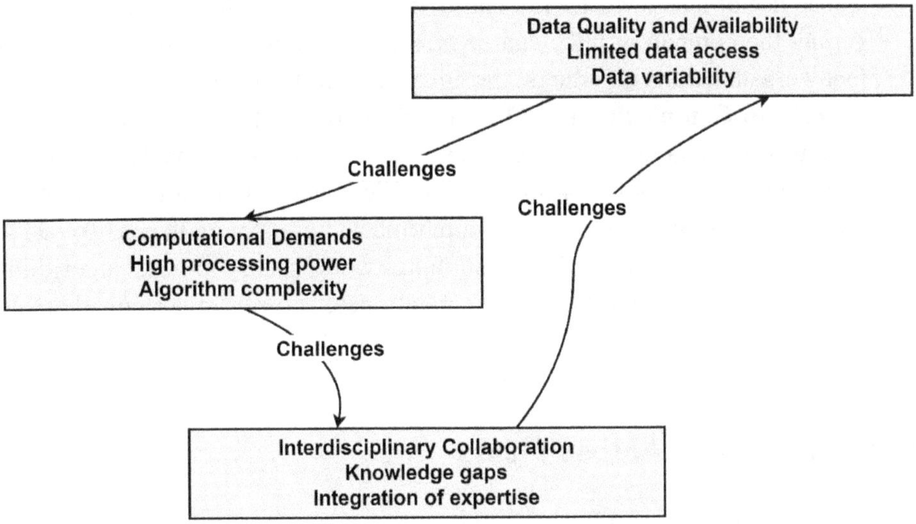

Data Quality and Availability

- Data Variability and Quality: Weather conditions, sensor limits, and calibration mistakes can mess up remote sensing data, causing changes in how good the data is. ML systems need top-notch, steady information to make smart choices and do good analysis. It's super important but often tricky to make sure the data is spot-on and trustworthy. This means checking, confirming, and keeping an eye on quality, which gets even harder when dealing with all sorts of different environments.
- Getting Data and Having It Available: How easy it is to get satellite pictures and information from sensors on the ground isn't the same everywhere. This makes it tough to watch over the environment in all places. When there's not enough data in some areas or for certain things, it's hard to make and check if our computer programs are working right. To fix this, we need to:

Computational Demands

- **Processing Power and Storage**: ML algorithms and deep learning models need a ton of computing power and storage to crunch through massive amounts of remote sensing data. You gotta have beefy HPC systems and cloud platforms to handle the heavy lifting and manage all that data in real-time or close to it. But here's the catch: the price tag for all this tech can be a real pain in the neck for some companies, which puts a damper on the widespread use of fancy ML tricks. The costs of setting up the infrastructure and getting the computing juice can be a real showstopper, making it tough for many organizations to jump on the advanced ML bandwagon.

- **Algorithm Complexity and Optimization**: To create and improve ML algorithms for remote sensing, you need to pick the right models, adjust settings, and handle computer power well. Tricky algorithms often need lots of training and testing, which takes more time and costs more to run. To tackle computer issues and boost how well things work, it's key to make algorithms run smoother, use multiple processors at once, and set up systems that can grow. This helps get past the hurdles of complex computations and makes everything run better in real-world use.

Working Together Across Fields

- **Integration of Expertise**: To make ML and remote sensing work well together, experts from different fields need to team up. Scientists, engineers, folks who know about the environment, and data nerds all have to work together. They need to share what they know and figure out how to do things the same way. This helps them create solid ways to keep an eye on things, understand tricky data, and turn what they find into stuff people can use. Setting up groups where everyone can work together making places where people can share what they know, and getting different sectors to partner up can help create a full picture of how to watch and take care of the environment.
- **Domain-Specific Knowledge**: People need to customize ML algorithms to tackle specific environmental issues and goals, like checking air quality, grouping land uses, or drawing habitat maps. To make sense of ML results, interpret them, and help make decisions, it's crucial to have knowledge about environmental processes, how ecosystems work, and the rules and laws in place. This know-how helps put ML outputs in context. To get past knowledge hurdles and make the most of these combined technologies, it's key to improve training programs, encourage research across different fields, and bring in experts to ML development teams.

Dealing with the problems of combining machine learning and remote sensing technology is key to improving how we watch over the environment and meet green goals. We need to make data better and easier to get, handle big computer needs, and get different fields to work together. These are the main things we need to do to use these tools to their full power. If we put money into building stuff, share data more, and get different groups to team up, we can get past roadblocks, use new ideas, and open doors for big changes in how we watch and take care of our planet. We still need to keep studying, come up with new tech, and have rules that help if we want to tackle big environmental issues and keep our natural resources safe for kids in the future.

FUTURE TRENDS AND OPPORTUNITIES

New technology, better sensors, and changing rules are shaping how we'll keep an eye on the environment in the future. When we mix machine learning with watching from far away, it could change how we look after nature, deal with environmental problems, and grow in a way that doesn't hurt the planet (Godwin Immanuel et al. 2024; Ramudu et al. 2023; Veeranjaneyulu, Boopathi, Kumari, et al. 2023; Veeranjaneyulu, Boopathi, Narasimharao, et al. 2023).

Figure 5. Future of environmental monitoring shaped emerging technologies

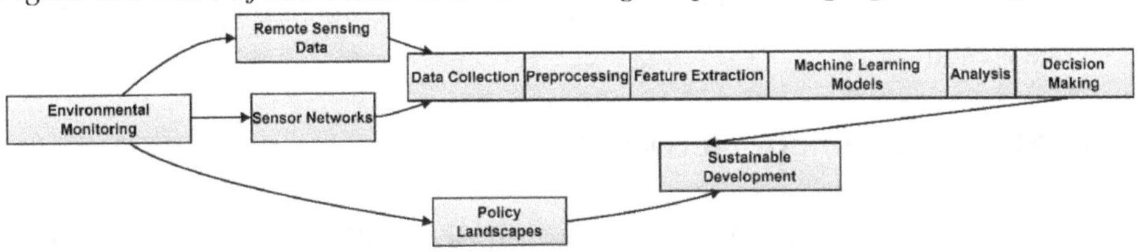

Figure 5 shows how environmental monitoring works. It starts with getting data from satellites and other sources. Then, people clean up this data and pick out the important parts. After that, computers use special programs to study the information. People look at what the computers find and make choices about how to protect nature. All of this helps us take better care of our planet. You can change some parts of this picture to fit how you use computers and satellite information for watching over the environment.

Emerging Technologies

- **Deep Learning**: Deep learning methods, like CNNs and RNNs, are causing a revolution in remote sensing data analysis. CNNs are really good at spotting things in images, which helps to make detailed maps of land cover and find objects in satellite pictures. RNNs work well to analyze data over time, which lets us guess what might happen with things like climate patterns and how ecosystems change. In the future, we'll see even better deep learning setups that will make models more accurate, able to handle more stuff, and easier to understand when it comes to keeping an eye on complex environmental stuff.

- **Internet of Things (IoT)**: IoT gadgets packed with sensors and hooked up through wireless networks, let us gather and watch environmental stuff in real-time. Mixing IoT with remote sensing makes things clearer in space and time helping out with smart farming, checking city air quality, and dealing with natural disasters. Looking ahead, we'll see more IoT-powered networks to keep an eye on the environment, data crunching spread out all over, and smart systems to help make choices. These systems will rely on AI to get the job done.
- **Cloud Computing**: Cloud computing platforms give you scalable storage, computing power, and ways to analyze data for handling tons of remote sensing info. Cloud-based ML models make it easier to access data, work together, and roll out environmental monitoring fixes worldwide. As cloud tech gets better, improvements in data safety real-time processing, and cheaper solutions will spark new ideas in remote sensing uses(Sundar et al., 2024; Tirlangi et al., 2024). This helps manage resources and deal with climate changes more.

Innovations in Sensor Networks

- **New Sensors**: New technology makes sensors smaller, cheaper, and easier to put in all sorts of places, even tough spots. These small sensors help us watch more areas, get more detailed information, and see what's happening right now for better environmental checks. Cool new sensors like multispectral, hyperspectral, and LiDAR can now look at more stuff in nature, from how wet the ground is and how healthy plants are to what's going on in the oceans and air.
- **AI and Sensor Fusion Working Together**: Smart computer programs team up with sensor fusion tricks to make data from lots of sensors work better together. When you mix information from cameras, radar, and sound detectors, you get a clearer picture of what's going on in the environment. It's more accurate, tougher, and gives you a fuller story. Looking ahead, we might see sensors that can change on their own, machines that gather data by themselves, and AI that helps keep sensors in good shape. All this means we'll have solid, top-notch environmental information to help make smart choices.

Policy Implications and Recommendations

- **Data Governance and Standards**: Setting up global rules and norms for data management has a big impact on how we share and use information from satellites and smart computer programs. We need clear ways to swap data, be open about what we're doing, and use these tools. When we make it easy for everyone to access environmental information, agree on how to share it, and protect people's privacy, we help folks work together, come up with new ideas, and make smart choices about watching over our planet.
- **Regulatory Frameworks and Compliance**: The rules need to keep up with all the new tech we're using to check on the environment. We should make laws that push for smart use of resources, cut down on pollution, and look after plants and animals. This helps us match up what we see from space and figure out with computers with what the big global agreements say we should do, like the Paris deal on climate change and the goals for making the world better for everyone.
- **Capacity Building and Education**: Putting money into programs to build skills, train people, and teach others creates a workforce that knows about remote sensing, machine learning, and environmental science. Encouraging teamwork across different fields, setting up ways to share knowledge, and working together on research helps come up with new ideas faster, fills in gaps in understanding data, and gives communities the power to use technology to watch over and protect the environment.

Case Studies: Success Stories in Machine Learning and Remote Sensing

Using remote sensing technology and machine learning together has shown good results in different places and climates. This combo brings new ways to tackle environmental problems and helps make development more sustainable. It's working well across many different areas.

- **Monitoring Deforestation in the Amazon Rainforest**: Deforestation in the Amazon basin poses a big threat to biodiversity and carbon storage. To keep an eye on deforestation activities in real-time, researchers now use satellite pictures along with smart computer programs like convolutional neural networks (CNNs). These systems look at the color patterns and shapes in the images to spot illegal tree cutting. This helps local officials and nature groups to take action. Thanks to this approach less trees are being cut down and forests are being looked after better. The systems check out the unique "fingerprints" of light and spatial arrangements to find where people are chopping down trees . This has helped to cut down on forest loss and make forest protection work better.

- **Urban Air Quality Management in Beijing China**: Beijing has big air pollution problems because of its fast growth and factories. Smart computer systems work with air checking networks to look at information from satellites, sensors on the ground, and weather data to guess how bad the pollution will be. These guesses help warn people and tell them how to stay healthy when the air is bad. This shows how smart technology can make city planning better and help make rules that clean up the air in crowded cities. The tech helps people who live there and those in charge make better choices to deal with the health risks that come from dirty air.
- **Precision Agriculture in the United States**: ML algorithms have an impact on crop management in US farming areas. These use satellite pictures and sensor networks to monitor soil water, crop health, and pests in real-time. This gives farmers useful tips, like when to water and where to spray. As a result, crops grow better, farmers use less stuff, and farming becomes more eco-friendly. This shows how ML is changing farming and helping make sure we have enough food even when the weathers unpredictable. These smart systems help farmers know what their fields need, making American agriculture more efficient and sustainable(Pachiappan et al., 2024; Venkateswaran et al., 2024).
- **Keeping an Eye on Sea Life in the Great Barrier Reef Australia**: The Great Barrier Reef, which UNESCO says is super special, is in trouble. It's dealing with a changing climate, corals turning white, and stuff that pollutes the water. To help out, people are using fancy tech with smart computer programs to check on how healthy the reef is, how warm the water gets, and if the water is clean enough. These systems look at pictures from space and information about the ocean to spot things that stress out the reef. This helps people come up with plans to protect all the cool sea creatures that live there. This project shows how smart tech and pictures from far away can team up to watch over important ocean spots and change plans as needed to keep them safe.

These examples show how machine learning and remote sensing work well together in different places and climates. By using cool tech and mixing different fields, people can tackle environmental problems, get better at dealing with climate change, and push for sustainable growth all over the world. To make these success stories bigger and hit global green goals, we need to keep coming up with new ideas, work together, and put money into data-based fixes.

CONCLUSION

Machine learning (ML) and remote sensing technology are changing how we keep an eye on and manage the environment. They're giving us new ways to tackle tricky problems in all sorts of places with different weather. In this chapter, we've looked at how ML and remote sensing are used, the hurdles they face, what's coming next, and some success stories in various environmental areas. ML makes it easier to monitor the environment more accurately, on a larger scale, and more accurately. These tools help with stuff like keeping city air clean, farming more, checking if forests are healthy, and protecting sea life. They do this by looking at pictures from satellites, data from internet-connected sensors, and past patterns. This helps people make smart choices to protect nature and grow in a way that doesn't harm the planet.

Even with all the new tech stuff, we still face big problems like making sure the data is good, dealing with all the computer power needed, and getting different experts to work together. To fix these issues, we need to put money into better data systems, teach people new skills, and create rules for using these technologies in a good way that works for everyone. New things like deep learning, IoT, and cloud computing could help us keep an eye on the environment better. Cool new ideas in sensor networks, making better algorithms, and setting up new rules will push the next big improvements in how we use remote sensing. If we team up with other countries and share what we know, we can start using data-driven answers to worldwide environmental problems faster.

ML combining with remote sensing gives us amazing chances to help the environment, deal with climate problems, and keep nature diverse. We need to keep studying, coming up with new ideas, and getting support from rules to use these tools . By using information from data and working together across different fields, we can make nature stronger, give everyone fair access to information about the environment, and help people all over the world take care of our planet. This team effort will have an impact on how we protect Earth and make it last. To implement these ideas, we need to analyze the data and consider how to put them into action. This will allow us to make a real difference in preserving our world for future generations.

REFERENCES

Asha, P., Natrayan, L., Geetha, B., Beulah, J. R., Sumathy, R., Varalakshmi, G., & Neelakandan, S. (2022). IoT enabled environmental toxicology for air pollution monitoring using AI techniques. *Environmental Research*, *205*, 112574. DOI: 10.1016/j.envres.2021.112574 PMID: 34919959

Baklanov, A., & Zhang, Y. (2020). Advances in air quality modeling and forecasting. *Global Transitions*, *2*, 261–270. DOI: 10.1016/j.glt.2020.11.001

Boopathi, S. (2023). Internet of Things-Integrated Remote Patient Monitoring System: Healthcare Application. In Dynamics of Swarm Intelligence Health Analysis for the Next Generation (pp. 137–161). IGI Global.

Boopathi, S. (2024). Advancements in Machine Learning and AI for Intelligent Systems in Drone Applications for Smart City Developments. In Futuristic e-Governance Security With Deep Learning Applications (pp. 15–45). IGI Global. DOI: 10.4018/978-1-6684-9596-4.ch002

Boopathi, S., Kumar, P. K. S., Meena, R. S., Sudhakar, M., & Associates. (2023). Sustainable Developments of Modern Soil-Less Agro-Cultivation Systems: Aquaponic Culture. In Human Agro-Energy Optimization for Business and Industry (pp. 69–87). IGI Global.

Boopathi, S., Lewise, K. A. S., Subbiah, R., & Sivaraman, G. (2022). Near-dry wire-cut electrical discharge machining process using water–air-mist dielectric fluid: An experimental study. Elsevier. *Materials Today: Proceedings*, *50*(5), 1885–1890. DOI: 10.1016/j.matpr.2021.08.077

Durairaj, M., Jayakumar, S., Karpagavalli, V., Maheswari, B. U., & Boopathi, S. (2023). Utilization of Digital Tools in the Indian Higher Education System During Health Crises. In *Multidisciplinary Approaches to Organizational Governance During Health Crises* (pp. 1–21). IGI Global. DOI: 10.4018/978-1-7998-9213-7.ch001

El Hafyani, H., Abboud, M., Zuo, J., Zeitouni, K., Taher, Y., Chaix, B., & Wang, L. (2024). Learning the micro-environment from rich trajectories in the context of mobile crowd sensing: Application to air quality monitoring. *GeoInformatica*, *28*(2), 177–220. DOI: 10.1007/s10707-022-00471-4

Ghahremanloo, M., Lops, Y., Choi, Y., & Yeganeh, B. (2021). Deep learning estimation of daily ground-level NO2 concentrations from remote sensing data. Journal of Geophysical Research: Atmospheres, 126(21), e2021JD034925.

Godwin Immanuel, D., Solaimalai, G., Chandrakala, B. M., Bharath, V. G., Singh, M. K., & Boopathi, S. (2024). Advancements in Electric Vehicle Management System: Integrating Machine Learning and Artificial Intelligence. In Advances in Web Technologies and Engineering (pp. 371–391). IGI Global. DOI: 10.4018/979-8-3693-1487-6.ch018

Ingle, R. B., Swathi, S., Mahendran, G., Senthil, T., Muralidharan, N., & Boopathi, S. (2023). Sustainability and Optimization of Green and Lean Manufacturing Processes Using Machine Learning Techniques. In Circular Economy Implementation for Sustainability in the Built Environment (pp. 261–285). IGI Global. DOI: 10.4018/978-1-6684-8238-4.ch012

Jeevanantham, Y. A., Saravanan, A., Vanitha, V., Boopathi, S., & Kumar, D. P. (2022). Implementation of Internet-of Things (IoT) in Soil Irrigation System. IEEE Explore, 1–5.

Jeya, R., Venkatakrishnan, G. R., Rengaraj, R., Rajalakshmi, M., Pradeep Mohan Kumar, K., & Boopathi, S. (2023). Water Resource Managements in Soil and Soilless Irrigation Systems Using AI Techniques. In Advances in Environmental Engineering and Green Technologies (pp. 245–266). IGI Global. DOI: 10.4018/979-8-3693-0338-2.ch014

Kamali Maskooni, E., Naghibi, S. A., Hashemi, H., & Berndtsson, R. (2020). Application of advanced machine learning algorithms to assess groundwater potential using remote sensing-derived data. *Remote Sensing (Basel)*, *12*(17), 2742. DOI: 10.3390/rs12172742

Kavitha, C., Varalatchoumy, M., Mithuna, H., Bharathi, K., Geethalakshmi, N., & Boopathi, S. (2023). Energy Monitoring and Control in the Smart Grid: Integrated Intelligent IoT and ANFIS. In Applications of Synthetic Biology in Health, Energy, and Environment (pp. 290–316). IGI Global.

Li, J., Garshick, E., Hart, J. E., Li, L., Shi, L., Al-Hemoud, A., Huang, S., & Koutrakis, P. (2021). Estimation of ambient PM2. 5 in Iraq and Kuwait from 2001 to 2018 using machine learning and remote sensing. *Environment International*, *151*, 106445. DOI: 10.1016/j.envint.2021.106445 PMID: 33618328

Méndez, M., Merayo, M. G., & Núñez, M. (2023). Machine learning algorithms to forecast air quality: A survey. *Artificial Intelligence Review*, *56*(9), 10031–10066. DOI: 10.1007/s10462-023-10424-4 PMID: 36820441

Pachiappan, K., Anitha, K., Pitchai, R., Sangeetha, S., Satyanarayana, T., & Boopathi, S. (2024). Intelligent Machines, IoT, and AI in Revolutionizing Agriculture for Water Processing. In Handbook of Research on AI and ML for Intelligent Machines and Systems (pp. 374–399). IGI Global.

Pitchai, R., Guru, K. V., Gandhi, J. N., Komala, C. R., Kumar, J. R. D., & Boopathi, S. (2024). Fog Computing-Integrated ML-Based Framework and Solutions for Intelligent Systems: Digital Healthcare Applications. In Technological Advancements in Data Processing for Next Generation Intelligent Systems (pp. 196–224). IGI Global. DOI: 10.4018/979-8-3693-0968-1.ch008

Prasad, M. S. C., Dhanalakshmi, M., Mohan, M., Somasundaram, B., Valarmathi, R., & Boopathi, S. (2024). Machine Learning-Integrated Sustainable Engineering and Energy Systems: Innovations at the Nexus. In Harnessing High-Performance Computing and AI for Environmental Sustainability (pp. 74–98). IGI Global. DOI: 10.4018/979-8-3693-1794-5.ch004

Rahimpour, A., Amanollahi, J., & Tzanis, C. G. (2021). Air quality data series estimation based on machine learning approaches for urban environments. *Air Quality, Atmosphere & Health*, *14*(2), 191–201. DOI: 10.1007/s11869-020-00925-4

Ramudu, K., Mohan, V. M., Jyothirmai, D., Prasad, D., Agrawal, R., & Boopathi, S. (2023). Machine Learning and Artificial Intelligence in Disease Prediction: Applications, Challenges, Limitations, Case Studies, and Future Directions. In Contemporary Applications of Data Fusion for Advanced Healthcare Informatics (pp. 297–318). IGI Global.

Reddy, T., RM, S. P., Parimala, M., Chowdhary, C. L., Hakak, S., Khan, W. Z., & others. (. (2020). A deep neural networks based model for uninterrupted marine environment monitoring. *Computer Communications*, *157*, 64–75. DOI: 10.1016/j.comcom.2020.04.004

Ren, X., Mi, Z., & Georgopoulos, P. G. (2020). Comparison of Machine Learning and Land Use Regression for fine scale spatiotemporal estimation of ambient air pollution: Modeling ozone concentrations across the contiguous United States. *Environment International*, *142*, 105827. DOI: 10.1016/j.envint.2020.105827 PMID: 32593834

Shen, H., Jiang, Y., Li, T., Cheng, Q., Zeng, C., & Zhang, L. (2020). Deep learning-based air temperature mapping by fusing remote sensing, station, simulation and socioeconomic data. *Remote Sensing of Environment*, *240*, 111692. DOI: 10.1016/j.rse.2020.111692

Singh, D., Dahiya, M., Kumar, R., & Nanda, C. (2021). Sensors and systems for air quality assessment monitoring and management: A review. *Journal of Environmental Management*, *289*, 112510. DOI: 10.1016/j.jenvman.2021.112510 PMID: 33827002

Sokhi, R. S., Moussiopoulos, N., Baklanov, A., Bartzis, J., Coll, I., Finardi, S., Friedrich, R., Geels, C., Grönholm, T., Halenka, T., Ketzel, M., Maragkidou, A., Matthias, V., Moldanova, J., Ntziachristos, L., Schäfer, K., Suppan, P., Tsegas, G., Carmichael, G., & Kukkonen, J. (2021). Advances in air quality research–current and emerging challenges. *Atmospheric Chemistry and Physics Discussion*, *2021*, 1–133. DOI: 10.5194/acp-2021-581

Subha, S., Inbamalar, T., Komala, C., Suresh, L. R., Boopathi, S., & Alaskar, K. (2023). A Remote Health Care Monitoring system using internet of medical things (IoMT). IEEE Explore, 1–6.

Sundar, R., Srikaanth, P. B., Naik, D. A., Murugan, V. P., Karumudi, M., & Boopathi, S. (2024). Achieving Balance Between Innovation and Security in the Cloud With Artificial Intelligence of Things: Semantic Web Control Models. In Advances in Web Technologies and Engineering (pp. 1–26). IGI Global. DOI: 10.4018/979-8-3693-1487-6.ch001

Syamala, M., Komala, C., Pramila, P., Dash, S., Meenakshi, S., & Boopathi, S. (2023). Machine Learning-Integrated IoT-Based Smart Home Energy Management System. In Handbook of Research on Deep Learning Techniques for Cloud-Based Industrial IoT (pp. 219–235). IGI Global. DOI: 10.4018/978-1-6684-8098-4.ch013

Tirlangi, S., Teotia, S., Padmapriya, G., Senthil Kumar, S., Dhotre, S., & Boopathi, S. (2024). Cloud Computing and Machine Learning in the Green Power Sector: Data Management and Analysis for Sustainable Energy. In Developments Towards Next Generation Intelligent Systems for Sustainable Development (pp. 148–179). IGI Global. DOI: 10.4018/979-8-3693-5643-2.ch006

Veeranjaneyulu, R., Boopathi, S., Kumari, R. K., Vidyarthi, A., Isaac, J. S., & Jaiganesh, V. (2023). Air Quality Improvement and Optimisation Using Machine Learning Technique. IEEE- Explore, 1–6.

Venkateswaran, N., Kiran Kumar, K., Maheswari, K., Kumar Reddy, R. V., & Boopathi, S. (2024). Optimizing IoT Data Aggregation: Hybrid Firefly-Artificial Bee Colony Algorithm for Enhanced Efficiency in Agriculture. *AGRIS On-Line Papers in Economics and Informatics*, *16*(1), 117–130. DOI: 10.7160/aol.2024.160110

Venkateswaran, N., Vidhya, R., Naik, D. A., Raj, T. M., Munjal, N., & Boopathi, S. (2023). Study on Sentence and Question Formation Using Deep Learning Techniques. In *Digital Natives as a Disruptive Force in Asian Businesses and Societies* (pp. 252–273). IGI Global. DOI: 10.4018/978-1-6684-6782-4.ch015

Yang, H., Kong, J., Hu, H., Du, Y., Gao, M., & Chen, F. (2022). A review of remote sensing for water quality retrieval: Progress and challenges. *Remote Sensing (Basel), 14*(8), 1770. DOI: 10.3390/rs14081770

Yuan, Q., Shen, H., Li, T., Li, Z., Li, S., Jiang, Y., Xu, H., Tan, W., Yang, Q., Wang, J., Gao, J., & Zhang, L. (2020). Deep learning in environmental remote sensing: Achievements and challenges. *Remote Sensing of Environment, 241*, 111716. DOI: 10.1016/j.rse.2020.111716

Chapter 15
Security in Internet of Things Applications Using Network-Attached Storage for Resource Usage Optimization

Antony Taurshia
https://orcid.org/0000-0001-9129-1859
Division of Data Science and Cyber Security, Karunya Institute of Technology and Sciences, Coimbatore, India

D. Ponmary Pushpa Latha
https://orcid.org/0000-0003-4000-0294
Division of Digital Sciences, Karunya Institute of Technology and Sciences (Deemed), Coimbatore, India

S. Princy Suganthi Bai
Department of MCA, Hindustan Institute of Technology and Science, (Deemed), Coimbatore, India

R. Vignesh
https://orcid.org/0009-0006-6065-987X
Division of Digital Sciences, Karunya Institute of Technology and Sciences (Deemed), Coimbatore, India

D. Joseph Pushpa Raj
https://orcid.org/0009-0006-6033-9825
Department of Computer Science and Engineering, PSN College of Engineering and Technology, Tirunelveli, India

DOI: 10.4018/979-8-3693-5448-3.ch015

ABSTRACT

The Internet of Things (IoT) enhances human society by enabling common objects to access the internet and make insightful conclusions, using technology elements like RFID, sensors, embedded devices, and service platforms. The IoT is enabled by certain elements, but they have limitations due to their limited processing power and resource availability. The constraints in IoT applications can compromise the security primitives that are fully utilized. The recommended method utilizes Network Attached Storage (NAS) to encrypt data before it leaves the local network, minimizing the strain on devices with limited capabilities. Cloud service platforms can be eliminated by securing data locally on the enterprise premises, eliminating the need for cloud-based services.

INTRODUCTION

The combination of physical objects or devices with sensors, software, and connectivity that allows them to share data with other devices or systems over the Internet is referred to as the Internet of Things IoT (Lin et al., 2017). It allows for the seamless interaction of multiple wearables and devices to create an interconnected ecosystem that can perform complex tasks with minimal human intervention IoT work using information provided by devices that together it brings real-time collection and analysis, enabling businesses and individuals to make data-driven decisions (Lin et al., 2017).

However, the IoT presents major challenges, such as security and privacy concerns, standards issues, and complex data management. IoT devices have different positions in consumer products Laptops, desktops, and smartphones connected to IoT applications can adopt more secure cryptographic primitives. Embedded systems with limited resources can accommodate tolerable cryptography. RFID tags and high-capacity sensor nodes can only accept hardware-based cryptographic primitives. Despite these challenges, IoT continues to grow rapidly, with estimates suggesting that by 2027 more than 41 billion IoT devices will be connected (Wang et al., 2022).

With the increasing data dependency and the growing concern for data security, individuals and organizations face the challenge of finding a storage solution that balances accessibility, security, and affordability. Cloud storage services offer convenience and remote access but raise concerns about data privacy and security (Cao et al., 2020). Therefore, there is a need for an alternative solution that combines the benefits of cloud storage, local control, and enhanced security while remaining cost-effective(Wang et al., 2022).

This work aims to address the problem of finding a storage solution that offers an alternative to cloud storage, prioritizes data security, and provides the benefits of local control and affordability. NAS is primarily used as a storage device where files are stored and shared by devices attached to the NAS via Ethernet (Kumar et al., 2021). NAS facilitates the working of stand stand-alone network with enhanced data control. Though NAS has its own set of limitations like network bottleneck, the performance of NAS can be enhanced through proper hardware, network, and file-sharing settings (Reed et al., 2000). The proposed work utilizes NAS as an alternative for edge computing to offer enhanced data security and privacy. The work further focuses on utilizing Raspberry Pi as a network-attached storage device, leveraging its low-cost and versatile nature to meet the requirements of IoT settings. By investigating the capabilities, implementation, and security features of Raspberry Pi NAS, the work intends to determine its effectiveness as a viable alternative to cloud and edge computing capabilities.

The Internet of Things (IoT) has emerged as a transformative paradigm, significantly enhancing human interaction with technology. It has enabled the integration of everyday objects into the digital world, facilitating seamless communication and data exchange. IoT encompasses a diverse range of technologies including Radio Frequency Identification (RFID), sensors, embedded systems, and service platforms. These components collectively enable the collection, transmission, and analysis of data from various sources, contributing to smarter environments and more efficient processes(Azrour et al., 2021).

Despite the numerous benefits offered by IoT, the technology also faces several challenges, particularly related to security and resource constraints. IoT devices typically have limited processing power and memory, which restricts their ability to implement robust security measures. This limitation poses a significant risk as the number of connected devices continues to grow, making IoT networks attractive targets for cyberattacks. Ensuring the security of IoT applications is critical to protect sensitive data and maintain the integrity and reliability of IoT systems(Lu et al., 2021).

One of the primary concerns in IoT security is the protection of data during transmission and storage. IoT devices frequently communicate over networks, often transmitting sensitive information that, if intercepted, could lead to serious security breaches. Traditional security measures, such as encryption, are computationally intensive and may not be feasible for resource-constrained IoT devices. Therefore, there is a need for innovative approaches to secure data without overburdening these devices(Sadhu et al., 2022).

Network Attached Storage (NAS) offers a promising solution to address these security challenges. NAS is a file-level storage architecture that enables data storage and retrieval from centralized disk capacity. By leveraging NAS, data can be encrypted before it leaves the local network, thus offloading the processing burden from individual IoT devices. This approach ensures that data is protected at rest and in transit without compromising the performance of the devices. The integration of NAS into IoT systems can also reduce the dependency on cloud services for data security. While cloud platforms provide robust security features, they introduce additional latency and dependency issues(Mousavi et al., 2021). By securing data locally on the enterprise premises using NAS, organizations can achieve lower latency and greater control over their data. This approach not only enhances security but also optimizes resource usage, leading to more efficient IoT operations.

In this paper, we propose a method that utilizes NAS for data encryption in IoT applications. This method aims to alleviate the security constraints imposed by the limited processing capabilities of IoT devices. By encrypting data at the NAS level before it is transmitted, we can ensure that even resource-constrained devices can operate securely. Additionally, this approach minimizes the reliance on external service platforms, thereby reducing latency and enhancing data control. The following sections will delve deeper into the enabling technologies of IoT, the limitations of current security measures, and the proposed NAS-based solution. We will explore the architecture and implementation of the proposed method, highlighting its advantages in terms of security and resource optimization. Furthermore, we will discuss the potential impact of this approach on the broader IoT ecosystem and future directions for research and development(Jurcut et al., 2020).

Thus, the IoT has revolutionized the way we interact with the world, but it also brings significant security challenges due to the limited resources of its devices. By leveraging NAS for data encryption, we can mitigate these challenges and enhance the overall security and efficiency of IoT applications. This paper aims to provide a comprehensive analysis of this approach, offering valuable insights for researchers and practitioners in the field of IoT security.

Background

The Internet of Things (IoT) has revolutionized how devices interact and share data, enabling smart environments through connected sensors, RFID, and embedded systems. However, the limited processing power and resources of these devices pose significant security challenges. Traditional security measures can strain these devices, making it difficult to ensure data privacy and integrity.

Scope

This study explores the use of Network Attached Storage (NAS) to enhance security and optimize resource usage in IoT applications. By leveraging NAS for data encryption and storage within local networks, the solution aims to reduce the dependency on external cloud services, ensuring better data control and privacy. The research focuses on the integration of NAS with IoT devices, evaluating performance and security improvements.

Objectives

- To investigate the security limitations of IoT devices and propose a NAS-based solution.
- To demonstrate how NAS can offload encryption tasks, preserving device resources.
- To assess the performance impact of NAS on data handling in IoT systems.
- To provide a framework for implementing NAS in IoT environments to enhance data privacy and security.

SYSTEM FUNCTIONS OF IoT WORKING

The IoT system functions through the collaboration of several effective technological components and service platforms comprised(Jurcut et al., 2020; Khan et al., 2022). The IoT system functions can be visualized using Mermaid code, illustrating the interaction between various components and service platforms within an IoT ecosystem (Figure 1).

Radio Frequency Identification (RFID): Radio waves are mainly used by RFIDs to identify and track items. It is a low-cost, disposable contactless smart card consisting of an RFID tag, reader, and antenna. The electronic tag carries a chip that contains a unique identification number and attributes corresponding to the device it was tagged. The antenna located in the chip is used for communicating and transmitting information to the reader. When the antenna comes within the range of the reader, it is powered by the reader to transmit information. Depending on their energy source, RFID tags can be classified as active, passive, or semi-passive (Agrawal et al., 2023; Koshariya et al., 2023). RFID tags are efficient enough that they can be connected to even routine objects making them smart and communicable. For example, RFID tags are used for the electronic verification of parking permits in smart parking systems. It is also used for tracking and locating objects in applications like supply chain management, asset management, and healthcare.

Figure 1. System Functions of IOT working

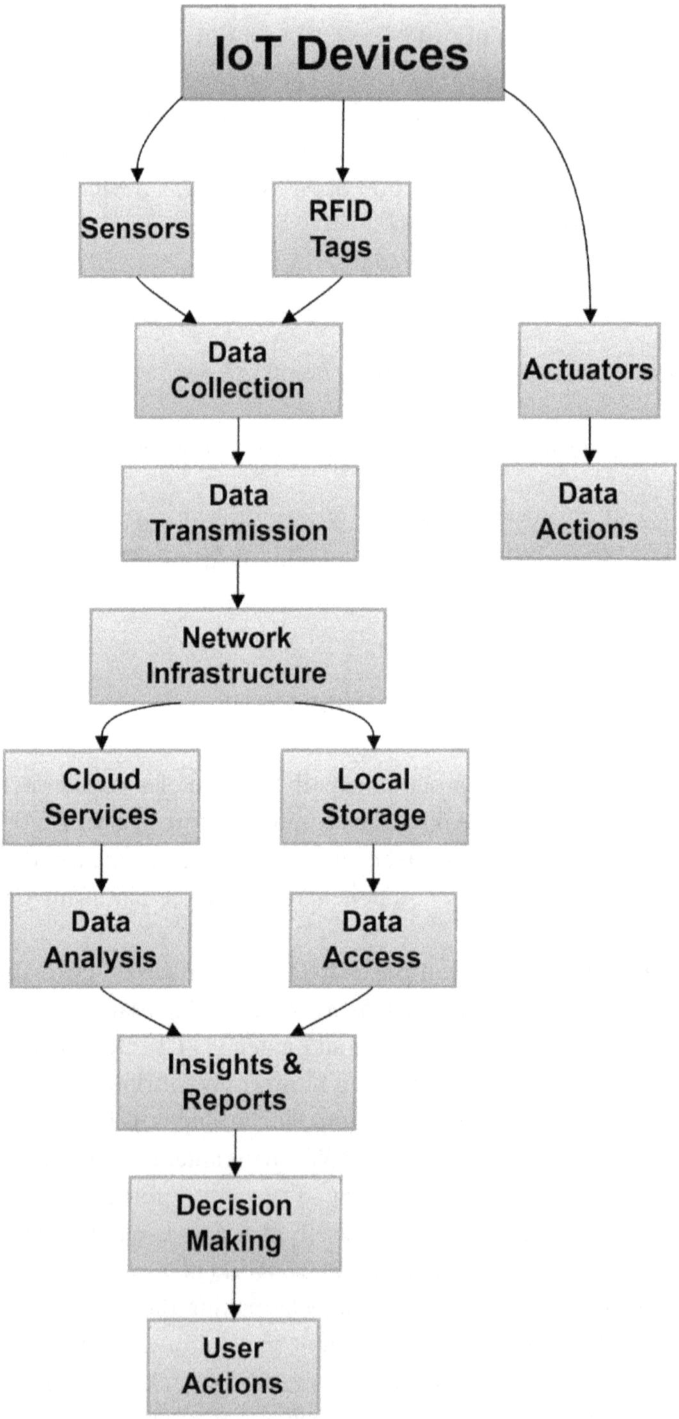

Sensor Networks: When RFID is accountable for object identification purposes, sensor networks are responsible for sensing and gathering information related to the environment like temperature, pressure, humidity, etc. Wireless Sensor Networks (WSN) comprise a network of a finite number of sensor nodes, responsible for tracking and monitoring device status, controlled by a central node. The sensors consist of an energy source, microprocessor, electronic circuit, and radio transceiver with antenna. Wireless Sensor and Actuator Network (WSAN) is an advanced form of WSN, where sensors sense and gather environmental information while actuators perform appropriate actions based on the sensed information. RFID sensor network (RSN) is also a form of sensor network where RFID and sensor work collaboratively to track the status information of devices.

Lightweight Communication Protocols: The communication protocols in IoT are distinct from the standard networks due to limited resource availability. For connectivity, low power wireless personal area networks (LoWPAN), IEEE 802.15.4, ZigBee, IEEE 802.11, and Z-wave are utilized as lightweight communication protocols. In the application layer, protocols like Constrained Application Protocol (CoAP) and Message Queue Telemetry Transport (MQTT) are utilized. Internet communications are often addressed using Internet Protocol version 4 (IPv4). However, a protocol like IPv6 with more address space is needed since more devices are being linked to the internet.

Cloud Service Platform: IoT devices are ubiquitous in nature, so it needs a platform for providing ubiquitous service. Cloud is one such platform for researchers that is capable of providing services like unlimited computing and storage capabilities to end-users anywhere and anytime. Software as a service (SaaS), Platform as a service (PaaS), and Infrastructure as a service (IaaS) are the three fundamental cloud services. But the integration of the cloud with IoT demands a new list of services leading to a new paradigm of 'everything as a service'(Nanda et al., 2024). Despite the services provided by the cloud, there are certain limitations like constrained network bandwidth and high latency in providing service due to its centralized nature. The cloud also poses greater limitations in providing data privacy and security to real-time IoT applications.

Fog/Edge/Cloudlet: Due to cloud constraints, new paradigms must be developed that can service end users by remaining near the network's edge. A cloudlet is a mobile user-friendly "data centre in a box.". It is the middle of the three-tiered architecture forming the middle element while the cloud is the upper part and end devices form the lower part. Several architectures for cloudlet are proposed including VM-based cloudlet, Ad hoc cloudlet, and Elastic Cloudlet(Dhanalakshmi et al., 2024; Pitchai et al., 2024). However, accessing a cloudlet requires using a Wi-Fi access point. Hence a cloudlet cannot support ubiquitous computing since its coverage area is small. Cloudlet is also incapable of supporting resource provisioning and service scalability. A novel paradigm called fog computing holds the potential to extend cloud services to the edge of the network. Similar to fog computing, edge computing offers services to end devices by being in close proximity to them. However, fog computing is implemented in network edges, a single or few hops away from the edge, while edge computing is placed in edge devices. Additionally, edge computing views the network edge as a segregated platform, while fog computing views the network edge as a decentralized platform that distributes decentralized services from the cloud to the edge. Though edge computing seems like a promising solution to the emerging Internet of Everything (IoE) by working close to the source of data, it has its limitations(Das et al., 2024; Maguluri et al., 2023). Since Edge holds distributed control, it fails to provide data and location privacy. Since edge computing is done at the edge of a network in feeble resource IoT devices it cannot adopt computationally heavy security primitives. Edge also struggles to implement identity authentication and access control over the data due to its distributed security management.

The Internet of Things (IoT) functions through a complex interplay of various technological components and service platforms, each contributing to the seamless integration and operation of connected devices. These components can be broadly categorized into four main areas: sensing, communication, data processing, and action. The efficient functioning of an IoT system relies on the coordination and collaboration of these areas, enabled by advanced technologies and robust service platforms.

Sensing and Data Collection: The first crucial function of an IoT system is sensing, which involves the collection of data from the physical environment. This is achieved through a variety of sensors embedded in IoT devices. Sensors can detect and measure a wide range of physical parameters, such as temperature, humidity, light, motion, and pressure. These devices convert physical stimuli into electrical signals, which can then be processed and analyzed(Jurcut et al., 2020; Omolara et al., 2022).

- **Types of Sensors**: Sensors used in IoT can be broadly classified into passive and active sensors. Passive sensors, such as thermistors and photoresistors, detect changes in environmental conditions without requiring an external power source. Active sensors, like ultrasonic sensors and radar sensors, emit signals and measure the response to determine various environmental factors (Venkateswaran et al., 2024).
- **RFID and Embedded Devices**: Radio Frequency Identification (RFID) tags and embedded devices play a critical role in IoT systems by enabling object identification and data collection. RFID tags use electromagnetic fields to automatically identify and track tags attached to objects, while embedded

devices are microcontroller-based systems that execute specific tasks within an IoT ecosystem.

Communication and Networking: Once data is collected, it needs to be transmitted to other devices or centralized systems for further processing. This is where communication and networking come into play. IoT systems rely on a variety of communication protocols and networking technologies to ensure reliable and efficient data transfer.

- **Communication Protocols**: Common communication protocols used in IoT include Wi-Fi, Bluetooth, Zigbee, LoRaWAN, and cellular networks. These protocols differ in terms of range, data rate, power consumption, and complexity. The choice of protocol depends on the specific requirements of the IoT application.
- **Networking Technologies**: IoT systems utilize different networking topologies, such as star, mesh, and hybrid networks, to connect devices. These networks can be local (e.g., Local Area Networks - LAN) or wide-area networks (e.g., the Internet). Networking technologies ensure that data collected by sensors is transmitted to gateways, which aggregate and forward the data to cloud platforms or edge devices(Devi et al., 2024; Verma et al., 2024).

Data Processing and Analysis: Data processing is a critical function in IoT systems, transforming raw data into meaningful insights. This process involves several stages, including data aggregation, storage, processing, and analysis.

- **Edge and Fog Computing**: Edge computing refers to processing data near the source of data generation, reducing latency and bandwidth usage. Fog computing extends this concept by creating a distributed computing infrastructure that processes data at intermediate nodes between the edge and the cloud.
- **Cloud Computing**: Cloud platforms provide scalable storage and powerful processing capabilities for IoT data. They offer services for data analytics, machine learning, and artificial intelligence, enabling complex data analysis and real-time decision-making(Rahamathunnisa et al., 2023; Tirlangi et al., 2024).
- **Big Data Analytics**: IoT generates massive amounts of data, which requires big data technologies for effective processing and analysis. Techniques such as data mining, machine learning, and predictive analytics are employed to extract valuable insights from the collected data.

Action and Automation: The final function of an IoT system is to take action based on the analyzed data. This involves triggering automated responses or sending commands to actuators and other devices to perform specific tasks.

- **Actuators**: Actuators are devices that convert electrical signals into physical actions, such as motors, valves, and relays. They interact with the physical environment by controlling mechanisms, adjusting settings, or performing operations based on the instructions received from the IoT system.
- **Automation**: IoT systems enable automation of processes through predefined rules and algorithms. For instance, in a smart home, the system can automatically adjust the thermostat settings based on the ambient temperature and user preferences. In industrial IoT, automation can optimize production processes, improve efficiency, and reduce human intervention.
- **Feedback Loops**: IoT systems often incorporate feedback loops where the actions taken by actuators influence subsequent data collection and analysis. This continuous loop ensures that the system adapts to changing conditions and operates optimally.

Integration with Service Platforms: Service platforms play a crucial role in the overall functionality of IoT systems(Agrawal et al., 2023; Karthik et al., 2023). These platforms provide essential services such as device management, data storage, analytics, and security. They also offer application development frameworks and APIs that facilitate the creation of IoT applications.

- **Device Management**: Service platforms provide tools for registering, monitoring, and managing IoT devices. They ensure that devices are functioning correctly, firmware updates are applied, and any issues are promptly addressed.
- **Data Management**: Platforms offer scalable storage solutions for handling the vast amounts of data generated by IoT devices. They provide mechanisms for data aggregation, transformation, and retrieval, ensuring that data is readily available for analysis.
- **Security Services**: Ensuring the security of IoT systems is paramount. Service platforms offer security features such as data encryption, access control, and threat detection to protect IoT networks and data.

Thus, the effective functioning of IoT systems relies on the seamless integration of sensing, communication, data processing, and action components, supported by robust service platforms. This collaborative approach enables IoT to deliver valuable insights, automate processes, and create smarter environments, ultimately enhancing the quality of life and driving innovation across various industries.

SECURITY THREATS TO IoT APPLICATIONS

The degree of data impact, network accessibility, scope, scalability, heterogeneity, repetition, and user participation can all be used to categories IoT application domains. The security aspects of IoT applications that must be guaranteed are availability, privacy, nonrepudiation, integrity, authorization, and confidentiality(Abba Ari et al., 2024; Bediya & Kumar, 2021).

Smart Home: Smart home systems provide comfort and safety to the residents of a house by using smart, automated, and ubiquitous information and communication technology services. The components of a smart home system are as follows: i) Home Entertainment System; ii) Home Appliances, Lighting, and Climate Control System; iii) Home Communication System; and iv) Home Security System. Security risks to a smart home include traffic analysis, masquerade, replay, denial of service, and eavesdropping, among others. Through these assaults, malicious individuals can obtain personal information about a resident, and robbers can exploit sensor data to gain entry. It is important to take the right precautions while implementing a system to protect residents' privacy and security(Syamala et al., 2023).

Intelligent Transport System (ITS): ITS involves electronic, wireless, and automated technologies that help to improve the efficiency and convenience of surface transportation. The Advanced Vehicle Control and Safety System (AVCSS), Advanced Traveller Information System (ATIS), and Advanced Traffic Management System (ATMS) are examples of ITS systems. The possible security threats in ITS systems include jamming, congestion, security and spectrum sharing, Sybil attack, DoS attack, flooding attack, masquerading, and replay attack. The small delay in basic safety messages can lead to fatal accidents in ITS systems. Confidentiality is not critical in some ITS systems whereas its crucial in some systems like electronic toll collection. Attacks on confidentiality include eavesdropping, brute force, and man-in-the-middle attack(Gopal et al., 2024; Revathi et al., 2024).

Smart Grid: Smart home systems provide comfort and safety to the residents of a house by using smart, automated, and ubiquitous information and communication technology services.: According to the Electric Power Research Institute (EPRI), a smart grid is an updated electricity delivery system that can automatically optimize, monitor, and safeguard the operation of all of its interconnected components(S., 2024; Saravanan et al., 2024). These elements range from industrial users and building automation systems to energy storage installations, end-user consumers and their thermostats, electric vehicles, appliances, and other household devices. Virtual

Power Plants, micro-grids, smart meter, and communication technologies are the main components of smart grid technology. Man-in-the-middle attacks in SCADA, denial-of-service (DoS) attacks and fake data injection attacks in smart meter, jammer attacks in micro-grids, and man-in-the-middle attacks in electric vehicles are among the security risks in the smart grid. Other possible security threats include physical attack, tampering, eavesdropping, worms, forgery, impersonation, message modification, insider attack, and replay attack. Physical security in a smart grid is difficult as some of the devices used cannot support cryptography and provision for updates and patches. SCADA, a major supporting technology of the smart grid is also prone to several attacks.

The Internet of Things (IoT) represents a significant advancement in technology, enabling a wide range of applications across various domains. These applications can be categorized based on the degree of data impact, network accessibility, scope, scalability, heterogeneity, repetition, and user participation. However, the proliferation of IoT devices introduces numerous security threats that must be addressed to ensure the reliability and trustworthiness of IoT systems. The primary security aspects that must be guaranteed include availability, privacy, nonrepudiation, integrity, authorization, and confidentiality. This section explores the main security threats to IoT applications and the necessary measures to mitigate these risks(Abba Ari et al., 2024).

Data Impact and Privacy Threats: Data impact in IoT applications refers to the significance and sensitivity of the data being collected, processed, and transmitted. IoT devices often handle personal, financial, and health-related information, making privacy a critical concern.

- Data Breaches: Unauthorized access to sensitive data can result in significant privacy violations. Attackers can exploit vulnerabilities in IoT devices to steal personal information, which can then be used for identity theft, financial fraud, or blackmail.
- Eavesdropping: Interception of data during transmission can lead to unauthorized disclosure of confidential information. IoT devices often use wireless communication, which is susceptible to eavesdropping attacks if not properly encrypted.

Network Accessibility and Availability Threats: Network accessibility in IoT refers to the ability of devices to communicate over various network types. Ensuring availability means that the IoT services are operational and accessible when needed (Agrawal et al., 2023; Veeranjaneyulu et al., 2023).

- Distributed Denial of Service (DDoS) Attacks: IoT devices are often targeted for DDoS attacks, where multiple compromised devices are used to overwhelm a network or service, rendering it unavailable to legitimate users.
- Device Hijacking: Attackers can gain control of IoT devices, using them to launch further attacks or disrupt services. This can lead to significant downtime and operational losses.

Scope and Scalability Threats: The scope and scalability of IoT applications refer to the extent of their deployment and their ability to scale to large numbers of devices.

- Scalability Issues: As the number of IoT devices grows, managing security becomes increasingly complex. Ensuring that all devices are secure and updated can be challenging, leading to potential vulnerabilities.
- Propagation of Infections: In a large-scale IoT deployment, malware or viruses can spread quickly from one device to another, causing widespread disruptions.

Heterogeneity and Integrity Threats: IoT applications are characterized by heterogeneity, involving diverse devices and technologies. Ensuring the integrity of data and devices in such a diverse environment is crucial.

- Firmware and Software Vulnerabilities: Different devices running varied firmware and software versions can have unpatched vulnerabilities that attackers can exploit to compromise system integrity.
- Data Tampering: Integrity threats involve unauthorized modification of data. Attackers can alter the data being transmitted or stored by IoT devices, leading to false readings or malicious actions.

Repetition and Nonrepudiation Threats: Repetition in IoT refers to the continuous and repetitive nature of data collection and actions performed by devices. Nonrepudiation ensures that actions or transactions cannot be denied after they have been executed.

- Replay Attacks: Attackers can capture and retransmit legitimate data or commands, causing devices to repeat actions maliciously. This can lead to unauthorized operations or data corruption.
- Lack of Accountability: Ensuring nonrepudiation is difficult in IoT environments due to the distributed nature of the devices. Without proper logging

and auditing mechanisms, it can be challenging to prove that a specific device performed a particular action.

User Participation and Authorization Threats: User participation in IoT applications often involves interaction with devices and services, requiring robust authorization mechanisms to ensure that only legitimate users have access to specific functionalities.

- Weak Authentication: Many IoT devices use weak authentication mechanisms, such as default passwords, which can be easily guessed or cracked by attackers. This can lead to unauthorized access and control over the devices.
- Privilege Escalation: Attackers can exploit vulnerabilities to gain higher levels of access than intended, allowing them to perform unauthorized actions or access sensitive data.

Mitigation Strategies: To address these security threats, several strategies and best practices can be implemented:

- Encryption: Implementing strong encryption protocols for data transmission and storage ensures that even if data is intercepted, it remains unreadable to unauthorized parties.
- Regular Updates and Patching: Ensuring that all IoT devices are regularly updated and patched to fix known vulnerabilities is crucial for maintaining security.
- Strong Authentication Mechanisms: Using multi-factor authentication and unique, strong passwords for devices can prevent unauthorized access.
- Network Security: Implementing network security measures such as firewalls, intrusion detection systems, and secure communication protocols can protect against various network-based attacks.
- Access Control: Implementing strict access control policies ensures that only authorized users and devices can access specific functionalities and data.
- Monitoring and Logging: Continuous monitoring and logging of device activity can help detect and respond to security incidents promptly, ensuring nonrepudiation and accountability.
- Segmentation: Network segmentation can limit the spread of attacks and contain potential breaches to a smaller subset of devices.

Security threats to IoT applications are diverse and multifaceted, stemming from the inherent characteristics of IoT systems, including their wide-ranging scope, scalability, and heterogeneity. Ensuring the security of IoT applications requires a comprehensive approach that addresses privacy, availability, integrity, nonrepudiation, authorization, and confidentiality. By implementing robust security measures and best practices, organizations can mitigate these threats and harness the full potential of IoT technology while safeguarding their systems and data.

PROPOSED WORK: USE OF NAS FOR SECURITY IN IoT APPLICATIONS

The proposed work uses NAS to optimize the usage of resources in IoT devices. The work also provides enhanced security by eliminating the limitations of edge by providing localized control over data thereby providing data privacy(Abba Ari et al., 2024; Ande et al., 2020; Sadhu et al., 2022).

Figure 2. Integrating NAS into the IoT ecosystem

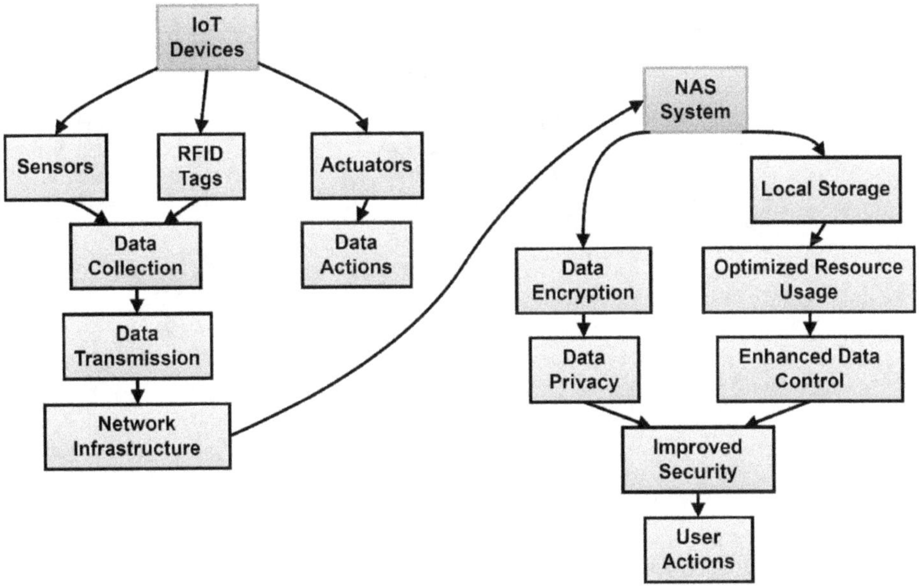

Figure 2 illustrates how integrating NAS into the IoT ecosystem can address resource limitations and enhance security, offering a more controlled and private data management solution. Figure 3 shows the architecture of the proposed work. IoT applications' resource-constrained devices are linked to NAS, which serves as a sink and encrypts the data these devices produce before it leaves the network.

As well as the data that are meant for the NAS-connected devices are first received by the NAS sink for decryption and then transferred to the respective devices. Traditionally NAS is used for file storage and sharing and acts as a file server. The proposed work utilizes the use of NAS to eliminate the limitations of edge security. NAS acts as a sink to the IoT devices where data can be aggregated and security primitives are implemented in NAS to secure the data before it leaves the local network.

Figure 3. The architecture of the proposed work

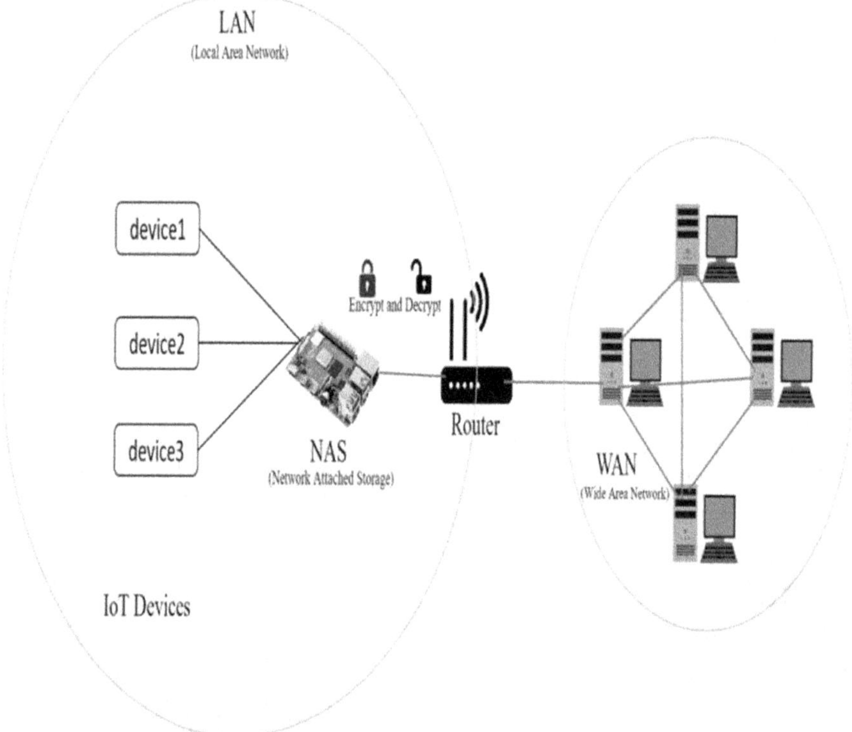

IMPLEMENTATION

Configuration Details

Raspberry Pi 4 Model B: The core of the NAS solution is the Raspberry Pi 4 Model B, a powerful single-board computer known for its versatility and performance. This model features a 1.5 GHz quad-core ARM Cortex-A72 CPU, making it well-suited for handling the encryption and decryption processes required for secure data storage. The 4GB RAM capacity ensures smooth operation and efficient multitasking, essential for a NAS setup serving multiple devices within a network(Abba Ari et al., 2024).

SanDisk Micro SD Card (16GB): Storage for the operating system and essential software is provided by a SanDisk Micro SD Card with a capacity of 16GB. While this storage is primarily used for the OS and applications, it is crucial to note that additional external storage devices (such as USB hard drives) are typically employed to handle the bulk of the data storage in a NAS configuration. The choice of a high-quality, reliable Micro SD card ensures the stability and performance of the system.

Operating System: Raspberry Pi OS: The Raspberry Pi NAS runs on Raspberry Pi OS with desktop and recommended software, providing a robust and user-friendly environment for managing and accessing stored data. Key details of the operating system configuration are as follows:

- **Release Date**: May 3rd, 2023
- **System**: 32-bit
- **Kernel Version**: 6.1
- **Debian Version**: 11 (Bullseye)
- **Size**: 2,701MB

The Raspberry Pi OS, based on Debian, is specifically optimized for Raspberry Pi hardware, ensuring compatibility and performance. The 32-bit system provides a stable platform, while the 6.1 kernel version incorporates the latest improvements and security updates. Debian 11 (Bullseye) is renowned for its reliability and extensive software repository, making it an excellent choice for a NAS setup.

Additional Software: The OS comes with a suite of recommended software, including tools for network management, file sharing, and security. These tools are crucial for setting up and maintaining the NAS, allowing users to easily configure network settings, manage shared folders, and implement security measures such as encryption and user authentication.

This configuration provides a solid foundation for a cost-effective, secure, and locally controlled NAS solution. The Raspberry Pi 4 Model B, combined with a reliable SanDisk Micro SD card and the robust Raspberry Pi OS, ensures efficient and secure data storage and management. This setup offers users greater control, privacy, and flexibility compared to traditional cloud storage services, making it an ideal choice for personal and small-scale enterprise applications.

Analysis of Encryption and Decryption Times for Different File Sizes

Encryption and decryption are critical processes for securing data, especially in IoT applications where data integrity and confidentiality are paramount. The time taken for these processes can vary significantly based on the file size and the encryption algorithm used. Here, we analyze the encryption and decryption times for three different file sizes: 508 MB, 34 MB, and 2.7 MB. Understanding these times helps in assessing the performance and efficiency of encryption mechanisms in IoT systems(Azrour et al., 2021; Bediya & Kumar, 2021; Khan et al., 2022; Mousavi et al., 2021). The figure 4 depicts an analysis of encryption and decryption times for various file sizes.

Case (a): File Size 508 MB

- **Encryption Time**: 5 minutes 41 seconds (341 seconds)
- **Approximate Time per MB**: 0.6701 seconds
- **Decryption Time**: 5 minutes 50 seconds (350 seconds)
- **Approximate Time per MB**: 0.688 seconds

For a 508 MB file, the encryption process takes 341 seconds, averaging approximately 0.6701 seconds per MB. The decryption process takes slightly longer at 350 seconds, averaging approximately 0.688 seconds per MB. This slight difference in time can be attributed to the overhead involved in decrypting and verifying the data.

The encryption and decryption times are relatively consistent, indicating that the process scales linearly with the file size for larger files. The slight variance in time per MB suggests that while both processes are efficient, decryption may involve additional verification steps, causing a marginal increase in time.

Case (b): File Size 34 MB

- **Encryption Time**: 10 seconds
- **Approximate Time per MB**: 0.294 seconds
- **Decryption Time**: 6.5 seconds
- **Approximate Time per MB**: 0.191 seconds

For a 34 MB file, the encryption process is significantly faster, taking only 10 seconds, which averages approximately 0.294 seconds per MB. The decryption process is even quicker, taking 6.5 seconds, averaging approximately 0.191 seconds per MB. This substantial reduction in time per MB compared to the 508 MB file can be due to the smaller data size, allowing the system to handle the encryption and decryption processes more efficiently.

The notable difference in time per MB for encryption and decryption suggests that for smaller files, the system's overhead is less pronounced, making both processes much faster. The decryption time per MB is significantly lower, indicating that the verification steps required for decryption are not as time-consuming for smaller files.

Figure 4. Analysis of Encryption and Decryption Times for Different File Sizes

```
pi@raspberrypi:~/Bookshelf $ time ccrypt -d File.pdf.cpt
Enter decryption key:

real	5m50.052s
user	0m38.940s
sys	0m1.247s
```

```
pi@raspberrypi:~/Bookshelf $ time ccrypt Book.pdf
Enter encryption key:
Enter encryption key: (repeat)

real	0m10.085s
user	0m1.823s
sys	0m0.238s
```

```
pi@raspberrypi:~/Bookshelf $ time ccrypt File.pdf
Enter encryption key:
Enter encryption key: (repeat)

real	5m41.191s
user	0m41.334s
sys	0m3.004s
```

```
pi@raspberrypi:~/Bookshelf $ time ccrypt -d Book.pdf.cpt
Enter decryption key:

real	0m6.543s
user	0m1.890s
sys	0m0.150s
```

```
pi@raspberrypi:~/Bookshelf $ time ccrypt Image.jpg
Enter encryption key:
Enter encryption key: (repeat)

real	0m8.310s
user	0m0.208s
sys	0m0.020s
```

```
pi@raspberrypi:~/Bookshelf $ time ccrypt -d Image.jpg
Enter decryption key:

real	0m4.495s
user	0m0.204s
sys	0m0.020s
```

Case (c): File Size 2.7 MB

- **Encryption Time**: 8 seconds
- **Approximate Time per MB**: 2.96 seconds
- **Decryption Time**: 4 seconds
- **Approximate Time per MB**: 1.481 seconds

For a 2.7 MB file, the encryption time is 8 seconds, averaging approximately 2.96 seconds per MB. The decryption time is 4 seconds, averaging approximately 1.481 seconds per MB. The encryption time per MB is significantly higher for this small file compared to the larger files. This suggests that there is a substantial fixed overhead involved in the encryption process that becomes more apparent with smaller files.

Similarly, the decryption time per MB is higher than that for larger files, though not as pronounced as the encryption time. The fixed overhead in the decryption process is less impactful but still noticeable. This overhead can include initial setup, important generation, and other preparatory steps that do not scale down proportionately with smaller file sizes.

Comparative Analysis

The analysis reveals several important points:

- **Linear Scaling for Larger Files**: For larger files like 508 MB, the encryption and decryption times scale linearly with file size. The time per MB remains relatively constant, indicating efficient processing.
- **Efficiency with Smaller Files**: For smaller files like 34 MB, the encryption and decryption processes are significantly more efficient, with much lower time per MB. This efficiency is due to reduced overhead and faster processing capabilities for smaller data sets.
- **Overhead Impact on Very Small Files**: For very small files like 2.7 MB, the fixed overhead involved in encryption and decryption becomes more noticeable. The time per MB is higher due to the initial setup and preparation steps that do not decrease proportionately with the smaller file size.

Implications for IoT Applications

In IoT applications, where devices often have limited processing power and storage capabilities, understanding the encryption and decryption times for different file sizes is crucial. Efficient encryption and decryption processes are essential to maintain data security without significantly impacting device performance.

- **Optimization for File Size**: IoT systems should optimize encryption and decryption processes based on typical file sizes encountered. For larger data transfers, ensuring linear scalability is significant, while for smaller data packets, minimizing overhead can enhance efficiency.
- **Balancing Security and Performance**: While strong encryption is necessary for data security, balancing it with performance considerations ensures that IoT devices can operate effectively without excessive delays.

By understanding and addressing these factors, IoT applications can achieve robust security while maintaining optimal performance, ensuring reliable and secure operation in various.

RESULTS AND DISCUSSION

The performance of encryption and decryption processes in IoT applications is influenced by several factors. This section analyzes the results from our experiments, considering the total average encryption and decryption times for various file types and sizes, operating systems, Raspberry Pi configurations, and memory card speeds(S., 2024; Upadhyaya et al., 2024; Venkateswaran et al., 2024).

Factors Affecting Encryption and Decryption Times

- File Type: Different file types can influence the encryption and decryption times due to their inherent structure and complexity. For example, text files, binary files, and multimedia files may vary in how they are processed by the encryption algorithm. Text files, typically smaller and less complex, may take less time compared to larger multimedia files that require more processing power.
- File Size: As demonstrated in the data, file size plays a significant role in the time taken for encryption and decryption. Larger files, such as the 508 MB file, take considerably longer compared to smaller files. This linear increase in time with file size is expected due to the increased amount of data that needs to be processed.
- Operating System: The choice of operating system can affect the performance of cryptographic operations. Different operating systems manage resources and handle processes uniquely. For instance, Linux-based systems like Raspbian, commonly used on Raspberry Pi, are known for their efficiency in handling background processes and resource management, which can influence the speed of encryption and decryption.

- Raspberry Pi Configuration: The hardware configuration of the Raspberry Pi, including the model, CPU speed, and available RAM, directly impacts the performance of encryption and decryption processes. Higher-end models with faster processors and more RAM can handle cryptographic operations more efficiently, reducing the time required.
- Memory Card Speed: The speed of the memory card used in the Raspberry Pi is another critical factor. Faster memory cards allow quicker read and write operations, which are essential for efficient encryption and decryption. A slow memory card can become a bottleneck, significantly increasing the time required for these processes.

Average Encryption and Decryption Times

- **Total Average Encryption Time for 1 MB**: $\approx$ 1.3083 Seconds
- **Total Average Decryption Time for 1 MB**: $\approx$ 0.787 Seconds

The results show a noticeable difference between the average times for encryption and decryption. Encryption, on average, takes about 1.3083 seconds per MB, while decryption is faster, averaging 0.787 seconds per MB. This disparity can be attributed to several factors, including the overhead involved in encrypting data securely, which often requires more computational steps compared to decryption.

The analysis of the results indicates that optimizing encryption and decryption times in IoT applications requires a multifaceted approach. Understanding the factors affecting these times helps in identifying areas for improvement(Venkateswaran et al., 2024).

- Optimizing File Handling: Efficiently handling different file types and sizes can reduce the time required for encryption and decryption. Implementing pre-processing steps to standardize file formats and sizes before encryption can streamline the process.
- Choosing the Right Operating System: Selecting an operating system that is optimized for performance and resource management can significantly enhance cryptographic operations. For IoT devices, lightweight and efficient OS choices like Raspbian or other Linux distributions are preferable.
- Upgrading Raspberry Pi Configuration: Investing in higher-end Raspberry Pi models with better CPU performance and more RAM can lead to substantial improvements in encryption and decryption times. Additionally, configuring the Raspberry Pi for optimal performance, such as overclocking (with caution) and ensuring efficient cooling, can further enhance processing speeds.

- Using High-Speed Memory Cards: Utilizing high-speed memory cards can drastically reduce read and write times, which are crucial for efficient encryption and decryption. Opting for Class 10 or UHS-I/UHS-II memory cards ensures faster data transfer rates.
- Implementing Lightweight Cryptographic Primitives: As previously discussed, using lightweight cryptographic algorithms can reduce the computational burden on IoT devices, leading to faster encryption and decryption times. These algorithms are specifically designed for resource-constrained environments, making them ideal for IoT applications.

The study of encryption and decryption times reveals significant insights into the factors influencing these processes in IoT applications. By addressing the factors such as file type, file size, operating system, hardware configuration, and memory card speed, it is possible to optimize cryptographic operations, ensuring that IoT devices can maintain high levels of security without compromising performance. Implementing these optimizations can lead to more efficient, secure, and reliable IoT systems.

Figures 5 depict the average time taken for encryption and decryption for variable file sizes. Though the time taken is higher for the present configuration it can be reduced with higher storage size. Implementing lightweight cryptographic primitives is also an alternate solution.

Figure 5. Average time taken for encryption and decryption for variable file sizes

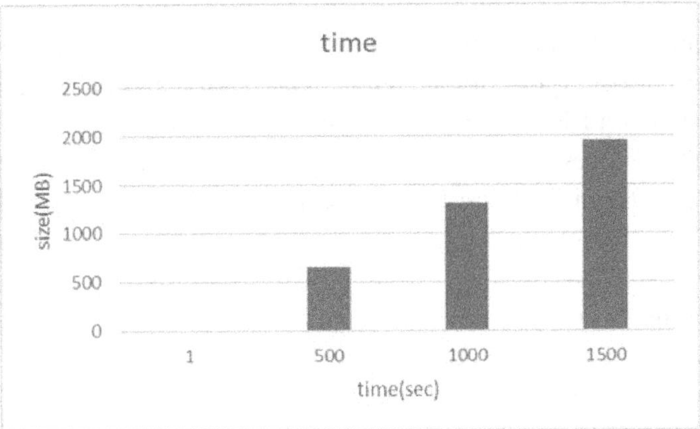

Time for encryption

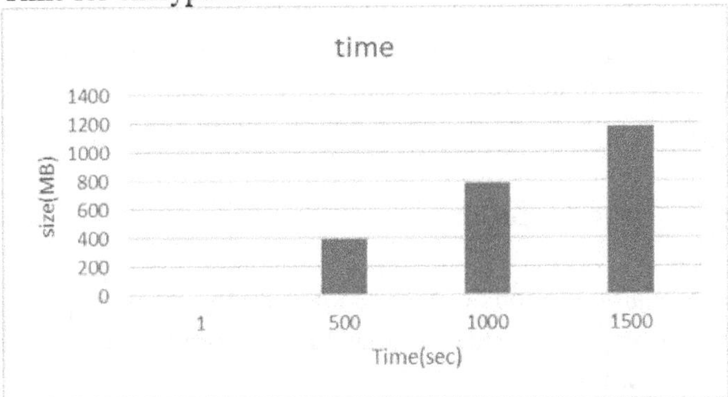

Time for decryption

Optimization of Encryption and Decryption Times for IoT Applications

Figure 5 illustrates the average time taken for encryption and decryption processes across variable file sizes. The data shows that while the current configuration results in relatively higher times for both encryption and decryption, these times can be optimized through various approaches, including increasing storage capacity and implementing lightweight cryptographic primitives. This section delves into these optimization strategies and their implications for IoT applications(Ali et al., 2024; Boopathi, 2023, 2024; Kavitha et al., 2023; Malathi et al., 2024).

Current Configuration Analysis

The current configuration demonstrates that as file sizes increase, the time taken for encryption and decryption also increases. This trend is expected due to the linear relationship between file size and processing time for cryptographic operations(-Boopathi, 2023). However, the relatively high times observed in the figure 5 suggest that the present system might not be fully optimized for performance. This can be attributed to several factors:

- **Processing Power**: IoT devices often have limited processing capabilities, which can slow down cryptographic operations.
- **Storage Capacity**: Limited storage can affect the performance of encryption and decryption processes, especially if there is insufficient space for efficient data handling.
- **Algorithm Efficiency**: The cryptographic algorithms currently in use might not be the most efficient for the given hardware configuration, leading to increased processing times.

Increasing Storage Capacity

One of the primary solutions to reduce encryption and decryption times is to increase the storage capacity of IoT devices. Larger storage sizes can enhance performance in several ways:

- **Improved Data Handling**: More storage allows for better management of data during cryptographic operations. It can enable the use of faster, more efficient temporary storage mechanisms, reducing the overall time taken for encryption and decryption.
- **Enhanced Buffering**: Increased storage can support larger buffer sizes, which can streamline data processing and reduce the need for frequent read/write operations. This can significantly cut down on the time required for cryptographic tasks.
- **Parallel Processing**: With more storage, it becomes feasible to implement parallel processing techniques, where multiple chunks of data are encrypted or decrypted simultaneously. This can drastically improve the speed of cryptographic operations.

Implementing Lightweight Cryptographic Primitives

Another effective approach to optimizing encryption and decryption times is the use of lightweight cryptographic primitives. These are specifically designed to provide strong security with minimal computational and storage overhead, making them ideal for resource-constrained IoT devices(Boopathi, 2024; Kavitha et al., 2023; Malathi et al., 2024; Venkateswaran et al., 2023).

- **Efficiency**: Lightweight cryptographic algorithms are tailored to be efficient on devices with limited resources. They require less processing power and memory, resulting in faster encryption and decryption times.
- **Reduced Energy Consumption**: IoT devices often run on battery power, making energy efficiency a critical factor. Lightweight cryptographic primitives consume less energy, thereby extending the battery life of IoT devices while still ensuring data security.
- **Scalability**: These primitives are designed to be scalable, allowing them to handle varying file sizes efficiently. They can maintain consistent performance even as the size of the data being encrypted or decrypted changes.

Practical Examples of Lightweight Cryptographic Primitives

Several lightweight cryptographic algorithms can be implemented to optimize IoT security(Boopathi & Khang, 2023; Mohanraj et al., 2024):

- **Advanced Encryption Standard (AES) in Lightweight Mode**: AES is widely used, but in lightweight mode, it can be optimized for faster processing on limited hardware.
- **Elliptic Curve Cryptography (ECC)**: ECC provides strong security with shorter crucial lengths, reducing the computational burden compared to traditional algorithms like RSA.
- **Speck and Simon**: These are block ciphers designed by the NSA for lightweight applications. They are specifically optimized for speed and efficiency on constrained devices.

Figure 5 highlights the need for optimizing encryption and decryption times in IoT applications. By increasing storage capacity and implementing lightweight cryptographic primitives, the performance of these critical security processes can be significantly enhanced.

Increasing Storage Capacity:

- Improved Data Handling
- Enhanced Buffering
- Parallel Processing

Implementing Lightweight Cryptographic Primitives:

- Efficiency
- Reduced Energy Consumption
- Scalability

Examples:

- AES in Lightweight Mode
- Elliptic Curve Cryptography (ECC)
- Speck and Simon

These strategies not only reduce the time required for encryption and decryption but also ensure that IoT devices can operate securely and efficiently. This balance between security and performance is essential for the widespread adoption and reliable functioning of IoT systems in various applications.

CONCLUSION

Overall, the Raspberry Pi NAS solution offers a cost-effective, secure, and locally controlled storage option that can be accessed by multiple devices within a network. It presents an attractive alternative to cloud storage services, providing individuals and organizations with greater privacy, control, and flexibility over their data. The research can also be extended further by implementing lightweight security primitives for authentication and access control using NAS. With the right setup and configuration, Raspberry Pi NAS can serve as a powerful storage and security solution that meets the demands of today's data-centric world. The Raspberry Pi Network Attached Storage (NAS) solution offers a compelling alternative to traditional cloud storage services, providing a cost-effective, secure, and locally controlled storage option accessible by multiple devices within a network. This approach enhances privacy and control, addressing concerns associated with third-party cloud storage services. By maintaining data within the local network, individuals and organizations can ensure greater security and flexibility, adapting the storage solution to their specific needs and preferences.

One of the main advantages of the Raspberry Pi NAS solution is its cost-effectiveness. Leveraging the affordability of Raspberry Pi hardware, users can set up a robust storage system without incurring the significant expenses typically associated with commercial NAS devices or cloud storage subscriptions. This makes it an ideal choice for small businesses, educational institutions, and home users who require reliable storage solutions on a budget. The security benefits of the Raspberry Pi NAS are also noteworthy. By encrypting data before it leaves the local network, the system mitigates the risks associated with data transmission over the internet. This local encryption ensures that sensitive information remains protected, reducing the potential for unauthorized access and data breaches. Furthermore, the use of NAS allows for customizable security configurations, enabling users to implement additional measures tailored to their specific security requirements.

In addition to encryption, the research highlights the potential for extending the Raspberry Pi NAS solution by incorporating lightweight security primitives for authentication and access control. These enhancements can further strengthen the security framework, ensuring that only authorized users can access the stored data. By integrating such security mechanisms, the Raspberry Pi NAS can evolve into a comprehensive storage and security solution capable of meeting the demands of today's data-centric world. With the right setup and configuration, the Raspberry Pi NAS can serve as a powerful tool for managing and securing data. Its flexibility and adaptability make it suitable for a wide range of applications, from personal data storage to enterprise-level solutions. As the need for secure and reliable storage continues to grow, the Raspberry Pi NAS presents a viable and attractive option, offering users greater control, privacy, and peace of mind.

REFERENCES

Abba Ari, A. A., Ngangmo, O. K., Titouna, C., Thiare, O., Mohamadou, A., & Gueroui, A. M. (2024). Enabling privacy and security in Cloud of Things: Architecture, applications, security & privacy challenges. *Applied Computing and Informatics, 20*(1/2), 119–141. DOI: 10.1016/j.aci.2019.11.005

Agrawal, A. V., Magulur, L. P., Priya, S. G., Kaur, A., Singh, G., & Boopathi, S. (2023). Smart Precision Agriculture Using IoT and WSN. In *Handbook of Research on Data Science and Cybersecurity Innovations in Industry 4.0 Technologies* (pp. 524–541). IGI Global. DOI: 10.4018/978-1-6684-8145-5.ch026

Ali, M. N., Senthil, T., Ilakkiya, T., Hasan, D. S., Ganapathy, N. B. S., & Boopathi, S. (2024). IoT's Role in Smart Manufacturing Transformation for Enhanced Household Product Quality. In *Advanced Applications in Osmotic Computing* (pp. 252–289). IGI Global. DOI: 10.4018/979-8-3693-1694-8.ch014

Ande, R., Adebisi, B., Hammoudeh, M., & Saleem, J. (2020). Internet of Things: Evolution and technologies from a security perspective. *Sustainable Cities and Society, 54*, 101728. DOI: 10.1016/j.scs.2019.101728

Azrour, M., Mabrouki, J., Guezzaz, A., & Kanwal, A. (2021). Internet of things security: Challenges and key issues. *Security and Communication Networks, 2021*(1), 5533843. DOI: 10.1155/2021/5533843

Bediya, A. K., & Kumar, R. (2021). A novel intrusion detection system for internet of things network security. [JITR]. *Journal of Information Technology Research, 14*(3), 20–37. DOI: 10.4018/JITR.2021070102

Boopathi, S. (2023). Securing Healthcare Systems Integrated With IoT: Fundamentals, Applications, and Future Trends. In *Dynamics of Swarm Intelligence Health Analysis for the Next Generation* (pp. 186–209). IGI Global.

Boopathi, S. (2024). Sustainable Development Using IoT and AI Techniques for Water Utilization in Agriculture. In *Sustainable Development in AI, Blockchain, and E-Governance Applications* (pp. 204–228). IGI Global. DOI: 10.4018/979-8-3693-1722-8.ch012

Boopathi, S., & Khang, A. (2023). AI-Integrated Technology for a Secure and Ethical Healthcare Ecosystem. In *AI and IoT-Based Technologies for Precision Medicine* (pp. 36–59). IGI Global. DOI: 10.4018/979-8-3693-0876-9.ch003

Cao, K., Liu, Y., Meng, G., & Sun, Q. (2020). An Overview on Edge Computing Research. *IEEE Access : Practical Innovations, Open Solutions*, *8*, 85714–85728. DOI: 10.1109/ACCESS.2020.2991734

Das, S., Lekhya, G., Shreya, K., Shekinah, K. L., Babu, K. K., & Boopathi, S. (2024). Fostering Sustainability Education Through Cross-Disciplinary Collaborations and Research Partnerships: Interdisciplinary Synergy. In *Facilitating Global Collaboration and Knowledge Sharing in Higher Education With Generative AI* (pp. 60–88). IGI Global.

Devi, J. V., Argiddi, R. V., Renuka, P., Janagi, K., Hari, B. S., & Boopathi, S. (2024). Study on Integrated Neural Networks and Fuzzy Logic Control for Autonomous Electric Vehicles. In *Advances in Web Technologies and Engineering* (pp. 104–127). IGI Global. DOI: 10.4018/979-8-3693-1487-6.ch006

Dhanalakshmi, M., Tamilarasi, K., Saravanan, S., Sujatha, G., Boopathi, S., & Associates. (2024). Fog Computing-Based Framework and Solutions for Intelligent Systems: Enabling Autonomy in Vehicles. In *Computational Intelligence for Green Cloud Computing and Digital Waste Management* (pp. 330–356). IGI Global.

Gopal, M., Lurdhumary, J., Bathrinath, S., Priya, A. P., Sarojwal, A., & Boopathi, S. (2024). Energy Harvesting and Smart Highways for Sustainable Transportation Infrastructure: Revolutionizing Roads Using Nanotechnology. In *Principles and Applications in Speed Sensing and Energy Harvesting for Smart Roads* (pp. 136–165). IGI Global. DOI: 10.4018/978-1-6684-9214-7.ch005

Jurcut, A., Niculcea, T., Ranaweera, P., & Le-Khac, N.-A. (2020). Security considerations for Internet of Things: A survey. *SN Computer Science*, *1*(4), 1–19. DOI: 10.1007/s42979-020-00201-3

Karthik, S., Hemalatha, R., Aruna, R., Deivakani, M., Reddy, R. V. K., & Boopathi, S. (2023). Study on Healthcare Security System-Integrated Internet of Things (IoT). In *Perspectives and Considerations on the Evolution of Smart Systems* (pp. 342–362). IGI Global.

Kavitha, C., Varalatchoumy, M., Mithuna, H., Bharathi, K., Geethalakshmi, N., & Boopathi, S. (2023). Energy Monitoring and Control in the Smart Grid: Integrated Intelligent IoT and ANFIS. In *Applications of Synthetic Biology in Health, Energy, and Environment* (pp. 290–316). IGI Global.

Khan, A. A., Laghari, A. A., Shaikh, Z. A., Dacko-Pikiewicz, Z., & Kot, S. (2022). Internet of Things (IoT) security with blockchain technology: A state-of-the-art review. *IEEE Access : Practical Innovations, Open Solutions*, *10*, 122679–122695. DOI: 10.1109/ACCESS.2022.3223370

Koshariya, A. K., Kalaiyarasi, D., Jovith, A. A., Sivakami, T., Hasan, D. S., & Boopathi, S. (2023). AI-Enabled IoT and WSN-Integrated Smart Agriculture System. In *Artificial Intelligence Tools and Technologies for Smart Farming and Agriculture Practices* (pp. 200–218). IGI Global. DOI: 10.4018/978-1-6684-8516-3.ch011

Kumar, R., Nagaraj, A., Paul, B., & Dixit, S. P. (2021). Network-Attached Storage: Data Storage Applications. *Turkish Journal of Computer and Mathematics Education V Ol.12 No, 12*(12), 2385–2396.

Lin, J., Yu, W., Zhang, N., Yang, X., Zhang, H., & Zhao, W. (2017). A Survey on Internet of Things: Architecture, Enabling Technologies, Security and Privacy, and Applications. *IEEE Internet of Things Journal, 4*(5), 1125–1142. DOI: 10.1109/JIOT.2017.2683200

Lu, J., Shen, J., Vijayakumar, P., & Gupta, B. B. (2021). Blockchain-based secure data storage protocol for sensors in the industrial internet of things. *IEEE Transactions on Industrial Informatics, 18*(8), 5422–5431. DOI: 10.1109/TII.2021.3112601

Maguluri, L. P., Arularasan, A., & Boopathi, S. (2023). Assessing Security Concerns for AI-Based Drones in Smart Cities. In *Effective AI, Blockchain, and E-Governance Applications for Knowledge Discovery and Management* (pp. 27–47). IGI Global. DOI: 10.4018/978-1-6684-9151-5.ch002

Malathi, J., Kusha, K., Isaac, S., Ramesh, A., Rajendiran, M., & Boopathi, S. (2024). IoT-Enabled Remote Patient Monitoring for Chronic Disease Management and Cost Savings: Transforming Healthcare. In *Advances in Explainable AI Applications for Smart Cities* (pp. 371–388). IGI Global.

Mohanraj, G., Krishna, K. S., Lakshmi, B. S., Vijayalakshmi, A., Pramila, P. V., & Boopathi, S. (2024). Optimizing Trust and Security in Healthcare 4.0: Human Factors in Lightweight Secured IoMT Ecosystems. In *Lightweight Digital Trust Architectures in the Internet of Medical Things (IoMT)* (pp. 52–72). IGI Global. DOI: 10.4018/979-8-3693-2109-6.ch004

Mousavi, S. K., Ghaffari, A., Besharat, S., & Afshari, H. (2021). Security of internet of things based on cryptographic algorithms: A survey. *Wireless Networks, 27*(2), 1515–1555. DOI: 10.1007/s11276-020-02535-5

Nanda, A. K., Sharma, A., Augustine, P. J., Cyril, B. R., Kiran, V., & Sampath, B. (2024). Securing Cloud Infrastructure in IaaS and PaaS Environments. In *Improving Security, Privacy, and Trust in Cloud Computing* (pp. 1–33). IGI Global. DOI: 10.4018/979-8-3693-1431-9.ch001

Omolara, A. E., Alabdulatif, A., Abiodun, O. I., Alawida, M., Alabdulatif, A., & Arshad, H. (2022). The internet of things security: A survey encompassing unexplored areas and new insights. *Computers & Security*, *112*, 102494. DOI: 10.1016/j.cose.2021.102494

Pitchai, R., Guru, K. V., Gandhi, J. N., Komala, C. R., Kumar, J. R. D., & Boopathi, S. (2024). Fog Computing-Integrated ML-Based Framework and Solutions for Intelligent Systems: Digital Healthcare Applications. In *Technological Advancements in Data Processing for Next Generation Intelligent Systems* (pp. 196–224). IGI Global. DOI: 10.4018/979-8-3693-0968-1.ch008

Rahamathunnisa, U., Sudhakar, K., Murugan, T. K., Thivaharan, S., Rajkumar, M., & Boopathi, S. (2023). Cloud Computing Principles for Optimizing Robot Task Offloading Processes. In *AI-Enabled Social Robotics in Human Care Services* (pp. 188–211). IGI Global. DOI: 10.4018/978-1-6684-8171-4.ch007

Reed, B. C., Chron, E. G., Burns, R. C., & Long, D. D. E. (2000). *A UTHENTICATING N ETWORK -A TTACHED S TORAGE W E PRESENT AN ARCHITECTURE FOR NETWORK - AUTHENTICATED DISKS THAT ENCRYPTION. O UR SYSTEM PROVIDES NETWORK CLIENTS WITH DIRECT*. 49–57.

Revathi, S., Babu, M., Rajkumar, N., Meti, V. K. V., Kandavalli, S. R., & Boopathi, S. (2024). Unleashing the Future Potential of 4D Printing: Exploring Applications in Wearable Technology, Robotics, Energy, Transportation, and Fashion. In *Human-Centered Approaches in Industry 5.0: Human-Machine Interaction, Virtual Reality Training, and Customer Sentiment Analysis* (pp. 131–153). IGI Global.

S., B. (2024). Advancements in Optimizing Smart Energy Systems Through Smart Grid Integration, Machine Learning, and IoT. In *Advances in Environmental Engineering and Green Technologies* (pp. 33–61). IGI Global. DOI: 10.4018/979-8-3693-0492-1.ch002

Sadhu, P. K., Yanambaka, V. P., & Abdelgawad, A. (2022). Internet of things: Security and solutions survey. *Sensors (Basel)*, *22*(19), 7433. DOI: 10.3390/s22197433 PMID: 36236531

Saravanan, S., Khare, R., Umamaheswari, K., Khare, S., Krishne Gowda, B. S., & Boopathi, S. (2024). AI and ML Adaptive Smart-Grid Energy Management Systems: Exploring Advanced Innovations. In *Principles and Applications in Speed Sensing and Energy Harvesting for Smart Roads* (pp. 166–196). IGI Global. DOI: 10.4018/978-1-6684-9214-7.ch006

Syamala, M., Komala, C., Pramila, P., Dash, S., Meenakshi, S., & Boopathi, S. (2023). Machine Learning-Integrated IoT-Based Smart Home Energy Management System. In *Handbook of Research on Deep Learning Techniques for Cloud-Based Industrial IoT* (pp. 219–235). IGI Global. DOI: 10.4018/978-1-6684-8098-4.ch013

Tirlangi, S., Teotia, S., Padmapriya, G., Senthil Kumar, S., Dhotre, S., & Boopathi, S. (2024). Cloud Computing and Machine Learning in the Green Power Sector: Data Management and Analysis for Sustainable Energy. In *Developments Towards Next Generation Intelligent Systems for Sustainable Development* (pp. 148–179). IGI Global. DOI: 10.4018/979-8-3693-5643-2.ch006

Upadhyaya, A. N., Saqib, A., Devi, J. V., Rallapalli, S., Sudha, S., & Boopathi, S. (2024). Implementation of the Internet of Things (IoT) in Remote Healthcare. In *Advances in Medical Technologies and Clinical Practice* (pp. 104–124). IGI Global. DOI: 10.4018/979-8-3693-1934-5.ch006

Veeranjaneyulu, R., Boopathi, S., Kumari, R. K., Vidyarthi, A., Isaac, J. S., & Jaiganesh, V. (2023). Air Quality Improvement and Optimisation Using Machine Learning Technique. *IEEE- Explore*, 1–6.

Venkateswaran, N., Kiran Kumar, K., Maheswari, K., Kumar Reddy, R. V., & Boopathi, S. (2024). Optimizing IoT Data Aggregation: Hybrid Firefly-Artificial Bee Colony Algorithm for Enhanced Efficiency in Agriculture. *AGRIS On-Line Papers in Economics and Informatics*, *16*(1), 117–130. DOI: 10.7160/aol.2024.160110

Venkateswaran, N., Kumar, S. S., Diwakar, G., Gnanasangeetha, D., & Boopathi, S. (2023). Synthetic Biology for Waste Water to Energy Conversion: IoT and AI Approaches. *Applications of Synthetic Biology in Health. Energy & Environment*, •••, 360–384.

Verma, R., Christiana, M. B. V., Maheswari, M., Srinivasan, V., Patro, P., Dari, S. S., & Boopathi, S. (2024). Intelligent Physarum Solver for Profit Maximization in Oligopolistic Supply Chain Networks. In *AI and Machine Learning Impacts in Intelligent Supply Chain* (pp. 156–179). IGI Global. DOI: 10.4018/979-8-3693-1347-3.ch011

Wang, C., Wang, D., Duan, Y., & Tao, X. (2022). Secure and Lightweight User Authentication Scheme for Cloud-Aided Internet of Things. *IEEE Transactions on Information Forensics and Security*, *18*, 2961–2976. DOI: 10.1109/TIFS.2023.3272772

Chapter 16
Sustainable Solutions for Autonomous Vehicles Using Deep Learning and IoT

Shantala Devi Patil
https://orcid.org/0009-0003-6032-5923
School of Computer Science and Engineering, REVA University, Bengaluru, India

Jerzy Ryszard Szymański
https://orcid.org/0000-0002-9493-8424
Department of Electrical Drives and Industrial Electronics, Casimir Pulaski University of Radom, Poland

Marta Żurek-Mortka
https://orcid.org/0000-0002-4804-8172
Department of Control Systems, Lukasiewicz Research Network Institute for Sustainable Technologies, Poland

Mithileysh Sathiyanarayanan
https://orcid.org/0009-0003-1456-2709
MIT Square, London, UK

ABSTRACT

The rise of self-driving cars shows promise to make getting around easier, but also brings big challenges when it comes to being eco-friendly. This paper lays out a plan to create lasting solutions for self-driving cars by combining deep learning, the internet of things (IoT), and car networks. This approach aims to boost efficiency cut

DOI: 10.4018/979-8-3693-5448-3.ch016

down on energy use, and make driving safer and more productive. IoT allows cars, roads, and checkpoints to share info in real time, which helps streamline travel and reduce harm to the environment. The car network ensures that self-driving cars stay and linked to all parts of their ecosystem. This paper looks at how these technologies can work together in the self-driving car industry to tackle green issues like using less energy, lowering carbon output, and using materials more . It zeros in on how this tech can shape what's to come.

INTRODUCTION

Self-driving cars also called autonomous vehicles (AVs) or driverless vehicles, are changing how we think about transportation. Regular cars need people to drive them, but AVs use advanced sensors, actuators, and AI systems to see what's around them and drive on their own. People are working on AV technology because it could make roads safer, help those who can't drive or don't have easy access to transportation, and make getting around more efficient. AVs use a mix of sensors like radar, lidar, cameras, and GPS to check their surroundings. These sensors create a live all-around view of what's near the car letting it spot lane lines, traffic signs, people walking, cyclists, and other vehicles (SZELE & KISGYÖRGY, 2018).

At the heart of how AVs work is the AI-powered decision-making system. This system processes sensor data to make driving decisions as they happen. Machine learning algorithms help AVs learn from huge amounts of data they collect while driving. This helps them get better at handling tricky traffic situations and spotting possible dangers. This learning is key to making AVs safer and more reliable over time. We group AVs into levels from 0 (no automation) to 5 (full automation). Level 1 and 2 AVs still need some human help, like steering or controlling speed. Levels 3, 4, and 5 mean the car can drive itself more and more, with less or no human help in certain situations or in all driving conditions (Chehri & Mouftah, 2019).

Major players in the self-driving car world include car makers, tech firms that focus on AI and sensors, government regulators, city planners, and insurance companies. These groups need to work together to tackle tech issues set rules, build needed structures, and get people to accept self-driving cars. Self-driving cars can do more than just make life easier for drivers. They can help society in big ways. These cars might cut down on traffic jams by finding the best routes and driving close together at steady speeds. This helps cars use less fuel and move through air more . Also, as more people use self-driving cars, we could see fewer crashes caused by human mistakes. Right now, these kinds of crashes are a top reason for road deaths around the world (Bahamonde-Birke et al. 2018).

But the big-scale rollout of AVs brings up issues and things to think about. Safety is still the top worry, as AVs need to show they're more trustworthy and secure than regular cars to win public confidence and get the green light from regulators. Moral questions, like how to decide in life-or-death situations also create tricky problems that need careful thought and discussion. Fitting AVs into our current transport setup means we need to update road designs, traffic control systems, and ways for vehicles to talk to each other. This is to make sure AVs and normal cars can work together . Also strong cyber defenses are key to keep AVs safe from online threats that could mess with how they work and their safety (Lim & Taeihagh, 2018).

This means self-driving cars are a game-changing idea that could change how we get around in the future. As tech gets better and rules catch up getting more people to use these cars will mean solving technical, legal, ethical, and social problems. This will help make the most of their ability to improve safety make it easier for people to move around, and be better for the environment in both cities and rural areas. When creating and using self-driving cars, it's essential to think about sustainability. We need to consider how these cars affect the environment, economy, and society. Self-driving cars have the potential to cause a revolution in how we move by cutting down on greenhouse gases, using energy more, and reducing the harm transportation does to the environment (Gružauskas et al. 2018).

A key reason to include sustainability in AV development is to lessen how much transportation harms the environment. Regular cars with gas engines play a big role in causing air pollution and releasing greenhouse gases. AVs give us chances to switch to cleaner more efficient power sources, like electric and hydrogen-powered vehicles. AVs can also cut down on fuel use and emissions by finding the best routes and ways of driving. This helps to improve air quality and fight climate change (Kovačić et al. 2022).

Financial viability is also a crucial factor to weigh when developing self-driving cars. At first, vehicles with autonomous technology might cost more than regular cars. However, they can save money in the long run. These savings come from needing less upkeep using fuel more, and cutting down on healthcare costs because of fewer crashes. What's more, self-driving cars can boost productivity and make economies more competitive. They do this by cutting down on travel time expenses related to traffic jams, and the need for fossil fuels (Kovačić et al. 2022).

Social sustainability aims to make sure AV technology helps all of society. This includes fair access to transportation safer roads, and better mobility for groups like the elderly and people with disabilities. AVs could make mobility more democratic by offering cheap and reliable transport in city and country areas where public transit might be scarce. To build sustainability into AV development, we need to look at the big picture. This means thinking about impacts over the whole lifecycle using resources, and being ready for future changes like climate shifts and growing cities. Key players like governments, industry chiefs, city planners, and community groups have a big part to play. They help shape the rules and plans that push for sustainable AV use(Du 2016).

Hurdles still exist in making AVs sustainable. We need to upgrade infrastructure, create rules that cut emissions and save energy, and get people to accept new tech. We must teach and inform people to build trust and tackle worries about AV safety, cyber threats, and privacy. So, making AVs sustainable isn't just a choice - it's a must. When we bake sustainability into how we design roll out, and run AVs, we can create a transport system that's good for the planet, makes money, and treats everyone (Li et al. 2015). As AV tech keeps growing, we have a chance to change how we move around while pushing forward global sustainability aims for those who come after us.

Objectives

The goals of making self-driving cars more eco-friendly are to cut down on pollution by lowering emissions and using energy better, to save money through cheaper running costs and better output, and to make transportation fair by offering easy-to-use and safe ways to get around. By putting sustainability first, self-driving car tech wants to build a transport system that can handle climate changes, make the air cleaner, and solve people's travel problems. These goals push new ideas in how we design, control, and build for self-driving cars to create a future where getting around isn't just quick and safe, but also good for the planet and society.

TECHNOLOGICAL FOUNDATIONS

Deep Learning in Self-Driving Cars

Deep learning has a key influence on the progress of self-driving cars allowing them to see, understand, and act in their environment much like human drivers. This branch of machine learning uses artificial neural networks (ANNs) to process huge datasets, spot patterns, and make choices on its own. For self-driving cars deep learning algorithms are essential to detect objects, plan routes, and make decisions, which boosts safety, productivity, and dependability on the road (Wang et al. 2022).

Neural Networks and Perception: Deep learning in AVs revolves around convolutional neural networks (CNNs). These networks stand out in image recognition and computer vision tasks. CNNs handle visual data from cameras on AVs. They spot objects like pedestrians, cyclists, road signs, and other vehicles. These networks can tell different types of objects apart. They also figure out where objects are in 3D space. By using past data and real-time observations, CNNs can guess how these objects will move next (Devi et al., 2024).

Object Detection and Recognition: Object detection has a crucial role in deep learning for AVs, as it helps identify and track objects in complex settings in real-time. Deep learning models can spot and classify important objects with great accuracy by looking at data from many sensors like cameras, radar, and lidar. This skill is key for AVs to drive through different and changing traffic situations lowering risks and making sure they interact with other people on the road.

Path Planning and Control: Deep reinforcement learning (DRL) is another branch of deep learning that boosts self-driving cars' decision-making and navigation skills. DRL algorithms figure out the best driving strategies through practice fine-tuning routes, speeds, and moves to hit specific goals like getting to a place and .

Simulation and Training: Teaching deep learning models for self-driving cars needs big datasets with all kinds of driving situations and weather. Virtual platforms play a key part to make fake data and test AV systems in computer worlds before putting them on actual streets. These fake worlds let developers check rare cases see how well things work in different situations, and make their systems better without putting AVs at risk on the road.

Challenges and Future Directions: Even with major steps forward, rolling out deep learning-based AV systems on a large scale still faces hurdles. Problems like performing well in all kinds of weather, understanding tricky traffic situations, and making quick choices in dangerous scenarios need more study and work. Looking ahead, we need to make deep learning models work better and be easier to understand, combine data from different sensors more, and boost ways for AVs to learn from each other's experiences. This will help AVs get smarter and safer over time.

This means deep learning has a revolutionary impact on self-driving car technology. It gives vehicles cutting-edge abilities to see, decide, and control themselves. Neural networks CNNs and DRL algorithms, allow self-driving cars to move through tricky settings on their own. This makes driving safer more effective, and better for people using the roads. As scientists keep studying this field, putting deep learning into self-driving systems could change how we get around. It might create smart flexible vehicles that can handle the ins and outs of today's city and country roads.

Internet of Things (IoT) in AV Ecosystems

The Internet of Things (IoT) has a big influence on boosting the abilities and productivity of self-driving cars by enabling smooth talks between vehicles, roads, and the whole transport system. Adding IoT to self-driving cars means putting in linked devices, sensors, and computer programs that let data move and decisions happen right away. This part looks at how IoT changes self-driving car systems to improve connections, safety, and how well they work (Du 2016; Mora et al. 2020).

Figure 1. Internet of Things (IoT) in AV Ecosystems

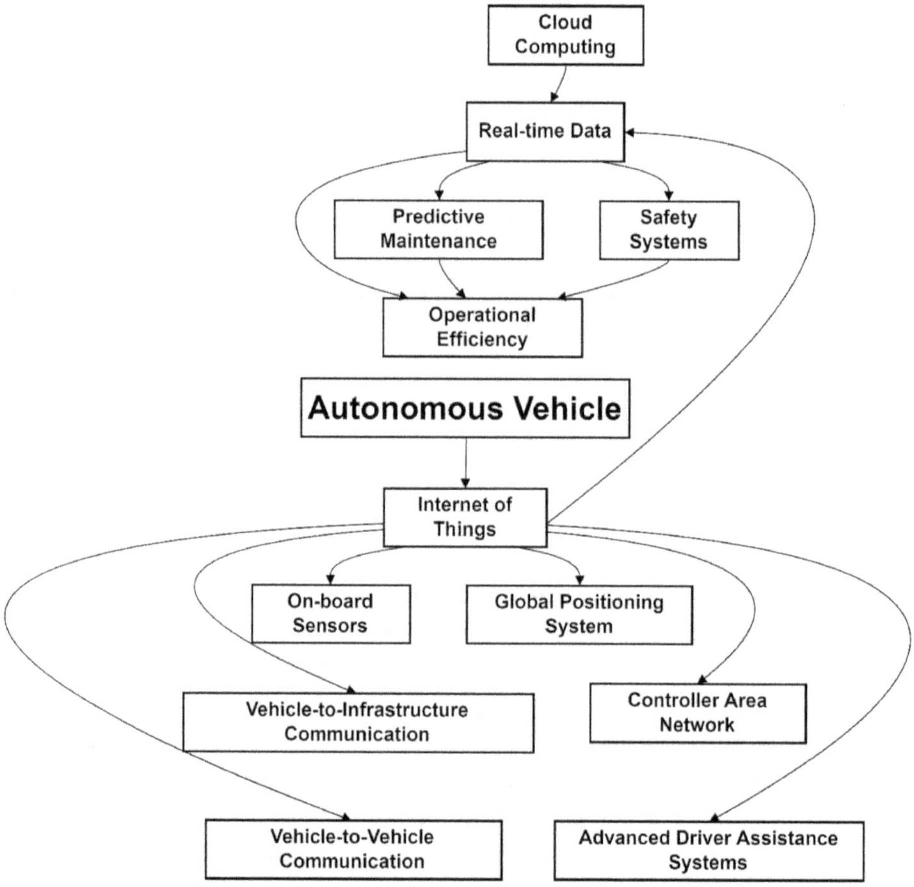

The Internet of Things (IoT) improves connections, safety, and how well self-driving car systems work, as shown in Figure 1.

Connectivity and Communication: IoT allows self-driving cars to talk to each other (V2V), to road equipment (V2I), and to online systems (V2C) creating a full Vehicle-to-Everything (V2X) network. This connection helps AVs get up-to-the-minute info on road status, traffic flow, and weather forecasts letting them make smart choices on their own. V2X chat cuts down on traffic jams, makes traffic move better, and boosts safety by giving early heads-up about possible dangers or crashes.

Sensor Fusion and Data Processing: IoT in AVs uses sensor fusion methods to combine data from different sensors like radar, lidar, cameras, and GPS. These sensors create a detailed all-around view of what's near the vehicle, which is key to navigate, spot objects, and avoid obstacles. IoT systems process this data right away using edge computing and cloud-based analysis. This allows AVs to understand complex traffic situations and change how they act based on what they see (Malathi et al. 2024; Pachiappan et al. 2024; Rahamathunnisa et al. 2024; Upadhyaya et al. 2024).

Predictive Maintenance and Efficiency: AVs with IoT capabilities gain from predictive upkeep features where sensors keep an eye on vehicle parts in real-time to spot oddities or wear. This forward-thinking strategy enables timely fixes and stops breakdowns making sure AVs run at their best. Also IoT analysis has an impact on route planning and vehicle routing cutting down on energy use and lowering running costs by using resources .

Cybersecurity and Data Privacy: As self-driving cars get more connected, keeping them safe from hackers becomes a big deal. IoT security steps include coding secrets safe talk channels, and strong ways to check who's who to keep data safe and private. Putting tough security rules in place is key to lower risks from possible cyber threats. This helps make sure self-driving cars run and people trust them(Nanda et al. 2024).

Automotive Networks and Connectivity

Automotive networks form the backbone of reliable operations and communication in self-driving cars. These networks include both car-to-car and car-to-infrastructure communication systems. This part looks at how automotive networks play a key role in self-driving tech focusing on how they connect how reliable they are, and how they keep people safe (Lim & Taeihagh 2018).

Figure 2. Operation and communication of autonomous vehicles (AVs)

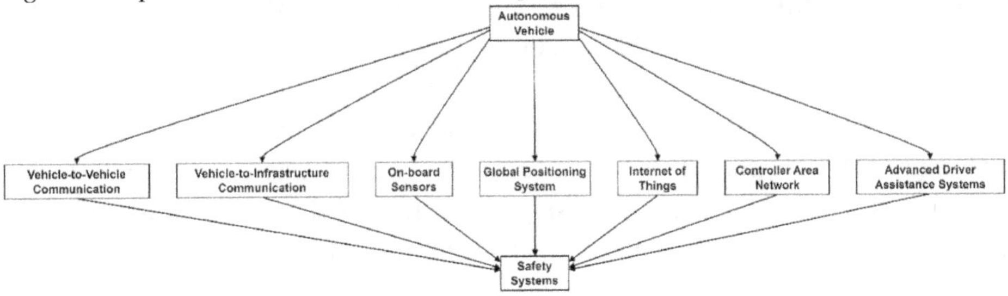

Automotive networks are key to self-driving cars. They allow cars to talk to each other and to road systems. This helps make sure the cars are safe, reliable, and stay connected, as Figure 2 shows.

CAN Bus and Ethernet Architectures: The Controller Area Network (CAN Bus) serves as a standard communication protocol in both old and new vehicles for communication within the vehicle. Self-driving cars combine CAN Bus with Ethernet setups to support faster data transmission and handle the growing need for bandwidth in sensor data and multimedia apps. Ethernet allows AV subsystems to exchange data in real-time, which makes them more responsive and easier to scale in complex self-driving car settings.

5G and Next-Generation Connectivity: New automotive networks use 5G tech to boost how self-driving cars talk to each other. They do this with super-fast super-reliable connections. 5G networks help V2X communication letting self-driving cars interact right away with other cars, people walking, and smart city stuff. This connection matters a lot for self-driving features like working together to control speed (CACC) and avoiding crashes at intersections. In these cases, making quick choices is key to stay safe and keep things running (Agrawal et al. 2023; Sreedhar et al. 2024).

Over-the-Air (OTA) Updates and Fleet Management: Car networks make OTA updates possible. These updates let car makers change AV software and firmware from afar to add new features, boost performance, and fix security weak spots. Fleet management systems use car networks to keep an eye on vehicle health, plan maintenance better, and make sure AV fleets follow the rules. This central control makes operations more effective and cuts down on idle time. As a result, AVs in use last longer and work more (Durairaj et al., 2023; Veeranjaneyulu et al., 2023).

Redundancy and Safety Critical Systems: Car networks put safety first by building backup systems and fail-safe features into self-driving car designs. This ensures key safety systems work well in all kinds of driving situations. Backup communication routes and spare systems lower the chances of communication breakdowns or cyber-attacks keeping self-driving cars working . Rules like ISO 26262 control how car makers develop these crucial safety systems. This makes sure self-driving cars meet tough safety standards before they hit public roads.

So IoT integration and automotive networks form the base for developing and running self-driving cars. They boost connectivity, safety, and productivity in AV systems. As tech keeps improving, using IoT features and strong car networks will be key to unlocking AVs' full power to change transport for a safer greener future.

CHALLENGES IN AV SUSTAINABILITY

Energy Consumption and Efficiency

Self-driving cars (AVs) could change transport by cutting energy use through better driving patterns and more efficient vehicles. But we need to tackle several issues to maximize their sustainability in terms of energy use(Muthukumar et al. 2024):

Battery Technology and Range: AVs use electric powertrains to cut down on emissions and use less fossil fuels. Better battery tech is key to boost energy density, speed up charging, and let cars go further. Right now, the big issues are how much batteries cost and weigh, plus the need for more charging stations to support lots of electric AVs (Satav et al., 2023).

Energy Optimization Algorithms: Self-driving cars use complex algorithms to save energy through planning routes controlling speed, and predicting future conditions. These algorithms look at things like traffic, weather forecasts, and road layouts to use less energy while still getting around . Some challenges include handling lots of real-time data and needing accurate models to predict the best way to use energy in different driving situations (Kumara et al. 2023; Nishanth et al. 2023; Vanitha et al. 2023).

Charging Infrastructure: The shift to electric AVs requires a strong charging network that can handle high-volume rapid-charging needs. Issues include setting up charging stations in cities and rural areas managing grid capacity to deal with peak loads, and creating standards for different charging systems to work together. Solving these problems is key to ensure AV users and fleet operators have easy access to charging options.

Environmental Impact and Emissions

AVs aim to cut down on environmental harm by reducing greenhouse gas emissions and making air cleaner(Chandrika et al. 2023). Yet, we need to tackle several environmental issues to achieve sustainable transportation:

Life Cycle Emissions: AVs have an impact on the environment throughout their entire existence. This includes when they're made, used, and thrown away. Sure, AVs might cut down on exhaust fumes while they're on the road, but we can't forget about other things. We need to think about how making batteries affects the planet where the electricity comes from, and what happens when we get rid of these cars. To get the full picture of how AVs affect the environment, we need to study their whole life cycle. This will show us where we can make AVs more eco-friendly (Boopathi, 2022; Maguluri et al., 2023; Sampath, 2021).

Traffic Congestion and Induced Demand: Self-driving cars might make people drive more because it costs less and it's easier to get around. This is called induced demand, and it can lead to using more energy and creating more pollution if we don't handle it well. To lessen the environmental impact of more driving from self-driving cars, we need to use strategies like charging for busy roads, sharing rides, and planning cities better (Sonia et al. 2024).

Material Resource Management: Making AV parts like batteries, sensors, and AI processors needs a lot of raw materials and rare earth elements. To cut down on harming the environment and using up limited resources, it's key to manage resources in a way that lasts. This means doing things like recycling and using circular economy ideas. To push forward AV tech that can last, we need breakthroughs in material science and better ways to handle supply chains.

Infrastructure Requirements

Adding self-driving cars to our current transportation setup creates hurdles in tech rollout, rules and regulations, and city planning (Chandrika et al. 2023; Maguluri et al. 2023):

Roadway Upgrades and Smart Infrastructure: Self-driving cars need smart infrastructure with sensors, communication systems, and real-time data analysis to improve navigation and safety. Updating existing roads and highways with smart tech is tricky due to funding issues, the need for teamwork among different groups, and worries about privacy when collecting and using data. To develop and launch smart infrastructure that supports self-driving cars, governments, businesses, and research groups must work together.

Regulatory and Policy Frameworks: Rules for AV rollout differ worldwide, which has an impact on tech breakthroughs and market uptake. Hurdles include setting up safety rules deciding who's responsible for AV crashes protecting data privacy, and making sure AV communication systems work together. To create flexible rules that spark new ideas while keeping people safe and protected, teamwork between rule-makers, industry bigwigs, and advocacy groups is key.

Public Acceptance and Behavior Change: For AVs to catch on, people need to trust and accept self-driving tech. Roadblocks include worries about safety online security, job losses in transport, and tricky choices in decision-making software. To build public trust and push for greener travel habits, we need education, outreach, and openness in AV progress.

This means tackling AV sustainability issues needs a broad strategy that covers new tech caring for the environment, and team efforts in management. When key players focus on using less energy, cutting down on emissions, and putting money into better infrastructure, they can speed up the shift to greener transport run by self-driving cars.

INTEGRATED FRAMEWORK FOR SUSTAINABLE AV SOLUTIONS

Integration of Deep Learning IoT, and Automotive Networks

The combination of deep learning Internet of Things (IoT), and automotive networks creates a framework that works together to improve the sustainability, productivity, and safety of self-driving cars as shown in Figure 3. This part looks at how these technologies come together to build a strong environment for long-lasting self-driving car solutions (Devi et al. 2024).

Figure 3. Integration of Deep Learning IoT, and Automotive Networks to AVs

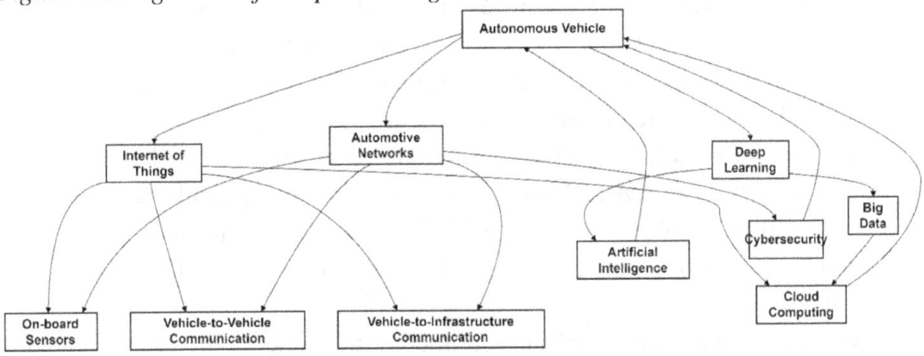

When deep learning, IoT, and automotive networks join forces, they build a framework that works together for self-driving cars. This teamwork makes the cars work better, run more, and stay safer (Boopathi 2023; KAV et al., 2023).

Deep Learning in AV Perception and Decision-Making: Deep learning algorithms have a big impact on how AVs see and understand their surroundings making them as sharp and quick as humans. Neural networks learn from huge sets of sensor data, including radar, lidar, and camera feeds, to spot objects, recognize lanes, and predict traffic patterns in real-time. As these models keep learning from data, they help AVs get better at driving through tricky city streets spotting road dangers before they happen, and making smart choices on their own (Sangeetha et al. 2023; Ugandar et al. 2023; Venkateswaran et al. 2023).

IoT and Real-Time Data Exchange: IoT connects AVs to an ecosystem where cars talk to each other (V2V), to roadside structures (V2I), and to cloud systems (V2C). This connection allows cars to share data about road states, traffic updates, and environmental factors. This improves how AVs understand their surroundings and make choices. IoT systems help combine sensor data . This lets AVs change their routes on the fly, use energy better, and work with traffic control systems to cut down on jams and pollution(Agrawal et al., 2023; Venkateswaran et al., 2024).

Automotive Networks for Reliable Communication: Automotive networks form the foundation to enable secure and reliable communication between AVs and infrastructure parts. Technologies like Controller Area Network (CAN Bus) and Ethernet setups support fast data transfer allowing AV subsystems to share key information . Backup communication routes and safety systems make sure AVs stay connected and work even in tough driving situations or when facing cyber threats.

Sustainability Benefits of Integration: Combining deep learning, IoT, and automotive networks has an impact on AV operations in several eco-friendly ways:

- **Energy Efficiency**: Deep learning algorithms have an influence on AVs' driving behavior and energy use based on real-time data inputs. IoT-enabled energy management systems change vehicle routes and speed profiles to use less fuel and cut carbon emissions.
- **Emissions Reduction**: By improving traffic flow and cutting down on stop-and-go driving patterns, AVs with built-in technologies can lower emissions compared to regular cars. IoT-enabled predictive maintenance keeps AVs running at their best further cutting down on environmental effects (Maguluri et al. 2023).
- **Infrastructure Optimization**: Automotive networks help set up smart infrastructure boosting AVs' ability to navigate and reducing the need for expensive road expansions. Integrated traffic management systems make roads safer and work better supporting efforts to create sustainable city transport.

Case Studies and Demonstrations: Real-world examples show how an all-in-one system for long-lasting self-driving car solutions works well:

- **Smart City Projects**: Cities like Singapore and Helsinki are rolling out self-driving car fleets with internet connections to make public transit easier to use and cut down on city traffic jams. Smart computer programs find the best routes and times for these cars, while car networks help them talk to each other and everything around them. This leads to better traffic control and faster emergency help.

- **New Ways to Manage Fleets**: Shipping companies use self-driving cars with internet connections to find the best delivery routes, use less fuel, and spend less money running their business. Smart computer models guess what people will want and help decide where to send the cars. This helps make the whole supply chain better for the environment.

Future Directions and Innovations: In the coming years, progress in AI, IoT, and car networks will keep pushing the boundaries of eco-friendly self-driving car solutions:

- **AI-driven Decision Making**: Smarter AI systems will help self-driving cars better handle changing traffic, navigate tricky areas, and focus on safety and getting around .
- **5G Connectivity**: The next wave of car networks using 5G will allow for super-fast communication and open doors to new features in self-driving cars, like controlling them from far away and using see-through displays.
- **Regulatory Frameworks**: New rules and guidelines will shape how we roll out green self-driving car tech making sure it follows environmental laws and helps people trust driverless transport.

This means that combining deep learning, IoT, and automotive networks creates a strong system to develop sustainable self-driving car solutions. Using these technologies together helps stakeholders speed up the use of self-driving cars while tackling issues like energy savings, cutting emissions, and improving infrastructure. To use sustainable self-driving car tech to shape the future of transport, teamwork across industries, universities, and government agencies is crucial.

Benefits of Integration for Sustainability

The rise of self-driving cars gives us a chance to build a green transport system that saves energy, cuts down on pollution, and looks after our planet. To create lasting self-driving car solutions, we need to bring together new tech, rules, and better roads in a way that makes them all work together (Muthukumar et al., 2024). When we join up different parts of making self-driving cars, it can help make things more eco-friendly, as Figure 4 shows.

Figure 4. Benefits of Integration for Sustainability in development of autonomous vehicles (AVs)

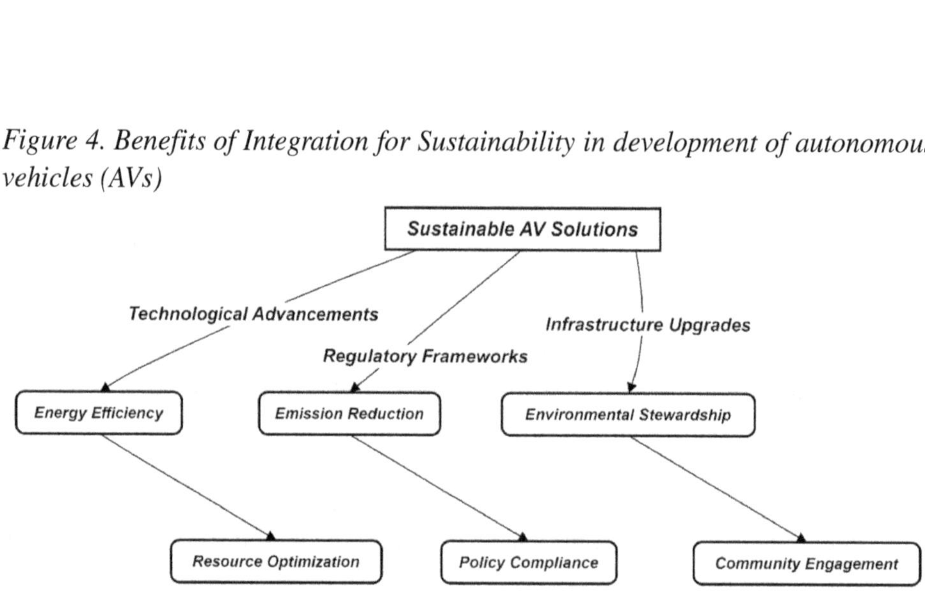

Technological Integration

Energy Efficiency and Optimization: Combining cutting-edge AI algorithms with real-time sensor data allows AVs to use energy more during operation. AI systems study traffic patterns, weather conditions, and road layouts to adjust driving behaviors, like speeding up and slowing down, to save the most energy. By cutting down on wasted energy, AVs lower overall fuel use and carbon emissions helping the environment.

Hybrid Propulsion Systems: Adopting electric and hybrid propulsion systems in AVs boosts sustainability by cutting down on fossil fuel use and lowering tailpipe emissions. Battery electric vehicles (BEVs) and plug-in hybrid electric vehicles (PHEVs) can drive with zero emissions, which is helpful in cities where air quality is a big concern. Also better battery tech, like solid-state batteries and fast-charging stations, helps electric AVs go further and charge quicker making them more practical and attractive to use.

Environmental Impact Reduction

Emissions Reduction Strategies: Self-driving cars help cut down on greenhouse gases and air pollution by making traffic move better easing traffic jams, and driving in ways that are good for the environment. Tech like cooperative adaptive cruise control (CACC) and platooning let these cars travel close together, which cuts down on air resistance and saves fuel. What's more keeping cars in good shape before problems happen and hooking them up to the power grid allows self-driving cars to play a part in keeping the energy grid stable and responding to power needs. This makes the whole energy system work better and more .

Lifecycle Assessment and Sustainable Practices: Lifecycle assessment methods help to guide sustainable AV design, manufacturing, and end-of-life practices. AVs that focus on recyclability, resource efficiency, and circular economy principles have less impact on the environment and save natural resources. Automakers, tech companies, and green groups work together to come up with new ideas for eco-friendly materials, energy-saving parts, and green manufacturing. This teamwork makes sure AVs meet global green goals.

Social and Economic Benefits

Mobility Accessibility and Equity: Self-driving cars make it easier for underserved groups to get around. This includes older folks, people with disabilities, and those living in rural areas. Driverless ride-sharing services and short-distance transport options help fill in gaps in transportation. This gives people more ways to travel and breaks down barriers. Making sure everyone has fair access to self-driving tech helps bring people together and creates job opportunities. This supports strong communities and growth that lasts.

Economic Efficiency and Cost Savings: AVs have an impact on the economy through savings in operating costs better productivity, and lower infrastructure upkeep costs. Systems that manage autonomous fleets make the best use of vehicles, cut down on idle time, and improve logistics, which boosts efficiency and profits (N. et al. 2024; Ravisankar et al. 2023). What's more, AVs help the economy grow by sparking new ideas in smart city tech digital infrastructure, and self-driving service industries, which leads to new jobs and economic value.

Benefits of Integration for Sustainability

Integrating sustainable practices into AV development and rollout has several main advantages (Devi et al. 2024):

Environmental Stewardship: Using less energy cutting down on emissions, and adopting green technologies allows integrated sustainable AV solutions to help fight climate change and boost air quality. AVs are key to shifting to a low-carbon transportation future matching global environmental sustainability goals and legal requirements.

Technological Innovation and Advancement: The push for sustainability leads to new tech in AV design, propulsion systems, and smart infrastructure fixes. Money put into research for energy-saving tech AI that makes things work better, and ways to use more renewable energy speeds up tech growth. This makes car and transport companies around the world more competitive.

Public Health and Quality of Life: Using fewer gas-powered cars and choosing greener ways to get around helps people stay healthier. It means less dirty air and less stress from traffic. AVs make roads safer, cut down on traffic jams, and make cities nicer places to live. This builds stronger healthier communities.

Regulatory Compliance and Market Acceptance: Following sustainable practices when developing AVs helps companies meet tough environmental rules and sustainability benchmarks. When companies tackle environmental and social issues head-on, it boosts market approval, builds trust with customers, and gives stakeholders more faith in AV tech. This paves the way to have AVs used and rolled out in many different global markets.

A complete system for sustainable self-driving car solutions doesn't just tackle environmental issues. It also boosts cost-effectiveness, pushes for fair treatment, and sparks new tech ideas. By tapping into progress in energy savings, ways to cut emissions, and fair transport options different groups can team up to shape a future where self-driving cars are key to creating sustainable transport systems around the world. Ongoing teamwork between rule-makers, industry bigwigs, and local groups is crucial to unlock the full promise of self-driving tech to advance sustainability and build a tough inclusive transport setup for the kids of tomorrow.

CASE STUDIES AND SIMULATIONS

Energy Efficiency Optimization

Energy efficiency optimization in self-driving cars involves using cutting-edge tech and smart operation plans to cut down on energy use while driving. A few real-world examples show good ways to make self-driving cars more energy-efficient (Mageswari et al. 2024; Naveeenkumar et al. 2024; Paul et al., 2024):

Advanced Driver Assistance Systems (ADAS): Real-world examples show that ADAS helps to save energy by aiding self-driving cars with adaptive cruise control, staying in lanes, and smart braking. These systems use data from sensors and AI algorithms to predict traffic and change the car's speed and how it speeds up or slows down. This cuts down on wasted energy and makes the car use fuel better.

Route Optimization and Predictive Analytics: Studies simulate how route optimization algorithms and predictive analytics affect energy efficiency. AVs can pick the most fuel-efficient routes and driving tactics by looking at past traffic trends, weather forecasts, and road conditions. Results show big cuts in fuel use and travel time showing how data-driven choices can benefit AV operations.

Regenerative Braking and Energy Recovery: Research looks at putting regenerative braking systems in AVs. These systems capture energy during slowing down and store it in batteries to use later. Models show how much more efficient these braking systems are pointing out how they could save energy and be better for the environment than old-style brakes.

Emissions Reduction Strategies

Emissions reduction strategies aim to cut greenhouse gas emissions and boost air quality through new technologies and better operations (Domakonda et al. 2022; Samikannu et al. 2022):

Electric and Hybrid Propulsion Systems: Studies look at how electric and hybrid propulsion systems in AVs help the environment. These systems stop tailpipe emissions and cut the need for fossil fuels. Computer models show how much emissions drop and check if it's possible to use more electric vehicles (EVs) in AV fleets. The results show these changes could help the environment.

Eco-Driving and Traffic Management: Studies simulate how eco-driving methods and smart traffic systems cut down emissions. Self-driving cars with eco-driving software make the best use of speeding up slowing down, and controlling speed to save fuel and reduce emissions. The results show less CO2 and air pollution proving that clever mobility solutions help create greener transport.

Vehicle-to-Grid (V2G) Integration: Research looks into how V2G can work with self-driving cars allowing energy to flow both ways between vehicles and the power grid. Computer models check the environmental good that V2G tech can do, like smoothing out peak power use and fitting in more renewable energy. This helps keep the grid steady and lowers emissions from self-driving car use.

Simulation Results and Analysis

Simulation results give us useful insights into how well energy efficiency and emissions reduction strategies work in AV operations:

Quantitative Performance Metrics: Simulation studies measure performance metrics, including energy use per mile/kilometer, CO2 emissions per trip, and fuel efficiency gains compared to regular vehicles. These metrics set benchmarks to evaluate the environmental and economic effects of green AV technologies.

Scenario Analysis and Sensitivity Testing: Simulation models run scenario analysis and sensitivity tests to check how well energy efficiency and emissions reduction strategies hold up under different driving conditions and operational scenarios. Sensitivity analysis spots key factors that affect performance outcomes such as changes in weather, traffic jams, and how drivers behave.

Policy Implications and Recommendations: Simulation results have an influence on policy implications and suggestions to encourage sustainable AV deployment and adoption. Findings stress the need for regulatory incentives, infrastructure investments, and public-private partnerships to advance energy efficiency and reduce emissions in AV ecosystems.

Case studies and simulations show how strategies to optimize energy efficiency and cut emissions can transform AV operations.

FUTURE DIRECTIONS AND INNOVATIONS IN AV SUSTAINABILITY

The future of self-driving cars has an influence on new tech, rules, and what's happening in the industry. AI better batteries, and mixing sensor data will make these cars safer and work better. People making laws will set goals for less pollution and put money into new roads and such. New trends like Mobility as a Service and self-driving trucks will lead to new ideas for getting around that don't harm the planet. Figure 5 shows what's next and what's new for self-driving cars.

Figure 5. Future Directions and Innovations in AV Sustainability

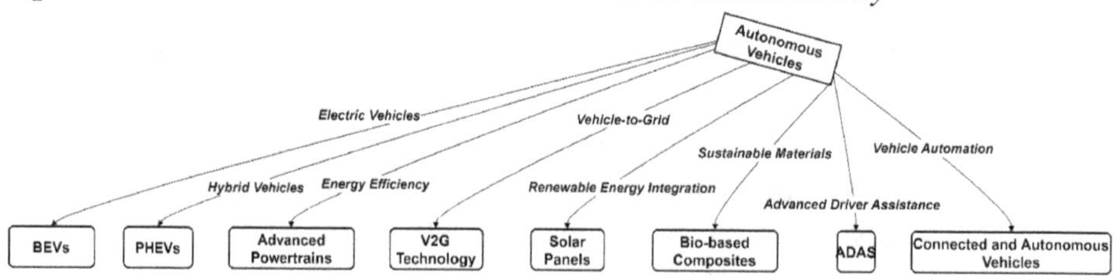

New Tech to Make Self-Driving Cars Last

AI and Machine Learning Breakthroughs: Future self-driving cars will use cutting-edge AI and machine learning systems to boost fuel efficiency, plan better routes, and predict maintenance needs. AI-powered simulations and real-time data analysis will allow these vehicles to adjust to changing traffic, use less energy, and run more overall.

New Energy Storage Tech: The next wave of self-driving cars will feature new energy storage tech, like solid-state batteries and ultra-capacitors, to drive further, charge faster, and store more power. These innovations will help more people choose electric and hybrid engines cutting down on fossil fuel use and reducing emissions over the car's lifetime.

Sensor Fusion and Perception Systems: AVs of the future will combine cutting-edge sensor fusion tech. This blend of radar, lidar, cameras, and AI-driven perception systems will give a clear all-around view of the surroundings.

Policy and Regulatory Considerations

Standardization and Interoperability: Rules will focus on making AV tech standard and able to work together. This ensures different makers and areas can use the same systems. New laws will set up safety rules, protect against cyber threats, and safeguard data privacy. These steps aim to build trust with users and other key players.

Emissions Reduction Targets: Governments and regulatory bodies will set bold targets to cut emissions for AV fleets. This will encourage companies to adopt zero-emission vehicles and green mobility options. New policies, like tax breaks, support for EV charging stations, and systems to trade emissions, will speed up the shift to cleaner transportation.

Urban Planning and Investments: Leaders will put money into smart cities and connected transport networks to help AVs succeed. They'll work on mixing AVs with public transit making roads safer, and using smart systems to manage traffic. These steps aim to improve traffic flow and reduce jams.

ETHICAL IMPLICATIONS AND PRIVACY CONCERNS

Introduction: The AVs epitomize technology at the pinnacle of integration, harnessing both deep learning and IoT to disrupt the way people and goods are moved. Even as these technologies promise safety, efficiency, and convenience time after time, a lot of ethical issues and privacy concerns tag along. This essay, therefore, provides a debate on some challenges such as ethical dilemmas in decision-making, privacy, and large-scale consequences for society as a whole (Giri et al., 2024; Godwin Immanuel et al., 2024; Muthukumar et al., 2024; Naresh et al., 2024; Rajan et al., 2024).

Ethical Implications of Autonomous Vehicles

Dilemmas in Decision-Making One of the crucial ethical dilemmas concerning AVs is making decisions, particularly in those involving causing harm. Deep learning algorithms are the roots of decision-making for AVs, which consequently enable them to operate in an ecological environment. These systems may, however, find themselves in situations where they have to make life-or-death decisions-what is often referred to as the "trolley problem." For instance, an AV may be faced with a decision between running over a pedestrian or putting its passengers at risk by swerving into oncoming traffic. The ethical dilemma concerns how the AV would prioritize lives and whether such decisions should be preprogrammed by the manufacturers or dynamically adjusted by the vehicle.

Bias and Fairness Other ethical issues with AVs are related to biases that may be incorporated into deep learning models of AVs. Their models, after all, are learnt from huge datasets and hence bound to pick up biases based on race, gender, or socioeconomic status. When left unaccounted for, these may lead to certain groups being treated inequivalently-misidentifying pedestrians from minority communities, for instance-or making suboptimal decisions that will affect them disproportionately. Hence, fairness in AV decision-making has to be ensured by scrutiny of training data, developing algorithms that identify and overcome bias.

Accountability and Responsibility Who should be held responsible in the case of an accident involving an AV is a tough call. In the case of conventional vehicles, the driver was solely responsible, but with AVs, responsibility lines blur between the driver and the manufacturer, and even the software developer. If an AV makes a wrong decision resulting in damage, it becomes a legal and ethical issue regarding liability: is it the owner of the vehicle, or is it the software developer, or perhaps the manufacturer? As more AVs hit the roads, clear frameworks for accountability are needed.

Privacy Concerns in Autonomous Vehicles

Data Collection and Surveillance AVs require huge amounts of data to operate effectively: GPS data, sensor readings, video, and communication with other vehicles and infrastructure. Adding IoT into the mix amplifies these levels to real-time data sharing and analysis. This also creates a host of privacy-related issues. AVs can observe not only the movements and behavior of passengers but even their conversations, building detailed profiles that could be misused if accessed by unauthorized parties. The potential for mass surveillance by government entities or private corporations is a considerable risk to privacy linked with AVs (Boopathi et al., 2024; Chandrika et al., 2023; Gowri et al., 2023; Yupapin et al., 2023).

Data Security Risks An in-depth study by KPMG also mentions that large volumes of data collected by AVs are equally vulnerable to certain types of cyberattacks. Depending on the sophistication of the hack, this could range from accessing sensitive information or manipulating vehicle controls to causing disruption on a larger scale, given that they affect the IoT networks which the AVs rely on. The security of such data is of utmost importance for ensuring the privacy of users and the safety of the operation of the AVs. Strong encryption methods, secure warehousing of data, and round-the-clock monitoring will be needed to keep these threats at bay.

Informed Consent and Data Ownership A related but distinct privacy question involves the issues of informed consent and data ownership. Users of AVs may have only a partial understanding of the degree to which data is gathered about them, much less how such data is utilized. In the case of AV, informed consent, where users agree to collect and use their data, is hard to apply since it forms part of continuous and pervasive data collection. Additionally, ownership of the AV data is ambiguous: ownership by either the vehicle owner, the maker, or a third-party service provider. Thus, establishment of clear regulations and policies will provide a complete description of who owns the data, and allow owners to control personal information.

Societal Implications and the Way Forward

Public Trust and Acceptance Ethical and privacy concerns regarding the AV directly relate to the level of trust the public has. For widespread deployment, there needs to be trust in the safety of AVs, fairness in the decisions made by them, and protection of privacy. This will require transparency on the part of manufacturers and policymakers regarding such concerns, strict testing of the AV systems for defects, and strong legal frameworks that take due care of such considerations (Boopathi et al., 2024; Devi et al., 2024; Giri et al., 2024).

Regulatory and Ethical Frameworks Accordingly, there should be the development of regulatory frameworks that have the capacity to solve ethical and privacy challenges of AVs. This includes setting regulations on ethical decision-making for AVs, standards on the protection of data and privacy, and also mechanisms that attribute responsibility. The processes of designing and developing AVs should have ethical considerations that ensure these cars operate within the norms of society.

Deep learning and IoT have huge potential in the application of autonomous vehicles, but simultaneously, bring immense ethical and privacy challenges. As AVs grow in permeation, solving these issues concurrently becomes vital, with the assurance that beneficial promises of this technology are acceded to without losing ethical ground or letting down personal privacy. Indeed, thoughtful regulation, transparency, and continuing public engagement can chart the way forward with AVs in ways that promote innovation and build trust.

CONCLUSION

To wrap up, the chapter on sustainable solutions for self-driving cars shows how combining cutting-edge tech new rules, and industry breakthroughs can make transportation greener more cost-effective, and fairer for everyone. As AI gets smarter, sensors work better together, and energy storage improves, self-driving cars are set to change how we move around cities and handle shipping. These improvements allow self-driving cars to use less energy cut down on pollution, and stay safer by seeing what's around them and predicting what might happen next.

Policy and regulatory factors have a big impact on AV sustainability's future. They focus on making things standard, work together, and cut emissions. Governments play a key part to encourage people to use zero-emission vehicles. They also help pay for new infrastructure and team up with others to push sustainable travel options forward. The industry is changing. We now see Mobility-as-a-Service (MaaS) platforms and self-driving delivery systems. These show a move towards joined-up travel solutions that put efficiency easy access, and caring for the environment first. Car makers, tech companies, and city planners need to work together. This helps create a strong transport system that meets everyone's needs while being kind to the environment.

So ongoing research new ideas, and teamwork between public and private sectors will help overcome obstacles and make the most of AV technology. By focusing on sustainability and pushing tech boundaries, AVs can play a key part in creating a sustainable connected future where getting around is safe, productive, and open to everyone.

REFERENCES

Agrawal, A. V., Magulur, L. P., Priya, S. G., Kaur, A., Singh, G., & Boopathi, S. (2023). Smart Precision Agriculture Using IoT and WSN. In *Handbook of Research on Data Science and Cybersecurity Innovations in Industry 4.0 Technologies* (pp. 524–541). IGI Global. DOI: 10.4018/978-1-6684-8145-5.ch026

Boopathi, S. (2022). An investigation on gas emission concentration and relative emission rate of the near-dry wire-cut electrical discharge machining process. *Environmental Science and Pollution Research International, 29*(57), 86237–86246. DOI: 10.1007/s11356-021-17658-1 PMID: 34837614

Boopathi, S., Ananth, J., Raghuram, K. S., Kistan, A., Velumani, M., & Gopinathan, R. (2024, January). *6342734- Machine for Analysis of the resistance of welded parts by the electrical resistance process.*

Chandrika, V., Sivakumar, A., Krishnan, T. S., Pradeep, J., Manikandan, S., & Boopathi, S. (2023). Theoretical Study on Power Distribution Systems for Electric Vehicles. In *Intelligent Engineering Applications and Applied Sciences for Sustainability* (pp. 1–19). IGI Global. DOI: 10.4018/979-8-3693-0044-2.ch001

Devi, J. V., Argiddi, R. V., Renuka, P., Janagi, K., Hari, B. S., & Boopathi, S. (2024). Study on Integrated Neural Networks and Fuzzy Logic Control for Autonomous Electric Vehicles. In *Advances in Web Technologies and Engineering* (pp. 104–127). IGI Global. DOI: 10.4018/979-8-3693-1487-6.ch006

Durairaj, M., Jayakumar, S., Karpagavalli, V., Maheswari, B. U., & Boopathi, S. (2023). Utilization of Digital Tools in the Indian Higher Education System During Health Crises. In *Multidisciplinary Approaches to Organizational Governance During Health Crises* (pp. 1–21). IGI Global. DOI: 10.4018/978-1-7998-9213-7.ch001

Giri, R., Neeraja, B., Alhamad, A. M., Mamidi, P. L., Somasundaram, R., & Boopathi, S. (2024). Human-Centered Design Approach to Autonomous Electric Vehicles: Revolutionizing Mobility. In Futuristic Technology for Sustainable Manufacturing (pp. 91-109). IGI Global.

Godwin Immanuel, D., Solaimalai, G., Chandrakala, B. M., Bharath, V. G., Singh, M. K., & Boopathi, S. (2024). Advancements in Electric Vehicle Management System: Integrating Machine Learning and Artificial Intelligence. In *Advances in Web Technologies and Engineering* (pp. 371–391). IGI Global. DOI: 10.4018/979-8-3693-1487-6.ch018

Gowri, N. V., Dwivedi, J. N., Krishnaveni, K., Boopathi, S., Palaniappan, M., & Medikondu, N. R. (2023). Experimental investigation and multi-objective optimization of eco-friendly near-dry electrical discharge machining of shape memory alloy using Cu/SiC/Gr composite electrode. *Environmental Science and Pollution Research International*, *30*(49), 1–19. DOI: 10.1007/s11356-023-26983-6 PMID: 37126160

Maguluri, L. P., Ananth, J., Hariram, S., Geetha, C., Bhaskar, A., & Boopathi, S. (2023). Smart Vehicle-Emissions Monitoring System Using Internet of Things (IoT). In *Handbook of Research on Safe Disposal Methods of Municipal Solid Wastes for a Sustainable Environment* (pp. 191–211). IGI Global.

Muthukumar, R., Pratheep, V. G. G., Sultanuddin, S. J., Kunduru, K. R., Kumar, P., & Boopathi, S. (2024). Leveraging Fuel Cell Technology With AI and ML Integration for Next-Generation Vehicles: Empowering Electric Mobility. In A Sustainable Future with E-Mobility: Concepts, Challenges, and Implementations (pp. 312-337). IGI Global.

Naresh, M., Subhahan, D. A., Narula, V. N., Roslin, D. K., Narula, S., & Boopathi, S. (2024). Edge Computing and Machine Learning Integration for Autonomous Electrical Vehicles. In *Solving Fundamental Challenges of Electric Vehicles* (pp. 99–127). IGI Global., DOI: 10.4018/979-8-3693-4314-2.ch005

Rajan, T. S., Rex, C. R. E. S., Shastri, D. S., Naidu, G., Senthil Kumar, C., & Boopathi, S. (2024). Study on Environmental and Social Impacts Through Electric Vehicles. In *Solving Fundamental Challenges of Electric Vehicles* (pp. 352–383). IGI Global., DOI: 10.4018/979-8-3693-4314-2.ch013

Sampath, B. (2021). *Sustainable Eco-Friendly Wire-Cut Electrical Discharge Machining: Gas Emission Analysis.*

Satav, S. D., Lamani, D., Harsha, K., Kumar, N., Manikandan, S., & Sampath, B. (2023). Energy and Battery Management in the Era of Cloud Computing: Sustainable Wireless Systems and Networks. In *Sustainable Science and Intelligent Technologies for Societal Development* (pp. 141–166). IGI Global.

Veeranjaneyulu, R., Boopathi, S., Kumari, R. K., Vidyarthi, A., Isaac, J. S., & Jaiganesh, V. (2023). Air Quality Improvement and Optimisation Using Machine Learning Technique. *IEEE- Explore*, 1–6.

Venkateswaran, N., Kiran Kumar, K., Maheswari, K., Kumar Reddy, R. V., & Boopathi, S. (2024). Optimizing IoT Data Aggregation: Hybrid Firefly-Artificial Bee Colony Algorithm for Enhanced Efficiency in Agriculture. *AGRIS On-Line Papers in Economics and Informatics*, *16*(1), 117–130. DOI: 10.7160/aol.2024.160110

Yupapin, P., Trabelsi, Y., Nattappan, A., & Boopathi, S. (2023). Performance improvement of wire-cut electrical discharge machining process using cryogenically treated super-conductive state of Monel-K500 alloy. *Iranian Journal of Science and Technology. Transaction of Mechanical Engineering*, *47*(1), 267–283. DOI: 10.1007/s40997-022-00513-0

Chapter 17
Wireless Sensor Networks (WSNs)–Integrated Machine Learning Algorithms for Water Resource Management

Ashay Devidas Shende
https://orcid.org/0000-0002-2082-1311
Department of Civil Engineering, K.D.K. College of Engineering, Nagpur, India

J. Bibiana Jenifer
https://orcid.org/0009-0005-9396-0450
Department of Information Science and Engineering, New Horizon College of Engineering, Bengaluru, India

Gururaj L. Kulkarni
Department of Computer Science and Engineering, Vardhaman College of Engineering, Hyderabad, India

Amar Choudhary
https://orcid.org/0000-0003-3865-2680
Department of Electronics and Communication Engineering, Alliance College of Engineering and Design, Bengaluru, India

Aparajita Mukherjee
https://orcid.org/0009-0009-1259-7864
Department of Computer Science and Engineering, IEM Newtown Campus, Kolkata, India

Sampath Boopathi
https://orcid.org/0000-0002-2065-6539
Department of Mechanical Engineering, Muthayammal Engineering College, Namakkal, India

ABSTRACT

The integration of Wireless Sensor Networks (WSNs) with advanced machine learning algorithms is revolutionizing water resource management. This chapter delves into

DOI: 10.4018/979-8-3693-5448-3.ch017

the synergistic application of these technologies to address critical challenges in monitoring and managing water resources. It begins with an introduction to WSN technology, emphasizing its role in real-time data collection for water quality monitoring, irrigation systems, and flood detection. The chapter then explores various machine learning algorithms, such as Support Vector Machines (SVM), Convolutional Neural Networks (CNN), and Recurrent Neural Networks (RNN), highlighting their applications in predictive analytics and anomaly detection. The chapter discusses successful implementations of WSNs and machine learning in water distribution and proactive infrastructure maintenance, emphasizing the importance of accurate data interpretation for informed decision-making in real-time data analysis techniques.

INTRODUCTION

Water resources management is an important aspect of sustainable development. It would involve a plan, development, and management of water resources that cater to present and future human needs while protecting the natural environment. This discipline encompasses a very wide area; therefore, it addresses activities that include water supply and quality control to flood protection and ecosystem conservation. Proper management in water resources has become critical due to the continuous increase of the world population and alteration of hydrological regimes as a consequence of climate variability and change (Rosero-Montalvo et al., 2020a) .

Management of water resources deals with providing clean water in a reliable manner for uses such as domestic, agriculture, industry, and recreation. This involves planning and building infrastructures like dams, reservoirs, pipelines, and treatment plants. Management of these resources takes into consideration the equilibrium among the competing uses and also protection of water quality. Industrial waste, agricultural runoff, and urbanization all contribute various contaminants to water supplies. Advanced treatment processes and strict regulations have been implemented to offset these impacts in an effort to have safe water (Padmanaban & others, 2021).

The second critical aspect in which water resources are managed involves the management of water quantity to avoid shortage and excess. Water scarcity results from over-extraction of groundwater, inequitable use of water, and long droughts. To address these issues, different measures are implemented with respect to the conservation of water, utilization of other water supplies including desalination and recycled water, and the use of efficient irrigation methods. On the other hand, excess water is regulated, particularly for areas that are experiencing flood, through proper method of flood control. This will involve constructing levees, floodways, and detention basins in addition to early warning systems that will reduce the threat of flooding and a targeted emergency response plan in case there is flooding (Krishnan et al., 2022a).

Ecosystem conservation: Aquatic ecosystems have much value and need to be conserved appropriately. For human civilization, generally speaking, a well-functioning aquatic ecosystem can also provide enriched biodiversity, clean water, and opportunities for recreation. These areas necessitate natural flow regimes, habitat restoration, and control of invasive species. Introduction of environmental considerations into the management plans of water allows passing ecological health in rivers, lakes, and wetlands to future generations. It is impossible to overestimate the role of technology in the management of water resources. New sensor networks, data analytics, and remote sensing innovations have changed the way water resources are monitored and managed. WSNs make it possible to collect data in real time, with regards to water quality, flow rates, and reservoir levels, among others. Such information would support decision-making and forecasting for the future (Ahmed et al., 2021). All these capabilities, of course, go further, with machine learning algorithms analyzing complex data sets, spotting patterns, and predicting problems to allow proactive management. Other significant concerns of water resource management include policy and governance issues.

IWRM stands for a holistic approach that advances coordinated development and management of water, land, and other related resources. It expresses a participatory stakeholder involvement, multidisciplinary approach, and consideration of social, economic, and environmental factors. Successful implementation of IWRM calls for appropriate governance structures, unquestionable legal frameworks, and efficient funding (Shaikh et al., 2021). It is thus dynamic, multidimensional, and requires a judicious blend of technical expertise, novelty in technologies, and appropriate policy frameworks. The quest for good water management practices would go on and on to become increasingly critical amidst global challenges of swelling populations and further intensification of climate change. Regarding this, the use of advanced technologies, stakeholders' cooperation, and environmental sustainability can guarantee that water is available and adequate for future generations. Integration of WSN with advanced machine learning algorithms pushes the edges in the domain of water resource management (Anh Khoa et al., 2021).

Climate change, population increase, and industrial demands have increasingly pressurized global water resources; the need for innovative and efficient management strategies becomes even a greater priority. This chapter will look into the integrative potential of machine learning integrated into WSNs in transforming water resource management into effective data collection, analysis, and decision-making methodologies. It covers the orderly planning, development, and monitoring of water resources to cater to diverse human and ecological needs (Rahaman & Azharuddin, 2022).

In conventional water resource management, everything is based on manual data collection and empirical methods only. All these approaches are insufficient in tackling the modern water systems' complexity and dynamics. WSNs have transformed data collection, as they can provide real-time and high-resolution monitoring of water parameters such as quality, flow rate, and level. They are a network of sensors that are usually distributed in space and communicate via wireless communication links to continuously stream data. This could be quite essential in timely and informed decisions. Machine Learning is a subset of Artificial Intelligence that deals with a set of algorithms, which learn from data to make a prediction (Palermo et al., 2022a).

The Machine Learning applied to water resource management uses large volumes of data created by WSNs to determine the pattern, forecast future trends, and detect anomalies. This capability is very useful in water quality management, optimization of water distribution, and prediction and mitigation of natural disasters such as floods and droughts. With regard to these aspects, the most significant application of WSNs in water resource management relates to water quality monitoring. Sensors deployed in water bodies will constantly measure certain parameters like pH, turbidity, dissolved oxygen, and presence of contaminants. The collected data is transmitted to centralized systems for real-time processing. Machine learning algorithms applied over this data detect deviations from normal conditions (Krishnan et al., 2022b). Anomaly detection, for example, may identify potential pollution events and enable rapid responses to minimize environmental and public health hazards. Predictive models can forecast the trends in water quality based on historical data and support the implementation of policies and taking remedial measures in advance. Besides water quality, WSNs integrated with machine learning contribute much to improving the standard of water distribution systems.

Urban water supply systems are very intricate and usually plagued with several adversities such as leakages, bursts, and inefficiency. Sensors that are integrated with infrastructure monitor flow rates and pressures, feeding data into machine learning algorithms that detect leaks and optimize distribution. Predictive maintenance models could predict the exact dates of probable failures, thus enabling proactive repairs and hence reducing water loss. This will not only enhance efficiency in the distribution process but also ensures a more reliable supply to the consumer. Rosero-Montalvo et al., 2020b Another important area for which the integration of WSNs with machine learning becomes a very valuable technique is that of flood management (Rosero-Montalvo et al., 2020b).

Floods present one of the most natural threats to human life and property. Traditional systems of flood monitoring are not in a position to provide resolution and timeliness to undertake proper flood management or studies. WSNs will be able to provide real-time data regarding rainfall, river levels, and soil moisture. Machine learning algorithms analyze such data and predict the flood event well in advance with greater accuracy. The early warning systems based on these technologies issue warnings much in advance for timely evacuation and other mitigation procedures. The benefits of WSN and machine learning integration in the management of water resources will be derailed by a number of challenges (Gupta et al., 2020).

Data should be good and reliable; else erroneous prediction and decisions from incomplete faulty data ensue. Similarly, guaranteeing the robustness of security in sensor networks will prevent data breach and maintain system integrity. Besides, most of these technologies need multi-level investment and technical skills that might be a challenge for resource-poor regions. This, therefore, calls for convergence of advanced machine learning algorithms and WSNs for better management of this resource (Palermo et al., 2022b).

Meeting these demands, the new approaches will markedly enhance the capabilities of monitoring and prediction and, therefore, management by offering real-time high-resolution data and analytical tools. These are new solutions that will become imperative in these times of increasing environmental and social pressures if the sustainability of this scarce resource is to be guaranteed. The intention of the chapter that follows is to provide an overall view of those technologies, their applications, and their potential for having a profound impact on water resource management.

Background

Water resource management has become an increasingly critical concern as more people demand water, while climate change continues to deteriorate the existing scarce resource. Conventional methods for monitoring and managing water resources fall short in mitigating a wide array of modern challenges such as real-time data collection, predictive analytics, and proactive infrastructure maintenance. Such limitations are overcome by the incorporation of advanced machine learning algorithms with WSNs. WSNs further allow monitoring of various water parameters continuously in real time, which can be conducted at a granular level, availing information needed for proper management. By integrating a network with machine learning algorithms like SVM, CNN, and RNN, higher-order predictive analytics and anomaly detection are enabled, expanding the decision space manifold.

Scope

It describes the advance integration between Wireless Sensor Networks and advanced machine learning algorithms that have been very helpful in improving water resource management. It discusses the practical use of WSNs in real-time data collection in monitoring functions such as water quality, irrigation, and flood detection. This chapter also discusses the application of machine learning techniques like SVM, CNN, and RNN in predictive analytics and anomaly detection. Apart from that, it extends several cases study on successful implementation and also focuses on how accurate data interpretation can lead to informed decisions so that water resources can be proactively managed.

Objectives

- Explore the integration of Wireless Sensor Networks (WSNs) with advanced machine learning algorithms in water resource management.
- Demonstrate the application of these technologies in real-time monitoring, predictive analytics, and anomaly detection.
- Highlight the benefits of enhanced data collection and analysis for water quality, distribution efficiency, and flood management.
- Discuss the challenges and solutions in implementing WSNs and machine learning.
- Present case studies and future trends to provide a comprehensive understanding of the potential and impact of these innovative technologies on sustainable water resource management.

WIRELESS SENSOR NETWORKS (WSNS) IN WATER MANAGEMENT

Wireless Sensor Networks (WSNs) are wireless sensors that record environmental conditions like temperature, humidity, pressure, and chemical compositions, transmitting data to a central system for analysis and decision-making, making them crucial in water management for real-time monitoring and resource management(Krishnan et al., 2022b; Rosero-Montalvo et al., 2020b).

Figure 1. Typical operations in a typical WSN node

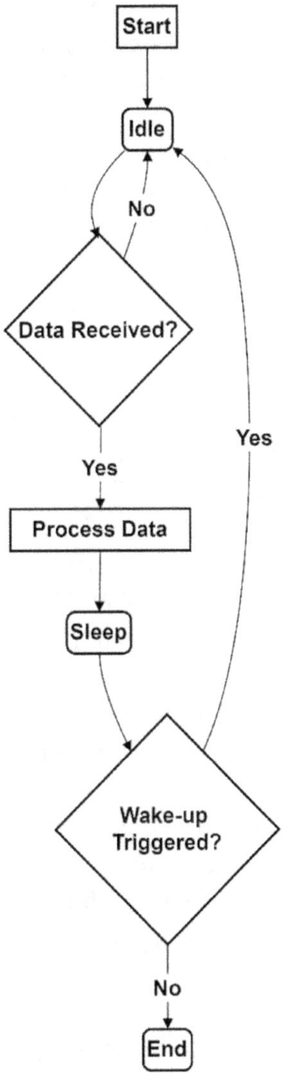

Figure 1 represents a simplified flow of operations in a typical WSN node, emphasizing states like idle, data processing, and power management (sleep and wake-up). Adjustments can be made based on specific functionalities and protocols used in different WSN applications, such as those in water management for environmental monitoring.

The important components of WSNs:

- **Sensors**: These are the devices that detect and measure specific physical or chemical parameters. In water management, sensors can monitor water quality (e.g., pH, turbidity, dissolved oxygen), flow rates, pressure, and levels. They convert the measured parameters into electrical signals that can be processed and transmitted.
- **Nodes**: A sensor node, also known as a mote, is an individual unit in the network that consists of a sensor, a microcontroller, a communication module, and a power source. The microcontroller processes the data collected by the sensor and manages communication with other nodes and the central system.
- **Communication Module**: This component enables wireless communication between nodes and the central system. It typically uses radio frequency (RF) communication, but other methods like infrared or optical communication can also be employed. The communication module ensures the reliable transmission of data across the network.
- **Power Source**: Sensor nodes are often powered by batteries, but energy-harvesting techniques (e.g., solar, kinetic) are increasingly used to extend the operational life of the network. Efficient power management is crucial to ensure the longevity and reliability of WSNs.
- **Central System (Sink Node)**: The central system, or sink node, collects data from all sensor nodes in the network. It serves as the main hub for data aggregation, processing, and analysis. This system can be a dedicated server, a cloud-based platform, or any other computational resource capable of handling the data load.
- **Network Topology**: WSNs can be organized in various topologies, including star, tree, and mesh. The choice of topology depends on the specific application and deployment environment. Mesh topology, for instance, offers high reliability and scalability by allowing multiple communication paths between nodes.

Applications of WSNs in Water Resource Management

WSNs have a wide range of applications in water resource management, significantly improving the efficiency, accuracy, and timeliness of data collection and analysis(More et al., 2020).

- **Water Quality Monitoring**: WSNs are extensively used to monitor the quality of water in rivers, lakes, reservoirs, and distribution systems. Sensors measure parameters such as pH, turbidity, dissolved oxygen, and the presence of contaminants. This real-time monitoring allows for the early detection of pollution events and helps in maintaining safe water quality standards. For example, sensors can detect sudden changes in turbidity that might indicate contamination from industrial discharge or agricultural runoff.

- **Irrigation Management**: In agriculture, WSNs optimize irrigation systems by providing real-time data on soil moisture, temperature, and humidity. This information enables precision irrigation, where water is applied only when and where it is needed, reducing water waste and enhancing crop yields. Automated irrigation systems can adjust water delivery based on sensor data, ensuring efficient water use and conserving resources.
- **Flood Detection and Management**: WSNs play a crucial role in flood monitoring and management. Sensors placed in rivers, dams, and catchment areas can monitor water levels and flow rates. When integrated with weather data, these systems can predict flood events and provide early warnings to communities, enabling timely evacuations and preventive measures. This application is particularly vital in flood-prone regions, where early detection can save lives and reduce property damage.
- **Urban Water Supply Management**: In urban areas, WSNs help in managing water distribution networks. Sensors monitor flow rates and pressure within pipelines, detecting leaks and bursts that can lead to significant water losses. By identifying these issues early, utilities can carry out targeted repairs, reducing downtime and improving the reliability of water supply systems. Additionally, real-time monitoring helps in balancing supply and demand, ensuring efficient water distribution across the network.
- **Wastewater Management**: WSNs are also applied in wastewater treatment plants to monitor the quality of effluents before they are discharged into water bodies. Sensors measure parameters such as chemical oxygen demand (COD), biological oxygen demand (BOD), and levels of harmful substances. This data helps in optimizing treatment processes and ensuring that discharged water meets environmental regulations.
- **Hydrological Monitoring**: WSNs are used to monitor various hydrological parameters such as rainfall, groundwater levels, and river flows. This data is essential for water resource planning and management, particularly in regions experiencing water scarcity. By understanding hydrological patterns, managers can develop more effective strategies for water allocation and conservation.

Hence, Wireless Sensor Networks (WSNs) are transforming water resource management by providing detailed, real-time data that supports efficient and sustainable practices. From monitoring water quality and managing irrigation to detecting floods and maintaining urban water supplies, WSNs enable proactive and informed decision-making, ultimately contributing to the sustainable management of water resources.

MACHINE LEARNING ALGORITHMS IN WATER RESOURCE MANAGEMENT

Figure 2. Machine Learning Algorithms in Water Resource Management

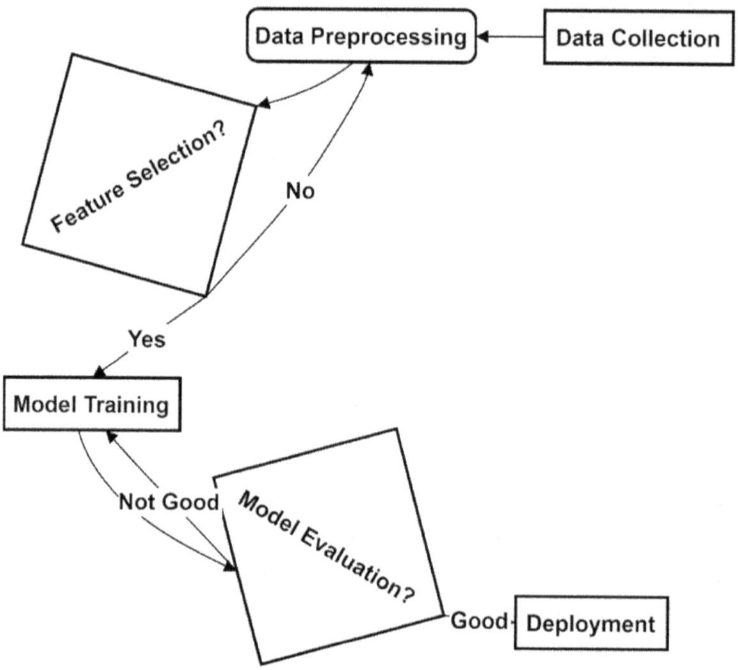

Figure 2 shows a typical workflow using machine learning algorithms for improving water resource management. Model Training uses algorithms like regression, decision trees, and neural networks to meet data needs. Adjustments can be made to include more detailed steps based on application complexity and algorithms used.

Machine learning (ML) is a subset of artificial intelligence (AI) that enables systems to learn from data, identify patterns, and make decisions with minimal human intervention. It involves the use of algorithms that can process vast amounts of data to generate insights, predict future events, and improve decision-making processes. In water resource management, machine learning offers significant potential by enhancing the ability to analyze complex datasets collected from various sources, including Wireless Sensor Networks (WSNs), satellite imagery, and historical records. By leveraging ML algorithms, water managers can optimize resource use, predict and mitigate risks, and ensure the sustainable management of water systems(Agrawal et al., 2023; Koshariya et al., 2023).

Machine learning is categorized into three main types: supervised learning, unsupervised learning, and reinforcement learning. Supervised learning involves training a model on labeled data, where the algorithm learns to predict outputs from input data. Unsupervised learning deals with unlabeled data, identifying patterns and relationships within the data. Reinforcement learning focuses on training models to make decisions by rewarding desired behaviors and penalizing undesired ones. In the context of water resource management, supervised learning is commonly used for predictive analytics, unsupervised learning for clustering and anomaly detection, and reinforcement learning for optimizing management strategies (Kumar et al., 2023; Pachiappan et al., 2024).

Types of Machine Learning Algorithms

Several machine learning algorithms are particularly relevant to water resource management. Among them, Support Vector Machines (SVM), Convolutional Neural Networks (CNN), and Recurrent Neural Networks (RNN) stand out for their specific capabilities and applications(Sreedhar et al., 2024; Tirlangi et al., 2024; Verma et al., 2024).

1. Support Vector Machines (SVM)

Support Vector Machines (SVM) are supervised learning models used for classification and regression tasks. SVMs work by finding the hyperplane that best separates data points of different classes in a high-dimensional space. This hyperplane maximizes the margin between the classes, ensuring robust classification.

In water resource management, SVMs can be applied to a variety of tasks. For instance, they can classify water quality data into categories such as safe, moderate, or polluted based on various parameters like pH, turbidity, and dissolved oxygen. SVMs can also be used in flood prediction models, where they classify weather and hydrological data to predict the likelihood of flood events. The ability of SVMs to handle high-dimensional data and their robustness to overfitting make them suitable for these complex classification tasks.

2. Convolutional Neural Networks (CNN)

Convolutional Neural Networks (CNN) are a type of deep learning algorithm primarily used for image and spatial data analysis. CNNs consist of multiple layers, including convolutional layers that automatically learn to identify spatial hierarchies and features within the data. These networks are particularly effective at recognizing patterns and structures in images and spatial datasets.

In water resource management, CNNs are valuable for analyzing satellite imagery and remote sensing data. For example, CNNs can be used to monitor changes in water bodies, detect algal blooms, and assess the impact of droughts on vegetation. By processing large amounts of spatial data, CNNs can identify subtle changes and patterns that are critical for effective water resource management. Additionally, CNNs can aid in mapping flood extents and monitoring land use changes that affect water resources.

3. Recurrent Neural Networks (RNN)

Recurrent Neural Networks (RNN) are a class of neural networks designed for sequential data processing. RNNs have loops that allow information to be passed from one step of the sequence to the next, making them suitable for time-series analysis and tasks where context and temporal dependencies are crucial.

In water resource management, RNNs are employed for time-series forecasting and predictive modeling. For example, RNNs can be used to predict river flow rates, reservoir levels, and precipitation patterns based on historical data. This predictive capability is essential for managing water supplies, planning for droughts, and anticipating flood events. RNNs can also be used in groundwater level prediction, where they analyze past water level data to forecast future trends, helping in sustainable groundwater management.

Applications in Water Resource Management

The integration of machine learning algorithms in water resource management offers numerous benefits. By analyzing data collected from WSNs and other sources, ML algorithms can provide insights that enhance decision-making processes. For instance, predictive models can help in efficient water allocation, ensuring that water is distributed based on demand forecasts and availability. Anomaly detection algorithms can identify unusual patterns in water quality or flow rates, prompting early interventions to prevent potential issues(Boopathi, 2024b; Pachiappan et al., 2024; Zekrifa et al., 2023).

Moreover, machine learning can improve the efficiency of water treatment processes by optimizing operational parameters based on real-time data. In flood management, ML algorithms can enhance early warning systems by providing accurate and timely predictions, allowing for proactive measures to protect communities. Additionally, machine learning can support the development of smart irrigation systems that adjust water delivery based on soil moisture and weather conditions, promoting water conservation and improving agricultural productivity.

Thus, the application of machine learning algorithms in water resource management holds immense potential for transforming how we monitor, predict, and manage water systems. By leveraging the capabilities of SVMs, CNNs, and RNNs, water managers can make more informed decisions, optimize resource use, and enhance the sustainability of water resources. As the challenges associated with water resource management continue to evolve, the integration of advanced machine learning techniques will be crucial in ensuring a resilient and sustainable water future.

Benefits of Machine Learning in Water Management

Machine learning (ML) offers substantial benefits in water management by enhancing decision-making processes, optimizing resource use, and improving overall system efficiency. One key benefit is predictive analytics, where ML algorithms analyze historical and real-time data to forecast water availability, demand patterns, and potential risks such as floods or water quality issues. This capability enables proactive planning and timely interventions to mitigate adverse impacts(Boopathi, 2024a; Veeranjaneyulu et al., 2023; Zekrifa et al., 2023).

ML also enhances the accuracy of water quality monitoring by detecting anomalies and predicting pollutant levels based on sensor data. By automating data analysis, ML reduces reliance on manual labor and improves the speed and precision of decision-making. Additionally, ML algorithms optimize water distribution systems by identifying inefficiencies, detecting leaks, and adjusting supply based on demand forecasts and environmental conditions. Moreover, ML contributes to sustainable water management by promoting water conservation strategies through smart irrigation systems and efficient use of water resources. Overall, the integration of ML in water management not only enhances operational efficiency but also supports environmental sustainability and resilience in the face of changing climatic conditions and growing water demands.

INTEGRATING WSNS AND MACHINE LEARNING

Methodologies for Integration

Integrating Wireless Sensor Networks (WSNs) with machine learning (ML) algorithms in the context of water resource management involves several methodologies that enhance data collection, analysis, and decision-making capabilities. Figure 3 outlines a comprehensive approach to integrating WSNs with ML algorithms in water resource management, highlighting iterative processes of data collection, analysis, model development, and decision support. Adjustments can be made to incorporate specific ML techniques, such as supervised learning for predictive modeling or unsupervised learning for anomaly detection, based on the application's requirements and data characteristics(Glady et al., 2024; Saravanan et al., 2024).

- **Data Collection and Preprocessing**: The integration begins with data collection from WSNs, which capture real-time measurements of water parameters such as quality, flow rates, and levels. ML algorithms require clean, structured data for accurate analysis, necessitating preprocessing steps such as data cleaning, normalization, and feature extraction. This ensures that the data fed into ML models is relevant and consistent, enhancing the reliability of predictions and insights derived from the data.
- **Model Selection and Training**: Once data preprocessing is complete, the next step involves selecting appropriate ML models based on the specific application and data characteristics. Commonly used models include Support Vector Machines (SVM), Convolutional Neural Networks (CNN), and Recurrent Neural Networks (RNN), each offering unique capabilities for tasks such as classification, regression, anomaly detection, and time-series forecasting. ML models are trained using historical data to learn patterns and relationships, enabling them to make predictions and decisions based on new incoming data from WSNs.
- **Real-Time Analytics and Decision Support**: The integration facilitates real-time analytics, where ML algorithms continuously analyze incoming data from WSNs to detect anomalies, predict future trends, and generate actionable insights. For example, anomaly detection algorithms can identify sudden changes in water quality or unexpected variations in flow rates, prompting immediate responses such as adjusting water treatment processes or activating flood control measures. Decision support systems powered by ML provide water managers with data-driven recommendations and scenarios to optimize resource allocation and mitigate risks effectively.

- **Feedback Loop and Model Updating**: Continuous improvement is crucial in integrating WSNs and ML. The feedback loop involves monitoring the performance of ML models over time, comparing predictions with actual outcomes, and updating models as new data becomes available. This iterative process ensures that ML algorithms remain accurate and adaptive to changing environmental conditions and operational requirements.

Figure 3. Integrating Wireless Sensor Networks (WSNs) with machine learning (ML) algorithms in water resource management

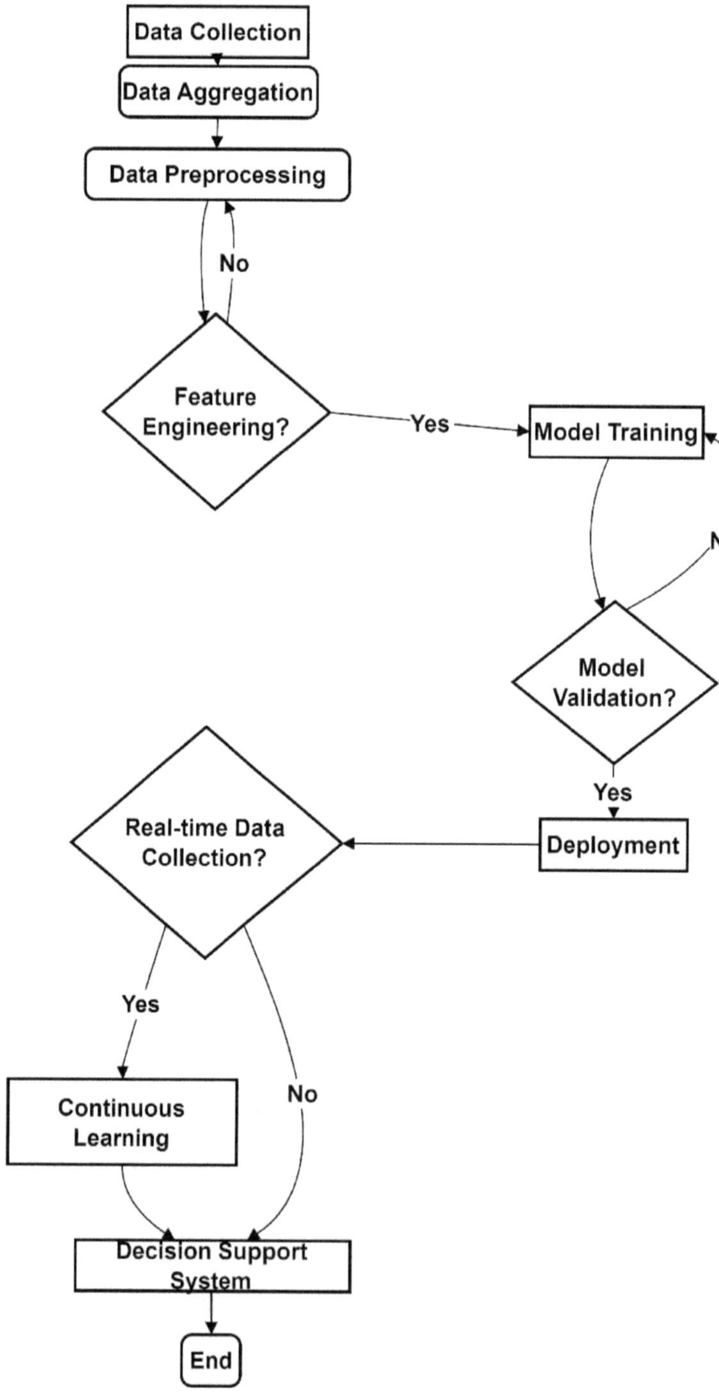

Challenges and Solutions

Despite its potential benefits, integrating WSNs and ML in water management faces several challenges that require innovative solutions(Agrawal et al., 2023):

- **Data Quality and Reliability**: WSNs may encounter data inaccuracies due to sensor malfunctions, environmental interference, or communication errors. Ensuring data quality and reliability involves deploying calibrated sensors, implementing redundancy in data collection, and using advanced signal processing techniques to filter out noise and anomalies.
- **Scalability and Network Management**: Managing large-scale WSN deployments can be complex, requiring efficient network protocols, data aggregation techniques, and power management strategies to extend sensor node longevity. Scalable ML algorithms that can process large volumes of data in real-time are essential for maintaining network performance and reliability.
- **Security and Privacy Concerns**: Protecting sensitive data transmitted through WSNs from cyber-attacks and unauthorized access is critical. Implementing encryption protocols, secure communication channels, and access control mechanisms ensures data confidentiality and integrity. Additionally, complying with data privacy regulations such as GDPR or CCPA safeguards user rights and promotes trust in the system.
- **Interdisciplinary Collaboration and Expertise**: Successful integration of WSNs and ML requires collaboration between water engineers, data scientists, and domain experts. Interdisciplinary teams can address complex challenges by combining domain knowledge with technical expertise to develop tailored solutions that meet specific operational needs and regulatory requirements.
- **Cost and Resource Constraints**: Deploying and maintaining WSNs with ML capabilities can be costly, particularly in resource-constrained regions. Implementing cost-effective sensor technologies, leveraging open-source ML frameworks, and exploring partnerships with academic institutions and industry stakeholders can help mitigate financial barriers and enhance technology adoption.

Hence, integrating WSNs with ML offers unprecedented opportunities to revolutionize water resource management by enabling data-driven decision-making, enhancing operational efficiency, and ensuring environmental sustainability. By addressing challenges related to data quality, scalability, security, interdisciplinary collaboration, and cost-effectiveness, organizations can unlock the full potential of these technologies to safeguard water resources for future generations. As advancements in sensor technology and machine learning continue, the synergy between WSNs and ML promises to drive innovation and resilience in global water management practices.

REAL-TIME MONITORING AND DATA ANALYSIS

Real-time monitoring and data analysis play a crucial role in modern water resource management, enabling timely decision-making, proactive interventions, and efficient resource utilization. This section explores techniques for real-time data collection, machine learning models for data analysis, and the interpretation and application of data in water management(Dhanya et al., 2023; Ramudu et al., 2023).

Techniques for Real-Time Data Collection

Real-time data collection in water management relies on advanced sensor technologies and network infrastructure to capture and transmit data continuously. Key techniques include:

- **Wireless Sensor Networks (WSNs)**: Deployed throughout water systems, WSNs consist of interconnected sensors that monitor various parameters such as water quality, flow rates, and environmental conditions. These sensors communicate wirelessly, providing real-time data streams to centralized systems for analysis.
- **Remote Sensing**: Satellite imagery and aerial drones equipped with sensors capture high-resolution data over large geographical areas. Remote sensing techniques are valuable for monitoring changes in water bodies, detecting land use patterns, and assessing environmental impacts on water resources.
- **Internet of Things (IoT) Devices**: IoT devices such as smart meters and connected infrastructure sensors collect data on water consumption, pipeline conditions, and operational performance. These devices enable continuous monitoring and feedback on system operations in real-time.
- **Data Telemetry and Communication**: Efficient data telemetry protocols and communication networks (e.g., cellular, satellite) ensure reliable data transmission from sensors to centralized servers or cloud-based platforms. Low-power communication protocols optimize energy usage in WSNs, extending operational lifetimes.

Machine Learning Models for Data Analysis

Machine learning (ML) models are instrumental in analyzing vast volumes of real-time data collected from various sources in water management (Godwin Immanuel et al., 2024; Naresh et al., 2024). ML models and their applications are:

- **Supervised Learning Models**: **Regression**: Predicts continuous variables such as water flow rates or pollutant concentrations based on historical data. **Classification**: Identifies water quality levels (e.g., safe, contaminated) or operational states (e.g., normal, abnormal) using labeled datasets.
- **Unsupervised Learning Models**: **Clustering**: Groups similar data points to identify patterns or anomalies in water quality or usage patterns. **Anomaly Detection**: Flags unusual events or deviations from normal behavior, such as sudden changes in water flow indicating leaks or contamination.
- **Deep Learning Models**: **Convolutional Neural Networks (CNN)**: Analyze spatial data from satellite imagery to monitor changes in water bodies, detect algal blooms, or assess land use impacts. **Recurrent Neural Networks (RNN)**: Process sequential data (e.g., time-series of water level fluctuations) for predicting floods, droughts, or optimizing hydropower generation.
- **Ensemble Learning**: **Random Forests, Gradient Boosting**: Combine multiple ML models to improve prediction accuracy and robustness in complex water management scenarios.

Interpretation and Application of Data

Interpreting and applying real-time data in water management involves several critical steps:

- **Data Visualization**: Visual tools such as dashboards, graphs, and maps facilitate the interpretation of complex data sets, enabling stakeholders to identify trends, anomalies, and actionable insights at a glance.
- **Decision Support Systems (DSS)**: Integrating ML outputs into DSS provides water managers with actionable recommendations and scenarios based on real-time data analysis. DSS can optimize water allocation, prioritize maintenance tasks, and mitigate risks such as floods or water contamination.
- **Predictive Analytics**: ML-driven predictive models forecast future water demand, quality trends, or environmental impacts, enabling proactive planning and resource allocation. For example, predicting seasonal water demand fluctuations helps utilities optimize supply strategies and avoid shortages.

- **Operational Optimization**: ML algorithms optimize operational parameters in water distribution networks, reducing energy consumption, minimizing leakage, and improving overall system efficiency. Smart irrigation systems adjust watering schedules based on soil moisture and weather forecasts, conserving water resources in agriculture.
- **Environmental Monitoring and Compliance**: Real-time data analysis ensures compliance with regulatory standards for water quality and environmental protection. ML models detect non-compliance events promptly, triggering corrective actions and minimizing environmental risks.

Real-time monitoring and data analysis are transforming water resource management by providing actionable insights, improving operational efficiency, and enhancing sustainability. Through advanced sensor technologies, IoT devices, and ML models, organizations can harness the power of data to make informed decisions, predict future trends, and mitigate risks in water systems. As technology continues to evolve, the integration of real-time monitoring with sophisticated data analytics will play a pivotal role in addressing global water challenges and ensuring the long-term health and resilience of water resources.

PREDICTIVE MAINTENANCE AND DECISION SUPPORT SYSTEMS

Predictive Maintenance Strategies

Predictive maintenance (PdM) strategies leverage advanced analytics and machine learning (ML) to predict equipment failures before they occur, optimizing maintenance schedules, reducing downtime, and extending asset lifespan. In water resource management, where operational reliability is critical, predictive maintenance strategies are essential for ensuring continuous service delivery and efficient resource utilization(Puranik et al., 2024; Sharma et al., 2024).

Condition Monitoring: Continuous monitoring of equipment conditions using sensors and IoT devices allows for real-time data collection on factors such as pump performance, pipeline integrity, and water quality parameters. Condition-based maintenance predicts failures based on deviations from normal operating conditions, enabling preemptive repairs and minimizing disruptions.

Data-Driven Analytics: Machine learning algorithms analyze historical and real-time data to detect patterns and anomalies indicative of impending failures. Supervised learning models, such as regression and classification algorithms, predict equipment degradation or failure based on factors like vibration analysis, temperature fluctuations, and operational stress levels.

Predictive Modeling: Statistical models and predictive analytics forecast equipment performance and maintenance needs based on historical data trends and environmental factors. Time-series analysis and probabilistic models, like Bayesian networks, quantify risks and prioritize maintenance tasks according to criticality and impact on operations.

Integrated Systems Approach: Combining predictive maintenance with asset management systems and enterprise resource planning (ERP) platforms optimizes resource allocation, schedules maintenance activities, and manages spare parts inventory effectively. Integration ensures seamless data flow between maintenance teams, operations personnel, and management, facilitating informed decision-making and resource optimization.

Development of Decision Support Systems

Decision support systems (DSS) enhance decision-making processes by providing stakeholders with actionable insights, scenario analysis, and predictive modeling outputs based on real-time data and predictive maintenance strategies. In water resource management, DSS are pivotal for optimizing operational efficiency, ensuring regulatory compliance, and mitigating risks(Tirlangi et al., 2024).

Figure 4. Development of Decision Support Systems (DSS) in water resource management

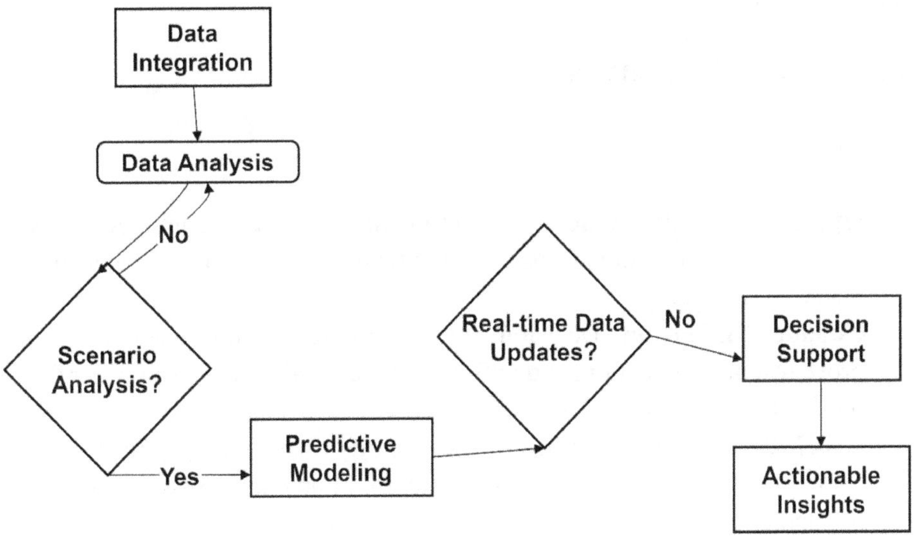

Figure 4 outlines the workflow for developing Decision Support Systems (DSS) in water resource management, emphasizing data integration, analysis, scenario evaluation, predictive modeling, and real-time decision support, emphasizing continuous data updates for operational efficiency and risk management.

Real-Time Data Integration: DSS integrate data from WSNs, IoT devices, and remote sensing technologies to provide a comprehensive view of water system operations. Data aggregation, cleansing, and normalization ensure data integrity and reliability, enabling accurate analysis and decision support.

Predictive Analytics: ML algorithms within DSS forecast water demand, monitor water quality trends, and predict environmental impacts. Models such as neural networks and decision trees analyze complex datasets to anticipate system behavior, optimize resource allocation, and identify potential risks in advance.

Scenario Analysis and Optimization: DSS simulate scenarios and evaluate alternative courses of action based on predictive models and operational constraints. What-if analysis assesses the impact of policy changes, infrastructure upgrades, or environmental factors on water system performance, guiding long-term planning and investment decisions.

Risk Management and Resilience: DSS quantify risks associated with water supply disruptions, infrastructure failures, and environmental hazards. Risk-based decision support prioritizes mitigation strategies, emergency response planning, and resilience measures to safeguard water resources and ensure continuity of service.

Stakeholder Collaboration: Interactive dashboards, visualization tools, and collaborative platforms facilitate communication and data sharing among stakeholders, enhancing transparency, accountability, and community engagement in water management initiatives.

Benefits and Challenges

Benefits:

- **Optimized Maintenance**: Predictive maintenance minimizes downtime, reduces operational costs, and extends asset lifespan through timely interventions.
- **Enhanced Efficiency**: Decision support systems streamline workflows, improve resource allocation, and optimize operational processes in water distribution, treatment, and conservation.
- **Risk Mitigation**: ML-driven analytics identify risks, forecast trends, and prioritize preventive measures to mitigate environmental impacts and ensure regulatory compliance.

- **Sustainability**: Integrated systems promote sustainable water management practices by conserving resources, reducing energy consumption, and enhancing system resilience.

Challenges:

- **Data Integration Complexity**: Integrating diverse data sources and ensuring data compatibility and integrity pose technical challenges.
- **Model Accuracy**: ML model accuracy depends on data quality, model complexity, and the dynamic nature of water system operations.
- **Cost and Implementation**: Initial investment in sensor networks, ML infrastructure, and staff training may be prohibitive for some organizations.
- **Regulatory and Privacy Concerns**: Compliance with data protection regulations and privacy policies, particularly with sensitive environmental data, requires stringent protocols and governance frameworks.

Predictive maintenance strategies and decision support systems are crucial for improving water resource management by enhancing operational efficiency, optimizing practices, and ensuring sustainable use. Utilizing advanced analytics, machine learning algorithms, and integrated systems, these strategies can anticipate equipment failures, mitigate risks, and prioritize investments. As technology evolves, these systems will play a vital role in achieving resilient, adaptive, and efficient global water management systems.

CHALLENGES AND OPPORTUNITIES

Water resource management is critical for sustainable development. The integration of advanced machine learning (ML) algorithms with Wireless Sensor Networks (WSNs) offers promising solutions for efficient water resource management. However, this integration faces significant challenges. This paper discusses these challenges and explores the opportunities presented by advanced ML in conjunction with WSNs(Agrawal et al., 2023; Koshariya et al., 2023). Water resources are under increasing pressure due to population growth, climate change, and industrial activities. Effective management of these resources is essential to ensure sustainability and accessibility. WSNs, combined with advanced ML algorithms, have emerged as powerful tools for monitoring and managing water resources(Rahamathunnisa et al., 2024; Venkateswaran et al., 2023). This integration allows for real-time data collection, analysis, and decision-making, enhancing the efficiency and effectiveness of water management practices.

Challenges

Data Quality and Quantity: One of the primary challenges is ensuring the quality and quantity of data collected by WSNs. ML algorithms require large datasets to train accurate models. Incomplete, noisy, or inaccurate data can lead to poor model performance. Additionally, the deployment of sensors in remote or harsh environments can result in data loss or corruption.

Energy Consumption: WSNs are typically deployed in remote areas where power sources are limited. Advanced ML algorithms, especially deep learning models, require significant computational power, leading to increased energy consumption. Balancing the energy efficiency of WSNs with the computational demands of ML algorithms is a critical challenge.

Scalability: As the number of sensors in a WSN increases, the volume of data generated can become overwhelming. ML algorithms must be scalable to handle large datasets efficiently. Ensuring that algorithms can process data in real-time without significant delays is essential for effective water resource management.

Integration and Interoperability: Integrating ML algorithms with WSNs involves combining hardware, software, and communication protocols. Ensuring interoperability between different components can be complex. Additionally, the integration process must be seamless to avoid disruptions in data collection and analysis.

Security and Privacy: Data collected by WSNs can be sensitive, particularly when monitoring water quality for human consumption. Ensuring the security and privacy of this data is crucial. ML algorithms must be designed to protect against cyber threats and unauthorized access.

Opportunities

- **Predictive Analytics:** ML algorithms can analyze historical data to predict future water demand, availability, and quality. Predictive models can help water managers make informed decisions, such as optimizing water distribution, scheduling maintenance, and planning for droughts or floods.
- **Anomaly Detection:** Advanced ML algorithms can detect anomalies in water quality or flow rates, indicating potential issues such as leaks, contamination, or equipment failures. Early detection of anomalies allows for prompt corrective actions, minimizing the impact on water resources and infrastructure.
- **Optimization of Water Usage:** ML algorithms can optimize water usage by analyzing consumption patterns and identifying areas for improvement. For example, in agricultural settings, ML models can recommend optimal

irrigation schedules based on soil moisture data, weather forecasts, and crop requirements, reducing water wastage.

- **Real-time Monitoring and Decision Support:** WSNs enable real-time monitoring of water resources, while ML algorithms can provide decision support by analyzing data and generating actionable insights. This combination allows water managers to respond quickly to changing conditions, improving the overall efficiency of water resource management.
- **Adaptive Management Strategies:** ML algorithms can support adaptive management strategies by continuously learning from new data and updating models accordingly. This adaptability is crucial for managing dynamic and complex water systems, ensuring that management practices remain effective over time.

The integration of advanced ML algorithms with WSNs presents significant opportunities for improving water resource management. However, addressing the challenges related to data quality, energy consumption, scalability, integration, and security is essential for realizing the full potential of this technology. By overcoming these challenges, WSNs and ML algorithms can provide valuable tools for sustainable and efficient water resource management.

FUTURE TRENDS AND INNOVATIONS

Emerging Technologies in WSNs and Machine Learning

The future of water resource management is increasingly reliant on emerging technologies in Wireless Sensor Networks (WSNs) and Machine Learning (ML), paving the way for more efficient, sustainable, and resilient water systems(Ahmed et al., 2021; Krishnan et al., 2022b; Palermo et al., 2022a; Rosero-Montalvo et al., 2020a).

Figure 5. Emerging technologies in Wireless Sensor Networks (WSNs) and Machine Learning (ML) for water resource management

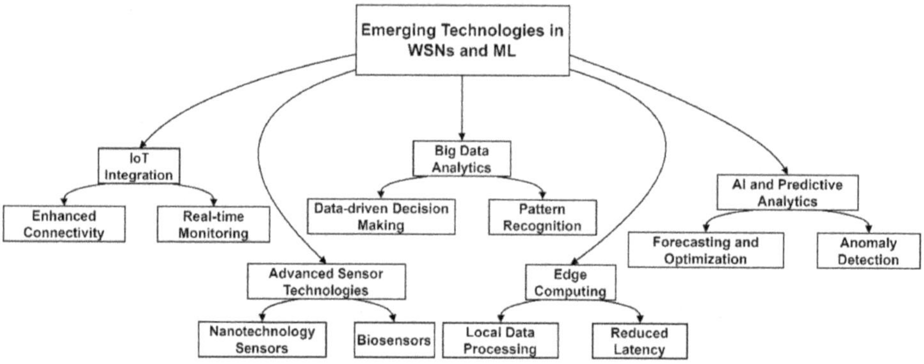

WSNs and ML are revolutionizing water resource management by enhancing real-time monitoring, predictive maintenance, and informed decision-making through advancements in sensor technologies, data analytics, and AI-driven insights (Figure 5).

- **Advanced Sensor Technologies**: Miniaturization and integration of sensors with enhanced functionalities are revolutionizing WSNs. Advanced sensors capable of measuring multiple parameters simultaneously, such as pH, turbidity, and heavy metal concentrations, enhance real-time monitoring capabilities and improve data accuracy in diverse environmental conditions.
- **Edge Computing and IoT Integration**: Edge computing brings computational capabilities closer to sensor nodes, reducing latency, and enabling real-time data processing and decision-making at the edge of the network. Integrated with IoT platforms, edge computing optimizes data transmission, minimizes bandwidth usage, and enhances system scalability in large-scale WSN deployments.
- **AI-Driven Data Analytics**: Machine Learning algorithms continue to evolve, leveraging advancements in deep learning, reinforcement learning, and federated learning to analyze complex datasets from WSNs. AI-driven analytics enable predictive modeling, anomaly detection, and optimization of water system operations, improving resource efficiency and sustainability.
- **Blockchain for Data Integrity**: Blockchain technology ensures data integrity, transparency, and security in WSNs by creating tamper-proof, decentralized ledgers of sensor data. Blockchain-based solutions enhance trust among stakeholders, facilitate secure data sharing, and support compliance with regulatory standards in water management.

- **Autonomous Systems and Robotics**: Autonomous underwater vehicles (AUVs) and aerial drones equipped with sensors and ML algorithms enable autonomous data collection, monitoring of remote or hazardous environments, and rapid response to water emergencies. Robotics enhance spatial data acquisition, environmental mapping, and infrastructure inspection, enhancing operational efficiency and reducing human intervention.

Potential Improvements and Innovations

- **Predictive and Prescriptive Analytics**: Future advancements in ML will focus on enhancing predictive and prescriptive analytics capabilities. Predictive models will forecast water demand with higher accuracy, anticipate environmental impacts, and optimize resource allocation based on dynamic operational conditions and climate variability.
- **Multi-Modal Sensor Fusion**: Integration of multiple sensor modalities, including optical, acoustic, and chemical sensors, enables comprehensive environmental monitoring and data fusion. Multi-modal sensor fusion enhances data richness, improves situational awareness, and supports holistic decision-making in complex water management scenarios.
- **Real-Time Adaptive Control Systems**: ML-driven adaptive control systems dynamically adjust operational parameters in real-time based on sensor data and predictive models. Adaptive systems optimize water distribution, treatment processes, and infrastructure management, responding to changing demands and mitigating risks of system failures.
- **Digital Twins and Simulation Modeling**: Digital twins simulate real-world water systems using virtual replicas based on sensor data, ML models, and historical trends. Simulation modeling enables scenario analysis, infrastructure planning, and optimization of maintenance strategies, promoting resilience and sustainability in water resource management.
- **Citizen Science and Community Engagement**: Leveraging crowdsourced data and participatory sensing initiatives empower communities to contribute to environmental monitoring efforts. Citizen science platforms integrate community observations with WSN data, enhancing data coverage, promoting environmental awareness, and fostering collaborative water stewardship.

The convergence of WSNs and ML is poised to revolutionize water resource management by introducing advanced technologies, predictive capabilities, and sustainable practices. As organizations embrace emerging technologies such as advanced sensors, edge computing, AI-driven analytics, and blockchain, they unlock new opportunities for improving operational efficiency, ensuring water quality, and enhancing resilience in the face of climate change and urbanization pressures.

Strategic Insights for Future Development in Water Resource Management

The future development of water resource management will be shaped by strategic insights that prioritize innovation, collaboration, and sustainability(Gupta et al., 2020; More et al., 2020).

- **Embrace Technological Advancements**: Continuously integrate emerging technologies such as AI, IoT, and blockchain to enhance monitoring capabilities, optimize resource allocation, and improve decision-making processes. Investing in research and development of advanced sensor networks and predictive analytics will drive operational efficiency and resilience in water systems.
- **Promote Data-Driven Decision-Making**: Foster a culture of data-driven decision-making by leveraging real-time data analytics and predictive modeling. Implement decision support systems that empower stakeholders with actionable insights for proactive management of water quality, supply, and infrastructure.
- **Enhance Interdisciplinary Collaboration**: Foster collaboration between water engineers, data scientists, policy-makers, and community stakeholders to address complex challenges in water management. Interdisciplinary approaches promote holistic solutions, innovation, and consensus-building for sustainable water practices.
- **Prioritize Resilience and Adaptation**: Build resilience against climate variability, population growth, and urbanization impacts by developing adaptive strategies and infrastructure. Incorporate resilience planning into water management frameworks to mitigate risks, enhance system reliability, and ensure long-term sustainability.
- **Promote Stakeholder Engagement and Transparency**: Engage communities through citizen science initiatives, educational outreach, and transparent communication strategies. Enhance public awareness of water issues, foster community participation in monitoring efforts, and promote accountability in decision-making processes.
- **Invest in Capacity Building and Training**: Equip water professionals with skills in emerging technologies, data analytics, and sustainable practices through training programs and knowledge exchange platforms. Build institu-

REFERENCES

Agrawal, A. V., Magulur, L. P., Priya, S. G., Kaur, A., Singh, G., & Boopathi, S. (2023). Smart Precision Agriculture Using IoT and WSN. In *Handbook of Research on Data Science and Cybersecurity Innovations in Industry 4.0 Technologies* (pp. 524–541). IGI Global. DOI: 10.4018/978-1-6684-8145-5.ch026

Ahmed, S. S., Bali, R., Khan, H., Mohamed, H. I., & Sharma, S. K. (2021). Improved water resource management framework for water sustainability and security. *Environmental Research*, *201*, 111527. DOI: 10.1016/j.envres.2021.111527 PMID: 34157270

Anh Khoa, T., Quang Minh, N., Hai Son, H., Nguyen Dang Khoa, C., Ngoc Tan, D., VanDung, N., Hoang Nam, N., Ngoc Minh Duc, D., & Trung Tin, N. (2021). Wireless sensor networks and machine learning meet climate change prediction. *International Journal of Communication Systems*, *34*(3), e4687. DOI: 10.1002/dac.4687

Boopathi, S. (2024a). Advancements in Machine Learning and AI for Intelligent Systems in Drone Applications for Smart City Developments. In *Futuristic e-Governance Security With Deep Learning Applications* (pp. 15–45). IGI Global. DOI: 10.4018/978-1-6684-9596-4.ch002

Boopathi, S. (2024b). Sustainable Development Using IoT and AI Techniques for Water Utilization in Agriculture. In *Sustainable Development in AI, Blockchain, and E-Governance Applications* (pp. 204–228). IGI Global. DOI: 10.4018/979-8-3693-1722-8.ch012

Dhanya, D., Kumar, S. S., Thilagavathy, A., Prasad, D., & Boopathi, S. (2023). Data Analytics and Artificial Intelligence in the Circular Economy: Case Studies. In *Intelligent Engineering Applications and Applied Sciences for Sustainability* (pp. 40–58). IGI Global.

Glady, J. B. P., D'Souza, S. M., Priya, A. P., Amuthachenthiru, K., Vikram, G., & Boopathi, S. (2024). A Study on AI-ML-Driven Optimizing Energy Distribution and Sustainable Agriculture for Environmental Conservation. In *Harnessing High-Performance Computing and AI for Environmental Sustainability* (pp. 1–27). IGI Global., DOI: 10.4018/979-8-3693-1794-5.ch001

Godwin Immanuel, D., Solaimalai, G., Chandrakala, B. M., Bharath, V. G., Singh, M. K., & Boopathi, S. (2024). Advancements in Electric Vehicle Management System: Integrating Machine Learning and Artificial Intelligence. In *Advances in Web Technologies and Engineering* (pp. 371–391). IGI Global. DOI: 10.4018/979-8-3693-1487-6.ch018

Gupta, A. D., Pandey, P., Feijóo, A., Yaseen, Z. M., & Bokde, N. D. (2020). Smart water technology for efficient water resource management: A review. *Energies*, *13*(23), 6268. DOI: 10.3390/en13236268

Koshariya, A. K., Kalaiyarasi, D., Jovith, A. A., Sivakami, T., Hasan, D. S., & Boopathi, S. (2023). AI-Enabled IoT and WSN-Integrated Smart Agriculture System. In *Artificial Intelligence Tools and Technologies for Smart Farming and Agriculture Practices* (pp. 200–218). IGI Global. DOI: 10.4018/978-1-6684-8516-3.ch011

Krishnan, S. R., Nallakaruppan, M., Chengoden, R., Koppu, S., Iyapparaja, M., Sadhasivam, J., & Sethuraman, S. (2022a). Smart water resource management using Artificial Intelligence—A review. *Sustainability (Basel)*, *14*(20), 13384. DOI: 10.3390/su142013384

Krishnan, S. R., Nallakaruppan, M., Chengoden, R., Koppu, S., Iyapparaja, M., Sadhasivam, J., & Sethuraman, S. (2022b). Smart water resource management using Artificial Intelligence—A review. *Sustainability (Basel)*, *14*(20), 13384. DOI: 10.3390/su142013384

Kumar, M., Kumar, K., Sasikala, P., Sampath, B., Gopi, B., & Sundaram, S. (2023). Sustainable Green Energy Generation From Waste Water: IoT and ML Integration. In *Sustainable Science and Intelligent Technologies for Societal Development* (pp. 440–463). IGI Global.

More, K. S., Wolkersdorfer, C., Kang, N., & Elmaghraby, A. S. (2020). Automated measurement systems in mine water management and mine workings–A review of potential methods. *Water Resources and Industry*, *24*, 100136. DOI: 10.1016/j.wri.2020.100136

Naresh, M., Subhahan, D. A., Narula, V. N., Roslin, D. K., Narula, S., & Boopathi, S. (2024). Edge Computing and Machine Learning Integration for Autonomous Electrical Vehicles. In *Solving Fundamental Challenges of Electric Vehicles* (pp. 99–127). IGI Global., DOI: 10.4018/979-8-3693-4314-2.ch005

Pachiappan, K., Anitha, K., Pitchai, R., Sangeetha, S., Satyanarayana, T., & Boopathi, S. (2024). Intelligent Machines, IoT, and AI in Revolutionizing Agriculture for Water Processing. In *Handbook of Research on AI and ML for Intelligent Machines and Systems* (pp. 374–399). IGI Global.

Padmanaban, K. (2021). A Novel Groundwater Resource Forecasting Technique for Cultivation Utilizing Wireless Sensor Network (WSN) and Machine Learning (ML) Model. [TURCOMAT]. *Turkish Journal of Computer and Mathematics Education*, *12*(2), 2186–2192.

Palermo, S. A., Maiolo, M., Brusco, A. C., Turco, M., Pirouz, B., Greco, E., Spezzano, G., & Piro, P. (2022a). Smart technologies for water resource management: An overview. *Sensors (Basel)*, *22*(16), 6225. DOI: 10.3390/s22166225 PMID: 36015982

Palermo, S. A., Maiolo, M., Brusco, A. C., Turco, M., Pirouz, B., Greco, E., Spezzano, G., & Piro, P. (2022b). Smart technologies for water resource management: An overview. *Sensors (Basel)*, *22*(16), 6225. DOI: 10.3390/s22166225 PMID: 36015982

Puranik, T. A., Shaik, N., Vankudoth, R., Kolhe, M. R., Yadav, N., & Boopathi, S. (2024). Study on Harmonizing Human-Robot (Drone) Collaboration: Navigating Seamless Interactions in Collaborative Environments. In *Cybersecurity Issues and Challenges in the Drone Industry* (pp. 1–26). IGI Global.

Rahaman, M. M., & Azharuddin, M. (2022). Wireless sensor networks in agriculture through machine learning: A survey. *Computers and Electronics in Agriculture*, *197*, 106928. DOI: 10.1016/j.compag.2022.106928

Rahamathunnisa, U., Sudhakar, K., Padhi, S., Bhattacharya, S., Shashibhushan, G., & Boopathi, S. (2024). Sustainable Energy Generation From Waste Water: IoT Integrated Technologies. In *Adoption and Use of Technology Tools and Services by Economically Disadvantaged Communities: Implications for Growth and Sustainability* (pp. 225–256). IGI Global.

Ramudu, K., Mohan, V. M., Jyothirmai, D., Prasad, D., Agrawal, R., & Boopathi, S. (2023). Machine Learning and Artificial Intelligence in Disease Prediction: Applications, Challenges, Limitations, Case Studies, and Future Directions. In *Contemporary Applications of Data Fusion for Advanced Healthcare Informatics* (pp. 297–318). IGI Global.

Rosero-Montalvo, P. D., López-Batista, V. F., Riascos, J. A., & Peluffo-Ordóñez, D. H. (2020a). Intelligent WSN system for water quality analysis using machine learning algorithms: A case study (Tahuando river from Ecuador). *Remote Sensing (Basel)*, *12*(12), 1988. DOI: 10.3390/rs12121988

Rosero-Montalvo, P. D., López-Batista, V. F., Riascos, J. A., & Peluffo-Ordóñez, D. H. (2020b). Intelligent WSN system for water quality analysis using machine learning algorithms: A case study (Tahuando river from Ecuador). *Remote Sensing (Basel)*, *12*(12), 1988. DOI: 10.3390/rs12121988

Saravanan, S., Khare, R., Umamaheswari, K., Khare, S., Krishne Gowda, B. S., & Boopathi, S. (2024). AI and ML Adaptive Smart-Grid Energy Management Systems: Exploring Advanced Innovations. In *Principles and Applications in Speed Sensing and Energy Harvesting for Smart Roads* (pp. 166–196). IGI Global. DOI: 10.4018/978-1-6684-9214-7.ch006

Shaikh, R. A. J., Naidu, H., & Kokate, P. A. (2021). Next-generation wsn for environmental monitoring employing big data analytics, machine learning and artificial intelligence. *Evolutionary Computing and Mobile Sustainable Networks: Proceedings of ICECMSN 2020*, 181–196.

Sharma, M., Sharma, M., Sharma, N., & Boopathi, S. (2024). Building Sustainable Smart Cities Through Cloud and Intelligent Parking System. In *Handbook of Research on AI and ML for Intelligent Machines and Systems* (pp. 195–222). IGI Global.

Sreedhar, P. S. S., Sujay, V., Rani, M. R., Melita, L., Reshma, S., & Boopathi, S. (2024). Impacts of 5G Machine Learning Techniques on Telemedicine and Social Media Professional Connection in Healthcare. In *Advances in Medical Technologies and Clinical Practice* (pp. 209–234). IGI Global. DOI: 10.4018/979-8-3693-1934-5.ch012

Tirlangi, S., Teotia, S., Padmapriya, G., Senthil Kumar, S., Dhotre, S., & Boopathi, S. (2024). Cloud Computing and Machine Learning in the Green Power Sector: Data Management and Analysis for Sustainable Energy. In *Developments Towards Next Generation Intelligent Systems for Sustainable Development* (pp. 148–179). IGI Global. DOI: 10.4018/979-8-3693-5643-2.ch006

Veeranjaneyulu, R., Boopathi, S., Narasimharao, J., Gupta, K. K., Reddy, R. V. K., & Ambika, R. (2023). Identification of Heart Diseases using Novel Machine Learning Method. *IEEE- Explore*, 1–6.

Venkateswaran, N., Kumar, S. S., Diwakar, G., Gnanasangeetha, D., & Boopathi, S. (2023). Synthetic Biology for Waste Water to Energy Conversion: IoT and AI Approaches. *Applications of Synthetic Biology in Health. Energy & Environment*, •••, 360–384.

Verma, R., Christiana, M. B. V., Maheswari, M., Srinivasan, V., Patro, P., Dari, S. S., & Boopathi, S. (2024). Intelligent Physarum Solver for Profit Maximization in Oligopolistic Supply Chain Networks. In *AI and Machine Learning Impacts in Intelligent Supply Chain* (pp. 156–179). IGI Global. DOI: 10.4018/979-8-3693-1347-3.ch011

Zekrifa, D. M. S., Kulkarni, M., Bhagyalakshmi, A., Devireddy, N., Gupta, S., & Boopathi, S. (2023). Integrating Machine Learning and AI for Improved Hydrological Modeling and Water Resource Management. In *Artificial Intelligence Applications in Water Treatment and Water Resource Management* (pp. 46–70). IGI Global. DOI: 10.4018/978-1-6684-6791-6.ch003

Conclusion

As we reach the culmination of Enhancing Data-Driven Electronics Through IoT, it is clear that this book represents a crucial step toward understanding the rapidly evolving relationship between electronics and the Internet of Things. The work presented across the various chapters not only demonstrates the transformative potential of IoT but also highlights the innovative strides being made across a diverse range of industries. By connecting data-driven insights with practical applications, this book illuminates the ways in which IoT is redefining the landscape of modern electronics. This volume brings together leading voices in the field, offering a rich blend of theoretical perspectives, technical challenges, and real-world case studies. The collaborative nature of this effort underscores the importance of interdisciplinary approaches as IoT technology continues to reshape industries, from healthcare and agriculture to urban infrastructure and beyond. At the heart of this book is a commitment to encouraging new ideas, encouraging further research, and inspiring innovative solutions that can meet the demands of a connected world. The exploration of key topics such as security, communication protocols, machine learning, and sustainable IoT-driven systems has opened new avenues for future exploration. Ultimately, the true success of this book lies not only in its ability to provide a thorough examination of current IoT practices but also in its capacity to spark curiosity, push boundaries, and motivate the next generation of engineers, researchers, and technologists. This journey through the intersections of IoT and electronics marks just the beginning of what is possible, as the field continues to expand and evolve.

I hope this book serves as a springboard for further discovery and contributes meaningfully to the ongoing discourse in IoT and electronics. Thank you for engaging with this work, and I look forward to seeing the innovations and advancements that it will inspire.

About the Contributors

Bhagwan Das is a renowned innovator, inventor, and academic leader with a passion for pioneering research in the fields of the Internet of Things (IoT), Artificial Intelligence (AI), and Data Science. With a rich career spanning academia and industry, Dr. Das has earned an esteemed reputation for his contributions to Information and Communication Technologies (ICT), where he continually drives innovation through cutting-edge research and practical applications. Currently serving as an Associate Professor and ICT Developer, Dr. Das brings a unique blend of technical expertise, problem-solving abilities, and leadership to his work, specializing in IoT systems, AI development, and data-driven solutions. His work is characterized by a strong commitment to advancing technological capabilities and creating sustainable, innovative solutions for modern challenges. Over the years, Dr. Das has been invited to share his insights at numerous international conferences and workshops, where his keynote speeches have inspired both peers and emerging scholars in the field. His thought leadership has reached audiences at prestigious events, including the International Conference on Pattern Recognition and Intelligent Control in China, and global forums such as the 1.5 Global Environmental Conference, among many others. Dr. Das has earned multiple awards for his innovative research, including patents and copyrights that reflect his inventive approach to ICT challenges. His professional qualifications as a Chartered Professional Engineer (CPEng) and Fellow of Engineers Australia (FIEAust) further solidify his standing as a respected figure in the global engineering community. Dr. Das has published extensively across leading platforms such as Scopus, Google Scholar, and ResearchGate, where his work continues to influence and shape the future of ICT and IoT innovation. His dedication to mentoring the next generation of engineers, combined with his remarkable ability to turn complex research into actionable solutions, makes him a key contributor to the field's ongoing evolution. Dr. Das's contributions extend far beyond the lab and classroom. His vision for technology-driven change, combined with his practical expertise, continues to have a lasting impact on the development of smarter, more connected systems that will drive the future of electronics and communication.

Engr Muhammad Zakir Shaikh is an accomplished and a young professional, currently serving as the Director of the National Center for Robotics, Automation and Artificial Intelligence at Mehran University of Engineering and Technology, Jamshoro. He is doing research on Fault Diagnosis of Rail Wheel Profiles via Deep Learning whose main aim is to reduce Railway Accidents across the globe and to improve the overall infrastructure and system of Railway Transportation. Currently, he is doing Cotutella(Joint PhD) after which he will be honored with two PhD Degrees. One in Electronics Engineering by Mehran University and the other PhD in Mechanical Engineering and Energy Efficiency by University of Malaga. He is originally a faculty member of Electronics Engineering and currently also working as a Business Development Manager at the Higher Education Commission's National Center for Robotics and Automation-Condition Monitoring Systems Lab at Mehran University of Engineering and Technology, Jamshoro. He is also working as President of QS ImpACT Pakistan to implement United Nation's Sustainable Development Goals along with various Charity and Voluntary works. Under his Leadership as the President of QS ImpACT Pakistan, He won the Overall Biggest Impact Award for the year 2022. Under his Leadership, Pakistan received the award of Best Council for the work on Sustainable Development Goals in the year 2022. Muhammad Zakir Shaikh is also a renowned motivational speaker. He has been a great motivation and a source of inspiration for the youth and many people across the globe. Notably, Muhammad Zakir Shaikh has represented Pakistan on the international stages, earning accolades such as the "I am a Mover, Shaker and Change-Maker" award and also represented the Youth and Pakistan at United Nations HQ in New York. He was the sole Pakistani recipient of the fully funded award to participate and represent Pakistan at the Global Youth Conference 2023 which was held in Casablanca, Morocco. Recently, he has been conferred with the PEC Engineering Excellence Award in the Category of Young Technopreneur for the year of 2023. He was also Honored with the Excellence Award for outstanding contributions and great services towards Technological Research, Modernizing Academic Learning, and Capacity Building in the Higher Education CENTRAL Project at Mae Fah Luang University, Chiang Rai, Thailand. He has also secured first Position in the International Unity 3D Competition held at the Technical University of Sofia, Bulgaria. He was also Co-PI of the project which has received funding of 1.5 Million from the Sindh Higher Education Commission under the Sindh Research Support Program. Besides that, he has already worked as a Research Associate to complete/implement the project titled "Design and Implementation of Intelligent Energy Efficient Industrial Process Control System Using Conveyor

Belts via Robotic Arm", a Fully Funded Project of the Pakistan Science Foundation. He has been declared as the Star Laureate 2011 for outstanding achievements in the field of Science and Technology by South Asian Publications. He has broken the historic record by getting Two gold medals simultaneously in 27th IEEEP All Pakistan Students Seminar for securing 1st position in one research paper and second position in the other paper. He is also working as the Focal Person of PRIDE (Public Relations Image Building for Development and Excellence) for positive image building and to highlight the contributions of individuals contributing towards society. He has expertise in Carbon Literacy training Programs and Actions related to Climate Change. He is also looking after all the media platforms of EU-funded Erasmus Plus Capacity Building Project CENTRAL, EU-funded Erasmus Plus ACTIVE Climate Action Project and EU-funded Erasmus Plus Project on Digital Twin Technology namely CATCH VR. Currently He is also working as a Researcher in EU Funded Capacity Building in Higher Education ACTIVE Climate Action Project at University of Malaga, Spain. Besides having several certifications, He has also won several awards and competitions. He is also a Professional member of PEC. He is also a member of IEEE and IEEEP. He received his Bachelor's degree in Electronics and Master's degree in Electronic Systems Engineering in 2013 and 2015 respectively from Mehran University of Engineering and Technology, Jamshoro. Besides being a great motivational speaker, He has won various debates and speech competitions and has been invited as a Guest Speaker and Keynote Speaker at various well-reputed platforms nationally and internationally. He has also many top-notch research publications in well-reputed National and International Journals. Besides that, he has been providing and teaching students for free and has done many charity works to help many people. Currently, Mr Zakir is working as a Researcher at University of Malaga, Spain. For his PhD research, he was selected as an Erasmus Plus PhD Exchange Scholar from Mehran UET to study and conduct his research at the University of Malaga, Spain for five months. He was also honored with the Academic Excellence Award for Outstanding Contributions in Entrepreneurship, Sustainable Future and Sustainable Development Goals in the year 2022. He has also been a Judge of QS Reimagine Education Awards. He is also the co-author of the Book Academia Reengineering-Renovating, Innovating, and Elevating the Higher Education Institutions.

Samreen Hussain is among the founding members of Sindh Higher Education Commission. Member Federal Higher Education Commission & First Female Elected member of Governing Body of Pakistan Engineering Council (PEC) in the discipline of Electronic Engineering. During her tenure of above Chancellor Aror University of Art, Architecture, Design & Heritage, she with her little team established a state of the Art Tent City for flood affectees during the unprecedented floods of 2022. The Tent City was solar powered, facilitated by RO Plant to use harvested rain water for drinking, make female Lavatories, School and Medical Care. Every tent was provided with pedestal fan, light, Cutlery, crockery, complete hygiene kit, school books and stationery. The tent city helped district Rohri to resume routine teaching and learning in 07 schools which were housing the flood affected families. She has authored two books (Academia Reengineering: Renovating, Innovating and Elevating the Higher Education Institutions and Women of Wisdom: Navigating the Path towards Women's Educational Excellence). Established Centre of Entrepreneurship Management and Emerging Technologies (CEMET) in Rohri as a pilot project to materialise the triple helix concept with the support of Sukkur Chamber of Commerce & Industries & District Council Sukkur. She had also established the Faculty of Information and Computing Science (I&CS) in District Central, Karachi in less than 06 months to utilize the infrastructure of a dormant school. She has the Unique record of being the first ever Female Vice Chancellor of any Engineering University in Pakistan ever in the history. She is the recipient of "PEC Engineers' Excellence Award - 2022, (Engineering Policy, Regulations and Advocacy)" and "National Engineering Excellence Award" by Institute of Engineers, Pakistan (IEP). She has received more than 200 citations in last two years. As a Convener Innovation & Entrepreneurship Committee Pakistan Engineering council she upgraded the division of PITC (Pakistan Innovation & Testing Centre) as Pakistan Innovation & Entrepreneurship Development Centre with one window operation to facilitate engineering products manufacturing industry through establishment of Pakistan Innovation, Testing, calibration & certification centre. She was also awarded with prestigious civil award (TI) on the diamond jubilee of the Pakistan recognizing her invaluable services in the field of education and research. Currently, She is serving as the Vice Chancellor of Dawood University of Engineering and Technology, Karachi.

Enrique Nava Baro finished his PhD thesis under supervision of Prof. J.M. Rebollar at Polytechnic University of Madrid, Spain, on the design of circular waveguide polarizer devices on microwaves/millimeter waves to be used on communication satellite antennas. As a result, a new developed computer software, based on a full solution of electromagnetic equations, was able to predict polarizer behaviour with enough accuracy to be used for industrial optimization and design purposes. This software was the only one available at that time and was transferred to relevant research institutions and companies on that field for industrial use. He serves as Associate Professor (Profesor Titular de Universidad) at University of Málaga in 1994, focusing his research work on interdisciplinary applications of digital signal and image processing, with many collaborations on medical, chemical, biological, underwater and industrial fields. Since 1995, Dr Nava is head of the

Andalusian research group 'Radiological Image Digital Processing' (TIC128), with eight post-doctoral researchers in Engineering and Medicine background. He has carried out several foreign research stays: Kurt Rossmann Lab (2000), at University of Chicago (USA), working on mammography computer-aided diagnosis algorithm development; at Institut für Technische Akustik (2011 and 2014) at RWTH (Aachen, Germany), working on advanced array processing techniques for location of acoustic sources; at University of New South Wales (UNSW@ADFA, 2012, Camberra, Australia) working in digital signal processing applied to biotremology in termites and ants; and on oceanographic research ships (Spanish Institute of Oceanography, IEO (MEDIAS campaign, 2015 and 2017), working on acoustics and digital image processing for fisheries research. Since some years ago, he had active collaborations in research with different national and international research groups: University of Malaga UMA Analytical Chemistry Group (pesticides in food quantification), Italian INFN-LNF group (X-ray microtomography new prototype), Romanian LAPI at Polytechnic University of Bucharest group (digital image processing), Prof. Moreno-Torres at UMA (acoustic phonetics applied to medical applications), several Pakistan universities (image and signal processing), Photonic Sensors and Algorithms (3D imaging) and LAV University of Cadiz group (underwater acoustics). In recent years, he has been actively involved in many projects for innovative teaching and internationalization, serving as Deputy Vice-Chancellor for Strategical Projects at University of Málaga.

***

Eshani Banik is pursuing BTech in Computer Science & Engineering at Brainware University, India.

Sampath Boopathi is an accomplished individual with a strong academic background and extensive research experience. He completed his undergraduate studies in Mechanical Engineering and pursued his postgraduate studies in the field of Computer-Aided Design. Dr. Boopathi obtained his Ph.D. from Anna University, focusing his research on Manufacturing and optimization. Throughout his career, Dr. Boopathi has made significant contributions to the field of engineering. He has authored and published over 225 research articles in internationally peer-reviewed journals, highlighting his expertise and dedication to advancing knowledge in his area of specialization. His research output demonstrates his commitment to conducting rigorous and impactful research. In addition to his research publications, Dr. Boopathi has also been granted one patent and has three published patents to his name. This indicates his innovative thinking and ability to develop practical solutions to real-world engineering challenges. With 17 years of academic and research experience, Dr. Boopathi has enriched the engineering community through his teaching and mentorship roles.

Archana Kedar Chaudhari has competed PhD in Biomedical Instrumentation from Savitribai Phule Pune University. She is currently working as Assistant Professor in Instrumentation Engineering Department, Vishwakarma Institute of Technology, Pune India. She has more than 20 years experience in Industry and Academia. She has several Indian patents published and granted. She has published many papers in referred international journals and conferences. Her filed of Interest include Image and Signal processing, Biomedical, Health Care, Internet of Things, Wireless Sensor Networks, Artificial Intelligence, Machine Learning and Deep Learning.

Sanchita Ghosh is a highly driven worker who holds a Master of Computer Applications degree. Her current position as a Technical Assistant at Brainware University allows her to specialize on IoT,NLP, AI, networking, cybersecurity, and cloud computing. Sanchita is passionate about using technology to solve complicated challenges, and she actively participates in projects and activities that promote innovation and progress the digital transformation. Her diversified skill set and dedication to continual learning make her an invaluable asset in the ever-changing field of information technology.

Noura Ibrahim Ganaou studied Computer Science at University of Sciences and Technology of Oran – Mohamed Boudiaf (USTO-MB/Algeria) where he received his License degree in 2015, then Masters in Computer Systems and Networks in 2017 at of Math and Computer Science, department of Computer Science. He pursuits his PhD studies from 2018 to 2021 at School of Doctoral School Exact Sciences and Faculty of Science department of Math and Computer Science, University of Abdou Moumouni (UAM) – Niamey/Niger and got a Doctorate degree in Computer Networks and Telecoms namely QoS assurance in distance learning environments. A part from professional carrier in computer science industry and published research works, he intervenes so far in academic institutes and universities in Niger as part-time temporary assistant professor with research interests in computer systems, networks, data science and artificial intelligence.

Allassane Issa Salaou studied Computer Science at University of Sciences and Technology of Oran – Mohamed Boudiaf (USTO-MB/Algeria) where he received his License degree in 2015, then Masters in Computer Systems and Networks in 2017 at of Math and Computer Science, department of Computer Science. He is currently pursuing PhD studies at Computer Science department of University of Oran 1 – Algeria in Computer Networks and Artificial Intelligence. Certified specialist in computer networks security and project management, his research interests focus in artificial intelligence namely event detection using ontology-based fuzzy logic and machine learning with published research works.

Veena Divya Krishnappa is an Assistant Professor in the Department of Electronics & Instrumentation Engineering at RV College of Engineering, Bengaluru, Karnataka, India. Her role involves teaching and researching in the field of electronics and instrumentation, contributing to the advancement of technology and engineering education. Her work at RV College of Engineering supports the development of future engineers and innovators.

Vidya M. J. is an Assistant Professor at the School of Computer Science and Engineering, RV University, Bengaluru, Karnataka, India. She specializes in computer science and engineering, contributing to the academic and research community through her role at RV University. Her expertise and academic contributions play a vital role in advancing knowledge and innovation within the field of computer science.

Harshada Bhushan Magar is a dedicated academic in the Department of Electronics and Telecommunication at AISSMS Institute of Information Technology, Pune, Maharashtra. With expertise in electronics and telecommunication, she is committed to delivering high-quality education and advancing research in the field. Harshada Bhushan Magar is recognized for her contributions to the academic community, where she mentors students and participates in innovative research projects. Her work focuses on bridging theoretical knowledge with practical applications, thereby preparing students for successful careers in the rapidly evolving electronics and telecommunication industry.

Osama A. Mahdi received his B.Sc. degree in Computer Science from Babylon University, Iraq, in 2009, his M.Sc. degree in Database Systems from Universiti Putra Malaysia in 2014, and his Ph.D. degree in Data Stream Analytics from La Trobe University in 2021. He is currently a full-time Lecturer in the Master of Data Analytics Program at the Melbourne Institute of Technology (MIT). His research interests include data stream mining (with a focus on concept drift), data analytics, and machine learning/deep learning.

Hitesh Mohapatra received the B.E. degree in Information Technology from Biju Patnaik University of Technology (BPUT), Odisha in 2006, and the MTech. Degree in CSE from Odisha University of Technology and Research (OUTR) University, Odisha in 2009. He received his Ph.D. in Computer Science & Engineering in 2021 from Veer Surendra Sai University of Technology (VSSUT),

Burla, India. He has contributed 30 SCI and Scopus indexed research papers, 21 international/national conferences and books on Software Engineering and C Programming respectively. He has 15 years of teaching experience both in industry and academia. He has served the research community with various capacities like session chair, technical chair, keynote speaker etc. His research interests include wireless sensor network, smart city, smart grid and smart water. Currently he is working as Associate professor at School of Computer Engineering, KIIT DU, Bhubaneswar, and Odisha.

T. Pravin is a dedicated professional in the Research and Development department at RSP Science Hub, Coimbatore, Tamil Nadu. He is actively involved in pioneering research initiatives and technological innovations. T. Pravin's work is characterized by a strong commitment to advancing scientific knowledge and developing cutting-edge solutions that address complex challenges. His contributions to the R&D sector at RSP Science Hub play a crucial role in driving innovation and fostering a culture of continuous improvement, making him a valuable asset to the scientific community in Coimbatore and beyond.

Mrutyunjay Padhiary is an Assistant Professor in the Department of Agricultural Engineering at Assam University, Silchar, India, with a deep academic background. His previous positions included significant teaching and research duties at the Faculty of Technology, UBKV in West Bengal, and Tribhuvan University in Nepal, as well as research experience at IIT Kharagpur. He has been awarded multiple awards, including an International NUS Research Fellowship and an excellent score on the 2001 GATE Exam, demonstrating his proficiency in precision agriculture. He has made substantial contributions to biofuel production, artificial intelligence vision, aquaculture, precision agriculture, agricultural automation, and artificial intelligence. He is currently working on several research projects, which range from autonomous surface vessels to precision agriculture all-terrain vehicles.

Piyal Roy is an accomplished academician in the field of Computer Science & Engineering. He has completed his BSc in Computer Science, MSc in Computer Science and MTech in Computer Science & Engineering. Currently, he is working as an Assistant Professor at Computer Science & Engineering department at Brainware University, India.

S.David Samuel Azariya is working as an Assistant Professor in the Department of Information Technology at Sona College of Technology. He is currently pursuing a Doctorate in Anna University, Chennai, Tamil Nadu, India. He has completed Diploma in Electronics and Communications and Bachelor of Engineering degree in Information Technology and Master of Engineering in Computer and

Communication Engineering. He is actively engaged in the research area of data visualization, computer vision, deep learning and IoT. He has published various research papers in Springer and IEEE conferences related to research domain. He is the member of IEI and ISTE professional society.

Rajasree P. M. is an Assistant Professor in the Department of Electronics & Instrumentation Engineering at RV College of Engineering, Bengaluru, Karnataka, India. She is involved in teaching and research within the field of electronics and instrumentation, focusing on advancing knowledge and practical applications in engineering. Her contributions at RV College of Engineering are integral to the academic development of students and the field.

S. C. Shamkuwar is an Assistant Professor in the Department of Mechanical Engineering at Vishwakarma Institute of Information Technology (VIIT), Pune, Maharashtra. With a strong foundation in mechanical engineering, he is dedicated to both teaching and research, contributing significantly to the academic and professional community. S. C. Shamkuwar is known for his commitment to educating future engineers, emphasizing the practical application of mechanical engineering principles. His involvement in research and innovative projects underscores his dedication to advancing the field and ensuring that his students are well-prepared for the challenges of the engineering industry.

Arpita Shome is working as assistance professor in the Computer Science & Engineering Department of Abacus Institute of Engineering and Management.

Siva Satya Sreedhar P is a faculty member in the Department of Information Technology at Seshadri Rao Gudlavalleru Engineering College, located in Gudlavalleru, Krishna district, Andhra Pradesh. With a robust background in information technology, he is committed to advancing both teaching and research in the field. Siva Satya Sreedhar P is known for his dedication to mentoring students, fostering innovation, and contributing to the academic community through research initiatives. His work plays a crucial role in shaping the future of IT education and preparing students to meet the challenges of the rapidly evolving tech industry.

V. Sujay is an academic and researcher in the Department of Computer Science and Engineering at Krishna University College of Engineering and Technology, Machilipatnam, Krishna, Andhra Pradesh. With a strong background in computer science, V. Sujay is dedicated to teaching and research in this dynamic field. His work focuses on equipping students with the essential knowledge and skills needed to thrive in the tech industry. Through innovative research and a commitment to academic excellence, V. Sujay contributes to the growth and development of both his students and the broader field of computer science and engineering.

V. Sathiyamoorthi is currently working as Lecturer in Computer Engineering Department at Government polytechnic college, Dharmapuri,Tamil Nadu, India. He was born on June 21, 1983, at Omalur in Salem District, Tamil Nadu, India. He received his Bachelor of Engineering degree in Information Technology from Periyar University, Salem with First Class. He obtained his Master of Engineering degree in Computer Science and Engineering from Anna University, Chennai with Distinction and secured 30th University Rank.He received his Ph.D degree from Anna University, Chennai in Web Mining. His areas of specialization include Web Usage Mining, Data Structures, Design and Analysis of Algorithm and Operating System. He has published many papers in International Journals and conferences. He has published many books and book chapters in various renowned international publishers. He has also participated in various National level Workshops and Seminars conducted by various reputed institutions.

Index

Symbols

A

B

C

D

E

F

G

H

I

L

M

N

O

P

R

S

T

U

V

W

www.ingramcontent.com/pod-product-compliance
Lightning Source LLC
LaVergne TN
LVHW080615300926
849162LV00005B/1173
* 9 7 9 8 3 6 9 3 5 4 4 8 3 *